SELF-CONSISTENT METHODS FOR COMPOSITES

SOLID MECHANICS AND ITS APPLICATIONS
Volume 148

Series Editor: G.M.L. GLADWELL
Department of Civil Engineering
University of Waterloo
Waterloo, Ontario, Canada N2L 3GI

Aims and Scope of the Series

The fundamental questions arising in mechanics are: *Why?*, *How?*, and *How much?* The aim of this series is to provide lucid accounts written by authoritative researchers giving vision and insight in answering these questions on the subject of mechanics as it relates to solids.

The scope of the series covers the entire spectrum of solid mechanics. Thus it includes the foundation of mechanics; variational formulations; computational mechanics; statics, kinematics and dynamics of rigid and elastic bodies: vibrations of solids and structures; dynamical systems and chaos; the theories of elasticity, plasticity and viscoelasticity; composite materials; rods, beams, shells and membranes; structural control and stability; soils, rocks and geomechanics; fracture; tribology; experimental mechanics; biomechanics and machine design.

The median level of presentation is the first year graduate student. Some texts are monographs defining the current state of the field; others are accessible to final year undergraduates; but essentially the emphasis is on readability and clarity.

For a list of related mechanics titles, see final pages.

Self-Consistent Methods for Composites

Vol.1: Static Problems

by

S.K. KANAUN

Instituto Tecnológico y de Estudios Superiores de Monterrey,
Campus Estado de México, México

and

V.M. LEVIN

Instituto Mexicano del Petroleo, México

Springer

A C.I.P. Catalogue record for this book is available from the Library of Congress.

ISBN 978-90-481-7694-6
ISBN 978-1-4020-6664-1 (e-book)

Published by Springer,
P.O. Box 17, 3300 AA Dordrecht, The Netherlands.

www.springer.com

Printed on acid-free paper

Contents

To our teachers:

Lazar Markovich Kachanov
and
Isaak Abramovich Kunin

Preface

The theory of heterogeneous materials has been intensively developed during the past few decades. The main reason for the interest of many researchers in this part of the mechanics of solids is the wide area of application of heterogeneous materials in modern material engineering. Self-consistent methods form a well-known branch of the theory of heterogeneous materials. In most books devoted to the mechanics and physics of heterogeneous media, the reader can find self-consistent solutions. But there are no books covering the entire spectrum of self-consistent methods in application to the calculation of static and dynamic properties of heterogeneous materials. This book has been written to cover this gap.

It is written for engineers because here they can find the equations for the effective properties of composites reinforced with various types of inclusions. The main advantage of self-consistent methods is that they give relatively simple equations for the effective parameters of composites. Such equations for static and dynamic properties of matrix composites reinforced with various types of inclusions, for porous media, media with cracks and other defects, for polycrystals, etc., are widely used in engineering practice, and many new self-consistent solutions are presented in the book.

This book is written also for scholars who wish to develop the theory of heterogenous media. In the book they will find the basic ideas and algorithms for the construction of self-consistent solutions. The book shows how these methods may be applied to composites with inclusions of complex structures, to problems of wave propagation, for calculation of higher statistical moments of physical fields in composites. Various ways for improving self-consistent solutions are proposed and discussed.

The book is divided into two volumes. The first volume is devoted to static problems of the mechanics and physics of heterogeneous materials. It starts from the solution of one-particle problems – physical fields in a homogeneous medium with an isolated inclusion. The inclusions that are important fillers for composite materials are considered here: ellipsoidal inclusions, multilayered spherical inclusions, hard flakes and ribbons, short and long fibers, cracks and cracklike inclusions. Effective static properties of composites with isolated inclusions are considered in the second part of this volume. Various self-consistent methods are used to obtain effective elastic, dielectric, thermo- and electroconductive properties of such composites. Predictions from different methods are compared with experimental data and exact solutions existing in the literature. Areas of application of the methods are indicated, and the character of possible errors is discussed.

The second volume is devoted to problems of wave propagation in heterogeneous materials. The velocities and attenuation coefficients of electromagnetic and elastic waves propagating in composites with various types of inclusions, and in polycrystals, are calculated in the framework of self-consistent methods.

The two volumes are almost independent.

The history of self-consistent methods is very long, and the number of works where these methods (explicitly or implicitly) have been applied to the solution of various problems of the mechanics and physics of heterogeneous media is overwhelming. It is impossible to mention all the authors who have made contributions to the subject. We therefore mention only the works that, in our opinion, contain either important new ideas or notable results. It is natural that the book reflects the authors' vision of self-consistent methods and is based mainly on the authors' works.

Our discussions of self-consistent methods with I. A. Kunin and his critique was an important driving force for the formation of our vision of self-consistent methods and of the principal ideas presented in the book. We had useful discussions on various problems considered in the book with D. Jeulin, V.G. Maz'ya, S.G. Mikhlin, V.A. Palmov, G.I. Petrashen, A.A. Vakulenko, and many others. We express our gratitude to all these people. We also deeply appreciate the corrections and suggestions made by the Series Editor, Professor Graham Gladwell, in the revision of the manuscript. The authors thank the Technological Institute of Higher Education of Monterrey (State of Mexico Campus), and Mexican Oil Institute, for supporting this book.

Mexico
June 2007

S.K. Kanaun
V.M. Levin

Notations

$\varepsilon_{ij}, \varepsilon_{\alpha\beta}, C_{ijkl}, C_{\alpha\beta\lambda\mu}, ...$	Lower case Greek or Latin indices are tensorial.
$\sigma_{ij} = C_{ijkl}\varepsilon_{kl}$	Summation with respect to repeating indices is implied.
$T_{(ij)kl} = \frac{1}{2}(T_{ijkl} + T_{jikl})$	Parentheses in indices mean symmetrization.
a, b, T or $\mathbf{a}, \mathbf{b}, \mathbf{T}, ...$	Nonindex notations for vectors and tensors.
$x, \mathbf{x}$	Point and vector of a point in 3D(2D)-space.
$\otimes$ $(\mathbf{a} \otimes \mathbf{b} \Rightarrow a_i b_j)$	Tensor product of vectors and tensors.
$\mathbf{a} \cdot \mathbf{b}$ $(\mathbf{a} \cdot \mathbf{b} = a_i b_i)$	Scalar product of vectors and tensors.
$\mathbf{a} \times \mathbf{b}$ $(\mathbf{a} \times \mathbf{b} \Rightarrow \epsilon_{ijk} a_j b_k)$	Vector product of vectors and tensors.
δ_{ij}	The Kronecker symbol.
ϵ_{ijk}	The Levi–Civita symbol.
$E^1, E^2, ..., E^6$	The basic 4 rank tensors (Appendix A.1).
$P^1, P^2, ..., P^6$	The basic 4 rank tensors (Appendix A.2).
$\nabla_i = \frac{\partial}{\partial x_i}$	The Nabla operator.
$\mathrm{def}\mathbf{u} \Rightarrow \nabla_{(i} u_{j)}$	Deformation operator.
$div\mathbf{T} \Rightarrow \nabla_i T_{ijk...}$	Divergence of a tensor field T.
$rot\mathbf{T} = \nabla \times \mathbf{T} \Rightarrow$ $\Rightarrow \epsilon_{ijk} \nabla_j T_{klm...}$	Rotor of a tensor field T.
$\triangle = \frac{\partial^2}{\partial x_1^2} + \frac{\partial^2}{\partial x_2^2} + \frac{\partial^2}{\partial x_3^2}$	The Laplace operator.
$f(k) =$ $= \int f(x) \exp(i\mathbf{k} \cdot \mathbf{x}) dx.$	The Fourier transform of a function $f(x)$.
k, $\mathbf{k}$	Point and vector of a point in the Fourier transform space.
$p.v. \int ...dx =$ $= \lim_{\delta \to 0} \int_{\lvert x \rvert > \delta} ...dx$	The Cauchy principal value of an integral.
$\delta(x)$	The Dirac delta-function.
$\Omega(x)$	Delta-function concentrated on a surface Ω.
V	Region in 3D(2D)-space.
$V(x) = 1$ if $x \in V$ $V(x) = 0$ if $x \notin V$	Characteristic function of a region V.
$\langle f(x) \rangle$	Mean value of a random function $f(x)$.
$\langle f(x) \vert x \rangle$	Mean value of $f(x)$ under the condition that $x \in V$.
$C^0, C,$ C^*	Tensors of elastic moduli of the matrix, inclusions, and the effective medium.
$C^1 = C - C^0$	Deviation of the elastic moduli in inclusions.
$\lambda_0, \mu_0;\ \lambda, \mu;$ $\lambda_*,\ \mu_*$	The Lame parameters of the matrix, inclusions, and the effective medium.

$K_0,\ K,\ K_*$	Bulk moduli of the matrix, inclusions, and the effective medium.
$E_0, E,\ E_*$	The Young moduli of the matrix, inclusions, and the effective medium.
$\nu_0, \nu,\ \nu_*$	The Poisson ratio of the matrix, inclusions, and the effective medium.
$K_1 = K - K_0$ $\mu_1 = \mu - \mu_0$ $\lambda_1 = \lambda - \lambda_0$	Deviation of properties in inclusions.

1. Introduction

The intensive development of the theory of heterogeneous media in the past few decades is related to the constantly increasing area of application of composite materials in modern industry. Nowadays, composites successfully compete with traditional homogeneous constructive materials like metals, homopolymers, alloys, and ceramics. Modern nanotechnologies open the possibilities for dictating the microstructure of composite materials and for creating materials with prescribed physical properties. The direct synthesis of composite materials requires comprehensive knowledge of the influence of details of their microstructure on macroproperties.

In this book, important classes of heterogeneous materials – the so-called matrix composites and polycrystals – are considered. The microstructure of matrix composites consists of a continuum phase (matrix) and a set of isolated inclusions homogeneously distributed inside the matrix. In applications, there are composites with polymer, metal, ceramic, and other types of matrices. Particles of quartz, sand, glass balls, caoutchouc globules, thin metallic flakes and ribbons, glass fibers, hard carbide whiskers, etc. are used as fillers in matrix composites. Traditional reinforcement elements are long glass and carbon fibers. Other important classes of heterogeneous materials that are within the scope of this book are polycrystals, porous media, and solids with cracks and other defects.

As a rule, the microstructure of the composite materials is stochastic, and the properties and sizes of the reinforcing particles are random, as are their spatial distributions. As a result, physical fields inside the composites are also random. An important problem of the theory of composites is to determine the mean values of these fields by nonrandom external loadings. This is the content of the so-called homogenization problem. The solution of this problem allows us to calculate effective properties of composite materials, and to change the latter for homogeneous ones with equivalent properties by the analysis of the global behavior of the composite constructions. Information about higher statistical moments of physical fields in the composites (multipoint correlation functions) is necessary for the description of structurally sensitive processes, like fracture and plastic flow in composites.

Physical fields considered in this book are described by linear differential or integral equations. Many problems of the mechanics and physics of composites are reduced to the solution of the following linear operator equation

$$u - Kc_1u = u_0. \tag{1.1}$$

Here $u(x)$ is an unknown function that presents the field in the composite medium, $c_1(x)$ is a known random function that describes the properties of the composite, $u_0(x)$ is a known nonrandom external field applied to the composite. K is a nonrandom operator that may be represented in the form of an integral operator

$$(Kf)(x) = \int K(x - x')f(x')dx', \tag{1.2}$$

where the kernel $K(x)$ is a generalized function. The solution $u(x)$ of equation (1.1) is a random function statistically dependent on $c_1(x)$. Presence of the product c_1u in (1.1) and dependence of u on c_1 make this problem statistically nonlinear.

If the function $c_1(x)$ is sufficiently small, the Neumann series of the solution of (1.1) converges

$$u = u_0 + Kc_1u_0 + Kc_1Kc_1u_0 + \cdots = \sum_{n=0}^{\infty}(Kc_1)^n u_0. \tag{1.3}$$

In order to calculate the mean value of the field $u(x)$, let us average both parts of this equation over the ensemble realizations of the random function $c_1(x)$. As the result, we find an equation for the mean field $\langle u(x)\rangle$ in the form

$$\langle u\rangle = u_0 + K\langle c_1\rangle u_0 + K\langle c_1Kc_1\rangle u_0 + K\langle c_1Kc_1Kc_1\rangle u_0 \ldots . \tag{1.4}$$

On the right-hand side of this equation, there appear an infinite number of multipoint correlation functions of the random field $c_1(x)$

$$\langle c_1(x)\rangle, \quad \langle c_1(x)c_1(x_1)\rangle, \quad \langle c_1(x)c_1(x_1)c_1(x_2)\rangle, \ldots \tag{1.5}$$

For actual composite materials, construction of the many-point correlation functions is not a trivial task. For instance, for a nonoverlapping random set of spherical inclusions, only the so-called Perkus–Yevick two-point correlation function is reliable up to high volume concentrations of inclusions. The construction of the three-point correlation function of a nonoverlapping set of spheres – a function that plays an important role in the molecular theory of liquids – was the goal of many research studies. Nevertheless, neither a good approximation to this function, nor efficient numerical algorithms for its calculations, can be found in the literature. Note that the composites that are interesting in applications usually consist of contrasting components, and the corresponding function $c_1(x)$ in (1.1) is not small. Therefore, one has to keep many terms in the series (1.4) in order to have reliable parameters for the mean fields in the composites. But technical difficulties in constructing multipoint correlation functions of the random function $c_1(x)$ that corresponds to an actual composite material are difficult to overcome.

In a number of works devoted to the mechanics of heterogeneous media, methods of approximate summations of series (1.4) were developed. The results of these works may be found, e.g., in [13], [102], [189]. By making some assumptions about the statistical properties of the random function $c_1(x)$ that essentially simplify the structure of the correlation functions (1.5) (perfectly disordered materials), the corresponding series can be summed, and finite equations for the mean fields and for the effective properties of the composites may be obtained. The materials with these properties are called unstructured blends, but the detailed area of application of this approximation remains unclear.

Another branch of the theory of composite materials was opened by the well-known work of Hashin and Strikman [58], where variational principles were used for the calculation of bounds for the effective properties of composites that consist of several different phases. This branch of the theory of composites was extensively developed during the past few decades, and its results are presented, e.g., in the books of Milton [148], Torquato [199], and series of surveys mentioned in these books. For composites with contrasting properties of the components, the variational bounds turn out to be very wide. But the variational formulation of the problems of the heterogeneous media is a solid basis for the construction of approximate solutions of the homogenization problem. Such approximations may be obtained by an appropriate choice of the so-called trial fields in the composite media. In static problems for composites with ellipsoidal inclusions, the simplest trial field (to be exact, the polarization field inside inclusions) may be taken as constant. Starting with the works of Walpole [207], [208], and then in a series of works by Willis and coauthors [215], [194], [195], [216], [196], this approach was successfully used for the solution of the homogenization problem in statics and for long wave propagation in composites.

A special field of studies was composites with a periodic distribution of components. The exact asymptotic theory of homogenization of periodic composites was developed in the 1970s and 1980s, and the results of this theory are presented in the books by Sanches-Palensia [180], Bakhvalov and Panasenko [8], Pobedrya [169], Zhikov, Koslov, and Oleinik [219]. In the framework of this theory, the homogenization problem was reduced to a standard problem for an elementary cell of the composite material, and numerical or analytical solutions of this problem allow calculation of the effective properties of the composite. Note that in the plane case, many exact results for the effective elastic properties of composites with regular microstructures were presented in the book by Grigoliuk and Filshtinsky [57] before the appearance of the asymptotic theory of homogenization.

Extensive growth of computer capacities during the past decades has opened the possibility of direct numerical simulation of the properties of composite materials. The finite element and other numerical methods are becoming efficient tools for studying linear and nonlinear effective properties of composite materials [152], [181]. But these simulations require powerful supercomputers and are not trivial tasks even now. The numerical simulations are specially difficult for wave propagation when the length of the propagating wave is comparable to, or shorter than, the sizes of the inclusions.

The application of self-consistent methods to the solution of the multiparticle problems in classical and quantum physics has a long history. For random structures, these problems have no exact solutions, and the main idea of self-consistent methods is to simplify a complex picture of interactions between many particles and to reduce a multiparticle problem to a problem for one particle. There is a situation in which the application of self-consistent methods is efficient: the field that acts on every particle in a random ensemble of interacting particles should not depend strongly on the specific configuration of this ensemble, but has to be a joint field of all the interacting particles. In the pioneering works of Clausius, Mossoti, Lorenz and Lorentz, Maxwell, and Rayleigh in the 19th century, self-consistent methods were used for calculation of the effective dielectric properties of composites with spherical particles. The equations they obtained were simple, and satisfactorily described experimental data in quasistatic (long wave) regions. (The historical survey of these works can be found in [111]). Later on, in the quantum theory of the nucleus (Hartree–Fock method), the theory of phase-transition (Weiss method, Landau theory) self-consistent schemes were successfully used to obtain some approximate solutions that turned out to be valid in many important cases. A survey of the application of self-consistent methods to the analysis of the physical properties of disordered crystalline solids can be found in [44].

The first self-consistent solutions in the mechanics of heterogeneous media were obtained in the middle of the 20th century in the works of Hershey [61], Kerner [95], Kröner [100], Budiansky [25], Hill [63]. In these works, self-consistent methods were used for calculation of effective elastic properties of polycrystals and matrix composites. Although nowadays there are numerous works where various modifications of self-consistent methods are applied to the solution of static and wave propagation problems of physics of mechanics of heterogeneous media, it is possible to point out two principal approaches to self-consistency: the effective field and the effective medium methods. These methods are based on different assumptions and, generally speaking, give different results when applied to the same composite medium. Both methods are based on two types of hypotheses: the first one reduces the multiparticle problem to the problem for one particle, and the second hypothesis is the condition of self-consistency (see below). It is necessary to emphasize that the hypotheses of the methods are introduced in a heuristic manner, and it is difficult to evaluate *ad hoc* the region of their validity. Only the comparisons of the predictions of the methods with experimental data or with exact solutions (when the latter exist) allow us to estimate the regions where the methods give reliable results.

Let us consider the dielectric properties of a composite medium containing a random set of spherical inclusions with volume concentration p. Let ε_0 be the dielectric permittivity of the matrix phase and ε that for the inclusions. The effective medium method (EMM) is based on two hypotheses: every inclusion may be considered as an isolated one on the medium with the effective properties of the composite; and that the field that acts on every inclusion is the external field applied to the medium (the condition of self-consisetncy). These hypotheses lead to an equation for the effective dielectric permittivity ε_* of the composite in the form

$$\varepsilon_* = \varepsilon_0 + p\frac{3(\varepsilon - \varepsilon_0)\varepsilon_*}{2\varepsilon_* + \varepsilon}. \tag{1.6}$$

This algebraic equation for the effective parameter ε_*, known as the Bruggeman formula, was obtained in 1935 in [23]. Its derivation is presented in detail in Volumes I and II of this book.

The effective field method (EFM) is based on two slightly different hypotheses: every inclusion in the composite behaves as an isolated one in the original matrix, but the local exciting field (effective field) that acts on this inclusion is the sum of the external field applied to the medium and the perturbation of this field by surrounding inclusions. In the simplest version of the method (the quasicrystalline approximation) the effective field is assumed to be constant and the same for all the inclusions. This method leads to an equation for the effective dielectric permittivity of the composite with spherical inclusions in the form

$$\varepsilon_* = \varepsilon_0 + \frac{3p(\varepsilon - \varepsilon_0)\varepsilon_0}{3\varepsilon_0 + (1-p)(\varepsilon - \varepsilon_0)}. \tag{1.7}$$

This is the so-called Maxwell Garnet equation for the effective permittivity that appeared in 1904 in [53]. The difference in the solutions of the EMM and EFM is clear seen from Equations (1.6), (1.7). The comparison of these solutions, and the areas of their applications are considered in Volumes 1 and 2 of the book.

For application of self-consistent methods, a heterogeneous medium should have specific features: a set of similar particles should be indicated in the medium. Such a particle may be an inclusion in the matrix composites, a pore in a porous medium, a crack in damage solids, crystal grain in the polycrystalline aggregate, etc. Any self-consistent method reduces the problem of interaction between many inclusions to a one-particle problem. This problem is the determination of the physical fields in an infinite homogeneous medium with an isolated inclusion by the action of an external field of a simple structure. Because of the importance of this problem for all the self-consistent methods, Part I of the book is devoted to its solution. Attention is focused on the following types of inclusions.

- Ellipsoidal inclusions and their limit forms
- Isotropic spherical or cylindrical inclusions with radial dependence of thermoelastic properties
- Thin inclusions with properties that are much harder or softer then the properties of the matrix
- Hard axisymmetric fibers

For ellipsoidal inclusions, the solution of the one-particle problem may be obtained in an explicit form on account of a remarkable polynomial conservation property: a polynomial external field produces a polynomial field of the same degree inside an ellipsoidal inclusion. This property is used in Chapter 3 to construct the solution of the elasticity problem for a homogeneous ellipsoidal inclusion and its particular forms: oblate and prolate spheroids, cylindrical inclusions, elliptical cracks.

Radially heterogenous spherical or cylindrical inclusions often occur in polymer–polymer and other types of composites. An efficient numerical algorithm for the solution of the elasticity and thermoelasticity problems for such inclusions for a constant external field is proposed in Chapter 3.

Chapter 4 is devoted to one-particle problems for thin inclusions. Inclusions in the form of hard flakes and ribbons are often used as fillers in composites, to increase their stiffness. On the other hand, thin soft inclusions and cracks are good models of damaged regions in solids. In considering thin inclusions in a homogeneous elastic medium, it is natural to reduce the exact elasticity problem to an approximate equation on the middle surface of the inclusion. To this end, asymptotic methods are used in Chapter 4 for the construction of the principal terms of the asymptotic expansion of the elastic fields in a medium with an inclusion. The solutions are phrased in terms of dimensionless parameters: the ratio δ_1 of minimal and maximal linear sizes of the inclusions; the ratio δ_2 of the property of the material of the matrix and the inclusion. The parameter δ_1 is always small, and the parameter δ_2 is small for inclusions softer than the medium and large for inclusions harder than the medium. For a thin soft (cracklike) inclusion, the method of matching of external (in the medium) and internal (in the inclusion) asymptotic expansions reduces the problem to a boundary value problem with the following conditions on the middle surface Ω of the inclusion

$$[u]_{\Omega} = b, \quad [n\sigma]_{\Omega} = 0, \quad n\sigma|_{\Omega} = \lambda b. \tag{1.8}$$

Here b is the vector of the jump $[u]$ of the displacement field u on Ω, $n\sigma$ is the normal component of the stress field on Ω. The parameter λ depends on the elastic moduli tensor C of the inclusion and its transverse dimension h

$$\lambda = \frac{nCn}{h}. \tag{1.9}$$

For a hard inclusion, the problem is reduced to the boundary value problem with other conditions on the middle surface Ω

$$[\theta\varepsilon]_\Omega = 0, \ \theta\varepsilon|_\Omega = \mu\overline{\sigma}, \ \partial\overline{\sigma}|_\Omega = -[n\sigma]_\Omega . \tag{1.10}$$

Here $\theta\varepsilon$ is the tangential component of the strain field on Ω (θ is the projector on the tangent plane), $\overline{\sigma}$ is the total force acting in the transverse section of the inclusion, ∂ is the gradient along Ω. The parameter μ is connected with the elastic properties and the transverse size of the inclusion:

$$\mu = \frac{\theta C^{-1}\theta}{h}. \tag{1.11}$$

In Chapter 4, these boundary value problems are reduced to the solution of integral equations on Ω. Explicit solutions for thin ellipsoidal inclusions are presented in that chapter.

In Chapter 5, a hard fiber that is a traditional reinforcement element in the composites is considered. The asymptotic analysis of this problem shows that the main component of the stress field inside the fiber is the axial stress σ_m. This component satisfies an ordinary differential equation of the second order:

$$\frac{d}{d\xi}\left(\alpha^2\frac{d}{d\xi}\sigma_m\right) - q^2\sigma_m = -q^2 C\varepsilon^0, \ q^2 = -\left(\delta_2\delta_1^2 \ln \delta_1\right)^{-1}. \tag{1.12}$$

Here ξ is the coordinate along the fiber axis, the function $\alpha(\xi)$ is proportional to the radius of the cross section of the fiber, ε^0 is the external strain field applied to the medium. The boundary conditions for this equation follow from minimization of the discrepancy of its solution and the solution of the exact integral equation of the problem. These conditions depend on the shapes of the ends of the fiber. Examples of the solutions of this equation for fibers of various forms are presented in Chapter 5.

In Chapter 6, fields with scalar potential are considered. Examples of such fields are temperature and electric fields in composite materials. The fields in an infinite homogeneous medium with an isolated inclusion of various forms are obtained in that chapter.

Chapters 7–9 are devoted to the solution of the homogenization problem in statics. Self-consistent methods are applied to the calculation of the effective statical properties of heterogeneous materials. In these chapters, composite materials with various types of inclusions, porous media, media with cracks and cracklike inclusions, are considered. The algorithm for the calculation of the effective properties based on the effective medium and effective field methods are derived, and explicit equations for elastic, thermoelastic, dielectric and electroconductive properties of composites are presented there.

Predictions of different self-consistent methods for composites with various types of inclusions are analyzed and compared. Experimental data existing in the literature are used to indicate regions of application of the various self-consistent solutions.

A substantial portion of Chapter 7 is devoted to the effective field method. This method reduces the homogenization problem to the construction of the mean local exciting field $\langle\sigma^*(x)|x\rangle$ that acts on every inclusion in the composite. Here $\sigma^*(x)$ is the random local exciting stress field, and $\langle\cdot|x\rangle$ is the mean under the condition that the point x belongs to the region occupied by the inclusions. In the simplest version of the method, quasicrystalline approximation , the mean $\langle\sigma^*(x)|x\rangle$ is assumed to be constant and the same for all the inclusions. The effective properties of composites with ellipsoidal inclusions, thin soft and hard inclusions, hard short and long fibers are obtained in Chapter 7 by this version of the method. Note that the equation for the effective parameters of composites with ellipsoidal inclusions obtained by this method coincides with the approximate solution of the homogenization problem on the basis of variational formulation if the polarization fields (trial fields) inside inclusions are taken as constant. The equation so obtained for the mean $\langle\sigma^*(x)|x\rangle$ includes a specific two-point correlation function of the random field of inclusions, and through this function, geometrical characteristics of the random field of inclusions influence the effective properties of the composites. In Chapter 7, composites with a periodic lattice of inclusions are considered in the framework of the effective field method, and its predictions are compared with exact solutions of the homogenization problem.

Chapter 8 is devoted to the improvement of predictions of the effective field method. The hypothesis of the quasicrystalline approximation is replaced by a more complex one. This leads to equations that include multipoint conditional means of the local exiting field that acts on every inclusion in the composite. The equation for the two point mean $\langle\sigma^*(x_1)|x_1, x_2\rangle$, where $\langle\cdot|x_1, x_2\rangle$ is the mean under the condition that points x_1 and x_2 are in different inclusions, gives more precise information about the structure of the local exciting fields. The equations for the mean $\langle\sigma^*(x_1)|x_1, x_2\rangle$ are obtained in Chapter 8. Information about geometrical properties of random set of inclusions in the form of two- and three-point correlation functions presents in these equations. The mean $\langle\sigma^*(x_1)|x_1, x_2\rangle$ allows us to evaluate higher statistical moments of elastic fields in composites. For rapidly changing external fields, the operator of the effective properties of the composites turns out to be non-local, and its properties are analyzed in that chapter.

Temperature and electric fields in matrix composites are considered in Chapter 9. Application of the EMM and EFM to the solution of the corresponding homogenization problems is presented in that chapter. An improved effective field method is developed for the calculation of the effective dielectric properties of composites with high volume concentrations of inclusions. The predictions of the method are compared with numerical solutions of the electrostatic homogenization problem for composites with spherical inclusions.

2. An elastic medium with sources of external and internal stresses

In this chapter, some results of the elasticity theory that are frequently used in the mechanics of heterogeneous materials are presented. Two types of sources of stress fields in a homogeneous media are considered: body forces as sources of external stress, and dislocation moments as the sources of internal stress. Integral representations of elastic fields arising from stress sources distributed in 3D-regions, on surfaces, and on lines are given and discussed. Discontinuities of the elastic fields on the borders of the regions containing the stress sources are indicated. Multipole expansions are used for presentation of the elastic fields far from the stress sources.

2.1 Medium with sources of external stresses

Let us consider a homogeneous elastic medium that occupies the whole 3D-space. The tensor of elastic moduli $C^0_{\alpha\beta\lambda\mu}$ of the medium is a positive constant four rank tensor symmetric over the first and second pairs of indices as well as over transposition of the pairs of indices. In what follows, the components of tensors are considered in Cartesian bases; summation with respect to repeated indices is implied.

One source of stress in an elastic medium is external body forces. Let the distribution of these forces in space be described by a vector function $q_\alpha(x)$, and let x be an arbitrary point of the medium with coordinates (x_1, x_2, x_3). We suppose that $q_\alpha(x)$ is a generalized function with a finite support ($q_\alpha(x) = 0$ outside some finite region). In the framework of the classical theory of elasticity, the displacement vector $u_\alpha(x)$ in the medium satisfies the following system of linear differential equations

$$(L_{\alpha\beta}u_\beta)(x) = q_\alpha(x)\,, \quad L_{\alpha\beta} = -\nabla_\lambda C^0_{\lambda\alpha\beta\mu}\nabla_\mu, \; \nabla_\alpha = \frac{\partial}{\partial x_\alpha}. \tag{2.1}$$

The strain $\varepsilon_{\alpha\beta}(x)$ and stress $\sigma_{\alpha\beta}(x)$ tensors in the medium are related to the vector $u_\alpha(x)$ by the equations

$$\varepsilon_{\alpha\beta}(x) = \mathrm{def}_\alpha u_\beta(x)\,, \quad \sigma_{\alpha\beta}(x) = C^0_{\alpha\beta\lambda\mu}\varepsilon_{\lambda\mu}(x)\,, \tag{2.2}$$

$$\mathrm{def}_\alpha u_\beta(x) = \nabla_{(\alpha}u_{\beta)}(x) = \frac{1}{2}\left(\nabla_\alpha u_\beta(x) + \nabla_\beta u_\alpha(x)\right). \tag{2.3}$$

The parentheses in indices mean symmetrization.

We can also write the system of equations for the elastic stress and strain fields in the form that is equivalent to (2.1)

$$\nabla_\alpha \sigma_{\alpha\beta}(x) = -q_\alpha(x)\,, \quad \sigma_{\alpha\beta}(x) = C^0_{\alpha\beta\lambda\mu}\varepsilon_{\lambda\mu}(x)\,, \quad \mathrm{Rot}_{\alpha\beta\lambda\mu}\varepsilon_{\lambda\mu}(x) = 0\,, \tag{2.4}$$

where the incompatibility operator Rot is determined by the equation

$$\mathrm{Rot}_{\alpha\beta\lambda\mu}\varepsilon_{\lambda\mu}(x) = \epsilon_{\alpha\nu\lambda}\epsilon_{\beta\rho\mu}\nabla_\nu\nabla_\rho\varepsilon_{\lambda\mu}(x)\,. \tag{2.5}$$

Here $\epsilon_{\alpha\nu\lambda}$ is the antisymmetric Levi–Civita tensor. The system (2.4) is called the system of the elasticity theory for external stresses [106].

The sources of the external stress include the forces applied at infinity. The corresponding strain and stress fields satisfy the homogeneous system of equations (2.4) ($q(x) = 0$). If the forces at infinity are absent, the solution of (2.1) vanishing at infinity is presented in the form:

$$u_\alpha(x) = \int G_{\alpha\beta}(x - x')q_\beta(x')dx'. \tag{2.6}$$

Here $G_{\alpha\beta}(x)$ is the Green function for the elastic displacements that satisfies the equation

$$L_{\alpha\beta}G_{\beta\lambda}(x) = \delta(x)\delta_{\alpha\lambda} \tag{2.7}$$

where $\delta(x)$ is the 3D-Dirac delta-function, $\delta_{\alpha\beta}$ is the Kronecker symbol, and $G(x) \to 0$ when $|x| \to \infty$.

It is known [131] that for an arbitrary anisotropic homogeneous medium, $G(x)$ is an even homogeneous function of the degree -1

$$G_{\alpha\beta}(x) = \frac{1}{|x|}g_{\alpha\beta}(n)\,, \quad n_\alpha = \frac{x_\alpha}{|x|} \tag{2.8}$$

Explicit forms of $G_{\alpha\beta}(x)$ may be obtained for an isotropic medium, a transversely isotropic medium, and a medium with hexagonal symmetry [131]. For of an isotropic medium with Lamé constants λ_0 and μ_0, the tensor $G_{\alpha\beta}(x)$ has the form

$$G_{\alpha\beta}(x) = \frac{1}{8\pi\mu_0|x|}\left[(2 - \kappa_0)\delta_{\alpha\beta} + \kappa_0 n_\alpha n_\beta\right]\,, \quad \kappa_0 = \frac{\lambda_0 + \mu_0}{\lambda_0 + 2\mu_0} \tag{2.9}$$

Let us consider the Green tensor for a transversely isotropic medium. If the direction of the isotropy axis is defined by a unit vector m, the tensor of elastic moduli C^0 of the medium is presented in the form

$$C^0 = k_0P^2 + 2m_0\left(P^1 - \frac{1}{2}P^2\right) + l_0(P^3 + P^4) + 4\mu_0P^5 + n_0P^6. \tag{2.10}$$

Here the tensors P^i are defined by the equations

$$P^1_{\alpha\beta\lambda\mu} = \theta_{\alpha)(\lambda}\theta_{\mu)(\beta}\,, \quad P^2_{\alpha\beta\lambda\mu} = \theta_{\alpha\beta}\theta_{\lambda\mu}\,, \tag{2.11}$$

$$P^3_{\alpha\beta\lambda\mu} = \theta_{\alpha\beta}m_\lambda m_\mu\,, \quad P^4_{\alpha\beta\lambda\mu} = m_\alpha m_\beta\theta_{\lambda\mu}, \tag{2.12}$$

$$P^5_{\alpha\beta\lambda\mu} = m_{(\alpha}\theta_{\beta)(\lambda}m_{\mu)}, \quad P^6_{\alpha\beta\lambda\mu} = m_\alpha m_\beta m_\lambda m_\mu\,. \tag{2.13}$$

$$\theta_{\alpha\beta} = \delta_{\alpha\beta} - m_\alpha m_\beta. \tag{2.14}$$

In this notation, k_0 is the plane strain bulk modulus for lateral dilatation without longitudinal extension, m_0 is the shear modulus for any transversal to the m direction, μ_0 is the longitudinal shear modulus, n_0 is the modulus of longitudinal uniaxial extension, l_0 is the associated cross-modulus. These five independent elastic moduli of a transversely isotropic medium are related to five components of the Voigt matrix c^0_{ij} commonly used in the literature [160] for the presentation of the elastic moduli of anisotropic solids

$$c^0_{11} = k_0 + m_0, \quad c^0_{12} = k_0 - m_0, \quad c^0_{13} = l_0, \tag{2.15}$$
$$c^0_{44} = \mu_0, \quad c^0_{33} = n_0, \quad c^0_{23} = c^0_{13}, \quad c^0_{55} = c^0_{44}, \quad c^0_{66} = m_0. \tag{2.16}$$

In a cylindrical coordinate system (ρ, φ, z) with the z-axis directed along the vector m , the tensor $g_{ik}(\theta, \varphi)$ in (2.8) is written in the form

$$g_{ik}(\theta, \varphi) = g_{\varphi\varphi}(\theta)e^\varphi_i e^\varphi_k + g_{\rho\rho}(\theta)e^\rho_i e^\rho_k + g_{zz}(\theta)e^z_i e^z_k$$
$$+ g_{\rho z}(\theta)(e^\rho_i e^z_k + e^z_i e^\rho_k), \tag{2.17}$$

where $e^\rho, e^\varphi, e^z = m$ are the basis vectors of the cylindrical coordinates system (ρ, φ, z)

$$g_{\varphi\varphi}(\theta) = \sum_{l=1}^{3} \frac{(q_l - p_l\xi_l)\sin^2\theta - p_l\cos^2\theta}{\sin^2\theta\sqrt{\xi_l\sin^2\theta + \cos^2\theta}}, \tag{2.18}$$

$$g_{\rho\rho}(\theta) = \sum_{l=1}^{3} \frac{q_l\sin^2\theta + p_l\cos^2\theta}{\sin^2\theta\sqrt{\xi_l\sin^2\theta + \cos^2\theta}}, \tag{2.19}$$

$$g_{\rho z}(\theta) = \sum_{l=1}^{3} \frac{r_l\cos\theta}{\sin\theta\sqrt{\xi_l\sin^2\theta + \cos^2\theta}}, \tag{2.20}$$

$$g_{\rho z}(\theta) = \sum_{l=1}^{3} \frac{s_l\cos\theta}{\sqrt{\xi_l\sin^2\theta + \cos^2\theta}}. \tag{2.21}$$

Here the coefficients p_l, q_l, r_l, s_l and ξ_1, ξ_2, ξ_3 depend on the elastic moduli of the medium

$$p_l = \frac{1}{\Delta_l}\left[-k_0(n_0 - \xi_l\mu_0) + (l_0+\mu_0)^2\right], \tag{2.22}$$

$$q_l = \frac{1}{\Delta_l}\left[(\mu_0 - \xi_l(k_0+m_0))(n_0 - \xi_l\mu_0) + \xi_l(l_0+\mu_0)^2\right], \tag{2.23}$$

$$r_l = \frac{1}{\Delta_l}(l_0+\mu_0)(\mu_0 - \xi_l m_0), \tag{2.24}$$

$$s_l = \frac{1}{\Delta_l}(\mu_0 - \xi_l(k_0+m_0))(\mu_0 - \xi_l m_0), \tag{2.25}$$

$$\Delta_l = 4\pi\mu_0 m_0(k_0+m_0)\prod_{\substack{j=1\\(j\neq l)}}^{3}(\xi_j - \xi_l), \quad \xi_1 = \frac{\mu_0}{m_0} \tag{2.26}$$

and ξ_2, ξ_3 are the roots of the quadratic equation

$$(k_0+m_0)\mu_0\xi^2 + \left[l_0^2 + 2l_0m_0 - n_0(k_0+m_0)\right]\xi + n_0\mu_0 = 0. \tag{2.27}$$

Let us consider examples of external stress sources.

Concentrated force. For a concentrated force, the density $q_\alpha(x)$ in (2.1) and (2.4) has the form

$$q_\alpha(x) = Q_\alpha\delta(x), \tag{2.28}$$

where Q_α is a constant vector (Fig. 2.1a).

Such a density corresponds to the concentrated force Q_α, applied at the origin ($x = 0$). It follows from (2.6) that the displacement field due to this source is defined by the equation

$$u_\alpha(x) = G_{\alpha\beta}(x)Q_\beta\,. \tag{2.29}$$

Force dipole. Let a concentrated force Qe_α act at the point $x = 0$ in the direction of a unit vector e_α, and an opposite force $(-Qe_\alpha)$ act at the point with coordinates $x_\alpha = he_\alpha$ (Fig. 2.1b). According to (2.28), (2.29) the displacement field due to action of these two forces has the form

$$u_\alpha(x) = [G_{\alpha\beta}(\mathbf{x}) - G_{\alpha\beta}(\mathbf{x} - h\mathbf{e})]Qe_\beta\,. \tag{2.30}$$

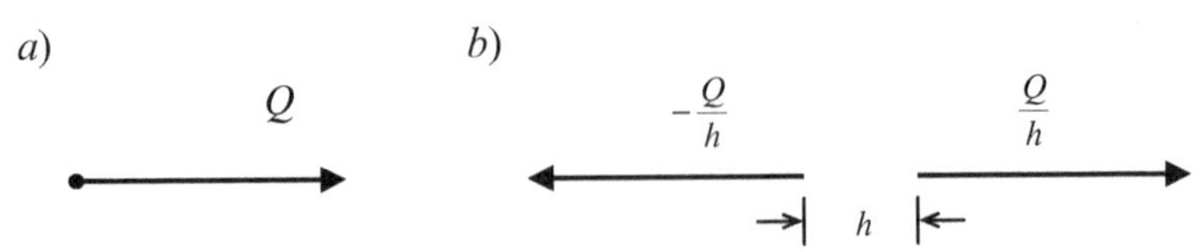

Fig. 2.1. Concentrated force (a), force dipole (b).

Suppose that $Q = h^{-1}Q^0$ and go to the limit $h \to 0$ in (2.30). As a result, we obtain the source that is called an elementary force dipole with intensity Q^0. The displacement field from such a source is

$$u_\alpha(x) = \lim_{h\to 0} \frac{1}{h}[G_{\alpha\beta}(\mathbf{x}) - G_{\alpha\beta}(\mathbf{x} - h\mathbf{e})]e_\beta Q^0 = \nabla_\lambda G_{\alpha\beta}(x) e_\lambda e_\beta Q^0 \,. \quad (2.31)$$

It follows from (2.6) and the definition of the derivatives of the generalized function $\delta(x)$ [54] that the body force $q(x)$ corresponding to the force dipole is defined by the equation

$$q_\alpha(x) = Q^0 e_\alpha e_\beta \nabla_\beta \delta(x) \,. \quad (2.32)$$

For such a body force, the displacement field $u(x)$ calculated from (2.6) coincides with (2.31).

Let the density of the external forces be defined by the equation

$$q_\alpha(x) = Q_{\alpha\beta} \nabla_\beta \delta(x) \,, \quad (2.33)$$

where $Q_{\alpha\beta}$ is a constant tensor. Note that the total force $\bar{Q}_\alpha$ of such a force distribution is equal to zero

$$\bar{Q}_\alpha = \int q_\alpha(x) dx = \int Q_{\alpha\beta} \nabla_\beta \delta(x) dx = 0 \,. \quad (2.34)$$

If $Q_{\alpha\beta}$ is a symmetric tensor, there exists an orthogonal basis $e_\alpha^{(i)}$ $(i = 1, 2, 3)$, where $Q_{\alpha\beta}$ is represented in the form

$$Q_{\alpha\beta} = Q^1 e_\alpha^{(1)} e_\beta^{(1)} + Q^2 e_\alpha^{(2)} e_\beta^{(2)} + Q^3 e_\alpha^{(3)} e_\beta^{(3)} \,. \quad (2.35)$$

Here Q^i $(i = I, 2, 3)$ are the eigenvalues of the tensor $Q_{\alpha\beta}$. As follows from (2.32), the source (2.33) is the sum of three elementary dipoles with intensities Q^1, Q^2, Q^3 acting along the main axes of the tensor $Q_{\alpha\beta}$.

The total moment M_α of the forces in (2.33) has the form

$$M_\alpha = \int \epsilon_{\alpha\beta\lambda} x_\beta Q_{\lambda\mu} \nabla_\mu \delta(x) dx = -\epsilon_{\alpha\beta\lambda} Q_{[\lambda\beta]} \,, \quad (2.36)$$

$$Q_{[\alpha\beta]} = \frac{1}{2}(Q_{\alpha\beta} - Q_{\beta\alpha}) \,. \quad (2.37)$$

Thus, the antisymmetric part of the tensor $Q_{\alpha\beta}$ ($Q_{[\alpha\beta]}$) defines the total moment of the source in (2.33). In what follows, the tensor $Q_{\alpha\beta}$ is called the generalized force moment. Such a source produces the following displacement field in the medium

$$u_\alpha(x) = \nabla_\lambda G_{\alpha\beta}(x) Q_{\beta\lambda} \,. \quad (2.38)$$

Let $q_{\alpha\beta}(x)$ be the density of the generalized force moments distributed in a finite region of 3D-space. The displacement field from such a source is defined by the equation

$$u_\alpha(x) = \int \nabla_\lambda G_{\alpha\beta}(x - x') q_{\beta\lambda}(x') dx', \tag{2.39}$$

that coincides with (2.38) when $q_{\alpha\beta}(x) = Q_{\alpha\beta}\delta(x)$. Using Gauss' theorem one can show that the displacement field in (2.39) coincides with the displacements from the following body force distribution

$$q_\alpha(x) = \nabla_\beta q_{\alpha\beta}(x) \,. \tag{2.40}$$

Thus, the density $q_\alpha(x)$ (2.40) is equivalent to the given density of the force moments $q_{\alpha\beta}(x)$. *Vice versa*, any body force distribution $q_\alpha(x)$ is equivalent to a distribution of some force moments. It follows from (2.40) that the density $q_{\alpha\beta}(x)$ of these moments is not determined uniquely by the vector $q_\alpha(x)$. For a given $q_\alpha(x)$, the tensor $q_{\alpha\beta}(x)$ is defined with the precision of the term $q^0_{\alpha\beta}(x)$ that satisfies the equation $\nabla_\beta q^0_{\alpha\beta}(x) = 0$ and is presented in the form $q^0_{\alpha\beta}(x) = rot_{\beta\lambda}\varphi_{\lambda\alpha}(x) = \epsilon_{\beta\lambda\mu}\nabla_\mu\varphi_{\lambda\alpha}(x)$, where $\varphi_{\lambda\alpha}(x)$ is an arbitrary function.

Taking two closely placed point dipoles with densities $Q_{\alpha\beta}/h$ and $-Q_{\alpha\beta}/h$, where h is the distance between the points of the dipole applications, and considering the limit $h \to 0$, we obtain a source of a more complicated structure (a quadrupole). Repeating this procedure we come to a sequence of sources that are called multipoles. The density of the body forces that corresponds to the multipole of the n-the order is

$$q_\alpha(x) = Q_{\alpha\beta_1\beta_2\ldots\beta_n}\nabla_{\beta_1}\nabla_{\beta_2}\ldots\nabla_{\beta_n}\delta(x) \,. \tag{2.41}$$

Sources of external forces distributed on surfaces and lines. Let Ω define a Lyapunov surface in 3D-space. We introduce the delta-function $\Omega(x)$ concentrated on such a surface. The function $\Omega(x)$ acts on any arbitrary smooth function $\psi(x)$ according to the equation [54]:

$$\int \psi(x)\Omega(x)dx = \int_\Omega \psi(x)d\Omega \,, \tag{2.42}$$

where the integral in the left-hand side is calculated over the whole 3D-space. The distribution of the body forces with density $Q_\alpha(x)$ on the surface Ω is described by the volume density $q_\alpha(x)$ in the form

$$q_\alpha(x) = Q_\alpha(x)\Omega(x) \,. \tag{2.43}$$

Such a distribution of body forces is called a simple layer concentrated on the surface Ω (Fig. 2.2). The displacement field corresponding to a simple layer is determined by the equation

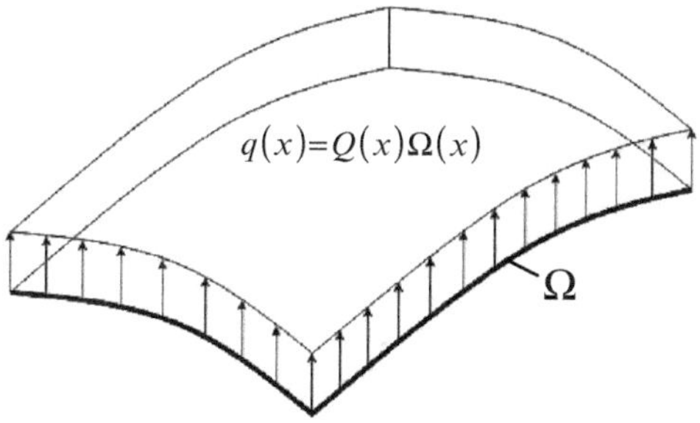

Fig. 2.2. Sources of external stresses distributed on surface Ω.

$$u_\alpha(x) = \int G_{\alpha\beta}(x-x')q_\beta(x')dx' = \int_\Omega G_{\alpha\beta}(x-x')Q_\beta(x')d\Omega' , \qquad (2.44)$$

that follows from (2.6), (2.42) and is called a potential of a simple layer in the static theory of elasticity.

Let force moments of density $Q_{\alpha\beta}(x)$ be distributed on the surface Ω. The corresponding volume density of these moments is

$$q_{\alpha\beta}(x) = Q_{\alpha\beta}(x)\Omega(x) . \qquad (2.45)$$

The displacement field produced by this source follows from (2.39) in the form

$$u_\alpha(x) = \int_\Omega \nabla_\lambda G_{\alpha\beta}(x-x')Q_{\lambda\beta}(x')d\Omega' . \qquad (2.46)$$

Note that for the density $Q_{\alpha\beta}(x)$ of a special kind

$$Q_{\alpha\beta}(x) = -C^0_{\alpha\beta\lambda\mu}n_\lambda(x)b_\mu(x) , \qquad (2.47)$$

the potential (2.46) is called a potential of a double layer in static elasticity. In (2.47), $n(x)$ is the normal to the surface Ω, $b(x)$ is a vector field on Ω.

Similar to the generalized function $\Omega(x)$ it is possible to introduce a generalized function $l(x)$ concentrated on a line l in 3D-space. The corresponding distributions of the forces and generalized moments on l are

$$q_\alpha(x) = Q_\alpha(x)l(x) , \quad q_{\alpha\beta}(x) = Q_{\alpha\beta}(x)l(x) . \qquad (2.48)$$

The displacement fields due to such sources of external stresses have the forms

$$u_\alpha(x) = \int_l G_{\alpha\beta}(x-x')Q_\beta(x')dl', \qquad (2.49)$$

$$u_\alpha(x) = \int_l \nabla_\lambda G_{\alpha\beta}(x-x')Q_{\beta\lambda}(x')dl'. \qquad (2.50)$$

Let us consider the strain and stress tensors in a homogeneous elastic medium with a volume distribution of generalized force moments. Supposing that the density of these moments $q_{\alpha\beta}(x)$ is a symmetric tensor, we obtain from (2.39)

$$\varepsilon_{\alpha\beta}(x) = -\int \mathrm{K}_{\alpha\beta\lambda\mu}(x-x')q_{\lambda\mu}(x')dx' \,, \tag{2.51}$$

$$\sigma_{\alpha\beta}(x) = C^0_{\alpha\beta\lambda\mu}\varepsilon_{\lambda\mu}(x) \,, \tag{2.52}$$

$$\mathrm{K}_{\alpha\beta\lambda\mu}(x) = -[\nabla_\alpha\nabla_\lambda G_{\beta\mu}(x)]_{(\alpha\beta)(\lambda\mu)} \,. \tag{2.53}$$

Because the function $G_{\alpha\beta}(x)$ has the singularity $|x|^{-1}$ when $x \to 0$, its second derivative $\mathrm{K}_{\alpha\beta\lambda\mu}(x)$ is a nonintegrable function with the singularity $|x|^{-3}$. To define the integral in (2.51) we consider $\mathrm{K}_{\alpha\beta\lambda\mu}(x)$ as a generalized function. The values of formally divergent integrals are calculated by a regularization procedure. In particular, the regularization of a generalized function that is the derivative of a regular (integrable) function was considered in [54]. It is shown in Appendix B.2 that the action of the generalized function $\mathrm{K}_{\alpha\beta\lambda\mu}(x)$ on a smooth finite tensor-function $\varphi_{\alpha\beta}(x)$ is defined by the equation

$$\int K_{\alpha\beta\lambda\mu}(x)\varphi_{\lambda\mu}(x)dx = \mathrm{A}_{\alpha\beta\lambda\mu}\varphi_{\lambda\mu}(0) + (\det a)\; p.v.\!\int \mathrm{K}_{\alpha\beta\lambda\mu}(ax)\varphi_{\lambda\mu}(ax)dx \,,$$

$$(ax)_\alpha = a_{\alpha\beta}x_\beta \,, \tag{2.54}$$

$$\mathrm{A}_{\alpha\beta\lambda\mu} = \frac{1}{4\pi}\int_{\Omega_1} \mathrm{K}^*_{\alpha\beta\lambda\mu}(a^{-1}k)d\Omega \,, \quad (a^{-1}k)_\alpha = a^{-1}_{\alpha\beta}k_\beta \,, \tag{2.55}$$

$$\mathrm{K}^*_{\alpha\beta\lambda\mu}(k) = \int \mathrm{K}_{\alpha\beta\lambda\mu}(x)\exp(ik\cdot x)dx \,, \quad k\cdot x = k_\alpha x_\alpha . \tag{2.56}$$

Here an arbitrary symmetric two-rank tensor $a_{\alpha\beta}$ determines a nondegenerate linear transformation of x-space, the symbol $p.v.\int \ldots.dx$ means the Cauchy principal value of the integral

$$p.v.\int \mathrm{K}(ax)\varphi(ax)dx = \lim_{\delta\to 0}\int_{|x|>\delta} \mathrm{K}(ax)\varphi(ax)dx. \tag{2.57}$$

$\mathrm{K}^*_{\alpha\beta\lambda\mu}(k)$ is the Fourier transform of the function $K_{\alpha\beta\lambda\mu}(x)$, Ω_1 is the surface of the unit sphere in the k-space of the Fourier transforms. According to (2.8) and (2.53) the equation for the tensor $\mathrm{K}^*_{\alpha\beta\lambda\mu}(k)$ has the form

$$\mathrm{K}^*_{\alpha\beta\lambda\mu}(k) = \left[k_\alpha k_\lambda G^*_{\beta\mu}(k)\right]_{(\alpha\beta)(\lambda\mu)} \,, \tag{2.58}$$

$$G^*_{\alpha\beta}(k) = L^{*-1}_{\alpha\beta}(k) \,, \quad L^*_{\alpha\beta}(k) = k_\lambda C^0_{\lambda\alpha\beta\mu}k_\mu \,. \tag{2.59}$$

Hence, each of the terms in the right-hand side of (2.54) is not determined uniquely, but the sum of these terms is unique and does not depend on the values of the tensor components. It follows from the uniqueness of the generalized function $K(x)$ defined in (2.53).

Let us return to equation (2.51) for the strain tensor. For any smooth function $q_{\alpha\beta}(x)$, the integral in (2.51) may be understood in the following sense (the indices are omitted for simplicity)

$$\varepsilon(x) = -\int \mathrm{K}(x-x')q(x')dx = -A^0 q(x) - p.v. \int \mathrm{K}(x-x')q(x')dx' \,, \quad (2.60)$$

where tensor A^0 is determined in (2.55) where $a_{\alpha\beta} = \delta_{\alpha\beta}$.

Note that the strain field $\varepsilon(x)$ in (2.51) satisfies the compatibility condition ($\mathrm{Rot}_{\alpha\beta\lambda\mu}\varepsilon_{\lambda\mu}(x) = 0$) for any finite density $q_{\alpha\beta}(x)$. This follows from the equation

$$\mathrm{Rot}_{\alpha\beta\lambda\mu}\mathrm{K}_{\lambda\mu\tau\rho}(x) = -\mathrm{Rot}_{\alpha\beta\lambda\mu}[\nabla_\mu \nabla_\rho G_{\lambda\tau}(x)]_{(\rho\tau)} \equiv 0 \,, \quad (2.61)$$

that can be checked directly using definition (2.5) of the operator *Rot*.

2.2 Medium with sources of internal stresses

Stresses in an elastic medium can exist in the absence of external forces. Such stresses are called internal, and their sources might be inhomogeneous temperature fields, plastic deformations, phase transitions that involve changing the parameters of the crystalline lattices, etc.

Let plastic deformation $m_{\alpha\beta}(x)$ take place in a finite region $V$of a homogeneous elastic medium. Because the region V is constrained by the surrounding material, there appear elastic strains $\varepsilon^e_{\alpha\beta}$ and stresses $\sigma_{\alpha\beta}$ in the medium. The total strain tensor $\varepsilon_{\alpha\beta}$, the sum of the elastic $\varepsilon^e_{\alpha\beta}$ and "inelastic" $m_{\alpha\beta}$ components, satisfies the compatibility equation

$$\mathrm{Rot}_{\alpha\beta\lambda\mu}\varepsilon_{\lambda\mu} = \mathrm{Rot}_{\alpha\beta\lambda\mu}(\varepsilon^e_{\lambda\mu} + m_{\lambda\mu}) = 0 \,. \quad (2.62)$$

According to Hooke's law, the stresses in the medium are determined by the elastic part $\varepsilon^e_{\alpha\beta}$ of the total strain tensor ($\sigma_{\alpha\beta} = C^0_{\alpha\beta\lambda\mu}\varepsilon^e_{\lambda\mu}$). In the absence of external forces, the stress tensor satisfies the equilibrium equation ($\mathrm{div}\sigma = 0$). Thus, the system of equations for the internal stresses may be written as follows:

$$\nabla_\beta \sigma_{\beta\alpha} = 0 \,, \quad \sigma_{\alpha\beta} = C^0_{\alpha\beta\lambda\mu}\varepsilon^e_{\lambda\mu} \,, \quad (2.63)$$

$$\mathrm{Rot}_{\alpha\beta\lambda\mu}\varepsilon^e_{\lambda\mu} = -\eta_{\alpha\beta} \,, \quad \eta_{\alpha\beta} = \mathrm{Rot}_{\alpha\beta\lambda\mu} m_{\lambda\mu} \,. \quad (2.64)$$

The tensor $\eta_{\alpha\beta}(x)$ in this equation is called the tensor of the dislocation density, and $m_{\alpha\beta}(x)$ is the tensor of the dislocation moments [106], [34], [35],

[36], [37]. If $\eta_{\alpha\beta}(x)$ is a finite integrable function, the solution of the system (2.63) may be presented in the form

$$\sigma_{\alpha\beta}(x) = \int Z_{\alpha\beta\lambda\mu}(x-x')\eta_{\lambda\mu}(x')dx', \tag{2.65}$$

where $Z_{\alpha\beta\lambda\mu}(x)$ is called the Green tensor for the internal stresses. The properties of this tensor are described in [106]. Taking into account the relation $\eta_{\alpha\beta}(x) = \mathrm{Rot}_{\alpha\beta\lambda\mu}m_{\lambda\mu}(x)$ and using Stokes' theorem (Appendix C), we obtain

$$\sigma_{\alpha\beta}(x) = \int \mathrm{S}_{\alpha\beta\lambda\mu}(x-x')m_{\lambda\mu}(x')dx', \tag{2.66}$$

$$\mathrm{S}_{\alpha\beta\lambda\mu}(x) = \mathrm{Rot}_{\alpha\beta\tau\rho}Z_{\tau\rho\lambda\mu}(x). \tag{2.67}$$

The function $S(x)$ in this equation may be presented also in terms of the Green function $G(x)$ defined in (2.8). Let a homogeneous medium be subjected to sources of external (body forces $q(x)$) and internal (dislocation moments $m(x)$) stresses. It follows from (2.4) and (2.63) that the stress field in the medium is the solution of the system

$$-\nabla_\beta\sigma_{\beta\alpha} = q_\alpha, \quad \mathrm{Rot}_{\alpha\beta\lambda\mu}\left(C^0\right)^{-1}_{\lambda\mu\nu\rho}\sigma_{\nu\rho} = -\eta_{\alpha\beta}. \tag{2.68}$$

Using the Green function $G_{\alpha\beta}(x)$ for the displacement and the Green function $Z_{\alpha\beta\lambda\mu}(x)$ for the internal stresses we can represent the equation for the stress tensor $\sigma_{\alpha\beta}(x)$ in the form

$$\sigma_{\alpha\beta}(x) = \int C^0_{\alpha\beta\lambda\mu}\nabla_\lambda G_{\mu\rho}(x-x')q_\rho(x')dx' + \int Z_{\alpha\beta\lambda\mu}(x-x')\eta_{\lambda\mu}(x')dx'. \tag{2.69}$$

Substituting in this equation the left-hand sides of equations for $q_\alpha(x)$ and $\eta_{\alpha\beta}(x)$ (2.68) and using Gauss's and Stokes's theorems, we obtain

$$\begin{aligned}\sigma_{\alpha\beta}(x) &= -C^0_{\alpha\beta\lambda\mu}\nabla_\lambda\int G_{\mu\rho}(x-x')\nabla'_\tau\sigma_{\tau\rho}(x')dx' \\ &\quad - \int Z_{\alpha\beta\lambda\mu}(x-x')\mathrm{Rot}_{\lambda\mu\rho\tau}(C^0)^{-1}_{\rho\tau\nu\delta}\sigma_{\nu\delta}(x')dx' = \\ &= -\int\Big[\nabla_\lambda C^0_{\alpha\beta\lambda\mu}\nabla_\tau G_{\mu\rho}(x-x') \\ &\quad + \mathrm{S}_{\alpha\beta\lambda\mu}(x-x')\left(C^0\right)^{-1}_{\lambda\mu\rho\tau}\Big]\sigma_{\rho\tau}(x')\,dx'.\end{aligned} \tag{2.70}$$

Comparing the left and right parts of this equation we obtain

$$-\nabla_\lambda C^0_{\alpha\beta\lambda\mu}\nabla_\tau G_{\mu\rho}(x) - \mathrm{S}_{\alpha\beta\lambda\mu}(x)(C^0)^{-1}_{\lambda\mu\rho\tau} = I_{\alpha\beta\rho\tau}\delta(x)\,, \tag{2.71}$$

where $I_{\alpha\beta\rho\tau} = \delta_{\alpha)(\rho}\delta_{\tau)(\beta}$ is the unit four rank tensor. Thus, the equation for the function $S_{\alpha\beta\lambda\mu}(x)$ may be written in the form

$$\mathrm{S}_{\alpha\beta\lambda\mu}(x) = C^0_{\alpha\beta\nu\rho}\mathrm{K}_{\nu\rho\tau\delta}(x)C^0_{\tau\delta\lambda\mu} - C^0_{\alpha\beta\lambda\mu}\delta(x)\,, \tag{2.72}$$

where the tensor $\mathrm{K}_{\alpha\beta\lambda\mu}(x)$ is defined in (2.53).

The function $S_{\alpha\beta\lambda\mu}(x)$, like $\mathrm{K}_{\alpha\beta\lambda\mu}(x)$ is generalized homogeneous function of the degree (−3). Its action on an arbitrary smooth finite function $m_{\lambda\mu}(x)$ is defined by the equation that follows from (2.54):

$$\int \mathrm{S}_{\alpha\beta\lambda\mu}(x)m_{\lambda\mu}(x)dx = D_{\alpha\beta\lambda\mu}m_{\lambda\mu}(0) + (\det a)\ p.v. \int \mathrm{S}_{\alpha\beta\lambda\mu}(ax)m_{\lambda\mu}(ax)dx\,, \tag{2.73}$$

$$D_{\alpha\beta\lambda\mu} = \frac{1}{4\pi}\int_{\Omega_1} \mathrm{S}^*_{\alpha\beta\lambda\mu}(a^{-1}k)d\Omega\,, \tag{2.74}$$

$$\mathrm{S}^*_{\alpha\beta\lambda\mu}(k) = C^0_{\alpha\beta\rho\tau}\mathrm{K}^*_{\rho\tau\delta\nu}(k)C^0_{\delta\nu\lambda\mu} - C^0_{\alpha\beta\lambda\mu}\,. \tag{2.75}$$

Here $\mathrm{S}^*_{\alpha\beta\lambda\mu}(k)$ is the Fourier transform of the function $S_{\alpha\beta\lambda\mu}(x)$, the other notation is the same as in (2.54).

Note that the field $\sigma_{\alpha\beta}(x)$ in (2.66) satisfies automatically the equation of equilibrium ($\nabla_\beta\sigma_{\beta\alpha} = 0$) for any finite integrable function $m_{\alpha\beta}(x)$. It is the consequence of the equation

$$\nabla_\beta \mathrm{S}_{\beta\alpha\lambda\mu}(x) = \nabla_\beta \mathrm{Rot}_{\beta\alpha\rho\tau}Z_{\rho\tau\lambda\mu} \equiv 0\,, \tag{2.76}$$

that follows from Equation (2.5) for the operator Rot.

Let us consider internal stresses that correspond to particular distributions of dislocation moments in a homogeneous elastic medium.

Constant density of dislocation moments. If $m_{\alpha\beta}(x) = m^0_{\alpha\beta}$ is a constant tensor in the whole space, the integral in equation (2.66) for the tensor $\sigma_{\alpha\beta}(x)$ formally diverges at $x = 0$ and $x = \infty$. In order to find the value of this integral we consider the following model problem. Let the internal stresses in the medium be generated by a temperature field $T(x)$. If the deformation is absent for $T = 0$, the density of the dislocation moments $m_{\alpha\beta}(x)$ that corresponds to the temperature $T(x)$ is

$$m_{\alpha\beta}(x) = \alpha^0_{\alpha\beta}T(x)\,, \tag{2.77}$$

where $\alpha^0_{\alpha\beta}$ is the tensor of the thermal expansion coefficients of the medium. For a constant temperature field ($T(x) = T = \text{const}$), the tensor $m_{\alpha\beta}$ is also constant. The thermal stresses in the medium depend on the condition at infinity. If there is no constraint at infinity, i.e., the medium can expand (contract) freely, a constant temperature does not produce internal stresses. Thus, if $m_{\alpha\beta}(x) = m^0_{\alpha\beta}$, the value of the integral in (2.66) should be zero:

$$\int S_{\alpha\beta\lambda\mu}(x - x')m^0_{\lambda\mu}dx' = 0 . \tag{2.78}$$

If the medium is completely constrained at infinity, the change of temperature does not affect the total strain in the medium ($\varepsilon_{\alpha\beta} = \varepsilon^e_{\alpha\beta} + m^0_{\alpha\beta} = 0$), and the internal stresses are

$$\sigma_{\alpha\beta} = C^0_{\alpha\beta\lambda\mu}\varepsilon^e_{\lambda\mu} = -C^0_{\alpha\beta\lambda\mu}m^0_{\lambda\mu} = -C^0_{\alpha\beta\lambda\mu}\alpha^0_{\lambda\mu}T . \tag{2.79}$$

From this equation and (2.66) it follows that for the constraint conditions at infinity ($\varepsilon = 0$), the following equation holds

$$\int \mathrm{S}_{\alpha\beta\lambda\mu}(x - x')m^0_{\lambda\mu}dx' = -C^0_{\alpha\beta\lambda\mu}m^0_{\lambda\mu} . \tag{2.80}$$

Equations (2.78) and (2.80) define the regularization of divergent integral (2.66) when $m_{\alpha\beta}$ = const. Note that there is not a unique natural regularization of this operator on constants, and the value of this integral depends on its physical meaning in specific problems of elasticity.

A consequence of the regularizations (2.78), (2.80), and (2.72) is the definition of the action of the integral operator with the kernel $\mathrm{K}(x)$ (2.53) on constants

$$\int \mathrm{K}(x - x')dx' = \begin{cases} 0, & \text{strain constraint at infinity} \\ (C^0)^{-1}, & \text{no strain constraint at infinity} \end{cases} \tag{2.81}$$

Note that this integral can be interpreted as the deformation of the medium under the action of force moments with constant density $q^0_{\alpha\beta} = -\delta_{\alpha\beta}$.

The absence of a unique regularization of the integral operators with the kernels $S(x)$ and $\mathrm{K}(x)$ on constants becomes clear if the convolution integral in (2.66) is calculated using the Fourier transform of the integrand functions

$$\int \mathrm{S}(x - x')m^0dx' = \int \mathrm{S}^*(k)m^0\delta(k)\exp(-ik \cdot x)dk = \mathrm{S}^*(0)m^0. \tag{2.82}$$

Here the equation

$$\int \exp(ik \cdot x)dx = (2\pi)^3\delta(k) \tag{2.83}$$

is taken into account [21]. It follows from (2.58) and (2.75) that the Fourier transform $\mathrm{S}^*(k)$ of the kernel $\mathrm{S}(x)$ is a homogeneous function of k of the degree 0, i.e.,

$$S^*(k) = S^*(m), \quad m_\alpha = \frac{k_\alpha}{|k|}. \tag{2.84}$$

Therefore, the value of the function $S^*(k)$ at $k = 0$ is not defined uniquely but depends on the direction in which the vector k tends to zero. The following equation is a consequence of (2.78), (2.80):

$$S^*(0) = \begin{cases} -C^0, & \text{strain constraint at infinity} \\ 0, & \text{no strain constraint at infinity} \end{cases}. \tag{2.85}$$

The function $K^*(k)$ is also a homogeneous function of the degree 0 with respect to k, and because of (2.81), we have

$$K^*(0) = \begin{cases} 0, & \text{strain constraint at infinity} \\ (C^0)^{-1}, & \text{no strain constraint at infinity} \end{cases}. \tag{2.86}$$

Homogeneous distribution of dislocation moments in a half-space. Let the density of the dislocation moments be constant in the half-space $x_3 > 0$

$$m_{\alpha\beta}(x) = m^0_{\alpha\beta} H(x_3) \quad H(x_3) = \begin{cases} 1, & x_3 > 0 \\ 0, & x_3 < 0 \end{cases}, \tag{2.87}$$

where x_1, x_2, x_3 are the Cartesian coordinates, and $H(x_3)$ is the Heaviside function.

In order to calculate the convolution integral in (2.66) we use the convolution theorem for Fourier transforms[1]:

$$\sigma(x) = \int S(x - x')m(x')dx'$$

$$= (2\pi)^{-3} \int S^*(k)\, m^0 H^*(k) \exp(-ik \cdot x)\, dk, \tag{2.88}$$

$$H^*(k_1, k_2, k_3) = (2\pi)^2 \delta(k_1)\delta(k_2)\left[\pi\delta(k_3) + ik_3^{-1}\right]. \tag{2.89}$$

The integral in the right-hand side of (2.88) has the value

$$\int S(x - x')m(x')dx' = \frac{1}{2}[S^*(0) + S^*(n)\, \mathrm{sign}(-n \cdot x)]m^0. \tag{2.90}$$

Here it is taken into account that $S^*(0, 0, k_3) = S^*(n)$, where n is the external normal to the region $x_3 > 0$ occupied by the dislocation moments. The value of the tensor $S^*(0)$ depends on the conditions at infinity (see (2.85)). But the jump of the stress tensor on the boundary of the region containing the dislocation moments does not depend on these conditions, and is determined by an equation that follows from (2.88) and (2.90):

$$[\sigma] = \sigma^+ - \sigma^- = -S^*(n)m^0. \tag{2.91}$$

[1] In what follows, we omit tensorial indices if it does not lead to misunderstanding.

Here σ^+ is the limit value of $\sigma(x)$ when the point x tends to the plane $x_3 = 0$ from the side of the external normal n, and σ^- is the value of $\sigma(x)$ when x tends to this plane from the opposite side. The equation

$$n_\alpha S^*_{\alpha\beta\lambda\mu}(n) = n_\alpha C^0_{\alpha\beta\rho\delta} n_\rho G^*_{\nu\delta}(n) n_\tau C^0_{\tau\nu\lambda\mu} - n_\alpha C^0_{\alpha\beta\lambda\mu} = 0\,, \tag{2.92}$$

follows from (2.58), (2.59) and (2.75). Therefore, the normal components of the stress tensor $n_\alpha \sigma_{\alpha\beta}(x)$ are continuous on the boundary of the region containing the dislocation moments

$$[n_\alpha \sigma_{\alpha\beta}] = n_\alpha S^*_{\alpha\beta\lambda\mu}(n) m^0_{\lambda\mu} \equiv 0\,. \tag{2.93}$$

Taking into account (2.51), (2.90), and (2.72), we represent the total strain tensor in the medium with distribution (2.87) of the dislocation moments in the form

$$\begin{aligned}\varepsilon(x) &= (C^0)^{-1}\sigma(x) + m(x) = \int \mathrm{K}(x - x')C^0 m^0 H(-n\cdot x')dx' \\ &= \frac{1}{2}\left[\mathrm{K}^*(0) + \mathrm{K}^*(n)C^0 \mathrm{sign}(-n\cdot x)\right] m^0.\end{aligned} \tag{2.94}$$

Thus, the jump of the strain tensor $\varepsilon(x)$ on the boundary of the half-space $(x_3 > 0)$ is defined by the equation

$$[\varepsilon] = -\mathrm{K}^*(n) C^0 m^0\,. \tag{2.95}$$

The Somigliana dislocation. Let V_h be a thin region in 3D-space such that one of its characteristic sizes is much smaller than other two, and Ω be the middle surface of V_h. We denote by $h(x)$ the transverse dimension of V_h along the normal $n(x)$ to Ω. Let us cut the region V_h from the medium and subject it to forces that produce plastic strains $\varepsilon^p(x)$ in V_h. Suppose that $\varepsilon^p(x)$ is constant along the normal to the surface Ω, and determined on this surface by the equation

$$\varepsilon^p_{\alpha\beta}(x) = \frac{1}{h(x)} b_{(\alpha}(x) n_{\beta)}(x),\ x \in \Omega\,. \tag{2.96}$$

Here $b(x)$ is a vector field given on Ω. Freezing the plastic strains, we subject the region V_h to surface forces that produce only elastic deformations and return the region V_h to its previous form. Keeping the applied forces, we put the material back into the medium and glue along the boundary of the region V_h. After that, we remove the external forces. As the result, internal stresses arise in the medium. Let us find the limit distribution of these stresses when $h \to 0$. The density of the dislocation moments $m(x)$ corresponding to such a source is determined by the equation

$$m_{\alpha\beta}(x) = \lim_{h\to 0} \frac{1}{h(x)} b_{(\alpha}(x) n_{\beta)} V_h(x) = b_{(\alpha}(x) n_{\beta)}(x) \Omega(x)\,, \tag{2.97}$$

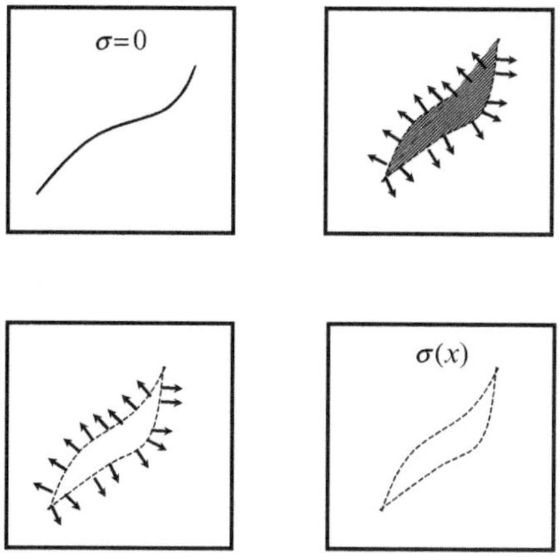

Fig. 2.3. The Somigliana dislocation.

where $V_h(x)$ is the characteristic function of the region V_h, $\Omega(x)$ is the generalized function concentrated on the middle surface of region V_h and determined in (2.42).

The density of dislocation moments in the form (2.97) is called the Somigliana dislocation. There is another simple interpretation of this source of internal stresses (see Fig. 2.3). Let us make a cut along a smooth nonclosed surface Ω in a homogeneous elastic medium, separate the edges of the cut by external forces in such a way that the points located on the opposed sides of the cut, and coinciding in the initial state, drift apart by a vector $b(x)$. We fill the void so obtained by the material of the medium and glue it to the void border. After removing the external forces, internal stresses appear in the medium. The corresponding density of the dislocation moments has the form (2.97). The stresses in the medium with the Somigliana dislocation are defined by the equation that follows from (2.66)

$$\sigma_{\alpha\beta}(x) = \int_{\Omega} \mathrm{S}_{\alpha\beta\lambda\mu}(x - x')n_\lambda(x')b_\mu(x')d\Omega' . \tag{2.98}$$

The Somigliana dislocation is connected with the elasticity problem for a crack in a homogeneous elastic medium. Let the medium be cut along a smooth nonclosed surface Ω and subjected to some external forces at infinity. As a result, the sides of the crack will be separated by a vector $b(x)$. We denote by $\sigma^0(x)$ the stress field that would have existed in the medium without the crack and by the same external forces. The stress field in the medium with the crack is the sum of the field $\sigma^0(x)$ and the field in (2.98):

$$\sigma_{\alpha\beta}(x) = \sigma^0_{\alpha\beta}(x) + \int_{\Omega} \mathrm{S}_{\alpha\beta\lambda\mu}(x - x')n_\lambda(x')b_\mu(x')d\Omega' . \tag{2.99}$$

Because of (2.76), the field $\sigma_{\alpha\beta}(x)$ satisfies the equilibrium equation $\mathrm{div}\sigma = 0$ in the whole medium, and using the condition on the crack surface we obtain the equation for the vector $b(x)$. If the edges of the crack are free from stresses, this condition is

$$n_\alpha(x)\sigma_{\alpha\beta}(x) = 0 \quad \text{when} \quad x \in \Omega.. \tag{2.100}$$

From this equation and (2.99) it follows that the equation for the vector $b(x)$ has the form

$$\int_\Omega T_{\alpha\beta}(x,x')b_\beta(x')d\Omega' = n_\beta(x)\sigma^0_{\beta\alpha}(x)\,, \quad x \in \Omega\,, \tag{2.101}$$

$$T_{\alpha\beta}(x,x') = -n_\lambda(x)\mathrm{S}_{\lambda\alpha\beta\mu}(x-x')n_\mu(x')\,. \tag{2.102}$$

This equation is considered in detail in Section 3.7.

Note that in deriving an integral equation of the crack problem, we use the double layer potential of static elasticity. This potential is determined in (2.46), and the displacement field in the medium with a crack is presented in the form

$$u_\alpha(x) = u^0_\alpha(x) - \int_\Omega \nabla_\lambda G_{\alpha\beta}(x-x')C^0_{\lambda\beta\mu\nu}n_\mu(x')b_\nu(x')d\Omega'\,, \tag{2.103}$$

where $u^0(x)$ is the displacement vector in the medium without the crack and by the same external loading. It is known that the double-layer potential is discontinuous, and the vector $u(x)$ in (2.103) has a jump on Ω that is equal to the potential density $b(x)$. The stress tensor that corresponds to the displacement field (2.103) has the form

$$\sigma_{\alpha\beta}(x) = \sigma^0_{\alpha\beta}(x) - C^0_{\alpha\beta\lambda\mu}\int_\Omega \nabla_\lambda\nabla_\nu G_{\mu\rho}(x-x')C^0_{\nu\rho\tau\delta}n_\tau(x')b_\delta(x')d\Omega'. \tag{2.104}$$

Taking into account (2.72) we see that the right-hand sides of (2.104) and (2.99) differ by the singular term $C^0_{\alpha\beta\lambda\mu}n_\lambda(x)b_\mu(x)\Omega(x)$ concentrated on the surface Ω. Outside Ω, the stress tensors in (2.99) and (2.104) coincide. This difference has the following explanation. If the solution of the crack problem is taken in the form of the double layer potential (2.103), the crack is modeled by the distribution of the force moments (dipoles) of density $Q_{\lambda\mu}(x) = -C^0_{\alpha\beta\lambda\mu}n_\lambda b_\mu(x)$ on Ω. As a result, the stress field in the medium contains a singular component proportional to $\Omega(x)$ and does not satisfy the equation of equilibrium ($\mathrm{div}\sigma = 0$) in the whole space. Equation (2.104) leads to the equations

$$\nabla_\beta\sigma_{\beta\alpha} = -q^s_\alpha\,, \quad q^s_\alpha(x) = \nabla_\lambda[C^0_{\lambda\alpha\beta\mu}n_\beta(x)b_\mu(x)\Omega(x)]\,, \tag{2.105}$$

where $q^s(x)$ is a generalized function concentrated on Ω. That is why modeling of a crack by the surface distribution of dislocation moments (2.97) is a more correct procedure. The stress field in (2.99) satisfies the equilibrium equation in the whole space, including the surface Ω, and does not contain singular components. The corresponding stress vector $t_\alpha(x) = n_\beta \sigma_{\beta\alpha}(x)$ is continuous on the surface Ω except, maybe, its border contour Γ. Because the limit values of the stress tensor in (2.104) and in (2.99) coincide when $x \to \Omega$, the equation for the vector $b(x)$ in (2.103) that follows from boundary condition (2.100) does not differ in essence from (2.101).

2.3 Discontinuities of elastic fields in a medium with sources of external and internal stresses

Let body forces with density $q(x)$be distributed in a region V whose boundary Ω is a closed Lyapunov surface. If $q_\alpha(x)$ is a smooth functions inside V, the displacement and strain fields in the medium are presented in the form of integrals with weak (integrable) singularities

$$u_\alpha(x) = \int_V G_{\alpha\beta}(x - x')q_\beta(x')dx', \tag{2.106}$$

$$\varepsilon_{\alpha\beta}(x) = \int_V \nabla_{(\alpha} G_{\beta)\mu}(x) q_\mu(x')dx'. \tag{2.107}$$

These integrals are continuous functions in the whole space.

If force moments of density $q_{\alpha\beta}(x)$ are distributed in the region V, the displacements and strains fields in the medium are presented by the integrals in (2.39) and (2.51):

$$u_\alpha(x) = \int_V \nabla_\lambda G_{\alpha\beta}(x - x')q_{\lambda\beta}(x')dx', \tag{2.108}$$

$$\varepsilon_{\alpha\beta}(x) = -\int_V \mathrm{K}_{\alpha\beta\lambda\mu}(x - x')q_{\lambda\mu}(x')dx'. \tag{2.109}$$

Here $u(x)$ is a continuous function, but $\varepsilon(x)$ is discontinuous on the surface Ω. To find the jump of $\varepsilon(x)$ on this surface, we present the integral in (2.109) in the form

$$\varepsilon(x) = -\int_V K(x - x')[q(x') - q(x)]dx' - \int_V K(x - x')dx' q(x)\,. \tag{2.110}$$

Let the function $q(x)$ outside V be defined by an arbitrary smooth continuation. For a smooth bounded function $q(x)$, the first integral in the right-hand side of (2.110) has a weak singularity and is continuous on Ω. Let us consider the limit of the second integral when x tends to a point $x_0 \in \Omega$ from outside or inside of the region V.

Let us introduce the Cartesian coordinates (y_1, y_2, y_3) with the origin at the point x_0 and the y_3-axis directed along the external normal $n(x_0)$ to Ω, and consider the limit value of the integral

$$J(y) = \int_V \mathrm{K}(y - y')dy' \tag{2.111}$$

when y tends to Ω from the side of the normal $n(x_0)$. Fix a point $y = y^0$ and introduce dimensionless coordinates $\xi_i = y_i/|y_0|$, $(i = 1, 2, 3)$.

Because $K(y)$ is a homogeneous function of the degree -3, we have

$$J(y) = J(\xi|y_0|) = \int \mathrm{K}(\xi - \xi')V(\xi')d\xi', \tag{2.112}$$

where $V(\xi)$ is the characteristic function of the region V in the coordinates ξ_i. If y^0 tends to zero, $\xi_0 = y_0/|y_0|$ is the unit vector of the direction along which the point y_0 tends to the origin. In the limit $y_0 \to 0$, the region $V(\xi)$ is transformed (in the coordinates ξ_i) into the half-space $\xi_3 < 0$, i.e. $V(\xi_1, \xi_2, \xi_3) \to H_1(\xi_1, \xi_2, \xi_3) = 1 - H(\xi_3)$, where $H(\xi_3)$ is the Heaviside function (Fig. 2.4). Thus, the following equation holds:

$$\begin{aligned} \lim_{y_0 \to 0} J(y_0) &= \int \mathrm{K}(\xi_0 - \xi')H_1(\xi')d\xi' \\ &= (2\pi)^{-3} \int \mathrm{K}^*(k)H_1^*(k)\exp(-ik \cdot \xi_0)dk\,. \end{aligned} \tag{2.113}$$

Taking into account equation (2.89) for $H^*(k)$ we obtain

$$J^+(0) = \lim_{y_0 \to 0} J(y_0) = \frac{1}{2}[\mathrm{K}^*(0) - \mathrm{K}^*(n)], \quad y_0 \notin V\,. \tag{2.114}$$

The same procedure gives us the limit of $J(y_0)$ when $y_0 \to 0$ from the inside of V :

$$J^-(0) = \lim_{y_0 \to 0} J(y_0) = \frac{1}{2}[\mathrm{K}^*(0) + \mathrm{K}^*(n)], \quad y_0 \in V\,. \tag{2.115}$$

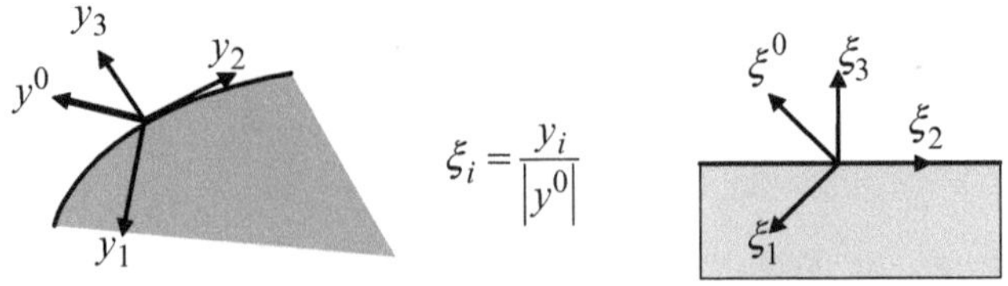

Fig. 2.4. Local coordinate systems on the border of the region containing stress sources, $|y^0| \to 0$, $|\xi^0| = 1$.

Hence, the jump of the integral $J(y)$ on the border Ω of the region V is

$$[J(0)] = J^{+}(0) - J^{-}(0) = -\mathrm{K}^{*}(n)\,. \tag{2.116}$$

Equation (2.110) leads to the equation for the jump of the potential $\varepsilon(x)$ in (2.109)

$$[\varepsilon(x_0)] = \varepsilon^{+}(x_0) - \varepsilon^{-}(x_0) = \mathrm{K}^{*}(n_0)q(x_0). \tag{2.117}$$

Here $n_0 = n(x_0)$ is the external normal to Ω at the point $x_0 \in \Omega$.

Let us consider a medium with dislocation moments distributed inside the region V. The tensor of the internal stresses $\sigma(x)$ in the medium is determined by equation (2.66), and may be presented in the form of the following sum:

$$\sigma(x) = \int_V \mathrm{S}(x-x')[m(x') - m(x)]dx' + \int_V \mathrm{S}(x-x')dx'm(x), \tag{2.118}$$

If $m(x)$ is a smooth bounded function inside V, the first term in this equation is continuous on the boundary Ω of the region V as the integral with a weak singularity. The second term has a jump on Ω, and the value of this jump follows from definition (2.72) of the function $\mathrm{S}(x)$, and Equation (2.117) for the jump of the integral in (2.109)

$$[\sigma(x_0)] = \sigma^{+}(x_0) - \sigma^{-}(x_0) = -\mathrm{S}^{*}(n_0)m(x_0)\,. \tag{2.119}$$

Here $\sigma^{+}(x_0)$ is the limit value of $\sigma(x)$ when $x \to x_0$ from the side of the external normal $n_0 = n(x_0)$ to Ω at the point x_0, $\sigma^{-}(x_0)$ is the same limit when x tends to x_0 from the opposite side. The function $\mathrm{S}^{*}(k)$ is defined in (2.75).

Let us now consider an elastic medium with sources of external stress concentrated on a smooth oriented surface Ω. The potential of a simple layer (2.44) defines the displacements in the medium by the action of body forces distributed on Ω :

$$u_\alpha(x) = \int_\Omega G_{\alpha\beta}(x-x')Q_\beta(x')d\Omega'\,. \tag{2.120}$$

If $Q_\beta(x)$ is a smooth function, the function $u(x)$ is continuous everywhere because the right-hand side of (2.120) is an integral with a weak singularity. The strain field $\varepsilon(x)$ corresponding to the displacement field (2.120) has the form

$$\varepsilon(x) = \int_\Omega \mathrm{def}\,[G(x-x')]\,Q(x')d\Omega' \tag{2.121}$$

and is discontinuous on Ω. def is the deformation operator defined in the notation list. To find the value of the jump of $\varepsilon(x)$ on Ω we present the potential $\varepsilon(x)$ in the form

$$\varepsilon(x) = \int_{\Omega} \operatorname{def}[G(x-x')]\,[Q(x')-Q(x)]d\Omega' + \int_{\Omega} \operatorname{def}[G(x-x')]\,d\Omega' Q(x) \tag{2.122}$$

and consider the limits of each term of this equation when $x \to \Omega$. The function $Q(x)$ outside Ω is determined by an arbitrary smooth continuation. The first term in (2.122) is continuous on Ω as the integral with a weak singularity. The limit value of the second integral

$$J_1(x) = \int_{\Omega} \operatorname{def}[G(x-x')]\,d\Omega' \tag{2.123}$$

is presented in the form

$$\lim_{x\to x_0\in\Omega} J_1(x) = \int_{\Omega\backslash\Omega_\delta(x_0)} \operatorname{def}[G(x-x')]\,d\Omega' + \int_{\Omega_\delta(x_0)} \operatorname{def}[G(x-x')]\,d\Omega'. \tag{2.124}$$

Here $\Omega_\delta(x_0)$ is a part of the surface Ω that is cut from it by the sphere of the radius δ centered at the point $x_0 \in \Omega$. Let us introduce the local coordinate system y_1, y_2, y_3 with the origin at the point $x_0 \in \Omega$ and the y_3-axis directed along the normal $n(x_0)$ to Ω. In these coordinates, the second integral in (2.124) takes the form

$$\lim_{x\to x_0\in\Omega} \int_{\Omega_\delta(x_0)} \operatorname{def}[G(x-x')]\,dx' = \lim_{y\to 0} \int_{\Omega_\delta(0)} \operatorname{def}[G(y-y')]\,d\Omega'. \tag{2.125}$$

In what follows, the parameter δ tends to zero. Therefore, $\Omega_\delta(0)$ may be considered as a plane disc defined by the equation $\Omega_\delta(0) = \{y_1^2 + y_2^2 < \delta^2,\ y_3 = 0\}$. Let us consider the integral

$$J^0(y) = \int_{\Omega_\delta(0)} \operatorname{def}[G(y-y')]\,d\Omega'. \tag{2.126}$$

Because $def G(y)$ is a homogeneous function of the degree -2, the value of this integral is independent of the radius δ of the disc of integration $\Omega_\delta(0)$. It allows us to write

$$J^0(y) = \lim_{\delta\to\infty} \int_{\Omega_\delta(0)} \operatorname{def}[G(y-y')]\,d\Omega' = \int \operatorname{def}[G(y-y')]\,\delta(y_3')dy'. \tag{2.127}$$

We take into account that in the limit $\delta \to \infty$ the surface $\Omega_\delta(0)$ is transformed into the plane $y_3 = 0$; the last integral is calculated over the 3D-space. Using the convolution property of this integral we find

$$J^0_{\alpha\beta\lambda}(y) = \frac{i}{2\pi} \int k_{(\alpha} G^*_{\beta)\lambda}(k)\delta(k_1)\delta(k_2)\exp(-iy \cdot k)dk$$
$$= \frac{1}{2}\delta_{3(\alpha} G_{\beta)\lambda}(0,0,1)\,\mathrm{sign}\,(y_3)\ . \tag{2.128}$$

In the invariant form, this equation is written as follows:

$$J^0_{\alpha\beta\lambda}(y) = \frac{1}{2} n_{(\alpha} G^*_{\beta)\lambda}(n)\,\mathrm{sign}\,(n \cdot y)\,, \tag{2.129}$$

where n is the normal to Ω at the point x_0.

Let δ tend to zero in (2.124). Taking into account (2.127)–(2.129) we obtain the limiting value of the integral $J_1(x)$ (J_1^+ is the limiting value from the side of the external normal, J_1^- - from the opposite side) in the form

$$J_1^{\pm}(x_0) = p.v. \int_\Omega \mathrm{def}\,[\,G(x_0 - x')]\, d\Omega' \pm \frac{1}{2}\Lambda(n_0)\,, \tag{2.130}$$

$$\Lambda_{\alpha\beta\lambda}(n_0) = n_{0(\alpha} G^*_{\beta)\lambda}(n_0)\ , \quad n_0 = n(x_0)\,. \tag{2.131}$$

The Cauchy principal value of the integral exists because $def\, G(x)$ is an odd function of x. From this equation and (2.122) follows that the limiting values of the potential (2.122) have the forms

$$\varepsilon^{\pm}(x_0) = p.v. \int_\Omega \mathrm{def}\,[G(x_0 - x')]\, Q(x')d\Omega' \pm \frac{1}{2}\Lambda(n_0)Q(x_0)\,. \tag{2.132}$$

Thus, the jump of this potential on Ω is determined by the equation

$$[\varepsilon_{\alpha\beta}(x_0)] = \Lambda_{\alpha\beta\lambda}(n_0)Q_\lambda(x_0)\,. \tag{2.133}$$

Let us consider the potential of a double layer in static elasticity (2.46), (2.47)

$$u_\alpha(x) = -\int_\Omega \nabla_\lambda G_{\alpha\beta}(x - x')C^0_{\lambda\beta\mu\nu} n_\mu(x') b_\nu(x') d\Omega'\,. \tag{2.134}$$

The properties of this potential are similar to the properties of the potential (2.121). The limiting values of this potential on the surface Ω are determined by the equation

$$u^{\pm}_\alpha(x_0) = -\int_\Omega \nabla_\lambda G_{\alpha\beta}(x_0 - x')C^0_{\beta\lambda\mu\nu} n_\mu(x') b_\nu(x') d\Omega' + \frac{1}{2} b_\alpha(x_0) \tag{2.135}$$

Here we take into account the equation

$$n_\lambda G^*_{\alpha\beta}(n) C^0_{\beta\lambda\mu\nu} n_\mu = \delta_{\alpha\nu}\,, \tag{2.136}$$

that follows from the definition (2.59) of $G^*(k)$.

Equations (2.132) and (2.135) hold for an arbitrary anisotropic medium. It follows from (2.135) that the jump of the potential of a double layer on the surface Ω is equal to the potential density $b(x)$.

2.4 Elastic fields far from the sources

Let body forces with density $q(x)$ be distributed inside a finite region V in a homogeneous elastic medium. If $q(x)$ is an integrable function inside V, the displacement field $u(x)$ outside V is an infinitely differentiable function presented in the form of the integral (2.6). The expansion of the Green function $G(x-x')$ in this integral in a Taylor series about the point ($x_0 \in V$, $x \notin V$) has the form

$$G_{\alpha\beta}(x-x') = \sum_{k=0}^{\infty} \frac{1}{k!}[\nabla_{\lambda_1}\nabla_{\lambda_2}...\nabla_{\lambda_k} G_{\alpha\beta}(x-x_0)] \\ \times (x_0-x')_{\lambda_1}(x_0-x')_{\lambda_2}...(x_0-x')_{\lambda_k}\,. \tag{2.137}$$

Substituting this expansion into the right-hand side of (2.6) we represent the displacement field outside V in the form of the following series:

$$u_\alpha(x) = \sum_{k=0}^{\infty} \nabla_{\lambda_1}\nabla_{\lambda_2}...\nabla_{\lambda_k} G_{\alpha\beta}(x-x_0) Q^k_{\beta\lambda_1\lambda_2...\lambda_k}(x_0)\,, \tag{2.138}$$

$$Q^k_{\beta\lambda_1\lambda_2...\lambda_k}(x_0) = \frac{(-1)^k}{k!}\int_V (x'-x_0)_{\lambda_1}(x'-x_0)_{\lambda_2}...(x'-x_0)_{\lambda_k} q_\beta(x')dx'. \tag{2.139}$$

The same result may be obtained if the function $q(x)$ in (2.6) is presented in the form of the following series of generalized functions:

$$q_\beta(x) = \sum_{m=0}^{\infty} Q^m_{\beta\lambda_1\lambda_2...\lambda_m}(x_0)\nabla_{\lambda_1}\nabla_{\lambda_2}...\nabla_{\lambda_m}\delta(x-x_0)\,. \tag{2.140}$$

Note, that the first term of this expansion

$$Q^0_\beta(x_0)\delta(x-x_0) = \int_V q_\beta(x')dx'\delta(x-x_0) \tag{2.141}$$

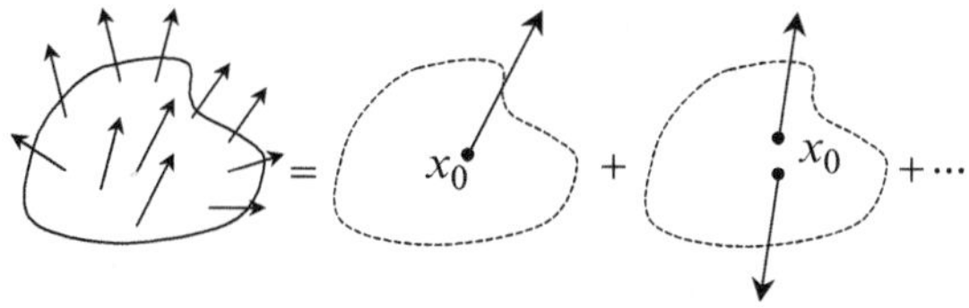

Fig. 2.5. Expansion of the distribution of body forces inside a region in the multipole series.

may be interpreted as the total force of the distribution $q_\alpha(x)$ applied to the point x_0. The second term in series (2.140) is a force dipole with total moment

$$Q^1_{\beta\lambda_1} = -\int_V (x' - x_0)_{\lambda_1} q_\beta(x')dx', \tag{2.142}$$

applied to the same point x_0 (Fig. 2.5). The other terms in (2.140) may be interpreted as multipoles of higher orders applied to x_0. The series (2.140) is called the expansion of the force distribution $q(x)$ inside V in the series over the multipoles concentrated at the point $x_0 \in V$. The series (2.140) should be understood as follows: convolution of the right and left parts of (2.140) with an arbitrary analytical function $\varphi(x)$ yields equality.

The k-th term of series (2.138) has the asymptotic $|x|^{-(k+1)}$ when $|x| \to \infty$. Therefore, the larger the distance from the point x to the region V the better is the convergence of this series. In order to calculate the elastic fields far from the region V one can take just the first few terms of the multipole expansion (2.140). Keeping three such terms, we represent the field $u(x)$ in the form

$$\begin{aligned} u_\alpha(x) &= G_{\alpha\beta}(x - x_0)Q^0_\beta + \nabla_\lambda G_{\alpha\beta}(x - x_0)Q^1_{\beta\lambda} \\ &+ \nabla_\lambda\nabla_\mu G_{\alpha\beta}(x - x_0)Q^2_{\beta\lambda\mu}\,. \end{aligned} \tag{2.143}$$

Here $G_{\alpha\beta}(x-x_0)Q^0_\beta$ is the principal term of the field $u(x)$ when $|x| \to \infty$.

Let the region V be an ellipsoid with semiaxes a_1, a_2, a_3 and the unit vectors of the principal axes $\mathbf{e}^{(1)}, \mathbf{e}^{(2)}, \mathbf{e}^{(3)}$. If the density q_α of the stress sources is constant in V, the first three coefficients Q^m in expansion (2.140) are

$$\begin{aligned} Q^0_\beta = vq_\beta,\ \ Q^1_{\alpha\beta} = 0,\ \ Q^2_{\beta\alpha\lambda} &= \frac{v}{10}(a_1^2 e^{(1)}_\alpha e^{(1)}_\beta \\ &+ a_2^2 e^{(2)}_\alpha e^{(2)}_\beta + a_3^2 e^{(3)}_\alpha e^{(3)}_\beta)q_\lambda,\ \ v = \frac{4}{3}\pi a_1 a_2 a_3\,. \end{aligned} \tag{2.144}$$

Suppose that the point x_0 is the center of the ellipsoid V. The displacement field far from such a source is described by the right-hand side of (2.143).

If the density $q(x)$ is a linear function inside the ellipsoid V: $q_\alpha(x) = p_{\alpha\beta}(x-x_0)_\beta V(x)$, the first three coefficients of the expansion of this function in the multipole series (2.140) are

$$Q^0_\beta = 0\,, \quad Q^1_{\beta\lambda} = -\frac{v}{5}(a_1^2 e^{(1)}_\lambda e^{(1)}_\gamma + a_2^2 e^{(2)}_\lambda e^{(2)}_\gamma + a_3^2 e^{(3)}_\lambda e^{(3)}_\gamma)p_{\beta\gamma}\,, \quad Q_{\beta\alpha\lambda} = 0\,. \tag{2.145}$$

The density of the force or dislocation moments distributed in a finite region V may be also expanded in a multipole series similar to (2.140). In this case, the strain $\varepsilon(x)$ and stress $\sigma(x)$ tensors far from the region V may be approximated by the first terms of these series

$$\varepsilon_{\alpha\beta}(x) = \sum_{m=0}^{\infty} \nabla_{\lambda_1}...\nabla_{\lambda_m} K_{\alpha\beta\lambda\mu}(x - x_0) C^0_{\lambda\mu\tau\rho} M_{\tau\rho\lambda_1...\lambda_m}(x_0)\,, \tag{2.146}$$

$$\sigma_{\alpha\beta}(x) = \sum_{m=0}^{\infty} \nabla_{\lambda_1}...\nabla_{\lambda_m} \mathrm{S}_{\alpha\beta\mu\nu}(x - x_0) M_{\mu\nu\lambda_1...\lambda_m}(x_0)\,, \tag{2.147}$$

$$M_{\alpha\beta\lambda_1...\lambda_m}(x_0) = \frac{(-1)}{m!}\int_V (x' - x_0)_{\lambda_1}...(x' - x_0)_{\lambda_m}\, m_{\alpha\beta}(x')\, dx'. \tag{2.148}$$

In conclusion, we consider elastic fields far from a thin region V_h with the middle surface Ω and the transverse size $h(x)$ ($x \in \Omega$). Let the sources of elastic fields be distributed inside V_h (Fig. 2.6). We introduce a local coordinate system with the origin at the point $x \in \Omega$ and the z-axis, directed along the normal $n(x)$ to Ω. For body forces with density $q(x)$ distributed in the region V_h, the displacement field may be presented as follows:

$$u(x) = \int_\Omega d\Omega' \int_{-h(x')/2}^{h(x')/2} G(x - x' - zn(x'))q(x' + zn(x'))dz\,. \tag{2.149}$$

If we expand the Green function $G(x - x' - zn(x'))$ in a Taylor series with respect to z ($x \notin V$)

$$G(x - x' - zn') = \sum_{k=0}^{\infty} \frac{z^k}{k!}\frac{\partial^k}{\partial(n')^k} G(x - x')\,, \tag{2.150}$$

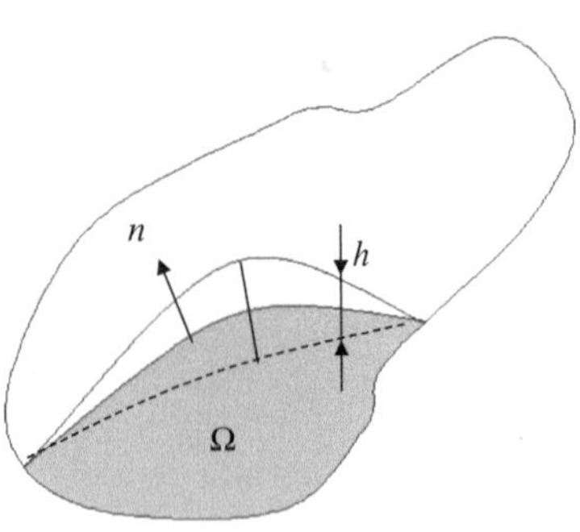

Fig. 2.6. A thin region V_h.

$$n' = n(x')\,, \frac{\partial}{\partial n'} = n'_\alpha \frac{\partial}{\partial (x'_\alpha)}\,. \tag{2.151}$$

and substitute this expansion in (2.149), the field $u(x)$ is presented in the form

$$u(x) = \sum_{k=0}^{\infty} \int_{\Omega} \left[\frac{\partial^k}{\partial (n')^k} G(x - x') \right] Q^{(k)}(x') d\Omega', \tag{2.152}$$

$$Q^{(k)}(x') = \frac{1}{k!} \int_{-h(x')/2}^{h(x')/2} q(x' + zn') z^k dz\,. \tag{2.153}$$

The series (2.152) corresponds to expansion of the density $q(x)$ in the following series:

$$q(x' + zn') = \sum_{k=0}^{\infty} \frac{\partial^k}{\partial (n')^k} [Q^{(k)}(x') \Omega(x')]\,, \tag{2.154}$$

where $\Omega(x)$ is the delta-function concentrated on the surface Ω (2.42).

The main term of expansion (2.152) is the potential of a simple layer, concentrated on the surface Ω with the density $Q^{(0)}(x)$:

$$Q^{(0)}(x) = \int_{-h(x)/2}^{h(x)/2} q(x + zn) dz\,. \tag{2.155}$$

2.5 Notes

The general structure of the Green tensor for an anisotropic homogeneous elastic medium was obtained in [131]. Regularization of the generalized functions that are the second derivatives of the Green function for displacements was defined in [54]. The regularization of operators similar to $\mathrm{K}(x)$ and $\mathrm{S}(x)$ on finite functions was defined in [48]. The method of regularization of the operators K and S on constants was proposed in [70]. The Green tensor for internal stresses in a homogeneous elastic medium was obtained in [105]. Fundamental results concerned the internal geometry and stresses in a medium with dislocation sources of various types may be found in [101], [46], [99]. The equations for the jumps of elastic fields on the borders of the regions containing the sources of external and internal stresses are presented in [94].

3. Equilibrium of a homogeneous elastic medium with an isolated inclusion

In this chapter, we consider problems of static elasticity and thermoelasticity for a homogeneous medium with an isolated inclusion (the one particle problem). These problems are reduced to the solution of integral equations for the stress and strain tensors in the region occupied by the inclusion. In many cases, integral formulations simplify the analysis of the tensor structure of the solutions of the one-particle problems. For an ellipsoidal inclusion or its limiting forms (elliptical crack and elliptical cylinder) and a polynomial external field, closed analytical solutions for the fields inside the inclusion may be constructed. Efficient numerical algorithms are proposed for the solution of elastic and thermoelastic problems for multilayered spherical and cylindrical inclusions subjected to a constant external field. Asymptotic analysis is used to obtain the principal terms in the solution of the elasticity problems for thin inclusions and hard slender fibers. Construction of these terms is reduced to integral equations on the middle surface of a thin inclusion, and to differential equations in the middle axis of the fiber.

3.1 Integral equations for a medium with an isolated inhomogeneity

Let us consider an infinite medium that contains an inhomogeneity (inclusion) in a finite region V. The tensor of the elastic moduli $C(x)$ of the medium is presented in the form

$$C(x) = C^0 + C^1(x)\,, \tag{3.1}$$

$$C^1(x) \neq 0 \ \text{ if } \ x \in V, \quad C^1(x) = 0 \text{ if } \ x \notin V. \tag{3.2}$$

Here C^0 is a constant tensor, and $C^1(x)$ is a finite piece-wise smooth tensor function. If the medium is subjected to body forces $q_\alpha(x)$, the displacement vector $u(x)$ of the points of the medium is the solution of the equation

$$L_{\alpha\beta}u_\beta(x) = q_\alpha(x)\,, \quad L_{\alpha\beta} = L^0_{\alpha\beta} + L^1_{\alpha\beta}\,, \tag{3.3}$$

$$L^0_{\alpha\beta} = -\nabla_\lambda C^0_{\lambda\alpha\beta\mu}\nabla_\mu\,, \quad L^1_{\alpha\beta} = -\nabla_\lambda C^1_{\lambda\alpha\beta\mu}(x)\nabla_\mu\,. \tag{3.4}$$

Let $u^0(x)$ be the displacement field in the medium without the inclusion under action of the same body forces and forces at infinity. This field satisfies the equation

$$(L^0_{\alpha\beta}u^0_\beta)(x) = q_\alpha(x)\,. \tag{3.5}$$

In what follows, the field $u^0(x)$ and the corresponding strain $\varepsilon^0(x) = \mathrm{def}u^0(x)$ and stress $\sigma^0(x) = C^0\varepsilon^0(x)$ fields are called the external fields.

The solution $u(x)$ of (3.3) is presented in the form

$$u(x) = u^0(x) + u^1(x)\,, \tag{3.6}$$

where $u^0(x)$ is the external field, and $u^1(x)$ is a perturbation of the displacement field due to the presence of the inclusion, $u^1(x) \to 0$ when $|x| \to \infty$. Substituting (3.6) into (3.3) we find the equation for the field $u^1(x)$:

$$(L^0 + L^1)u^1 = -L^1u^0\,. \tag{3.7}$$

Because $C^1(x)$ is a finite function, the right-hand side of this equation is also finite. Application of the operator $G^0 = (L^0)^{-1}$ and after that, the operator def to (3.7) leads to the equation

$$\varepsilon^1 + \mathrm{def}G^0L^1u^1 = -\mathrm{def}G^0L^1u^0 \quad \varepsilon^1 = \mathrm{def}u^1. \tag{3.8}$$

Taking into account equation (3.4) for the operator L^1, we can rewrite (3.8) in the form

$$\varepsilon^1(x) + (\mathrm{K}C^1\varepsilon)(x) = 0\,, \tag{3.9}$$

$$\mathrm{K} = -\mathrm{def}G^0\mathrm{def}, \quad \varepsilon = \mathrm{def}(u^0 + u^1)\,.$$

Adding to both sides of (3.9) the external strain field ε^0 gives the final equation for the tensor $\varepsilon(x)$ in the form

$$\varepsilon(x) + (\mathrm{K}C^1\varepsilon)(x) = \varepsilon^0(x). \tag{3.10}$$

In detail, this equation is written as follows:

$$\varepsilon_{\alpha\beta}(x) + \int \mathrm{K}_{\alpha\beta\lambda\mu}(x-x')C^1_{\lambda\mu\nu\rho}(x')\varepsilon_{\nu\rho}(x')dx' = \varepsilon^0_{\alpha\beta}(x)\,, \tag{3.11}$$

$$\mathrm{K}_{\alpha\beta\lambda\mu}(x) = -[\nabla_\alpha\nabla_\lambda G^0_{\beta\mu}(x)]_{(\alpha\beta)(\lambda\mu)}\,. \tag{3.12}$$

Note that the function $\mathrm{K}(x)$ coincides with the kernel of the operator K (2.53) that presents a strain field in a homogeneous medium with a volume distribution of generalized force moments. As mentioned in Section 2.1, the kernel $\mathrm{K}(x)$ is a homogeneous function of the degree -3. Action of such a function on a continuous finite function is defined in (2.54). The existence and uniqueness of the solution of (3.11) follows from the equivalence of this equation to the system of differential equations of elasticity theory (3.3).

Consider the problem of thermoelasticity, and suppose that the tensors of linear expansion coefficients of the medium $\alpha(x)$ is presented in the form

$$\alpha(x) = \alpha^0 + \alpha^1(x), \tag{3.13}$$

where $\alpha^1(x)$ is a piece-wise smooth finite function, α^0 is a constant tensor. The medium is subjected to body forces $q(x)$, some forces at infinity, and the known temperature field $T(x)$. In this case, the total strain field is the sum of the elastic ε^e and temperature $\alpha(x)T(x)$: deformations

$$\varepsilon(x) = \varepsilon^e(x) + \alpha(x)T(x).$$

The stress and elastic strain tensors in such a medium satisfy the system of equations that follow from (2.4) and (2.63), (2.64) when $m(x) = \alpha(x)T(x)$

$$\text{div}\sigma(x) = -q(x) \quad \sigma(x) = C(x)\varepsilon^e(x)\,, \tag{3.14}$$

$$\text{Rot}\varepsilon^e(x) = -\text{Rot}[\alpha(x)T(x)]\,. \tag{3.15}$$

This system may be rewritten in the form similar to Equations (2.4), (2.63), (2.64) for the stress $\sigma(x)$ and strain $\varepsilon(x)$ fields in a homogeneous medium C^0 with sources of external and internal stresses

$$\text{div}\sigma(x) = -q(x), \quad \sigma(x) = C^0\varepsilon'(x), \quad \text{Rot}\varepsilon'(x) = -\text{Rot}\, m(x)\,, \tag{3.16}$$

$$\varepsilon'(x) = \varepsilon^e(x) - B^1(x)\sigma(x)\,, \quad m(x) = B^1(x)\sigma(x) + \alpha(x)T(x)\,, \tag{3.17}$$

$$B^1(x) = B(x) - B^0, \; B(x) = C^{-1}(x), \quad B^0 = (C^0)^{-1}. \tag{3.18}$$

Here $\varepsilon'(x)$ is an auxiliary strain field.

Let $\sigma^0(x)$ be the external stress field that satisfies the equations

$$\text{div}\sigma^0(x) = -q(x)\,, \quad \sigma^0(x) = C^0\varepsilon^0(x)\,, \quad \text{Rot}\varepsilon^0(x) = 0, \tag{3.19}$$

and given conditions at infinity. Using Equation (2.66) for the tensor $\sigma(x)$ in the medium with the sources of internal stresses, we may represent the solution of (3.16) in the form

$$\sigma(x) = \sigma^0(x) + \int \mathrm{S}(x - x')m(x')dx'\,. \tag{3.20}$$

Here the function $\mathrm{S}(x)$ is defined in (2.72). Let us substitute in this equation the equation for the density $m(x)$ from (3.17). As a result, we arrive at the equation for the stress field in the medium with inhomogeneous elastic and thermoelastic characteristics

$$\sigma_{\alpha\beta}(x) - \int \mathrm{S}_{\alpha\beta\lambda\mu}(x-x')B^1_{\lambda\mu\tau\rho}(x')\sigma_{\tau\rho}(x')dx'$$

$$= \sigma^0_{\alpha\beta} + \int \mathrm{S}_{\alpha\beta\lambda\mu}(x-x')\alpha_{\lambda\mu}(x')T(x')dx', \tag{3.21}$$

$$\mathrm{S}_{\alpha\beta\lambda\mu}(x) = C^0_{\alpha\beta\nu\rho}\mathrm{K}_{\nu\rho\tau\delta}(x)C^0_{\tau\delta\lambda\mu} - C^0_{\alpha\beta\lambda\mu}\delta(x)\,. \tag{3.22}$$

When $T = 0$, the integral in the right-hand side of (3.21) vanishes, and this equation takes the form

$$\sigma(x) - \int \mathrm{S}(x-x')B^1(x')\sigma(x')dx' = \sigma^0(x)\,. \tag{3.23}$$

Using Hooke's law and representation (3.22) for $\mathrm{S}(x)$ we can show that this equation is equivalent to (3.11).

If the external field $\sigma^0(x)$ is equal to zero, (3.21) describes temperature stresses in a medium with inhomogeneous thermoelastic characteristics and a given temperature field $T(x)$.

Note important properties of the integral Equations (3.11) and (3.21):

1. Let the function $C^1(x)$ not be equal to zero in a finite region V (inclusion), where $V(x)$ is the characteristic function of V. Representing the strain tensor $\varepsilon(x)$ in the medium with the inhomogeneity in the form

$$\varepsilon(x) = \varepsilon^0(x) + \varepsilon^1(x)\,, \tag{3.24}$$

where $\varepsilon^1(x)$ is the perturbation of the strain field caused by the inclusion, we obtain from (3.11) the equation for $\varepsilon^1(x)$ in the form

$$\varepsilon^1(x) + \int \mathrm{K}(x-x')C^1(x')\varepsilon^1(x')V(x')dx'$$

$$= -\int \mathrm{K}(x-x')C^1(x')\varepsilon^0(x')V(x')dx'. \tag{3.25}$$

Here the relation $C^1(x) = C^1(x)V(x)$ is taken into account. It follows from this equation that the field $\varepsilon^1(x)$ depends only on the value of the external field $\varepsilon^0(x)$ inside the inclusion.

2. If the strain $\varepsilon^+(x) = \varepsilon(x)V(x)$ and stress $\sigma^+(x) = \sigma(x)V(x)$ fields inside V are known, the fields outside V are reconstructed for (3.11) and (3.23):

$$\varepsilon(x) = \varepsilon^0(x) - \int \mathrm{K}(x-x')C^1(x')\varepsilon^+(x')dx', \tag{3.26}$$

$$\sigma(x) = \sigma^0(x) + \int \mathrm{S}(x-x')B^1(x')\sigma^+(x')dx'. \tag{3.27}$$

The equations for the fields $\varepsilon^+(x)$ and $\sigma^+(x)$ inside the inclusion follow from (3.11) and (3.23) by multiplying them with $V(x)$:

$$\varepsilon^+(x) + \int \mathrm{K}^+(x,x')C^1(x')\varepsilon^+(x')dx' = \varepsilon^{0+}(x)\,, \tag{3.28}$$

$$\sigma^+(x) - \int \mathrm{S}^+(x,x')B^1(x')\sigma^+(x')dx' = \sigma^{0+}(x)\,, \tag{3.29}$$

$$\mathrm{K}^+(x,x') = V(x)\mathrm{K}(x-x')V(x')\,, \;\; \mathrm{S}^+(x,x') = V(x)\mathrm{S}(x-x')V(x')\,. \tag{3.30}$$

3. Let $C^1(x) \to -C^0$ or $C(x) = C^0 + C^1(x) \to 0$ (for a cavity). In this case, equation (3.11) for the strain tensor takes the form

$$\varepsilon^+(x) - \int K(x-x')C^0\varepsilon^+(x')dx' = \varepsilon^0(x)\,, \quad x \in V\,. \tag{3.31}$$

Taking into account that in the limit $C \to 0$, we have

$$B^1\sigma^+ = [C^{-1} - (C^0)^{-1}]C\varepsilon^+ \to \varepsilon^+, \tag{3.32}$$

The equation (3.23) for the stress tensor $\sigma(x)$ is transformed to the following one:

$$\sigma(x) = \sigma^0(x) + \int \mathrm{S}(x-x')\varepsilon^+(x')dx'\,. \tag{3.33}$$

Hence, after constructing the tensor $\varepsilon^+(x)$ from (3.31) we can find the stress field outside the cavity using (3.33).

Note that for a cavity, the tensor B^1 is singular, and uniqueness of the solutions of (3.3) and (3.31) is violated. To indicate all the solutions of (3.31) in this case, we present the kernel of the operator in the left-hand side of (3.31) in the form that follows from (3.22)

$$\mathrm{K}(x)C^0 = (C^0)^{-1}\mathrm{S}(x) + I\delta(x)\,. \tag{3.34}$$

As a result , (3.31) is transformed into the equation

$$\int \mathrm{S}(x-x')\varepsilon^+(x')dx' = C^0\varepsilon^0(x). \tag{3.35}$$

Equation (2.67) for $\mathrm{S}(x)$ implies, that the homogeneous Equation (3.35) is equivalent to the following one:

$$\int \mathrm{S}(x-x')\varepsilon^+(x')dx' = \int Z(x-x')\mathrm{Rot}\,\varepsilon^+(x')dx' = 0\,. \tag{3.36}$$

Here Stokes' theorem was used. Hence, every function that satisfies the compatibility equation $\mathrm{Rot}\varepsilon^+ = 0$ and is equal to zero outside V is a solution of the homogeneous equation (3.36). Further, for the strain field inside a cavity, we take the limiting solution of (3.11) when $C^1 \to -C^0$.

4. Let now $B^1(x) \to -B^0$, or $B \to 0$ inside V (absolutely rigid inclusion). In this case, equation (3.23) for the stress field in the medium with the inhomogeneity takes the form

$$\sigma(x) + \int \mathrm{S}(x - x')B^0\sigma^+(x')dx' = \sigma^0(x) \, , \quad x \in V \, . \tag{3.37}$$

Because in the limit $B(x) \to 0$ we have

$$C^1\varepsilon^+ = [B^{-1} - (B^0)^{-1}]B\sigma^+ \to \sigma^+, \tag{3.38}$$

equation (3.11) is transformed into the following one:

$$\varepsilon(x) = \varepsilon^0(x) - \int \mathrm{K}(x - x')\sigma^+(x')dx'. \tag{3.39}$$

This equation determines the strain field outside a rigid inclusion if the stress field $\sigma^+(x)$ inside it is known.

For $B = 0$, uniqueness of the solution of (3.37) is lost. Taking into account Equation (3.22) for the tensor $\mathrm{S}(x)$, we can represent the kernel of the operator in (3.37) in the form

$$\mathrm{S}(x)B^0 = C^0\mathrm{K}(x) - I\delta(x) \, . \tag{3.40}$$

Using this Equation, Equation (3.12) for $\mathrm{K}(x)$, and Gauss' theorem we can show that the homogeneous ($\sigma^0 = 0$) Equation (3.37) is equivalent to the following one:

$$\sigma(x) + \int \mathrm{S}(x - x')B^0\sigma^+(x')dx' = C^0 \int \mathrm{K}_{\alpha\beta\lambda\mu}(x - x')\sigma^+_{\lambda\mu}(x')dx'$$

$$= C^0 \int \nabla_\alpha G_{\beta\mu}(x - x')\nabla'_\lambda\sigma^+_{\lambda\mu}(x')dx' = 0 \, . \tag{3.41}$$

Thus, an arbitrary stress field inside V that satisfies the equilibrium equation $\nabla_\lambda\sigma^+_{\lambda\mu} = 0$ is the solution of the homogeneous Equation (3.37).

Further, the limiting solution of (3.29) when $B^1 \to -B_0$ is taken as the solution of Equation (3.37).

3.2 Conditions on the interface between two media

Let us consider a homogeneous elastic medium with an inclusion (inhomogeneity) that occupies a region V with a smooth boundary Ω. The material of the inclusion is homogeneous with the tensor of elastic moduli $C = C^0 + C^1$, where C^0 is the tensor of elastic moduli of the medium. If the external field $\varepsilon^0(x)$ is an arbitrary smooth function, the stress and strain tensors are continuous everywhere except on the boundary Ω of V. In order to find the jumps of the strain tensor $\varepsilon(x)$ on Ω we present the function $\varepsilon(x)$ in (3.11) in the form of the following sum:

$$\varepsilon(x) = \varepsilon^0(x) + \varepsilon^1(x)\,, \quad \varepsilon^1(x) = -\int_V \mathrm{K}(x - x')C^1\varepsilon(x')dx' \tag{3.42}$$

Here $\varepsilon^0(x)$ is a continuous function. Therefore, the jumps of the strain field coincide with the jumps of the potential $\varepsilon^1(x)$. Comparison of this equation with (2.51) shows that the potential $\varepsilon^1(x)$ can be interpreted as the strain field in the homogeneous elastic medium with the tensor of elastic moduli C^0 under the action of the generalized force moments of the density $q_{\alpha\beta}(x) = C^1_{\alpha\beta\lambda\mu}\varepsilon_{\lambda\mu}(x)$ inside V.

The jumps of the strain field on the boundary of the region containing distributed force moments were considered in Section 2.3. Let $n(x_0)$ be the external normal to Ω at the point $x_0 \in \Omega$. If $\varepsilon^+(x_0)$ is the limiting value of the strain tensor $\varepsilon(x)$ inside the inclusion when $x \to x_0$, and $\varepsilon^-(x)$ is the similar limit when x is outside V, then (2.117) and (3.42) give the equation for the jump of the strain field on Ω (see Fig. 3.1)

$$[\varepsilon(x_0)] = \varepsilon^-(x_0) - \varepsilon^+(x_0) = \mathrm{K}^*(n_0)C^1\varepsilon^+(x_0)\,. \tag{3.43}$$

Here the tensor $\mathrm{K}^*(n_0)$ is determined in (2.58).

Equation (3.43) implies that the limiting values of the strain tensor in the medium $\varepsilon^-(x_0)$ and in the inclusion $\varepsilon^+(x_0)$ at the point x_0 of the interface Ω are connected by the relation

$$\varepsilon^-(x_0) = [I + \mathrm{K}^*(n_0)C^1]\varepsilon^+(x_0)\,, \quad x_0 \in \Omega\,, \quad n_0 = n(x_0)\,. \tag{3.44}$$

Here I is the unit rank four tensor.

Let us turn to Equation (3.23) for the stress tensor $\sigma(x)$ in the medium with the inclusion, and present the function $\sigma(x)$ in the form

$$\sigma(x) = \sigma^0(x) + \sigma^1(x)\,, \quad \sigma^1(x) = \int_{\mathcal{V}} \mathrm{S}(x - x')B^1\sigma(x')dx', \tag{3.45}$$

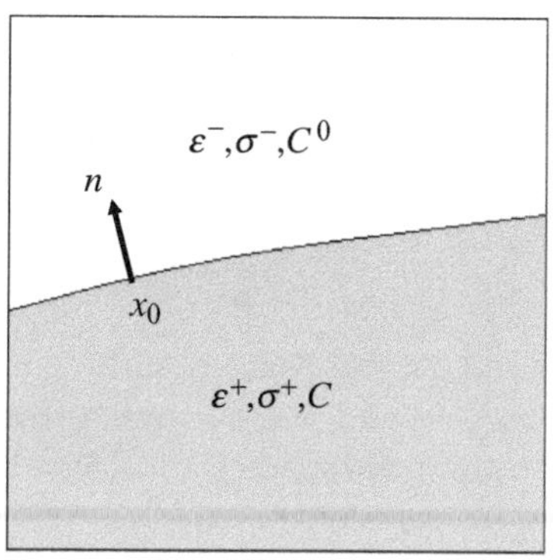

Fig. 3.1. The interface between two media.

where the external field $\sigma^0(x)$ is a continuous function. Equation (2.66) implies that the potential $\sigma^1(x)$ may be interpreted as the tensor of internal stresses in a homogeneous medium with the distribution of dislocation moments of the density $m(x) = B^1\sigma(x)$ inside V. It follows from (2.119) that the jump of this potential on Ω is determined by the equation

$$[\sigma(x_0)] = \sigma^-(x_0) - \sigma^+(x_\circ) = -\mathrm{S}^*(n_0)B^1\sigma^+(x_0)\,, \quad x_0 \in \Omega\,. \tag{3.46}$$

Here $\sigma^-(x_0)(\sigma^+(x_0))$ are the limiting values of the stress tensor in the medium (in the inclusion) when $x \to x_0 \in \Omega$, the tensor $\mathrm{S}^*(n)$ is determined in (2.75). Hence, the limiting values of the stress tensor outside and inside the region V at the point x_0 of the interface Ω are connected by the equation

$$\sigma^-(x_0) = [I - \mathrm{S}^*(n_0)B^1]\sigma^+(x_\circ)\,, \quad x_0 \in \Omega\,. \tag{3.47}$$

Note some consequences of these equations.

1. Introduce the projection operators Π and Θ on the normal $n(x_0)$ and the tangent plane to Ω at the point x_0

$$\Pi = \mathrm{I} - \Theta\,, \quad \mathrm{I}_{\alpha\beta\lambda\mu} = \delta_{\alpha)(\lambda}\delta_{\mu)(\beta}\,, \quad \Theta_{\alpha\beta\lambda\mu} = \theta_{\alpha)(\lambda}\theta_{\mu)(\beta}\,, \tag{3.48}$$

$$\theta_{\alpha\beta} = \delta_{\alpha\beta} - n_\alpha n_\beta\,, \quad \Pi_{\alpha\beta\lambda\mu}\Pi_{\lambda\mu\nu\rho} = \Pi_{\alpha\beta\nu\rho}. \tag{3.49}$$

Using these operators, we can split any symmetric two-rank tensor $q(x)$ on Ω into the "tangent" $\Theta_{\alpha\beta\lambda\mu}q_{\lambda\mu}$ and "normal" $\Pi_{\alpha\beta\lambda\mu}q_{\lambda\mu}$ components.

The definition of the tensor $\mathrm{K}^*(n)$ (2.58) implies

$$\Theta_{\alpha\beta\lambda\mu}(n)\mathrm{K}^*_{\lambda\mu\nu\rho}(n) \equiv 0\,. \tag{3.50}$$

This equation and (3.43) show that the tangential component $\theta\varepsilon$ of the strain tensor is continuous on the boundary Ω of the region V:

$$[\Theta_{\alpha\beta\lambda\mu}(n_0)\varepsilon_{\lambda\mu}(x_0)] = \Theta_{\alpha\beta\lambda\mu}(n_0)\varepsilon^-_{\lambda\mu}(x_0) - \Theta_{\alpha\beta\lambda\mu}(n_0)\varepsilon^+_{\lambda\mu}(x_0) = 0\,. \tag{3.51}$$

The definitions (2.75), (2.58) of the tensor $\mathrm{S}^*(n)$ implies the equality

$$n_\alpha \mathrm{S}^*_{\alpha\beta\lambda\mu}(n) = 0\,. \tag{3.52}$$

This equation and (3.46) show continuity of the normal component of the stress tensor on the surface Ω

$$[n_{0\alpha}\sigma_{\alpha\beta}(x_0)] = n_{0\alpha}\sigma^-_{\alpha\beta}(x_0) - n_{0\alpha}\sigma^+_{\alpha\beta}(x_0) = 0\,. \tag{3.53}$$

2. Equations (3.43) and (3.46) contain only the limiting values of the elastic fields at a point $x_0 \in \Omega$ and the normal vector n_0 to Ω at this point. It is shown in [110] that the conditions (3.43) and (3.46) take place on an arbitrary smooth surface Ω of discontinuity of the elastic moduli of a heterogeneous medium. In this case, the tensors C^0 and $C^0 + C^1$ play the role of

limiting values of elastic moduli of the medium at the point x_0 on Ω when the point x tends to x_0 from the side of normal (C^0) and from the opposite side ($C^0 + C^1$).

3. Equations (3.44) and (3.47) show that one does not need to calculate the integrals $\varepsilon^1(x)$ and $\sigma^1(x)$ (3.42) and (3.45) to find the stress and strain concentration in the medium, on the surface of the inclusion. One can find the values of $\varepsilon^-(x)$ and $\sigma^-(x)$ by multiplying the limiting values of the fields $\varepsilon^+(x)$ and $\sigma^+(x)$ at the points of the interface with the tensors $I + \mathrm{K}^*(n)C^1$ and $I - \mathrm{S}^*(n)B^1$, respectively. These tensors can be called the stress concentration coefficients

3.3 Ellipsoidal inhomogeneity

Let an inclusion with constant elastic moduli occupy an ellipsoidal region with surface defined by the equation

$$\frac{x_1^2}{a_1^2} + \frac{x_2^2}{a_2^2} + \frac{x_3^2}{a_3^2} = 1. \tag{3.54}$$

Here a_1, a_2, a_3 are of the ellipsoid semiaxes, and x_1, x_2, x_3 are the axes of the Cartesian coordinate system with the origin at the center of the ellipsoid.

In this case, the solution of Equation (3.28) has a remarkable property (the polynomial conservation theorem) [46], [109] (see Fig. 3.2).

If the external field $\varepsilon^0(x)(\sigma^0(x))$ is a polynomial of the degree m with respect to coordinates x_1, x_2, x_3 inside the ellipsoidal region V, the strain field $\varepsilon^+(x)$ (stress field $\sigma^+(x)$) inside the ellipsoid is also a polynomial of the degree no more than m.

The proof of this theorem given in [109] is presented below.

Considering Equation (3.11) for a homogeneous ellipsoidal inclusion

$$\varepsilon_{\alpha\beta}(x) + \int_V \mathrm{K}_{\alpha\beta\lambda\mu}(x - x')C^1_{\lambda\mu\nu\rho}\varepsilon_{\nu\rho}(x')dx' = \varepsilon^0_{\alpha\beta}(x), \tag{3.55}$$

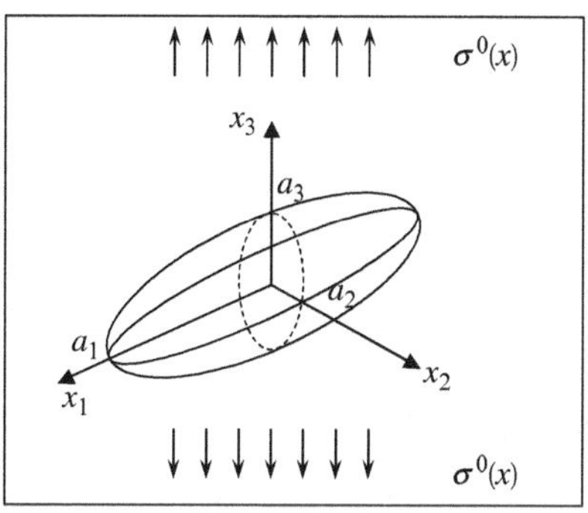

Fig. 3.2. Ellipsoidal inhomogeneity.

we suppose that its right-hand side is a polynomial of degree m. Because the linear transformation $y = a^{-1}x$ ($a_{\alpha\beta} = a_\alpha \delta_{\alpha\beta}$ (no sum with α !)) of the x-space converts the region V to the ball of a unit radius, and the right-hand side remains polynomial of the same degree, it is sufficient to give the proof of the theorem for the case when V is the ball $|x| \leq 1$.

Polynomial conservation holds if the operator K with kernel $\mathrm{K}(x)$ transforms a polynomial of degree m into a polynomial of the same degree. Moreover, in order to prove this property, it is sufficient to consider the action of the operator K on a homogeneous polynomial of degree m or on the tensor product $x^{(m)} = x_{\lambda_1} x_{\lambda_2} ... x_{\lambda_m}$.

The regular representation of the operator $\mathrm{K}(x)$ was determined in (2.60). When $q_{\alpha\beta}(x) = q^0_{\alpha\beta\lambda_1\lambda_2...\lambda_m} x_{\lambda_1} x_{\lambda_2} ... x_{\lambda_m}$, where q^0 is a constant tensor, we have

$$\int \mathrm{K}(x - x')q(x')dx' = A^0 q(x) + p.v. \int \mathrm{K}(x - x')q^0 x'^{(m)} dx', \tag{3.56}$$

Here A^0 is the constant tensor and the integral in the right-hand side is understood in the sense of a Cauchy principal value (2.57).

Because $\mathrm{K}(x)$ is a homogeneous function of the degree -3, it is presented in the form

$$\mathrm{K}(x - x') = \frac{\mathrm{K}^0(n)}{r^3}, \quad n = \frac{(x - x')}{r}, \quad r = |x - x'|, \tag{3.57}$$

where $\mathrm{K}^0(n)$ is an even function of n: $\mathrm{K}^0(-n) = \mathrm{K}^0(n)$.

It is known [149] that for the existence of the principal value of an integral, the function $\mathrm{K}^0(n)$ should satisfy the equation

$$\int_{\Omega_1} \mathrm{K}^0(n) d\Omega_n = 0, \tag{3.58}$$

where Ω_1 is the surface of a unit sphere.

Let us consider the integral in the right-hand side of (3.56). Substituting $x' = x + r\,n$ under the integral we obtain

$$p.v. \int_V \mathrm{K}(x - x') x'^{(m)} dx' = \sum_{k=0}^{m} x^{(m-k)} J^{(k)}(x), \tag{3.59}$$

$$J^{(k)}(x) = p.v. \int_V \mathrm{K}^0(n) n^{(k)} r^{k-3} dx', \quad x \in V. \tag{3.60}$$

Note that $J^{(0)} = 0$ because of property (3.58), and other integrals in (3.60) exist in the ordinary sense.

We put $dx' = r^2 dr d\Omega_1$ in (3.60) and integrate over the elemental cone with the origin at the point $x \in V$ (see Fig. 3.3). As a result, this integral takes the form

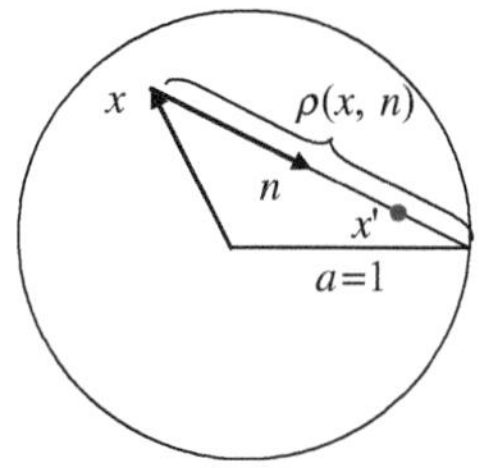

Fig. 3.3. To the proof of the polynomial conservation theorem.

$$J^{(k)}(x) = \frac{1}{k}\int_{\Omega_1} K^0(n)n^{(k)}\rho^k(x,n)d\Omega_1\,, \quad x \in V, \quad k = 1, 2, ..., m\,, \tag{3.61}$$

where $\rho(x,n)$ is the distance from the point x in the direction n to the surface of the unit sphere (Fig. 3.3)

$$\rho(x,n) = -(x \cdot n) + \sqrt{1-(x\cdot x)+(x\cdot n)^2}\,. \tag{3.62}$$

For ρ^k we have

$$\rho^k = \sum_{l=0}^{k}(-1)^l C_k^l (x\cdot n)^l[1-(x\cdot x)+(x\cdot n)^2]^{(k-l)/2}\,, \tag{3.63}$$

where C_k^l are the binomial coefficients. Because $K^0(n)$ is an even function of n, only those terms in this series satisfying $k + l = 2m$ contribute to the integral $J^{(k)}(x)$. These terms contain the root $\sqrt{1-(x\cdot x)+(x\cdot n)^2}$ with an even degree, and have the form

$$\sum_{t=0}^{\left[\frac{k}{2}\right]} A_{(k-2t)}(n)x^{(k-2t)}\,,$$

where $\left[\frac{k}{2}\right]$ is the integer part of $\frac{k}{2}$, $A_{(k-2t)}(n)$ are tensor coefficients. Therefore, $J^{(k)}(x)$ is a polynomial with respect to x of degree k with the tensor values

$$J^{(k)}(x) = \sum_{t=0}^{\left[\frac{k}{2}\right]} B_{(k-2t)}^{(k)} x^{(k-2t)}\,, \tag{3.64}$$

$$B_{(k-2t)}^{(k)} = \frac{1}{k}\int_{\Omega_1} K^0(n)n^{(k)}A_{(k-2t)}(n)d\Omega_1\,. \tag{3.65}$$

Finally, substituting (3.64) into (3.59) we obtain that the second term in (3.56) is a polynomial of degree m which terms are tensors of rank $(m + 4)$. This completes the proof of the polynomial conservation theorem for the ellipsoidal inclusion.

Because the generalized function $\mathrm{S}(x)$ in (3.29) has the same property as the function $\mathrm{K}(x)$, the polynomial conservation theorem may be proved on the basis of equation (3.29) for the stress tensor

$$\sigma(x) - \int\limits_V \mathrm{S}(x-x')B^1\sigma(x')dx' = \sigma^0(x). \tag{3.66}$$

It follows from the proof of the theorem that eigen-functions of the static problem of elasticity for an ellipsoidal inclusion are polynomials. The same property holds for other problems of mathematical physics for ellipsoidal regions (see Chapter 6). The symbol of the integral operators in these problems (the Fourier transform of the kernel) should be a homogeneous function of the degree 0.

Consider an algorithm for constructing solutions of equations (3.55) and (3.66) for an ellipsoidal inclusion and a polynomial right-hand side of an arbitrary degree. According to the polynomial conservation theorem, the action of the operator K on a homogeneous polynomial $x^{(m)}$ is defined by the equation

$$\int\limits_V K(x-x')x'^{(m)}dx' = \sum_{j=0}^{m} A_j^m x^{(j)}, \tag{3.67}$$

where A_j^m is a constant tensors of rank $(m+j+4)$. The detailed form of this equation is

$$\int\limits_V \mathrm{K}_{\alpha\beta\lambda\mu}(x-x')x'_{\nu_1}x'_{\nu_2}...x'_{\nu_m}dx'$$

$$= \sum_{j=0}^{m} A^m_{j\alpha\beta\lambda\mu\nu_1\nu_2...\nu_m\tau_1\tau_2...\tau_m} x_{\tau_1}x_{\tau_2}...x_{\tau_m}.. \tag{3.68}$$

In order to find the coefficients A_j^m let us differentiate both sides of (3.68) j times with respect to x, and put $x=0$ in the resulting equation. As the result, we obtain the coefficients A_j^m in the following symbolic form:

$$A_j^m = (-1)^j \int\limits_V [\nabla^{(j)}\mathrm{K}(x)]x^{(m)}dx\,, \qquad \nabla^{(j)}_{\lambda_1\lambda_2...\lambda_j} = \nabla_{\lambda_1}\nabla_{\lambda_2}...\nabla_{\lambda_j}. \tag{3.69}$$

When $j<m$, the integral in this equation converges absolutely. Because $\mathrm{K(x)}$ is an even function of x, the following equations hold:

$$A^{2p}_{2l-1} = 0 \quad \text{when} \quad l = 1,2,...,p, \quad \text{if} \quad m = 2p, \tag{3.70}$$

$$A^{2p+1}_{2l} = 0 \quad \text{when} \quad l = 0,1,...,p, \quad \text{if} \quad m = 2p+1\,. \tag{3.71}$$

When $j = m$, integral (3.69) can be considered as the action of the generalized function $[\nabla^{(m)}\mathrm{K}(x)]x^{(m)}$ on the function $V(x)$. Since the Fourier transform of $[\nabla^{(m)}\mathrm{K}(x)]x^{(m)}$ is a homogeneous function of the zero degree, integral (3.69) for $j = m$ takes the form (see Appendix B.2)

$$\mathrm{A}_m^m = \frac{1}{4\pi}\int_{\Omega_1} \mathrm{K}_m^*(a^{-1}k)d\Omega\,. \tag{3.72}$$

Here Ω_1 is the surface of a unit sphere in the k-space of the Fourier transforms,

$$\mathrm{K}_m^*(k) = (-1)^m\nabla^{(m)}[k^{(m)}\mathrm{K}^*(k)]\,, \tag{3.73}$$

and the function $\mathrm{K}^*(k)$ is defined in (2.58). The components of the tensor a in the basis of the ellipsoid principal axes x_1, x_2, x_3 have the forms (no sum with respect to α !)

$$a_{\alpha\beta} = a_\alpha\delta_{\alpha\beta}\,. \tag{3.74}$$

Let the external field $\varepsilon^0(x)$ be a polynomial of degree m

$$\varepsilon^0(x) = \sum_{j=0}^{m} d_j^0 x^{(j)}\,, \tag{3.75}$$

where d_j^0 is the tensor of rank $(2 + j)$. In this case, the solution of (3.55) has the form

$$\varepsilon(x) = \sum_{j=0}^{m} d_j x^{(j)}\,. \tag{3.76}$$

After substituting (3.75) and (3.76) into (3.55), and taking into account property (3.67) of the operator K, we obtain

$$\sum_{j=0}^{m} d_j x^{(j)} + \sum_{j=0}^{m}\left[\sum_{\ell=0}^{j} \mathrm{A}_\ell^j C^1 d_\ell x^{(\ell)}\right] = \sum_{j=0}^{m} d_j^0 x^{(j)}\,. \tag{3.77}$$

Equating the coefficients of powers of x we obtain a system of algebraic equations for the tensors d_j. Solution of this system is presented in the form of the following recurrence formulas:

$$d_m = \left[I_m + A_m^m C^1\right]^{-1} d_m^0\,, \tag{3.78}$$

$$d_j = \left[I_j + A_j^j C^1\right]^{-1}\left[d_j^0 - \sum_{\ell=j+1}^{m} A_\ell^j C^1 d_\ell\right], \tag{3.79}$$

$$j = m-1,\ m-2, ..., 0\,,$$

where I_j is the unit tensor of the rank $(j + 4)$

$$A_{2\ell-1}^{2j} = A_{2\ell}^{2j-1} = 0\,, \quad \ell,\ j = 1, 2, ..., m\,. \tag{3.80}$$

3.4 Ellipsoidal inhomogeneity in a constant external field

Let us consider in detail the case of a constant external fields. If $\varepsilon^0 = \text{const}$ in the region occupied by the inclusion, the strain field ε^+ inside the ellipsoidal inclusion is also constant and has the form

$$\varepsilon^+ = \Lambda^\varepsilon \varepsilon^0 \,, \quad \Lambda^\varepsilon = (I + A(a)C^1)^{-1}, \tag{3.81}$$

where I is the unit rank four tensor, and the tensor $A(a)$ is defined by Equation (3.72):

$$A(a) = A_0^0 = \frac{1}{4\pi}\int_{\Omega_1} \mathrm{K}^*(a^{-1}k)d\Omega \,. \tag{3.82}$$

If the external stress tensor σ^0 is constant, the stress tensor σ^+ inside the inclusion is also constant and has the value that follows from (3.66) in the form

$$\sigma^+ = \Lambda^\sigma \sigma^0 \,, \quad \Lambda^\sigma = (I - D(a)B^1)^{-1}, \tag{3.83}$$

$$D(a) = \frac{1}{4\pi}\int_{\Omega_1} \mathrm{S}^*(a^{-1}k)d\Omega = C^0 A(a) C^0 - C^0. \tag{3.84}$$

Let us note that Eshelby's tensor [45] $S_{ijkl}(a)$ is defined as $S_{ijkl} = A_{ijmn}(a)C^0_{mnkl}$.

For an isotropic medium, the constant tensors $A(a)$ and $D(a)$ have the symmetry of the ellipsoid and are defined by nine essential components. In the coordinate system of the ellipsoid principal axes, these components have the forms [109], [110]

$$A_{1111} = \frac{\varkappa_0}{8\pi\mu_0}[3J_{11} + (1-4\nu_0)J_1] \,, \quad A_{1122} = \frac{\varkappa_0}{8\pi\mu_0}(J_{21} - J_1) \,, \tag{3.85}$$

$$A_{1212} = \frac{\varkappa_0}{8\pi\mu_0}[J_{21} + J_{12} + (1-2\nu_0)(J_1 + J_2)] \,, \tag{3.86}$$

$$D_{1111} = 4\mu_0\varkappa_0\left[\frac{1}{8\pi}(3J_{11} + J_1) - 1\right], \tag{3.87}$$

$$D_{1122} = 4\mu_0\varkappa_0\left\{\frac{1}{16\pi}J_{21} + J_{12} - (1-4\nu_0)\,(J_1 + J_2) - \nu_0\right\}, \tag{3.88}$$

$$D_{1212} = 4\mu_0\varkappa_0\left\{\frac{1}{16\pi}[J_{21} + J_{12} + (1-2\nu_0)\,(J_1 + J_2)] - \frac{1-\nu_0}{2}\right\}. \tag{3.89}$$

Here ν_0 is the Poisson ratio of the medium, the coefficient $\varkappa_0$ is defined in (2.9), and J_p, J_{pq} are the 1D integrals

$$J_p = \frac{3}{2}v\int_0^\infty \frac{du}{(a_p^2+u)\Delta(u)}\,,\quad J_{pq} = \frac{3}{2}va_p^2\int_0^\infty \frac{du}{(a_p^2+u)(a_q^2+u)\Delta(u)}\,,\tag{3.90}$$

$$\Delta(u) = \left[(a_1^2+u)(a_2^2+u)(a_3^2+u)\right]^{1/2},\quad v = \frac{4}{3}\pi a_1 a_2 a_3\,,\quad p,q = 1,2.\tag{3.91}$$

Note that J_p and J_{pq} are expressed in terms of elliptic integrals. The other six components of the tensors A and D are obtained by cyclic transposition of the indices 1,2,3.

Let us consider an ellipsoid of revolution around the x_3-axis (spheroid) with the semiaxes $a_1 = a_2 = a, a_3$. In this case, the tensors A and D become transversally isotropic with the symmetry axes x_3. Let m be a unit vector along the x_3-axis. For the presentation of the rank four tensors of transversally isotropic symmetry, it is convenient to use the P-basis defined in (2.11)–(2.13) (see also Appendix A.2) . In this basis, the tensor A for an oblate spheroid ($a > a_3$) takes the form

$$A = A_1P^2 + A_2(P^1 - \frac{1}{2}P^2) + A_3(P^3+P^4) + A_5P^5 + A_6P^6,\tag{3.92}$$

The scalar coefficients $A_1, ..., A_6$ in this equation are:

$$A_1 = \frac{1}{2\mu_0}[(1-\kappa_0)f_\circ + f_1],\quad A_2 = \frac{1}{2\mu_0}[(2-\kappa_0)f_0 + f_1],\tag{3.93}$$

$$A_3 = -\frac{1}{\mu_0}f_1\,,\quad A_5 = \frac{1}{\mu_0}(1 - f_0 - 4f_1)\,,\tag{3.94}$$

$$A_6 = \frac{1}{\mu_0}[(1-\kappa_0)(1-2f_0) + 2f_1]\,,\tag{3.95}$$

$$f_0 = \frac{1-g}{2(1-\gamma^2)},\quad f_1 = \frac{\kappa_0}{4(1-\gamma^2)^2}\left[(2+\gamma^2)g - 3\gamma^2\right],\tag{3.96}$$

$$g = \frac{\gamma^2}{\sqrt{\gamma^2-1}}\,\mathrm{arctg}\sqrt{\gamma^2-1}\,,\quad \gamma = \frac{a}{a_3} > 1\tag{3.97}$$

The tensor Λ^ε in (3.81) for an oblate spheroid takes the following form:

$$\Lambda^\varepsilon = \Lambda_1^\varepsilon P^2 + \Lambda_2^\varepsilon\left(P^1 - \frac{1}{2}P^2\right) + \Lambda_3^\varepsilon P^3 + \Lambda_4^\varepsilon P^4 + \Lambda_5^\varepsilon P^5 + \Lambda_6^\varepsilon P^6,\tag{3.98}$$

$$\Lambda_1^\varepsilon = \frac{1}{2\Delta_\varepsilon}\left[1 + A_6(\lambda_1+2\mu_1) + 2A_3\lambda_1\right],\quad \Lambda_2^\varepsilon = (1+2A_2\mu_1)^{-1}\,,\tag{3.99}$$

$$\Lambda_3^\varepsilon = -\frac{1}{\Delta_\varepsilon}\left[2A_1\lambda_1 + 2A_3(\lambda_1+2\mu_1)\right],\tag{3.100}$$

$$\Lambda_4^\varepsilon = -\frac{1}{\Delta_\varepsilon}\left[2A_3(\lambda_1+\mu_1) + A_6\lambda_1\right],\tag{3.101}$$

$$\Lambda_5^\varepsilon = 2(1+A_5\mu_1)^{-1},\quad \Lambda_6^\varepsilon = \frac{2}{\Delta_\varepsilon}\left[\frac{1}{2} + 2A_1(\lambda_1+\mu_1) + A_3\lambda_1\right],\tag{3.102}$$

$$\Delta_\varepsilon = 2\left\{\left[\frac{1}{2} + 2A_1(\lambda_1+\mu_1) + A_3\lambda_1\right]\left[1 + A_6(\lambda_1+2\mu_1) + 2A_3\lambda_1\right]\right.$$
$$\left. - \left[2A_1\lambda_1 + 2A_3(\lambda_1+2\mu_1)\right]\left[2A_3(\lambda_1+\mu_1) + A_6\lambda_1\right]\right\}, \qquad (3.103)$$

where the coefficients $A_1, ..., A_6$ are defined in (3.93)–(3.97).

The equation for the tensor D in the same basis can be obtained from (3.83) using the multiplication table of the P-tensors (Appendix A.2)

$$D = D_1P^2 + D_2(P^1 - \frac{1}{2}P^2) + D_3(P^3 + P^4) + D_5P^5 + D_6P^6, \qquad (3.104)$$
$$D_1 = -\mu_0[4\kappa_0 - 1 - 2(3\kappa_0 - 1)f_0 - 2f_1], \qquad (3.105)$$
$$D_2 = -2\mu_0[1 - (2-\kappa_0)f_0 - f_1], \quad D_3 = -2\mu_\circ[(2\kappa_0 - 1)f_0 + 2f_1], \qquad (3.106)$$
$$D_5 = -4\mu_0(f_0 + 4f_1), \quad D_6 = -8\mu_0(\kappa_0 f_0 - f_1). \qquad (3.107)$$

In the same basis, the tensor Λ^σ in (3.83) is

$$\Lambda^\sigma = \Lambda_1^\sigma P^2 + \Lambda_2^\sigma\left(P^1 - \frac{1}{2}P^2\right) + \Lambda_3^\sigma P^3 + \Lambda_4^\sigma P^4$$
$$+ \Lambda_5^\sigma P^5 + \Lambda_6^\sigma P^6, \qquad (3.108)$$

$$\Lambda_1^\sigma = \frac{1}{2\Delta_\sigma}\left[1 - D_6b_6^1 - 2D_3b_3^1\right], \quad \Lambda_2^\sigma = \left(1 - D_2b_2^1\right)^{-1}, \qquad (3.109)$$
$$\Lambda_3^\sigma = \frac{1}{\Delta_\sigma}\left(2D_1b_3^1 + D_3b_6^1\right), \quad \Lambda_4^\sigma = \frac{1}{\Delta_\sigma}\left(2D_3b_1^1 + D_6b_3^1\right), \qquad (3.110)$$
$$\Lambda_5^\sigma = 2\left(1 - \frac{1}{4}D_5b_5^1\right)^{-1}, \quad \Lambda_6^\sigma = \frac{2}{\Delta_\sigma}\left[\frac{1}{2} - D_1b_1^1 - D_3b_3^1\right], \qquad (3.111)$$

$$\Delta_\sigma = 2\left\{\left[\frac{1}{2} - D_1b_1^1 - D_3b_3^1\right]\left[1 - D_6b_6^1 - 2D_3b_3^1\right]\right.$$
$$\left. - \left(2D_1b_3^1 + D_3b_6^1\right)\left(2D_3b_1^1 + D_6b_3^1\right)\right\}. \qquad (3.112)$$

In these equations, the coefficients $D_1, ..., D_6$ are defined in (3.105)–(3.107), and the coefficients b_i^1 are

$$b_1^1 = \frac{1}{3}\left[\frac{1}{3}\left(\frac{1}{K} - \frac{1}{K_0}\right) + \frac{1}{4}\left(\frac{1}{\mu} - \frac{1}{\mu_0}\right)\right], \quad b_2^1 = \frac{1}{2}\left(\frac{1}{\mu} - \frac{1}{\mu_0}\right), \qquad (3.113)$$
$$b_1^1 = \frac{1}{3}\left[\frac{1}{3}\left(\frac{1}{K} - \frac{1}{K_0}\right) - \frac{1}{2}\left(\frac{1}{\mu} - \frac{1}{\mu_0}\right)\right], \quad b_5^1 = \frac{1}{\mu} - \frac{1}{\mu_0}, \qquad (3.114)$$
$$b_6^1 = \frac{1}{3}\left[\frac{1}{3}\left(\frac{1}{K} - \frac{1}{K_0}\right) + \frac{1}{\mu} - \frac{1}{\mu_0}\right]. \qquad (3.115)$$

If $\gamma >> 1$, the main terms of the coefficients A_i and D_i $(i = 1-6)$ in (3.93)–(3.97) and (3.105)-(3.107) have the order γ^{-1} and take the forms

$$A_1 = \frac{\pi}{\gamma}\frac{2-\kappa_0}{16\mu_0}, \quad A_2 = \frac{\pi}{\gamma}\frac{4-\kappa_0}{16\mu_0}, \quad A_3 = -\frac{\pi}{\gamma}\frac{\kappa_0}{8\mu_0}, \tag{3.116}$$

$$A_5 = \frac{1}{\mu_0}\left[1 - \frac{\pi}{4\gamma}(1+2\kappa_0)\right], \quad A_6 = \frac{1}{\mu_0}\left[1 - \kappa_0 - \frac{\pi}{4\gamma}(2-3\kappa_0)\right], \tag{3.117}$$

$$D_1 = -\mu_0(4\kappa_0 - 1) + \frac{\pi\mu_0}{4\gamma}(7\kappa_0 - 2), \quad D_2 = -2\mu_0 + \frac{\pi\mu_0}{4\gamma}(4-\kappa_0), \tag{3.118}$$

$$D_3 = -\frac{\pi\mu_0}{2\gamma}(3\kappa_0 - 1), \quad D_5 = -\frac{\pi\mu_0}{\gamma}(1+2\kappa_0), \quad D_6 = -\frac{\pi\mu_0}{\gamma}\kappa_0. \tag{3.119}$$

In the limit $\gamma \to \infty$ (thin ellipsoidal disk), we obtain

$$A = \frac{1}{\mu_0}\left[P^5 + (1-\kappa_0)P^6\right], \tag{3.120}$$

$$D = -2\mu_0\left[P^1 + (2\kappa_0 - 1)P^2\right]. \tag{3.121}$$

For a very thin void ($C = 0$), $\Lambda^\sigma = 0$, and Λ^ε is defined by the equation

$$\Lambda^\varepsilon = \frac{\gamma}{\pi}\left[\frac{\lambda_0}{\mu_0}\frac{1}{\kappa_0}P^4 + \frac{8}{1+2\kappa_0}P^5 + \frac{1}{\kappa_0(1-\kappa_0)}P^6\right] + O(1). \tag{3.122}$$

For a thin rigid disk ($C \to \infty$), $\Lambda^\varepsilon = 0$, and the tensor Λ^σ is

$$\Lambda^\sigma = \frac{\gamma}{\pi}\left[\Lambda_1^\sigma P^2 + \Lambda_2^\sigma\left(P^1 - \frac{1}{2}P^2\right) + 2(1-2\kappa_0)\Lambda_1^\sigma P^3\right], \tag{3.123}$$

$$\Lambda_1^\sigma = \frac{2}{(2-\kappa_0)(4\kappa_0 - 1)}, \quad \Lambda_2^\sigma = \frac{8}{4-\kappa_0}. \tag{3.124}$$

If the inclusion is a prolate spheroid ($a < a_3$, $\gamma = a/a_3$, $\gamma < 1$), the tensors A and D are determined by the same Equations (3.92) and (3.104), where the function $g(\gamma)$ should be replaced by

$$g(\gamma) = \frac{\gamma^2}{2\sqrt{1-\gamma^2}}\ln\frac{1+\sqrt{1-\gamma^2}}{1-\sqrt{1-\gamma^2}}. \tag{3.125}$$

Fibers are often used as reinforcement elements of the composite materials, and a prolate spheroid ($\gamma << 1$) may be considered as model of such a fiber. For a prolate spheroid, equations for the tensors A and D have the same form (3.92) and (3.104), where only the main terms with respect to the small parameter γ should be kept in the equations for the functions f_0 and f_1

$$f_0 = \frac{1}{2}\left(1 + \gamma^2 - \gamma^2\ln\frac{2}{\gamma}\right), \quad f_1 = \frac{\kappa_0}{4}\left(2\gamma^2\ln\frac{2}{\gamma} - 3\gamma^2\right). \tag{3.126}$$

If the parameter γ tends to zero, we go to the equation for the tensors A and D that correspond to the inclusion in the form of an infinite circular cylinder

$$A = \frac{1}{4\mu_0}\left[(1-\kappa_0)P^2 + (2-\kappa_0)(P^1 - \frac{1}{2}P^2) + 2P^5\right], \tag{3.127}$$

$$D = -\mu_0\left[\kappa_0 P^2 + \kappa_0\left(P^1 - \frac{1}{2}P^2\right) + \right.$$
$$\left. + (2\kappa_0 - 1)\left(P^3 + P^4\right) + 2P^5 + 4\kappa_0 P^6\right]. \tag{3.128}$$

In this case, the components of the tensor Λ^ε are

$$\Lambda_1^\varepsilon = \frac{1}{2}\left[1 + \frac{1-\kappa_0}{\mu_0}(\lambda_1 + \mu_1)\right]^{-1}, \quad \Lambda_2^\varepsilon = \left(1 + \frac{2-\kappa_0}{2\mu_0}\mu_1\right)^{-1}, \tag{3.129}$$

$$\Lambda_3^\varepsilon = -\frac{1-\kappa_0}{2\mu_0}\lambda_1\left[1 + \frac{1-\kappa_0}{\mu_0}(\lambda_1 + \mu_1)\right]^{-1}, \quad \Lambda_4^\varepsilon = 0, \tag{3.130}$$

$$\Lambda_5^\varepsilon = 2\left(1 + \frac{\mu_1}{2\mu_0}\right)^{-1}, \quad \Lambda_6^\varepsilon = 1, \tag{3.131}$$

and the tensor Λ^σ is to be found from (3.128) and (3.83) or is expressed via the tensor Λ^ε

$$\Lambda^\sigma = C\Lambda^\varepsilon B^0. \tag{3.132}$$

For a spherical inclusion, the tensors A and D are isotropic and independent on the inclusion radius

$$A = \frac{1-\kappa_0}{9\mu_0}E^2 + \frac{5-2\kappa_0}{15\mu_0}\left(E^1 - \frac{1}{3}E^2\right), \tag{3.133}$$

$$D = \frac{4\mu_0(1-4\kappa_0)}{9}E^2 - \frac{2\mu_0(5+4\kappa_0)}{15}\left(E^1 - \frac{1}{3}E^2\right), \tag{3.134}$$

$$E^1_{\alpha\beta\lambda\mu} = I_{\alpha\beta\lambda\mu} = \delta_{\lambda)(\alpha}\delta_{\beta)(\mu}, \quad E^2_{\alpha\beta\lambda\mu} = \delta_{\alpha\beta}\delta_{\lambda\mu}. \tag{3.135}$$

The tensors $\Lambda^\varepsilon = \left(I + AC^1\right)^{-1}$ and $\Lambda^\sigma = (I - DB^1)^{-1}$ in (3.81), (3.83) that define the strain and stress fields inside the inclusion may be obtained using the the algebra of the E-tensors presented in Appendix A.1. In particular, for a spherical isotropic inclusion we have

$$\Lambda^\varepsilon = \frac{\lambda_0 + 2\mu_0}{3\lambda_1 + 2\mu_1 + 3(\lambda_0 + 2\mu_0)}E^2$$
$$+ \frac{15\mu_0(\lambda_0 + 2\mu_0)}{2\mu_1(3\lambda_0 + 8\mu_0) + 15\mu_0(\lambda_0 + 2\mu_0)}\left(E^1 - \frac{1}{3}E^2\right). \tag{3.136}$$

Note, that the tensors Λ^{ε} and Λ^{σ} exist for every $C^1$and B^1including a void or an absolutely rigid inclusion. In particular, for a void (pore) ($C^1 = -C^0$), Λ^{ε} takes the form

$$\Lambda^{\varepsilon} = \frac{\lambda_0 + 2\mu_\circ}{4\mu_0}\left[E^2 + \frac{60\mu_0}{9\lambda_0 + 14\mu_0}\left(E^1 - \frac{1}{3}E^2\right)\right], \tag{3.137}$$

and the tensor Λ^{σ} in (3.83) is zero.

For an absolutely rigid inclusion ($C^1 \to \infty$), the strain tensor ε^+ inside the inclusion is zero, and the tensor Λ^{σ} in (3.83) is defined by the equation

$$\Lambda^{\sigma} = \frac{\lambda_0 + 2\mu_\circ}{3\lambda_0 + 2\mu_0}\left[E^2 + \frac{15(3\lambda_0 + 2\mu_\circ)}{2(3\lambda_0 + 8\mu_0)}\left(E^1 - \frac{1}{3}E^2\right)\right]. \tag{3.138}$$

For a constant temperature field T, the equation for the temperature stresses in a medium with an inclusion follows from (3.21), and takes the form

$$\sigma(x) - \int_V \mathrm{S}(x - x')B^1\sigma(x')dx' = D(a)\alpha^1 T. \tag{3.139}$$

In this case, the stress tensor σ^+ inside the ellipsoidal inclusion is constant, and has the form

$$\sigma^+ = \Lambda^{\sigma}(a)D(a)\alpha^1 T. \tag{3.140}$$

For a spherical inclusion, this equation is

$$\sigma^+_{\alpha\beta} = -\frac{4\mu_0(3\lambda + 2\mu)}{4\mu_\circ + 3\lambda + 2\mu}\alpha_1 T\delta_{\alpha\beta}. \tag{3.141}$$

Using the equation for the jumps of the elastic fields on the border between the medium and the inclusion (Section 3.2), we obtain the limits of the strain and stress tensors in the medium, on the surface of the ellipsoid in the following forms

$$\varepsilon^-(x_0) = F^{\varepsilon}(n_0)\varepsilon^0, \quad \sigma^-(x_0) = F^{\sigma}(n_0)\sigma^0, \tag{3.142}$$

where $x_0 \in \Omega$, $n_0 = n(x_0)$ is the external normal to the interface Ω, F^{ε} and F^{σ} are the tensor coefficients of the strain and stress concentrations

$$F^{\varepsilon}(n_0) = (I + K^*(n_0)C^1)\Lambda^{\varepsilon}, \quad F^{\sigma}(n_0) = (I - \mathrm{S}^*(n_0)B^1)\Lambda^{\sigma}. \tag{3.143}$$

The stresses on the interface due to a homogeneous temperature field are defined by the same equation for $\sigma^-(x_0)$ in (3.142), where the tensor σ^0 is replaced by $D\alpha^1 T$.

3.5 Inclusion in the form of a plane layer

Let us consider the inhomogeneity in the form of a plane layer that is the limiting form of the ellipsoidal inclusion when two of its semiaxes tend to infinity (Fig. 3.4).

The stress σ^+ and strain ε^+ tensors inside the layer are defined by the same equations (3.81) and (3.83) where tensors A and D are equal to the corresponding limiting values of these tensors. Suppose that a_1 and a_2 tend to infinity, and the semiaxis a_3 is finite. Because the integrands in (3.82) and (3.83) are homogeneous of zero degree, we can write

$$\lim_{a_1, a_2 \to \infty} \mathrm{K}^*\left(\frac{k_1}{a_1}, \frac{k_2}{a_2}, \frac{k_3}{a_3}\right) = \mathrm{K}^*\left(0, 0, \frac{k_3}{a_3}\right) = \mathrm{K}^*(n) , \tag{3.144}$$

$$\lim_{a_1, a_2 \to \infty} \mathrm{S}^*\left(\frac{k_1}{a_1}, \frac{k_2}{a_2}, \frac{k_3}{a_3}\right) = \mathrm{S}^*\left(0, 0, \frac{k_3}{a_3}\right) = \mathrm{S}^*(n) , \tag{3.145}$$

where n is the normal to the layer surface. Thus, the functions $K^*(n)$ and $\mathrm{S}^*(n)$ can be moved from the integral signs in (3.82) and (3.84), and the equations for the tensors A and D take the form

$$A = \mathrm{K}^*(n) , \quad D = \mathrm{S}^*(n) . \tag{3.146}$$

Hence, for a constant external field, the stress σ^+ and strain ε^+ fields inside the layer are defined by the equations

$$\sigma^+ = (I - \mathrm{S}^*(n)B^1)^{-1}\sigma^0, \quad \varepsilon^+ = (I + \mathrm{K}^*(n)C^1)^{-1}\varepsilon^0. \tag{3.147}$$

Note, that the fields σ^- and ε^- outside the layer coincide with the external fields σ^0 and ε^0. To prove this statement, we consider the equation for the stress field outside the inclusion:

$$\sigma^-(x) = \sigma^0 + \int_V \mathrm{S}(x - x')dx'\, B^1\sigma^+, \tag{3.148}$$

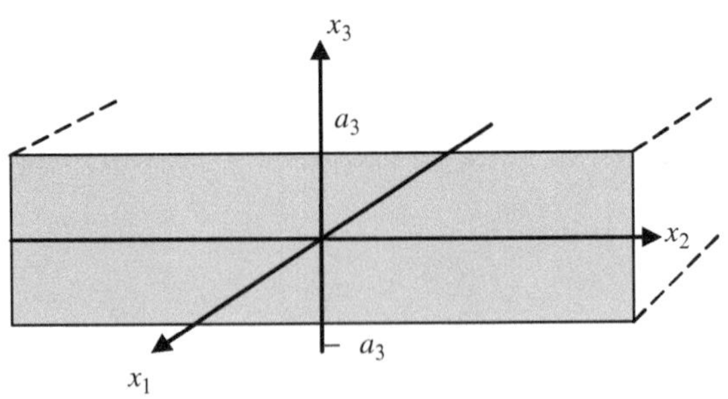

Fig. 3.4. Inclusion in the form of a plane layer.

where $\sigma^+$is the constant tensor and V is the layer of width $2a_3$. The integral in this equation is presented in the form

$$\int_V \mathrm{S}(x-x')dx' = \int_{-a_3}^{a_3} dx_3' \int_{-\infty}^{\infty}\int_{-\infty}^{\infty} \mathrm{S}(x_1-x_1', x_2-x_2', x_3-x_3')dx_1'dx_2' . \quad (3.149)$$

The internal integral in (3.149) may be written as the integral over the whole space:

$$J(x_1, x_2, x_3 - x_3') = \int_{-\infty}^{\infty}\int_{-\infty}^{\infty} \mathrm{S}(x_1 - x_1', x_2 - x_2', x_3 - x_3')dx_1'dx_2'$$

$$= \int\int_{-\infty}^{\infty}\int \mathrm{S}(x_1 - x_1', x_2 - x_2', x_3 - x_3' - z)\delta(z)dx_1'dx_2'dz . \quad (3.150)$$

Using the Fourier transform of the function $\mathrm{S}^*(k)$, we have

$$J(x_1, x_2, x_3 - x_3') =$$

$$= \frac{1}{2\pi}\int\int_{-\infty}^{\infty}\int \mathrm{S}^*(k_1, k_2, k_3)\ \delta(k_1)\delta(k_2)e^{-i[k_1x_1+k_2x_2+k_3(x_3-x_3')]}dk_1dk_2dk_3$$

$$= \mathrm{S}^*(n)\delta(x_3 - x_3') . \quad (3.151)$$

Substituting this result in (3.149) and in (3.148), we obtain that the stress field inside the layer is defined by the first equation (3.147), and the field outside coincides with σ^0, because

$$\int_{-a_3}^{a_3} \delta(x_3 - x_3')dx_3' = \begin{cases} 0, & \text{if } |x_3| > a_3 \\ 1, & \text{if } |x_3| < a_3 \end{cases} . \quad (3.152)$$

Thus, for a constant external field, the fields in the medium and the inclusion are defined by the equations

$$\sigma(x_3) = \sigma^0 + \mathrm{S}^*(n)B^1 \int_{-a_3}^{a_3} \delta(x_3 - x_3')\sigma(x_3')dx_3' , \quad (3.153)$$

$$\varepsilon(x_3) = \varepsilon^0 + \mathrm{K}^*(n)C^1 \int_{-a_3}^{a_3} \delta(x_3 - x_3')\sigma(x_3')dx_3' . \quad (3.154)$$

The strain and stress fields inside the layer are defined by Equations (3.147), and the fields in the medium coincide with the external fields

$$\sigma^- = \sigma^0, \quad \varepsilon^- = \varepsilon^0. \quad (3.155)$$

Note that the same equations hold if the external fields σ^0 and ε^0 depend only on the coordinate x_3.

3.6 Spheroidal inclusion in a transversely isotropic medium

Let the medium be transversely isotropic with the isotropy axis x_3, and let then contain a spheroidal inclusion with the same symmetry axis. In this case, for the calculation of the tensor $A(a)$, we use the equation (B.47) derived in Appendix B

$$A_{ijkl} = -\int_{\Omega} \left(e_m^r a_{mn} e_n^r\right)^{-1} a_{js} e_s^r \left[\nabla_l(\theta,\varphi) - e_l^r\right] g_{ik}(\theta,\varphi) d\Omega \bigg|_{(ij)(kl)} . \quad (3.156)$$

Here a_{mn} is the second-rank tensor, that depends on the inclusion shape and is defined by the expression

$$a_{mn} = \frac{1}{a_1^2} e_m^1 e_n^1 + \frac{1}{a_2^2} e_m^2 e_n^2 + \frac{1}{a_3^2} e_m^3 e_n^3, \quad (3.157)$$

$$\nabla_l(\theta,\varphi) = \frac{e_l^\varphi}{\sin\theta} \frac{\partial}{\partial\varphi} + e_l^\theta \frac{\partial}{\partial\theta}, \quad (3.158)$$

$\mathbf{e}^r, \mathbf{e}^\theta, \mathbf{e}^\varphi$ are the basis vectors of the spherical coordinate system with the origin in the center of the ellipsoid, and $\mathbf{e}^1, \mathbf{e}^2, \mathbf{e}^3$ are the unit vectors of the principal axes of the ellipsoid. Tensor $g_{ik}(\theta,\varphi)$ in (3.156) has the form (2.17)

$$g_{ik}(\theta,\varphi) = g_{\varphi\varphi}(\theta) e_i^\varphi e_k^\varphi + g_{\rho\rho}(\theta) e_i^\rho e_k^\rho + g_{zz}(\theta) e_i^z e_k^z + g_{\rho z}(\theta)(e_i^\rho e_k^z + e_i^z e_k^\rho), \quad (3.159)$$

where $\mathbf{e}^\rho, \mathbf{e}^\varphi, \mathbf{e}^z$ are the basis vectors of the cylindrical coordinate system $(\rho, \varphi, z = x_3)$ with the origin in the ellipsoid center

Using the connections between the basis vectors for the spherical and cylindrical coordinate systems:

$$\mathbf{e}^r = \mathbf{e}^\rho \sin\theta + \mathbf{e}^z \cos\theta, \quad \mathbf{e}^\theta = \mathbf{e}^\rho \cos\theta - \mathbf{e}^z \sin\theta, \quad (3.160)$$

we represent the Nabla operator $\nabla(\theta,\varphi)$ in the form

$$\nabla(\theta,\varphi) = \frac{\mathbf{e}^\rho}{\sin\theta} \frac{\partial}{\partial\varphi} + (\mathbf{e}^\rho \cos\theta - \mathbf{e}^z \sin\theta) \frac{\partial}{\partial\theta}. \quad (3.161)$$

For a spheroidal inclusion ($a_1 = a_2 = a$, a_3, $\gamma = a/a_3$) we have

$$a_{ij} = \frac{1}{a^2}\left(\theta_{ij} + \gamma^2 e_i^z e_j^z\right), \quad (3.162)$$

$$e_i^r a_{ij} e_j^r = \frac{1}{a^2}\left(\sin^2\theta + \gamma^2 \cos^2\theta\right), \quad (3.163)$$

where $\theta_{ij} = \delta_{ij} - e_i^z e_j^z$ is the projector on the isotropy plane.

Taking into account the equations

$$\int_0^{2\pi} e_i^\varphi e_j^\varphi d\varphi = \int_0^{2\pi} e_i^\rho e_j^\rho d\varphi = \pi\theta_{ij}, \tag{3.164}$$

$$\int_0^{2\pi} e_i^\rho e_j^\rho e_k^\rho e_l^\rho d\varphi = \frac{\pi}{4}\left(\theta_{ij}\theta_{kl} + \theta_{ik}\theta_{lj} + \theta_{il}\theta_{kj}\right), \tag{3.165}$$

$$\int_0^{2\pi} e_i^\rho e_j^\rho e_k^\varphi e_l^\varphi d\varphi = \frac{\pi}{4}\left(3\theta_{ij}\theta_{kl} - \theta_{ik}\theta_{lj} - \theta_{il}\theta_{kj}\right) \tag{3.166}$$

and integrating in (3.156) over the angles φ, we obtain

$$A_{\alpha\beta\lambda\mu} = \sum_{l=1}^{3} \int_0^{\pi} A^l_{\alpha\beta\lambda\mu}(\theta) \sin\theta d\theta. \tag{3.167}$$

Here the tensors $A^l_{\alpha\beta\lambda\mu}(\theta)$ in the P-basis are

$$A^l_{\alpha\beta\lambda\mu}(\theta) = -\frac{\pi}{2\Delta_l}\{(q_l - \xi_l^2 p_l)\xi_l \sin^2\theta \cdot P^2_{\alpha\beta\lambda\mu} + (2q_l$$
$$-\xi_l p_l)p_l \sin^2\theta\left(P^1_{\alpha\beta\lambda\mu} - \frac{1}{2}P^2_{\alpha\beta\lambda\mu}\right) + r_l\xi_l\left(\sin^2\theta - \gamma^2\cos^2\theta\right)(P^3_{\alpha\beta\lambda\mu}$$
$$+P^4_{\alpha\beta\lambda\mu}) + \left(2\gamma^2\left(2q_l - \xi_l p_l\right)\cos^2\theta - 2r_l\xi_l\left(\sin^2\theta - \gamma^2\cos^2\theta\right)\right.$$
$$\left.+2s_l\xi_l \sin^2\theta\right) P^5_{\alpha\beta\lambda\mu} + 4s_l\gamma^2\cos^2\theta \cdot P^6_{\alpha\beta\lambda\mu}\} \tag{3.168}$$

$$\Delta_l = \left(\xi_l \sin^2\theta + \cos^2\theta\right)^{3/2}\left(\sin^2\theta + \gamma^2\cos^2\theta\right), \quad l = 1, 2, 3. \tag{3.169}$$

The coefficients p_l, q_l, s_l, r_l, and ξ_l $(l = 1, 2, 3)$ are defined in (2.18)–(2.27). After integration over the angle θ in (3.167) we obtain the equation for the tensor A in the form

$$A_{\alpha\beta\lambda\mu} = A_1 P^2_{\alpha\beta\lambda\mu} + A_2\left(P^1_{\alpha\beta\lambda\mu} - \frac{1}{2}P^2_{\alpha\beta\lambda\mu}\right) + A_3(P^3_{\alpha\beta\lambda\mu} + P^4_{\alpha\beta\lambda\mu})$$
$$+ A_5 P^5_{\alpha\beta\lambda\mu} + A_6 P^6_{\alpha\beta\lambda\mu}, \tag{3.170}$$

$$A_1 = \frac{\pi}{2}\sum_{l=1}^{3}(q_l - \xi_l p_l) J_1^{(l)}, \quad A_2 = \frac{\pi}{2}\sum_{l=1}^{3}(2q_l - \xi_l p_l) J_1^{(l)}, \tag{3.171}$$

$$A_3 = -\frac{\pi}{2}\sum_{l=1}^{3} r_l\left(J_1^{(l)} - \gamma^2\xi_l J_2^{(l)}\right), \tag{3.172}$$

$$A_5 = \pi \sum_{l=1}^{3} \left[(2q_l - \xi_l p_l)\gamma^2 J_2^{(l)} - r_l \left(J_1^{(l)} - \gamma^2 \xi_l J_2^{(l)} \right) + s_l J_1^{(l)} \right], \tag{3.173}$$

$$A_6 = 2\pi \sum_{l=1}^{3} s_l \gamma^2 J_2^{(l)}. \tag{3.174}$$

Here $J_1^{(l)}$, $J_2^{(l)}$ are the functions

$$J_1^{(l)} = 2\lambda_l^2 \left[1 - \gamma^2 \lambda_l \xi_l \ln \left(\frac{\lambda_l + 1}{\lambda_l - 1} \right) \right], \tag{3.175}$$

$$J_1^{(l)} = 2\lambda_l^2 \left[\frac{1}{2} \lambda_l \ln \left(\frac{\lambda_l + 1}{\lambda_l - 1} \right) - 1 \right], \tag{3.176}$$

$$\lambda_l = \sqrt{\frac{1}{1 - \xi_l \gamma^2}} \tag{3.177}$$

For transition to the isotropic case in (3.170), let us introduce the variables τ, δ, η

$$\frac{k_0 - m_0}{k_0 + m_0} = \tau, \qquad \frac{m_0}{k_0 + m_0} = \frac{1}{2}(1 - \tau), \tag{3.178}$$

$$\frac{l_0}{k_0 + m_0} = \tau - \delta^2, \qquad \frac{n_0}{k_0 + m_0} = 1 + \eta\delta^2, \tag{3.179}$$

and put $\mu_0/m_0 = 1$. In these variables, the coefficients ξ_1, ξ_2, ξ_3 in (3.168), (3.169) take the forms

$$\xi_1 = 1, \quad \xi_2 = 1 - g\delta + q\delta^2, \quad \xi_3 = 1 + g\delta + q\delta^2, \tag{3.180}$$

$$g = \sqrt{\frac{(1+\tau)(2+\tau)}{1-\tau}}, \qquad q = \sqrt{\frac{1+\tau+\eta}{1-\tau}}. \tag{3.181}$$

Transition to the isotropic case corresponds to the limit $\delta \to 0$ in (3.170) and gives equation (3.92) for the tensor A.

When $\gamma \to 0$, we go to an infinite circular cylinder. In this case, $J_1^{(l)} = 2$ and the coefficients in (3.170) take the forms

$$A_1 = \pi \sum_{l=1}^{3} (q_l - \xi_l p_l), \quad A_2 = \pi \sum_{l=1}^{3} (2q_l - \xi_l p_l), \tag{3.182}$$

$$A_3 = -\pi \sum_{l=1}^{3} r_l, \quad A_5 = 2\pi \sum_{l=1}^{3} (s_l - r_l), \quad A_6 = 0. \tag{3.183}$$

The sums here can be easily calculated and give

$$A_1 = \frac{1}{4(k_0 + m_0)}, \quad A_2 = \frac{1}{4} \left(\frac{1}{k_0 + m_0} + \frac{1}{m_0} \right), \tag{3.184}$$

$$A_3 = 0, \quad A_5 = \frac{1}{2\mu_0}. \tag{3.185}$$

For an isotropic medium, $k_0 + m_0 = \lambda_0 + 2\mu_0$, $m_0 = \mu_0$, and these equations coincide with (3.127).

For a strongly oblate spheroid ($\gamma \to \infty$) the expressions for the integrals (3.175)–(3.177) are reduced to

$$J_1^l = \frac{\pi}{\gamma\sqrt{\xi_l}} + O\left(\frac{1}{\gamma^2}\right), \quad J_2^l = \frac{2}{\xi_l\gamma^2}\left(1 - \frac{\pi}{2\gamma\sqrt{\xi_l}}\right) + O\left(\frac{1}{\gamma^4}\right). \tag{3.186}$$

Substitution of these expressions into (3.171)–(3.174) gives the following formulas for the tensor A and its coefficients A_{1-6}

$$A = A^0 + \frac{\pi}{\gamma}A^1 + O\left(\frac{1}{\gamma^2}\right), \tag{3.187}$$

$$A_1^0 = A_2^0 = A_3^0 = A_4^0 = 0, \quad A_5^0 = \frac{1}{\mu_0}, \quad A_6^0 = \frac{1}{n_0}, \tag{3.188}$$

$$A_1^1 = \frac{n_0 + \mu_0\sqrt{\xi_2\xi_3}}{8(k_0 + m_0)\mu_0\left(\sqrt{\xi_2} + \sqrt{\xi_3}\right)\sqrt{\xi_2\xi_3}}, \tag{3.189}$$

$$A_2^1 = A_1^1 + \frac{1}{8\sqrt{\mu_0 m_0}}, \tag{3.190}$$

$$A_3^1 = A_4^1 = -\frac{l_0 + \mu_0}{4(k_0 + m_0)\mu_0\left(\sqrt{\xi_2} + \sqrt{\xi_3}\right)\sqrt{\xi_2\xi_3}}, \tag{3.191}$$

$$A_5^1 = -\frac{1}{4\mu_0}\left[\sqrt{\frac{m_0}{\mu_0}} + \frac{(k_0 + m_0)n_0 - l_0^2}{\mu_0\sqrt{(k_0 + m_0)n_0}\left(\sqrt{\xi_2} + \sqrt{\xi_3}\right)}\right]. \tag{3.192}$$

$$A_6^1 = -\frac{\mu_0\left(\xi_2 + \xi_3 + \sqrt{\xi_2\xi_3}\right) - n_0}{2n_0\mu_0\left(\sqrt{\xi_2} + \sqrt{\xi_3}\right)\sqrt{\xi_2\xi_3}}. \tag{3.193}$$

3.7 Crack in an elastic medium

Let us consider a homogeneous medium with a flattened cavity that occupies a finite region V with a smooth boundary. The integral equations for the medium with the cavity have the forms (3.31), (3.33):

$$\varepsilon(x) - \int \mathrm{K}(x - x')C^0\varepsilon^+(x')dx' = \varepsilon^0(x), \tag{3.194}$$

$$\sigma(x) - \int \mathrm{S}(x - x')\varepsilon^+(x')dx' = \sigma^0(x). \tag{3.195}$$

Here $\varepsilon^+(x) = \varepsilon(x)\, V(x)$ is the limiting value of the strain tensor inside the inclusion when its elastic moduli tend to zero. The stress field outside the cavity is calculated via $\varepsilon^+(x)$ from (3.195).

We consider the transition from a cavity to a crack, i.e., to a cut along a smooth surface Ω located inside V, and introduce a local Cartesian coordinate system at the point $x \in \Omega$ with the z-axis directed along the normal $n(x)$ to Ω, $h(x)$ be the transverse size of the cavity, and $z_1(x,h)$, $z_2(x,h)$ be the coordinates of the points of intersection of the z-axis with the boundary of V, $z_1, z_2 \to 0$ as $h \to 0$. For a fixed point $x \notin V$, the kernel $\mathrm{S}(x)$ in (3.195) is a smooth bounded function. Hence, the main term of the stress field outside the cavity has the form:

$$\sigma(x) = \sigma^\circ(x) + \int_\Omega \mathrm{S}(x - x')\bar{\varepsilon}(x', h)d\Omega' , \tag{3.196}$$

$$\bar{\varepsilon}(x,h) = \int_{z_1(x,h)}^{z_2(x,h)} \varepsilon^+(x + zh(x))dz , \quad x \in \Omega . \tag{3.197}$$

The function $\bar{\varepsilon}(x,h)$ may be interpreted as the coefficient of the principal term of the expansion of the function $\varepsilon^+(x)$ in a series of multipoles concentrated on Ω (Section 2.4):

$$\varepsilon^+(x)V(x) = \bar{\varepsilon}(x,h)\Omega(x) + \cdots \tag{3.198}$$

Here $\Omega(x)$ is the delta-function concentrated on Ω (2.42). It is important that all other terms of the multipole expansion of the function $\varepsilon^+(x)V(x)$ vanish as $h \to 0$, i.e., the total strain field $\varepsilon(x)$ contains a singular component proportional to $\Omega(x)$. Let the transverse size of the cavity h tend to zero, and the cavity be transformed into an infinitely thin cut (crack). On the crack surface, the stress $n_\alpha(x)\sigma_{\alpha\beta}(x)$ and displacement $u_\alpha(x)$ vectors satisfy the conditions

$$n_\alpha(x)\sigma_{\alpha\beta}(x) = 0, \quad [u_\alpha(x)] = b_\alpha(x), \quad x \in \Omega , \tag{3.199}$$

where $b(x)$ is the vector of the displacement jump or the opening of the crack which is to be found from the solution of the crack problem. In what follows we suppose that the crack faces do not contact. This condition imposes some restrictions on the external field $\sigma^0(x)$.

Since the displacement field is discontinuous on Ω, the strain tensor in the medium with the crack is the sum of the regular ε^r and singular ε^s components:

$$\varepsilon(x) = \varepsilon^r(x) + \varepsilon^s(x) , \quad \varepsilon^s_{\alpha\beta}(x) = n_{(\alpha}(x)b_{\beta)}(x)\Omega(x) . \tag{3.200}$$

From comparison of this equation with (3.198) it follows that in the limit $h \to 0$ the unwritten terms in (3.198) vanish, and the limiting value $\bar{\varepsilon}(x)$ of the function $\varepsilon^+(x,h)$ takes the form

$$\bar{\varepsilon}_{\alpha\beta}(x) = \lim_{h\to 0}\bar{\varepsilon}^{+}_{\alpha\beta}(x,h) = n_{(\alpha}(x)b_{\beta)}(x), \; x \in \Omega\,. \tag{3.201}$$

This equation and (3.197) show that in the limit $h \to 0$ the "normal" component of the tensor $\varepsilon^{+}(x)$ tends to infinity, whereas the function $\bar{\varepsilon}(x)$ is bounded.

Thus, taking into account (3.196), (3.201) and symmetry of the tensor $\mathrm{S}(x)$, we represent the stress field outside the crack in the form

$$\sigma_{\alpha\beta}(x) = \sigma^{0}_{\alpha\beta}(x) + \int_{\Omega} \mathrm{S}_{\alpha\beta\lambda\mu}(x-x')n_{\lambda}(x')b_{\mu}(x')d\Omega'. \tag{3.202}$$

From the first condition (3.199) we obtain the equation for the vector field $b(x)$ on Ω:

$$(T_{\alpha\beta}b_{\beta})(x) = \int_{\Omega} T_{\alpha\beta}(x,x')b_{\beta}(x')d\Omega' = n_{\beta}(x)\sigma^{0}_{\alpha\beta}(x)\,, \tag{3.203}$$

$$T_{\alpha\beta}(x,x') = -n_{\lambda}(x)\mathrm{S}_{\lambda\alpha\beta\mu}(x-x')n_{\mu}(x')\,. \tag{3.204}$$

Note, that the operator T in Equation (3.203) can be written as an integral operator with the kernel $T(x,x')$ only symbolically, because the corresponding integral formally diverges as $x \in \Omega$ for any smooth function $b(x)$ ($T(x,x')\sim|x-x'|^{-3}$ as $x \to x'$).

It is not difficult to see that the representation (3.202) coincides with the stress field in an elastic medium containing the Somigliana dislocation (2.99), and (3.203) is equivalent to Equation (2.101) that corresponds to the dislocation model of a crack.

Let us consider the generalized function $T(x,x')$ that is the kernel of integral (3.203), and for simplicity, let Ω be a region in the plane $x_3 = 0$. The function $b(\bar{x})$ ($\bar{x} = \bar{x}(x_1,x_2)$) is continued by zero outside Ω. In this case, T may be considered as the convolution operator with the generalized function $T(\bar{x})$, that is created by the generalized function (3.204). We have from (3.204)

$$T_{\alpha\beta}(x_1,x_2) = -\mathrm{S}_{3\alpha\beta3}(x_1,x_2,x_3)|_{x_3=0}\,. \tag{3.205}$$

The Fourier transform $T^{*}_{\alpha\beta}(k_1,k_2)$ of the function $T_{\alpha\beta}(x_1,x_2)$ follows from this equation in the form

$$T^{*}_{\alpha\beta}(k_1,k_2) = -\frac{1}{2\pi}\int_{-\infty}^{\infty} \mathrm{S}^{*}_{3\alpha\beta3}(k_1,k_2,k_3)dk_3\,, \tag{3.206}$$

where $\mathrm{S}^{*}(k)$ is determined in (2.75). This integral converges absolutely and defines an even homogeneous function of $\overline{k} = \overline{k}(k_1,k_2)$. Therefore, if $b(\bar{x})$ is a function from $\mathrm{S}(R^2)$-space (it consists of smooth functions vanishing at infinity faster than any negative degree of $|x|$), the action of the operator T on such a function is defined by the equation

$$(Tb)(\bar{x}) = \frac{1}{(2\pi)^2} \int T^*(\bar{k}) b^*(\bar{k}) e^{-i\bar{k}\cdot\bar{x}} d\bar{k} \,, \tag{3.207}$$

where the integral converges absolutely. This equation gives a regular representation of the operator T on the functions from $S(R^2)$ for the case of a plane region Ω.

If Ω is an arbitrary Lyapunov surface bounded by a smooth contour Γ, the regular representation of the operator in (3.203) is obtained in Appendix C.2 in the form

$$(T_{\alpha\beta} b_\beta)(x) = p.v. \int_\Omega T_{\alpha\beta}(x, x')[b_\beta(x') - b_\beta(x)] d\Omega' \\ - n_\nu(x) \oint \mathrm{rot}_{\lambda\beta} Z_{\nu\alpha\lambda\mu}(x - x') d\Gamma'_\beta b_\mu(x) \,. \tag{3.208}$$

The integrals here exist if the function $b(x)$ has a continuous derivative on Ω and is zero on the contour Γ.

For an arbitrary smooth surface Ω, the operator T is a generalized pseudo-differential operator with principal homogeneous symbol $T^*(k)$ defined in (3.206) is a homogeneous function of the degree 1. The existence and uniqueness of the solution of the equation $Tb = f$ are proved in [48]. It is shown in [48] that for a smooth right-hand side $f(x)$, the asymptotic form of this solution near Γ is

$$b(x) = \beta(x_0)\sqrt{r} + 0(r^{3/2}) \,, \tag{3.209}$$

where r is the distance from the point $x \in \Omega$ to $x_0 \in \Gamma$ along the normal to Γ, $\beta(x_0)$ is a smooth function along Γ.

Let us consider the asymptotic form of the stress field $\sigma(x)$ outside the crack, in the neighborhood of its edge Γ, and let y_1, y_2, y_3 be the local Cartesian coordinates at the point $x_0 \in \Gamma$. The y_3-axis is directed along the limit normal to Ω at the point x_0, and the y_2-axis is tangent to Γ. Then, the y_1-axis lies in the plane which is tangent to Ω at the point x_0 (Fig. 3.5).

Taking into account Equation (3.209), we have the following equation for the asymptotic form of the crack opening vector in the vicinity of the point $x_0 \in \Gamma$

$$b(y) = \beta(x_0)\sqrt{y_1} + 0(y_1^{3/2}) \,. \tag{3.210}$$

Using Equation (3.201), we can write the equation for the stress tensor at the point y with coordinates $y_1 = -r\cos\theta$, $y_2 = 0$, $y_3 = -r\sin\theta$, where r is the distance from point y up to the origin, θ is the polar angle in the plane (y_1, y_3) $(\xi_i = y_i/r)$

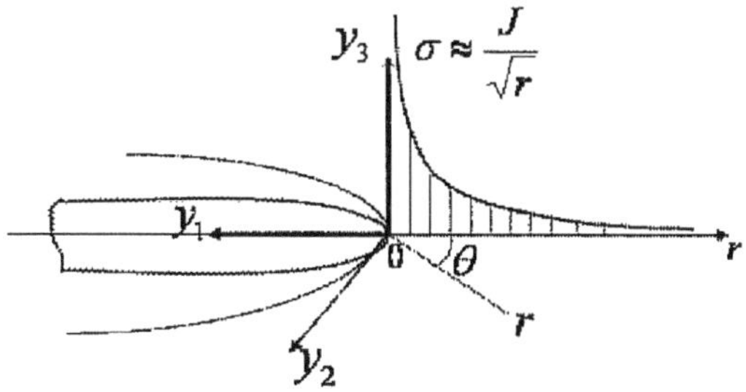

Fig. 3.5. Local coordinate system in the vicinity of the crack edge.

$$\sigma(y) = \frac{1}{\sqrt{r}} \int\limits_{\Omega(r)} \mathrm{S}\left(\cos\theta + \xi_1, \xi_2, \sin\theta + \xi_3\right) n\left(r\xi\right) \beta\left(x_0\right) \sqrt{\xi_1} d\Omega_\xi + O\left(\sigma^0\right) . \tag{3.211}$$

Here we take into account, that $\mathrm{S}(x)$ is an even homogeneous function of the degree -3. In the limit $r \to 0$, the integral multiplier at $r^{-1/2}$ in this equation has a finite value

$$J(\theta, x_0) = \lim \sqrt{r}\sigma(y), \quad r \to 0 . \tag{3.212}$$

Taking into account, that in the limit $r \to 0$ the surface $\Omega(r)$ in coordinates ξ_i is transformed into the semiplane ($\xi_3 = 0, \quad \xi_1 > 0$), we obtain the following equation for the components of the tensor $J(\theta, x_0)$:

$$J_{\alpha\beta}(\theta, x_0) = \mathrm{s}_{\alpha\beta\lambda\mu}(\theta) n_\lambda(x_0) \beta_\mu(x_0) , \tag{3.213}$$

$$\mathrm{s}(\theta) = \int\limits_0^\infty \sqrt{\xi_1} d\xi_1 \int\limits_{-\infty}^{\infty} \mathrm{S}(\cos\theta + \xi_1, \xi_2, \sin\theta) d\xi_2 , \tag{3.214}$$

where $n(x_0)$ is the limiting value of the normal to Ω at the point $x_0 \in \Gamma$.

Thus, the function $J(\theta, x_0)$ is presented in the form of the product of two terms. The first, $\mathrm{s}(\theta)n(x_0)$, is independent of the shape of the surface Ω and the external field, and it is determined only by the local orientation of the edge Γ of Ω at the point $x_0 \in \Gamma$. The second term (vector $\beta(x_0)$) is a functional of the whole surface Ω and the external field $\sigma^0(x)$.

Equation (3.212) implies

$$\sigma(y) = \frac{1}{\sqrt{r}} J(\theta, x_0) + O(\sigma^0), \tag{3.215}$$

and therefore, $J(\theta, x_0)$ may be called the stress intensity tensor coefficient.

The tensor $J(\theta, x_0)$ can be represented as a sum of three tensors that correspond to the three components of the vector $\beta(x_0)$ in the basis of the axes y_1, y_2, y_3 (do not sum over i!)

$$J = J^1 + J^2 + J^3 , \quad J^i_{\alpha\beta}(\theta, x_0) = s_{\alpha\beta\lambda i}(\theta) n_\lambda(x_0) \beta_i(x_0) . \tag{3.216}$$

Note that in elasticity theory and linear fracture mechanics, the asymptotic of the stress field in the neighborhood of the edge of a crack is characterized by the stress intensity factors K_I, K_{II}, K_{III} [28]. The connection of these factors with the components of the tensor $J^i(\theta, x_0)$ is given by the equations

$$K_I(x_0) = \sqrt{2\pi} J^3_{33}(0, x_0) , \quad K_{II}(x_0) = \sqrt{2\pi} J^1_{13}(0, x_0) , \tag{3.217}$$

$$K_{III}(x_0) = \sqrt{2\pi} J^2_{23}(0, x_0) . \tag{3.218}$$

It follows from this equation and (3.216) that with the accuracy to multipliers that depend on the elastic moduli of the medium, the stress intensity factors coincide with the components of the vector $\beta(x_0)$.

3.8 Elliptical crack

Let Ω be a plane elliptical cut in a homogeneous elastic medium. In this case, there exists the following analog of the polynomial conservation theorem.

If the external field $\sigma^0(x)$ applied to the elliptical crack is a polynomial of the degree m, the jump $b(x)$ of the displacement vector on the crack surface has the form

$$b(x_1, x_2) = B(x_1, x_2) \sqrt{1 - \left(\frac{x_1}{a_1}\right)^2 - \left(\frac{x_2}{a_2}\right)^2} , \tag{3.219}$$

where $B(x_1, x_2)$ is a polynomial of a degree no higher than m in coordinates x_1, x_2 connected with the principal axes of the ellipse; a_1, a_2 are its semiaxes.

In order to prove this statement, let us consider the elliptical crack as a limit of an ellipsoidal cavity with the semiaxes a_1, a_2, h when h tends to zero. According to (3.197), (3.201), the vector $b(x)$ is determined by the limiting value of the integral

$$\bar{\varepsilon}^+(\bar{x}, h) = \int_{-hz(\bar{x})}^{hz(\bar{x})} \varepsilon^+(x + \xi n) d\xi , \quad \bar{x} \in \Omega , \tag{3.220}$$

where ξ is the coordinate along the normal n to Ω,

$$z(\bar{x}) = \sqrt{1 - \left(\frac{x_1}{a_1}\right)^2 - \left(\frac{x_2}{a_2}\right)^2} . \tag{3.221}$$

From the polynomial conservation theorem (Section 3.3) follows that for the polynomial external field of degree m, the field $\varepsilon^+(\bar{x} + \xi n)$ inside the cavity is also a polynomial of a degree no higher than m with respect to the coordinates x_1, x_2, ξ . Thus, this field is presented in the form

$$\varepsilon^{+}(\bar{x}+\xi n) = \varepsilon^{(0)}(\bar{x}) + \varepsilon^{(1)}(\bar{x})\xi + \cdots + \varepsilon^{(m)}(\bar{x})\xi^{m}, \tag{3.222}$$

where $\varepsilon^{(k)}(\bar{x})$ is polynomial of a degree no higher than $m-k$ with respect to the coordinates x_1, x_2. After substituting this presentation in (3.220) and integrating over ξ, we find that the terms with odd degrees of ξ vanish, and the integrals of the terms with ξ^{2k} give terms that have the form $Q^{2k}(\bar{x})z(\bar{x})$, where Q^{2k} is a polynomial of degree not higher than $2k$. Hence, the tensor $\bar{\varepsilon}^{+}(x,h)$ is represented in the form

$$\bar{\varepsilon}^{+}(\bar{x},h) = P_m(\bar{x},h)z(\bar{x})\,, \tag{3.223}$$

where P_m is a polynomial of degree not more than m with respect to the coordinates x_1, x_2. Finally, setting $h=0$, we obtain from (3.201) the required Equation (3.219).

Property (3.219) means that for an elliptical crack, the operator T in equation (3.203) transforms a polynomial $B(\bar{x})$ of degree m multiplied by the function $z(\bar{x})$ (3.221) into a polynomial of the same degree m on Ω. In particular, if σ^0 is a constant tensor, $B_\alpha = \beta^0_\alpha$ is a constant vector that is the solution of the equation

$$T^0_{\alpha\beta}B^0_\beta = \sigma^0_{\alpha\beta}n_\beta\,, \tag{3.224}$$

where T^0 is the following constant tensor

$$T^0 = \int_\Omega T(\bar{x}-\bar{x}')z(\bar{x}')d\bar{x}'0 = \int_\Omega T(\bar{x})z(\bar{x})d\bar{x}$$

$$= p.v.\int T(\bar{x})[z(\bar{x})-z(0)]\,d\bar{x}\,. \tag{3.225}$$

Here the function $z(\bar{x})$ is continued by zero outside Ω; the Cauchy principal value of the integral is calculated over the whole plane x_1, x_2. Note that this integral exists in the ordinary sense.

Let us consider (3.224) for an isotropic medium. It follows from Equations (3.204), (2.72), (3.22) that in this case, the function $T(\bar{x})$ takes the form

$$T_{\alpha\beta}(\bar{x}) = -\frac{\mu_0}{4\pi(1-\nu_0)|\bar{x}|^3}\left[(1-2\nu_0)\,\delta_{\alpha\beta} + \nu_0\left(2n_\alpha n_\beta + \frac{3\bar{x}_\alpha\bar{x}_\beta}{|\bar{x}|^2}\right)\right]. \tag{3.226}$$

Let us substitute this equation into (3.225) and introduce the coordinates r, φ in the (x_1, x_2)-plane

$$x_1 = a_1 r\ \cos\varphi\,, \quad x_2 = a_2 r\ \sin\varphi\,. \tag{3.227}$$

In these coordinates, the integral T^0 takes the form

$$T^0_{\alpha\beta} = -\frac{\mu_0 a_1 a_2}{4\pi(1-\nu_0)}\int_0^\infty \frac{z(r)-1}{r^2}dr\int_0^{2\pi}\Big[(1-2\nu_0)\delta_{\alpha\beta} + \nu_0\left(2n_\alpha n_\beta + \frac{3m_{\alpha\beta}(\varphi)}{t^2(\varphi)}\right)\Big]\frac{d\varphi}{t^3(\varphi)}, \tag{3.228}$$

$$m_{11} = a_1^2\cos^2\varphi\,,\quad m_{22} = a_2^2\sin^2\varphi\,,\quad m_{12} = m_{21} = \frac{1}{2}a_1 a_2\sin 2\varphi\,, \tag{3.229}$$

$$m_{\alpha 3} = m_{3\alpha} = 0,\quad t^2(\varphi) = a_1^2\cos^2\varphi + a_2^2\sin^2\varphi\,, \tag{3.230}$$

$$z(r) = \sqrt{1-r^2},\quad r\le 1,\quad z(r) = 0,\quad r>1. \tag{3.231}$$

Calculating the integrals in (3.228) and solving Equation (3.224) with respect to B^0, we obtain (do not sum over α!)

$$B^0_\alpha = (T^0_{\alpha\beta})^{-1}\sigma^0_{\beta 3}\,,\quad T^0_{\alpha\beta} = T^0_\alpha\delta_{\alpha\beta}\,, \tag{3.232}$$

$$T^0_1 = \frac{a_2\mu_0}{2a_1^2(1-\nu_0)}\left[c_1 + \nu_0(c_2 - 2c_1)\right], \tag{3.233}$$

$$T^0_2 = \frac{a_2\mu_0}{2a_1^2(1-\nu_0)}\left[c_1 + \nu_0(c_3 - 2c_1)\right], \tag{3.234}$$

$$T^0_3 = \frac{a_2\mu_0 c_1}{2a_1^2(1-\nu_0)}\,,\quad c_1 = \frac{E(k)}{1-k}\,, \tag{3.235}$$

$$c_2 = c_1 - \frac{E(k)-K(k)}{k}\,,\quad c_3 = 3c_1 - c_2\,,\quad k \tag{3.236}$$

$$= 1-(a_2/a_1)^2,\quad a_1\ge a_2\,. \tag{3.237}$$

Here $K(k)$ and $E(k)$ are the complete elliptic integrals of the first and second kinds, respectively:

$$K(k) = \int_0^{\pi/2}\frac{d\theta}{\sqrt{1-k\sin^2\theta}},\quad E(k) = \int_0^{\pi/2}\sqrt{1-k\sin^2\theta}\,d\theta. \tag{3.238}$$

Let us consider the asymptotic values of the vector $b(x)$ on the contour of an elliptic crack. For a polynomial external field, this asymptotic form follows from (3.219) in the form

$$b(x) = \beta(x_0)\sqrt{r} + O(r^{3/2})\,, \tag{3.239}$$

$$\beta(x_0) = \sqrt{2}B(x_0)\left[\left(\frac{x_1^0}{a_1^2}\right)^2 + \left(\frac{x_2^0}{a_2^2}\right)^2\right]^{1/4}, \tag{3.240}$$

where r is the same as in (3.209), $B(x_0)$ is the value of the polynomial $B(x)$ at the point $x_0 \in \Gamma$.

The stress intensity tensor $J(\theta, x_0)$ for the elliptical crack follows from (3.213) in the form

$$J_{\alpha\beta}(\theta, x_0) = \sqrt{2}\pi_{\alpha\beta\lambda}B_\lambda(x_0)\left[\left(\frac{x_1^0}{a_1^2}\right)^2 + \left(\frac{x_2^0}{a_2^2}\right)^2\right]^{1/4}, \tag{3.241}$$

$$\pi_{\alpha\beta\lambda}(\theta) = \mathrm{s}_{\alpha\beta\lambda\mu}(\theta)n_\mu . \tag{3.242}$$

For an isotropic medium, from the definition (3.214) of the tensor $\mathrm{s}(\theta)$ we obtain the following equation for the components of the tensor $\pi(\theta)$ in the basis of the local coordinates with the origin at the point $x_0 \in \Gamma$ (the x_3-axis is directed along the normal to Γ, x_2 -along the tangent to Γ, x_1 - along the normal to Γ in the plane Ω)

$$\pi_{\alpha\beta\lambda} = \pi_{\alpha\beta\lambda}\,, \quad \pi_{1\beta\lambda} = \pi_{1\lambda\beta}\,, \tag{3.243}$$

$$\pi_{\alpha\alpha 3} = \pi_{i3j} = \pi_{ij3} = 0\,, \quad i \neq j\,, \quad i,j = 1,2\,, \tag{3.244}$$

$$\pi_{i33} = \frac{1-\nu_0}{2\nu_0}\pi_{33i}\,, \quad \pi_{33i} = \frac{\mu_0\nu_0}{2(1-\nu_0)}\cos\left[\frac{\pi}{2}(2-i) + \frac{\theta}{2}\right], \tag{3.245}$$

$$\pi_{ii2} = \frac{\mu_0}{4(1-\nu_0)}\cos\frac{\theta}{2}\left[1 - (-1)^i \sin\frac{\theta}{2}\sin\frac{3\theta}{2}\right], \tag{3.246}$$

$$\pi_{111} = -\frac{\mu_0}{4(1-\nu_0)}\sin\frac{\theta}{2}\left[2 + \cos\frac{\theta}{2}\cos\frac{3\theta}{2}\right], \tag{3.247}$$

$$\pi_{122} = \pi_{221} = \frac{\mu_0}{8(1-\nu_0)}\sin\theta\cos\frac{3\theta}{2}\,. \tag{3.248}$$

In the conclusion of this section, we consider the external field that compresses a crack. If the faces of the crack are closed by the external field, the force vector $t_\alpha(x) = n_\beta(x)\sigma_{\beta\alpha}(x)$ on the surface Ω is not equal to zero and conditions (3.199) are violated. The normal components of the stress field on the surface Ω depend on the contact forces of the faces of the crack. In particular, if the face interaction is described by Coulomb's law (the tangential component $t(x)$ of the force vector on Ω is proportional to its normal component), this condition takes the form

$$-t_\alpha(x)n_\alpha(x) = \chi\, t_\alpha(x)e_\alpha(x)\,, \quad e_\alpha(x) = \left|t^e\right|^{-1}t_\alpha^e\,, \tag{3.249}$$

$$t_\alpha^e = t_\alpha(x) - n_\beta(x)t_\beta(x)n_\alpha(x)\,, \quad x \in \Omega\,, \tag{3.250}$$

where χ is the Coulomb constant. The jump $b(x)$ of the displacement vector on the crack surface is located in the tangent plane to Ω, and its direction is defined by the unit vector $e_\alpha(x)$ in (3.249)

$$b_\alpha(x) = b(x)e_\alpha(x), \tag{3.251}$$

where $b(x)$ is a scalar function.

Because of (3.202), the force vector $t_\alpha(x)$ on the crack surface is represented in the form

$$t_\alpha(x) = n_\beta(x)\sigma^0_{\beta\alpha}(x) - \int_\Omega T_{\alpha\beta}(x,x')b_\beta(x')d\Omega' \,. \tag{3.252}$$

Substituting this equation into (3.249) and taking into account (3.251) we obtain the equation for the vector field $b_\alpha(x)$ on Ω. If the crack faces are partially in contact, the condition $t_\alpha(x) = 0$ holds on the part of Ω where the faces are separated, and condition (3.249) takes place on the other part that are in contact.

If the external field is polynomial and the faces of the crack are partially in contact, the property (3.219) of the solution is violated. However, if the contact takes place over the whole surface Ω, and the boundary condition (3.249) holds, the polynomial conservation property is true. If $n_\beta\sigma^0_{\beta\alpha}(x)$ is a polynomial of degree m on Ω, and the vector $b(x)$ has the form $b(x) = B(x)z(x)$, where $B(x)$ is also a polynomial of degree m, we obtain that the vector $t_\alpha(x)$ defined in (3.252) is a polynomial of degree m on Ω. Under this condition, (3.252) is transformed into the equivalence of two polynomials of the same degree. Equating the terms of the same degree in $\bar{x}$ and taking into account (3.251) we obtain a system of equations for the coefficients of the polynomial $B(\bar{x})$.

In particular, if $n_\beta\sigma^0_{\beta\alpha}$ is a constant vector, (3.249) and (3.251) lead to the following equations for the constant vector B_α:

$$-\sigma^0_{33} + T^0_{3\beta}B_\beta = \chi(\sigma^0_{3\alpha} - T^0_{\alpha\beta}B_\beta)e_\alpha, \quad \frac{\sigma^0_{13} - T^0_{1\beta}B_\beta}{\sigma^0_{23} - T^0_{2\beta}B_\beta} = \frac{B_1}{B_2}, \tag{3.253}$$

$$e_1 = \frac{B_1}{B_2}\left[1 + \left(\frac{B_1}{B_2}\right)^2\right]^{-1/2}, \quad e_2 = e_1\frac{B_2}{B_1}, \quad e_3 = B_3 = 0. \tag{3.254}$$

Here the tensor T^0 is defined in (3.225); the vector B should satisfy the condition on the closed part of the crack

$$t_\alpha n_\alpha = \sigma^0_{33} - T^0_{3\beta}\,B_\beta < 0\,. \tag{3.255}$$

If the crack faces are absolutely smooth, the boundary conditions on Ω take the form

$$t^e_\alpha = 0\,, \quad n_\beta b_\beta(x) = 0\,, \quad x \in \Omega\,. \tag{3.256}$$

In this case, the equation for the vector field $b(x)$ on Ω is derived by substituting the vector $t_\alpha(x)$ in the form (3.252) into (3.249). For an elliptical crack, property (3.219) holds, and for the calculation of the coefficients of the polynomial $B(x)$, it is possible to use the procedure described in Section 3.3.

3.9 Radially heterogeneous inclusion

Homogeneous inclusions in a homogeneous elastic medium form an idealized model of the microstructure of actual composite materials. As a rule, there exist the so-called transition zones (layers) on the interface between the matrix and inclusions. The properties of the layers differ from the properties of the matrix as well as of the inclusion. For polymer–polymer composites, such layers may occupy considerable portions of the volumes of the inclusions. In some cases, mechanical properties of the composite materials with strongly inhomogeneous inclusions turn out to be better than the properties of similar materials with homogeneous inclusions. In this section, we consider a spherical isotropic inclusion the elastic moduli of which depend on the distance from the center. This case has many important applications in material engineering. For the solution of this problem, we use the integral equations obtained in the previous sections. In contrast with traditional approaches based on differential equations of elasticity (see, e.g., [29]), the integral equations allow us to discover the explicit tensor structure of the strain field in the medium with the inclusion. For a spherical multi-layered inclusion, an efficient numerical algorithm is proposed for the construction of the solution.

3.9.1 Elastic fields in a medium with a radially heterogeneous inclusion

Let us consider an infinite homogeneous medium containing a spherical inclusion. The tensor $C(x)$ of the elastic moduli of the inclusion is a piece-wise smooth function of distance r from the center. The medium is subjected to a constant external strain field ε^0. Because of the linearity of the problem, the strain field $\varepsilon(x)$ in the medium with the inclusion has the form

$$\varepsilon(x) = \varepsilon^0 + \varepsilon^1(x)\,, \quad \varepsilon^1_{\alpha\beta}(x) = A_{\alpha\beta\lambda\mu}(x)\varepsilon^0_{\lambda\mu}\,, \tag{3.257}$$

where the rank four tensor $A(x)$ is symmetric with respect to the first and second pairs of indices, and $A(x) \to 0$ as $|x| \to \infty$. Substituting (3.257) into Equation (3.11) for the strain tensor and taking into account that ε^0 is an arbitrary tensor, we arrive at the equation for the tensor $A(x)$ in the form

$$A_{\alpha\beta\lambda\mu}(x) + \int_V \mathrm{K}_{\alpha\beta\tau\rho}(x-x')C^1_{\tau\rho\nu\delta}(x')A_{\nu\delta\lambda\mu}(x')dx'$$

$$= -\int_V \mathrm{K}_{\alpha\beta\tau\rho}(x-x')C^1_{\tau\rho\lambda\mu}(x')dx'. \tag{3.258}$$

Here V is the region occupied by the inclusion, $C^1(x) = C^1(|x|) = C(|x|) - C_0$ is the disturbance of the elastic moduli inside the inclusion,

C_0 is the tensor of the elastic moduli of the medium. In a short symbolic form, (3.258) is written as follows:

$$A(x) + (\mathrm{K}C^1 A)(x) = -(KC^1)(x)\,. \tag{3.259}$$

If the Lamé parameters $\lambda(r)$, $\mu(r)$ of the inclusion material are piece-wise smooth functions inside V, the tensor $C^1(r)$ has the form

$$C^1(r) = \lambda_1(r)E^2 + 2\mu_1(r)E^1, \tag{3.260}$$

$$\lambda_1(r) = \lambda(r) - \lambda_0\,, \quad \mu_1(r) = \mu(r) - \mu_0, \quad r = |x|\,. \tag{3.261}$$

Here λ_0, μ_0 are the Lamé parameters of the medium (matrix). For representation of the solution of (3.259) (the tensor A), we introduce the tensor basis of six linearly independent rank four tensors $E^i(n)$ built of a fixed unit vector n_α and the unit rank two tensor $\delta_{\alpha\beta}$:

$$E^1_{\alpha\beta\lambda\mu} = \delta_{\lambda)(\alpha}\delta_{\beta)\mu}\,, \quad E^2_{\alpha\beta\lambda\mu} = \delta_{\alpha\beta}\delta_{\lambda\mu}\,, \qquad E^3_{\alpha\beta\lambda\mu} = \delta_{\alpha\beta}n_\lambda n_\mu\,, \tag{3.262}$$

$$E^4_{\alpha\beta\lambda\mu} = n_\alpha n_\beta \delta_{\lambda\mu}\,, \quad E^5_{\alpha\beta\lambda\mu} = n_{(\alpha}\delta_{\beta)(\lambda}n_{\mu)}\,, \quad E^6_{\alpha\beta\lambda\mu} = n_\alpha n_\beta n_\lambda n_\mu\,. \tag{3.263}$$

These tensors form a closed algebra with respect to the product operation defined by the equation

$$(E^i E^j)_{\alpha\beta\lambda\mu} = E^i_{\alpha\beta\nu\rho}\, E^j_{\nu\rho\lambda\mu}\,. \tag{3.264}$$

The multiplication table of the tensors E^i is presented in Appendix A.1.

For the solution of (3.259), a special representation of the singular integral operator K in (3.259) is used. Let us introduce a spherical coordinate system (r, n) with the origin at the inclusion center ($r = |x|$, $n = x/|x|$ is the normal vector on the unit sphere Ω_1), and consider a piece-wise smooth finite function $f(r, n)$. The Mellin transform of this function with respect to the variable r is denoted by $f^*(\mathrm{s}, n)$. The direct and inverse Mellin transforms are defined by the following equations [201]:

$$f^*(s,n) = \int_0^\infty r^{s-1} f(r,n)dr, \;\; f\,(r,n) = \frac{1}{2\pi i}\int_{\tau-i\infty}^{\tau+i\infty} f^*\,(s,n)\,r^{-s}ds\,. \tag{3.265}$$

Here the parameter τ defines the position of the line of integration in the complex s-plane. It is shown in Appendix B.4 that the action of the operator K on the function $f(r, n)$ is defined by the equation

$$(\mathrm{K}f)(r,n) = \frac{1}{2\pi i}\int_{\tau-i\infty}^{\tau+i\infty} r^{-s}(\mathrm{K}_s f^*)(s,n)ds\,, \tag{3.266}$$

where the operator K_s acts on the Mellin transform $f^*(s, n)$ of the function $f(r, n)$ according to the equation

$$(\mathrm{K}_s f^*)(s,n) = (2\pi)^{-d} \exp\left(\frac{i\pi d}{2}\right) \Gamma(d-s)\Gamma(s) \times \int_{\Omega_1(d)} (-n\cdot m)^{-s} dm \int_{\Omega_1(d)} \mathrm{K}^*(m) f^*(s,l)(m\cdot l)^{s-d} dl. \quad (3.267)$$

Here d is the dimension of the space, n, m, l are vectors on the unit sphere Ω_1 in this space, $\mathrm{K}^*(m)$ is the symbol of the operator K that has the form (2.58), $\Gamma(s)$ is the Euler Gamma-function. A similar representation may be written for the operator S.

In the E-basis (3.262)-(3.263), the Fourier transforms of the kernels $\mathrm{K}(x)$ and $\mathrm{S}(x)$ of the integral operators K and S in (3.11) and (3.23) have the forms

$$\mathrm{K}^*(k) = \frac{1}{\mu_0}[E^5(m) - \kappa_0 E^6(m)] \ , \ m = \frac{k}{|k|} \ , \quad (3.268)$$

$$\mathrm{S}^*(k) = -2\mu_0[E^1 - (1-2\kappa_0)(E^2 - E^3(m) - E^4(m)) - 2E^5(m) + 2\kappa_0 E^6(m)], \quad (3.269)$$

$$\kappa_0 = \frac{\lambda_0 + \mu_0}{\lambda_0 + 2\mu_0}. \quad (3.270)$$

Let us find the result of action of the operator K_s in (3.267) on the elements of the E-basis (3.262)-(3.263) when $d = 3$. For calculation of the integrals in the right-hand side of (3.267), we use the following formulas for integrals over the angle coordinates on the unit sphere Ω_1 ($\mathrm{Re}\, s < 1$):

$$\int_{\Omega_1} (n\cdot m)^{-s} dm = I(s) \ , \ \ I(s) = 2\pi \frac{1+e^{-i\pi s}}{1-s} \ , \quad (3.271)$$

$$\int_{\Omega_1} (n\cdot m)^{-s} m_\alpha m_\beta dm = \frac{I(s)}{3-s} \left(\delta_{\alpha\beta} - s n_\alpha n_\beta\right) \ , \quad (3.272)$$

$$\int_{\Omega} (n\cdot m)^{-s} E^6(m) dm = \frac{I(s)}{(3-s)(5-s)} \left[E^2 + 2E^1 - s\left(E^3(n) + E^4(n) + 4E^5(n)\right) + s(2+s)E^6(n)\right]. \quad (3.273)$$

From these equations and (3.266) and (3.267) we obtain the results of the action of the operator K_s on the elements of the E-basis in the forms

$$\mathrm{K}_s E^1 = \frac{1}{\mu_0(3-s)(5-s)}[T^1 + (1-\kappa_0)T^3], \quad (3.274)$$

$$\mathrm{K}_s E^2 = \frac{1-\kappa_0}{\mu_0} \cdot \frac{1}{(3-s)} T^2, \tag{3.275}$$

$$\mathrm{K}_s E^3 = \frac{1-\kappa_0}{\mu_0} \left(\frac{1}{3-s} T^2 - \frac{1}{5-s} T^3 \right), \tag{3.276}$$

$$\mathrm{K}_s E^4 = -\frac{1-\kappa_0}{\mu_0} \cdot \frac{(2-s)}{s(3-s)} T^2, \tag{3.277}$$

$$\mathrm{K}_s E^5 = -\frac{1}{2\mu_0(3-s)(5-s)} \{(1-s)[T^1 + (1-\kappa_0)T^3] + (3-s)(1-\kappa_0)T^3\}, \tag{3.278}$$

$$\mathrm{K}_s E^6 = \frac{s-2}{\mu_0 s(2+s)} \left\{ (1-\kappa_0) \left[\frac{1}{3-s} T^2 - \frac{3-s}{5-s} T^3 \right] + \frac{2}{(3-s)(5-s)} \left[T^1 + (1-\kappa_0) T^3 \right] \right\}, \tag{3.279}$$

The right-hand sides of these equations depend on three linearly independent tensors T^i

$$T^1(s,n) = (E^1 - sE^5(n))(5-s) - T^3(s,n), \quad T^2(s,n) = E^2 - sE^4(n), \tag{3.280}$$

$$T^3(s,n) = E^2 + 2E^1 - s[E^3(n) + E^4(n) + 4E^5(n)] + s(s+2)E^6(n). \tag{3.281}$$

Note, that tensors T^i are the eigen-elements of the operator K_s

$$\mathrm{K}_s T^1 = \frac{1}{2\mu_0} T^1, \quad \mathrm{K}_s T^2 = \frac{1-\kappa_0}{\mu_0} T^2, \quad \mathrm{K}_s T^3 = \frac{1-\kappa_0}{\mu_0} T^3. \tag{3.282}$$

These equations follow from (3.274)–(3.281).

Let us find the solution of (3.258) in the form of a linear combination of tensors of the E-basis (3.262)–(3.263) with scalar coefficients that depend only on r. In this case, the product $C^1 A$ under the integral in the left-hand side of (3.258) can be represented as the following sum:

$$(C^1 A)(r,n) = \sum_{i=1}^{6} S_i(r) E^i(n), \tag{3.283}$$

where $S_i(r)$ are scalar functions of r. After substituting these equations into (3.259) and applying the Mellin transform to both sides of the resulting equation, we arrive at the equation for the tensor $A^*(s,n)$ (the Mellin transform of tensor $A(r,n)$ with respect to r)

$$A^*(s,n) + \sum_{i=1}^{6} S_i^*(s)(\mathrm{K}_s E^i)(s,n) = -(\mathrm{K}_s C^{1*})(s,n), \tag{3.284}$$

$$C^{1*}(s) = \lambda_1^*(s) E^2 + 2\mu_1^*(s) E^1. \tag{3.285}$$

Here $S_i^*(s)$ is the Mellin transform of scalar coefficients $S_i(r)$ in the expansion (3.283).

From (3.274) to (3.279) it follows that the tensors $\mathrm{K}_s E^i$ and $\mathrm{K}_s C^{1*}$ are linear combinations of three tensors T^j, defined in (3.280)-(3.281). Thus, it is natural to find the tensor $A^*(s,n)$ in the form of the same linear combination:

$$A^*(s,n) = \sum_{j=1}^{3} \alpha_j^*(s)\, T^j(s,n). \tag{3.286}$$

Here $\alpha_j^*(s)$ are scalar functions of the parameter s of the Mellin transform, and the r-representations of these functions are $\alpha_j(r)$ $(j = 1, 2, 3)$.

Because the multiplier $(-s)$ in the space of the Mellin transforms corresponds to the differential operation D [201]

$$D = r\frac{d}{dr} \tag{3.287}$$

in the r-space, the tensor $A(r,n)$ in the $r-$representation takes the form

$$\begin{aligned} A(r,n) &= [E^1 + E^5(n)D](5+D)\alpha_1(r) + [E^2 + E^4(n)D]\alpha_2(r) \\ &\quad + \left\{E^2 + 2E^1 + \left[E^3(n) + E^4(n) + 4E^5(n)\right]D + E^6(n)\, D(D-2)\right\} \\ &\quad \times(\alpha_3(r) - \alpha_1(r)) \end{aligned} \tag{3.288}$$

that follows from (3.286) and (3.280), (3.281).

Let us determine the coefficients $\alpha_j(r)$ in (3.288). After application the Mellin transform to (3.259) we obtain

$$A^*(s,n) + \sum_{i=1}^{6} \mathrm{K}_s \left(C^1 A\right)^* (s,n) = -(\mathrm{K}_s C^{1*})(s,n). \tag{3.289}$$

If we substitute the tensor $A^*(s,n)$ (3.286) and the tensor $A(r,n)$ (3.288) into (3.289) and take into account (3.280) and (3.282), we obtain an equation in which left and right-hand sides are linear combinations of the linearly independent tensors T^1, T^2, T^3. Equating the coefficients of the tensors T^1 and then the coefficients of T^3 on both sides of this equation give us two equations connecting the scalar functions $\alpha_1^*(s)$ and $\alpha_3^*(s)$:

$$\mu_0 s(s+2)(s-3)(s-5)\alpha_1^*(s) + \Phi_1^*(s) = -2s(s+2)\mu_1^*(s), \tag{3.290}$$

$$\mu_0 s(s+2)(s-3)(s-5)[\alpha_3^*(s) - (1-\kappa_0)\alpha_1^*(s)] + (1-\kappa_0)\Phi_3^*(s) = 0, \tag{3.291}$$

$$\Phi_1^*(s) = s(s+2)s_1^*(s) + \frac{1}{2}(s+2)(s-1)S_5^*(s) + 2(s-2)S_6^*(s), \tag{3.292}$$

$$\Phi_3^*(s) = (s-3)2\,(s+2)S_3^*(s) + (s+2)S_5^*(s) + 2\,(s-2)S_6^*(s). \tag{3.293}$$

Here $S_i^*(s)$ are the Mellin transforms of the scalar coefficients $S_i(r)$ in the expansion (3.283). After substitution (3.288) into (3.283) we find these coefficients in the forms

$$S_1 = 2\mu_1[(3+D)\alpha_1 + 2\alpha_3], \tag{3.294}$$

$$S_2 = \lambda_1[(3+D)\alpha_2 + (5+D)\alpha_3] + 2\mu_1(\alpha_2 + \alpha_3 - \alpha_1), \tag{3.295}$$

$$S_3 = \lambda_1(5+D)D\alpha + 2\mu_1 D(\alpha_3 - \alpha_1), \quad S_4 = 2\mu_1 D(\alpha_2 + \alpha_3 - \alpha_4), \tag{3.296}$$

$$S_5 = 2\mu_1 D[(1+D)\alpha_1 + 4\alpha_3], \quad S_6 = 2\mu_1 D(D-2)(\alpha_3 - \alpha_1). \tag{3.297}$$

The equivalence of the coefficients in front of tensor T^2 gives us an equation similar to (3.290) that contains α_2. But instead of α_2, it is more convenient to use the function

$$\beta(r) = 3\alpha_2(r) + (5+D)\alpha_3(r). \tag{3.298}$$

The equation for this function can be obtain by multiplying both sides of (3.284) with tensor E^2 from the right and taking into account the equations

$$AE^2 = (E^2 + E^4 D)\beta, \quad C^1 AE^2 = S_7 E^2 + S_8 E^1, \tag{3.299}$$

$$S_7 = \lambda_1(3+D)\beta + 2\mu_1\beta, \quad S_8 = 2\mu_1 D\beta. \tag{3.300}$$

The resulting equation contains only the function $\beta^*(s)$

$$\mu_0 s(s-3)\beta^*(s) - (1-\kappa_0)\Phi_2^*(s) = (1-\kappa_0)s[3\lambda_1^*(s) + 2\mu_1^*(s)], \tag{3.301}$$

$$\Phi_2^*(s) = sS_7^*(s) + (s-2)S_8^*(s). \tag{3.302}$$

After applying the inverse Mellin transform to (3.290), (3.291), and (3.301) we get the differential equations for the functions $\alpha_1, \alpha_3, \beta$ in the r-representation. By this transition, the Mellin transforms of all the functions should be changed for their r-representations (originals) and parameter $(-s)$ for the differential operator D (3.287). Finally, we obtain a system of three ordinary differential equations for the functions $\alpha_1(r), \alpha_3(r)$, and $\beta(r)$.

Let $\lambda_1(r), \mu_1(r)$ be finite functions with the piece-wise continuous second derivatives, and $d\lambda_1/dr = d\mu_1/dr = 0$ at $r = 0$. From (3.290) we obtain a system of two ordinary differential equations of the fourth order for the functions $\alpha_1(r)$ and $\alpha_3(r)$. An equation of the second order for the function $\beta(r)$ follows from (3.301). The right-hand sides and the coefficients of these equations are piece-wise continuous functions of r. A bounded solution of these equations should satisfy the following conditions:

$$D\alpha_i = D^2\alpha_i = 0, \quad i = 1, 3, \quad D\beta = 0 \quad \text{if} \quad r = 0 \tag{3.303}$$

$$\alpha_1, \alpha_3, \ \beta \to 0 \quad \text{if} \quad r \to \infty. \tag{3.304}$$

The first group of these conditions follows from continuity of the function $A(r, n)$ at $r = 0$, and the second group is the consequence of $A(r, n)$ vanishing at infinity.

3.9.2 Thermoelastic problem for a medium with a radially heterogeneous inclusion

Let us go to the solution of the thermoelastic problem for a medium with a radially heterogeneous inclusion, and assume that the temperature field T is constant. When $T = 0$, the medium is free from internal stresses. If $T \neq 0$, temperature stresses $\sigma(x)$ appear in the medium. In this case, the stress field $\sigma(x)$ satisfies Equation (3.21) with $\sigma^0(x) = 0$:

$$\sigma(x) - \int \mathrm{S}(x - x')B^1(x')\sigma(x')dx' = \int \mathrm{S}(x - x')\alpha(x')dx'T. \tag{3.305}$$

Here $\alpha(x)$ is the tensor of the thermal expansion coefficients of the medium with the inclusion. Similar to the tensor of elastic moduli $C(x)$, the tensor $\alpha(x)$ is a function of the distance r from the center of the spherical inclusion.

Let us introduce the tensors of the elastic strain $\varepsilon^e(x)$, the total strain $\varepsilon(x)$, and the strain perturbation $\varepsilon^{1T}(x)$ in the medium with the inclusion

$$\varepsilon^e(x) = C^{-1}(x)\sigma(x)\ , \quad \varepsilon(x) = \varepsilon^e(x) + \alpha(x)T, \tag{3.306}$$

$$\varepsilon^{1T}(x) = \varepsilon(x) - \alpha^0 T. \tag{3.307}$$

Here α^0 is the tensor of the thermal expansion coefficients of the medium, the tensor $\varepsilon^{1T}(x)$ vanishes when $|x| \to \infty$.

After various algebraic transformations we go from (3.305) to the integral equation for the function $\varepsilon^{1T}(x)$ in the form

$$\varepsilon^{1T}(x) + \int \mathrm{K}(x - x')C^1(x')\varepsilon^{1T}(x')dx'$$

$$= \int \mathrm{K}(x - x')C(x')\alpha^1(x')dx'T, \tag{3.308}$$

where $\alpha^1(x) = \alpha(x) - \alpha^0$ is zero outside the inclusion. To obtain this equation, we accept that the deformation of the medium is not constrained at infinity, and according to (2.81) the following equation holds:

$$\int \mathrm{K}(x - x')C^0\alpha^0 dx' = \alpha^0.. \tag{3.309}$$

In what follows, we take $T = 1$. Therefore, the stress and strain tensors obtained below should be multiplied with the temperature T in order to obtain the actual values of these tensors.

For an isotropic medium and inclusion, the tensors $C(x)$ and $\alpha(x)$ have the forms

$$C(x) = \lambda(r)E^2 + 2\mu(r)E^1, \quad \alpha_{\alpha\beta}(r) = \alpha(r)\delta_{\alpha\beta}\,. \tag{3.310}$$

Applying the Mellin transform to both sides of (3.308) and taking into account (3.265)–(3.267) we get the equation

$$\varepsilon^{*1T}(s,n) + \mathrm{K}_s(C^1\varepsilon^{1T})^*(s,n) = \mathrm{K}_s(C\alpha^1)^*(s,n)\,. \tag{3.311}$$

Due to (3.274)–(3.279), the right-hand side of this equation can be presented as follows:

$$C_{\alpha\beta\lambda\mu}(x)\alpha^1_{\lambda\mu}(x) = \gamma(r)\delta_{\alpha\beta}\,, \tag{3.312}$$

$$\gamma(r) = 3k(r)\alpha_1(r)\,, \quad k(r) = \lambda(r) + \frac{2}{3}\mu(r)\,, \quad \alpha_1(r) = \alpha(r) - \alpha^0, \tag{3.313}$$

$$\mathrm{K}_s(C\alpha^1)^*_{\alpha\beta}(s,n) = \frac{\gamma^*(s)}{(\lambda_0+2\mu_0)(3-s)}h_{\alpha\beta}(s,n), \tag{3.314}$$

$$h_{\alpha\beta}(s,n) = \delta_{\alpha\beta} - sn_\alpha n_\beta\,, \tag{3.315}$$

where $\gamma^*(s)$ is the Mellin transform of the function $\gamma(r)$.. The action of the operator K_s on the tensor $n\otimes n$ is proportional to the tensor $h(s,n)$

$$\mathrm{K}_s(n\otimes n) = \frac{(s-2)}{(\lambda_0+2\mu_0)s(3-s)}h(s,n). \tag{3.316}$$

Thus, the solution of (3.311) is to be found in the form

$$\varepsilon^{1T*}(s,n) = \beta^*_T(s)h(s,n)\,, \tag{3.317}$$

where $\beta^*_T(s)$ is the Mellin transform of the scalar function $\beta_T(r)$. It follows from this equation that function $\varepsilon^{1T}(r,n)$ in the r-representation is

$$\varepsilon^{1T}_{\alpha\beta}(r,n) = (\delta_{\alpha\beta} + n_\alpha n_\beta D)\beta_T(r)\,. \tag{3.318}$$

The product $C^1\varepsilon^{1T}$ in (3.311) takes the form

$$C^1_{\alpha\beta\lambda\mu}(r)\varepsilon^{1T}_{\lambda\mu}(r,n) = S_{1T}(r)\delta_{\alpha\beta} + S_{2T}(r)n_\alpha n_\beta\,, \tag{3.319}$$

$$S_{1T}(r) = [3k_1(r) + \lambda_1(r)D]\beta_T(r)\,, \quad k_1(r) = k(r) - k_0, \tag{3.320}$$

$$S_{2T}(r) = 2\mu_1(r)D\beta_T(r)\,, \quad k_0 = \lambda_0 + \frac{2}{3}\mu_0. \tag{3.321}$$

Substituting (3.317) and (3.319) into (3.311) and taking into account (3.314) and (3.316) we get an equation in which right and left-hand sides are proportional to the tensor $h(s,n)$. Equating coefficients we obtain

$$(\lambda_0+2\mu_0)(3-s)\beta^*_T(s) + \left[S^*_{1T}(s) - \frac{2-s}{s}S^*_{2T}(s)\right] = \gamma^*(s)\,. \tag{3.322}$$

After multiplying both parts of this equation with s and applying the inverse Mellin transform, we derive the differential equation for function $\beta_T(r)$ in the form

$$(\lambda_0+2\mu_0)(3+D)D\beta_T - L(D)\beta_T = D\gamma\,, \tag{3.323}$$

$$L(D) = D(3k_1 + \lambda_1 D) + 2(2 + D)\mu_1 D\,. \qquad (3.324)$$

From (3.318) and the properties of the tensor $\varepsilon^{1T}(r, n)$ follows that the function $\beta_T(r)$ is bounded at $r = 0$ and vanishes when $r \to \infty$. Under these conditions, equation (3.323) has a unique solution. When this solution is found, the temperature stresses and strains in the medium with the inclusion follow from (3.306) and (3.318).

3.10 Multilayered spherical inclusion

Let the elastic moduli and thermal expansion coefficients of the inclusion be piece-wise constant functions of r with jumps at the points $r = a_i$ ($i = 1, 2..., N$; $0 < a_1 < a_2 < ... < a_N$). In this case, the inclusion consists of a spherical kernel and $N - 1$ spherical layers with constant thermoelastic properties (Fig. 3.6). The functions $\alpha_1, \alpha_3, \beta$ and β_T in (3.288) and (3.318) determine the strains and stresses inside the inclusion and in the medium. The differential equations for these functions follow from (3.290)–(3.302) and (3.323) and are essentially simplified in the regions where the thermoelastic properties are constant; they take the forms

$$D(D-2)(D+3)(D+5)\alpha_j = 0\,, \quad j = 1, 3\,, \qquad (3.325)$$

$$D(D+3)\beta = 0\,, \quad D(D+3)\beta_T = 0\,. \qquad (3.326)$$

The general solutions of these equations inside the intervals $a_{i-1} < r < a_i$, ($i = 1, 2, ..., N+1$; $a_0 = 0$, $a_{N+1} = \infty$) are presented in the forms

$$\alpha_1(r) = Y_1^i + Y_2^i r^2 + Y_3^i r^{-3} + Y_4^i r^{-5}, \qquad (3.327)$$

$$\alpha_3(r) = Y_5^i + Y_6^i r^2 + Y_7^i r^{-3} + Y_8^i r^{-5}, \qquad (3.328)$$

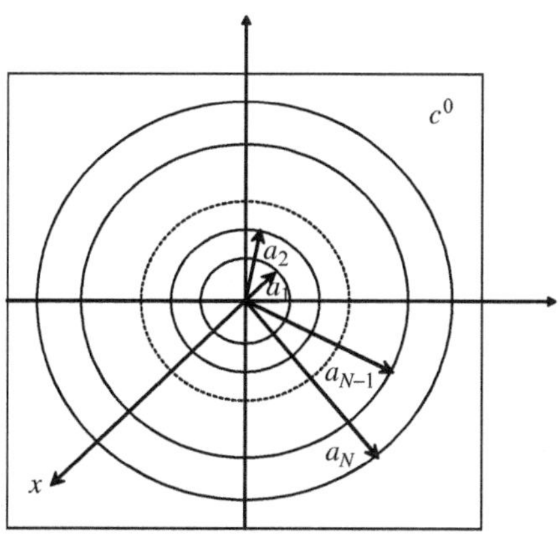

Fig. 3.6. Spherically layered inclusion.

$$\beta(r) = Y_9^i + Y_{10}^i r^{-3}, \quad \beta_T(r) = Y_{11}^i + Y_{12}^i r^{-3}. \tag{3.329}$$

Here Y_j^i are arbitrary constants. Hence, inside every layer, the solution of the problem involves 12 arbitrary constants.

Let us consider the behavior of the functions $\alpha_1, \alpha_3, \beta$ and β_T on the boundaries of the layers. Because the displacement vector $u(x)$ is continuous everywhere, the tensors $\varepsilon(r,n)$, $A(r,n)$ and $\varepsilon^{1T}(r,n)$ have no singular components (delta-functions and their derivatives). It follows from Equations (3.327)–(3.329), and (3.288) for $A(r,n)$ and (3.318) for $\varepsilon^{1T}(r,n)$ that the functions β and β_T are continuous, and α_1 and α_3 are continuous together with their first derivatives.

Let $\varphi(r)$ be a piece-wise constant function with jumps at the points $r = a_i$, $i = 1, 2, ..., N$, and zero when $r > a_N$. The Mellin transform (3.265) of $\varphi(r)$ has the form

$$\varphi^*(s) = -\frac{1}{s}\sum_{i=1}^{N}[\varphi]_i a_i^s, \tag{3.330}$$

where $[\varphi]_i$ is the jump of the function at the point a_i

$$[\varphi]_i = \varphi(a_i+0) - \varphi(a_i-0)\,, \tag{3.331}$$

$$\varphi(a_i \pm 0) = \lim_{\delta\to 0}\varphi(a_i \pm \delta)\,, \quad \delta > 0\,. \tag{3.332}$$

For a piece-wise constant function $\mu_1(r)$ and the functions $\alpha_j(r)$ ($j = 1, 3$) that are continuous together with their first derivatives, the following equations hold:

$$(\mu_1 D\alpha_j)^*(s) = -\sum_{i=1}^{N}[\mu_1\alpha_j]_i a_i^s - s(\mu_1\alpha_j)^*(s)\,, \tag{3.333}$$

$$(\mu_1 D^2\alpha_j)^*(s) = -\sum_{i=1}^{N}\left([\mu_1 D\alpha_j]_i - s[\mu_1\alpha_j]\right)a_i^s + s^2(\mu_1\alpha_j)^*(s)\,. \tag{3.334}$$

To prove these equation, we use the definition of the Mellin transform (3.265), and integration by parts.

On account of (3.330) and (3.333), (3.334) the function $\Phi_1^*(s)$ in (3.291) takes the form

$$\begin{aligned}\Phi_1^*(s) = 2\sum_{i=1}^{N}\{&(s+2)(s^2-8s+9)[\mu_1\alpha_1]_i - 4(s+2)[\mu_1\alpha_3]_i\\ &-(s^2-3s+6)[\mu_1 D\alpha_1]_i - 4(s-2)[\mu_1 D\alpha_3]_i\}\,a_i^s\\ &+ s(s+2)(s-3)(s-5)(\mu_1\alpha_1)^*(s)\,.\end{aligned} \tag{3.335}$$

Direct integration shows that the function $\alpha_1(r)$ in (3.327) satisfies the following equation:

$$s(s+2)(s-3)(s-5)(\mu\alpha_1)^*(s) = -\sum_{i=1}^{N}\{s^3[\mu\alpha_1]_i - s[\mu(6+D)\alpha_1]_i$$

$$+s\left[\mu\left(D^2+6D-1\right)\alpha_1\right]_i - \left[\mu\left(D^3+6D^2-D-30\right)\alpha_1\right]\}a_i^s\,. \qquad (3.336)$$

Substituting (3.335) into the first equation of (3.290) and taking into account (3.336) we go to the equation

$$\sum_{i=1}^{N}\{s^3\mu_0[\alpha_1]_i - s^2\mu_0[(D+6)\alpha_1]_i + s\left(\left[\mu\left(D^2+6D-1\right)\alpha_1\right]_i\right.$$

$$\left.-[\mu_1(7\alpha_1-3D\alpha_1+4\alpha_3+4D\alpha_3)]_i\right) - [\mu(D^3+6D^2-D-30)\alpha_1]_i$$

$$+2[\mu_1(4\alpha_3-4D\alpha_3-9\alpha_1+3D\alpha_1)]_i\}a_i^s = -2\sum_{i=1}^{N}(2+s)[\mu]_i a_i^s\,. \qquad (3.337)$$

Equating the multipliers of the linearly independent functions a_i^s on both sides of this equation, we get N equations. Each of them connects two polynomials of s. The terms of the same degree of s in these polynomials should coincide. From these conditions, we obtain the equations for the jumps of the function $\alpha_1(r)$ and its derivatives at points $r = a_i$ on the boundaries between the layers

$$[\alpha_1]_i = 0, \quad [D\alpha_1]_i = 0\,, \qquad (3.338)$$

$$[\mu D^2\alpha_1]_i = -2[\mu]_i - 3[\mu(2+D)\alpha_1]_i - 4[\mu(1+D)\alpha_3]_i\,, \qquad (3.339)$$

$$[\mu D^3\alpha_1]_i = 16[\mu]_i + [\mu(48+25D)\alpha_1]_i + 16[\mu(2+D)\alpha_3]_i\,. \qquad (3.340)$$

In the same way, from (3.290) and (3.301) we obtain similar equations for the jumps of the functions α_3, β and their derivatives on the boundaries of the layers

$$[\alpha_3]_i = 0\,, \quad [D\alpha_3]_i = 0\,, \qquad (3.341)$$

$$[(\lambda+2\mu)D^2\alpha_3]_i = -2[\mu]_i - 6[\mu(1+D)\alpha_1]_i - 4[\mu\alpha_3]_i - [(5\lambda+6\mu)D\alpha_3]_i\,, \qquad (3.342)$$

$$[(\lambda+2\mu)D^3\alpha_3]_i$$

$$= 16[\mu]_i + 24[\mu(2+D)\alpha_1]_i + 32[\mu\alpha_3]_i + [(25\lambda+42\mu)D\alpha_3]_i\,, \qquad (3.343)$$

$$[\beta]_i = 0\,, \quad [(\lambda+2\mu)D\beta]_i = -[3\lambda+2\mu]_i - [(3\lambda+2\mu)\beta]_i\,. \qquad (3.344)$$

The equations for the jumps of the function β_T follow from (3.319) and (3.322) in the form

$$[\beta_T]_i = 0\,, \quad [(3\lambda+2\mu)\beta_T]_i + [(\lambda+2\mu)D\beta_T]_i = [\gamma]_i\,. \qquad (3.345)$$

Equations (3.338)–(3.345) and the conditions for the functions $\alpha_1, \alpha_3, \beta$ and β_T at $r = 0$ and $r \to \infty$ give us a system of equations for the arbitrary constants Y_j^i in (3.327)–(3.329) that define the solution inside each layer. An efficient numerical algorithm for the calculation of these constants is considered below.

Let us introduce $(N+1)$ 12D-vectors Y^i with components which are the constants Y_j^i in the solution (3.327)–(3.329) inside the i-th layer, and $(N+1)$ vector functions $X^i(r)$ with the components

$$X_1^i = \alpha_1, \quad X_2^i = D\alpha_1, \quad X_3^i = D^2\alpha_1, \quad X_4^i = D^3\alpha_1\,, \tag{3.346}$$

$$X_5^i = \alpha_3, \quad X_6^i = D\alpha_3, \quad X_7^i = D^2\alpha_3\,, \quad X_8^i = D^3\alpha_3\,, \tag{3.347}$$

$$X_9^i = \beta, \quad X_{10}^i = D\beta, \quad X_{11}^i = \beta_T, \quad X_{12}^i = D\beta_T\,, \tag{3.348}$$

$$(\alpha_j = \alpha_j(r), \;\; j = 1,3, \;\; \beta = \beta(r), \;\; \beta_T = \beta_T(r), \;\; a_{i-1} < r < a_i)\,. \tag{3.349}$$

From (3.327) to (3.329) it follows that the vectors Y^i and X^i are connected by the equations

$$X^i(r) = H(r)Y^i, \quad Y^i = H^{-1}(r)X^i(r)\,, \tag{3.350}$$

$$H(r) = h_1(r) \otimes h_1(r) \otimes h_2(r) \otimes h_2(r)\,. \tag{3.351}$$

Here the symbol $\otimes$ means the direct (Cartesian) product of the matrices h_1 and h_2 that have the forms

$$h_1 = \begin{bmatrix} 1 & r^2 & r^{-3} & r^{-5} \\ 0 & 2r^2 & -3r^{-3} & -5r^{-5} \\ 0 & 4r^2 & 9r^{-3} & 25r^{-5} \\ 0 & 8r^2 & -27r^{-3} & -125r^{-5} \end{bmatrix}, \quad h_2 = \begin{bmatrix} 1 & r^{-3} \\ 0 & -3r^{-3} \end{bmatrix}. \tag{3.352}$$

Thus, if the vector $X^i(r)$ is known at the point r $(a_{i-1} < r < a_i)$, a unique vector Y^i, and therefore the solution of (3.327)-(3.329), may be found inside the i-th interval (layer).

If the value of vector $X^i(r)$ is known at the point $r = a_{i-1}+0$ (at the left end of the i-th interval), its value at the right end $r = a_i - 0$ may be found from (3.350), and takes the form

$$X^i(a_i) = R^i X^i(a_{i-1})\,, \quad R^i = H(a_i)H^{-1}(a_{i-1})\,, \tag{3.353}$$

where the matrix R^i may be called the matrix of transition through the layer.

Equations (3.338)–(3.345) imply that the vectors X^i and X^{i+1} at point $r = a_i$ on the boundary between the i-th and $(i+1)$-th layers are connected by the equation

$$X^{i+1}(a_i) = F^i + \Gamma^i X^i(a_i)\,. \tag{3.354}$$

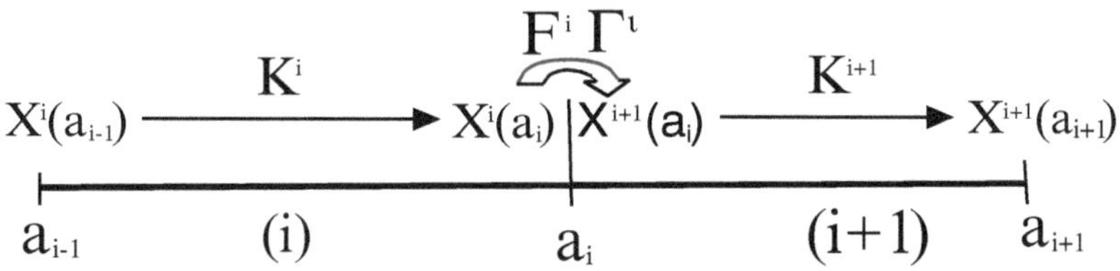

Fig. 3.7. Transition of the solution through the layers in the elasticity problem for a multilayered spherical inclusion.

The value of vector F^i and matrix Γ^i(the vector and matrix of transition through the border) can be reconstructed from (3.338)–(3.345). Explicit equations for these objects are given in Appendix D.1. Schematically, the process of transition of the solution through the layer and through the boundary between the i-th and $(i+1)$-th layers is shown in Fig. 3.7.

Let the vector $X^1(a_1)$ for the first layer (kernel) be known. Then, the vector $X^{i+1}(a_i)$ that defines the solution inside the $(i+1)$-th layer due to (3.353) and (3.354) is expressed via the vector $X^1(a_1)$ by the equation

$$X^{i+1}(a_i) = g^i + G^i X^1(a_1)\,, \quad i = 2, 3, ..., N, \tag{3.355}$$

$$g^1 = F^1, \ g^i = F^i + \sum_{j=1}^{i-1} \left(\prod_{k=1}^{j+1} Q^k \right) F^j, \tag{3.356}$$

$$G^i = \prod_{k=i}^{1} Q^k, \ \ Q^k = \Gamma^k R^k, \tag{3.357}$$

where R^1 is the unit matrix, and $R^k (k = 2, 3, ..., N)$ are the transition matrices defined in (3.353)

In order to find the vector $X^1(a_1)$ the conditions at $r = 0$ and $r = \infty$ should be taken into account. Because the solution is bounded at $r = 0$, Equations (3.327)–(3.329) for the first layer should contain only positive powers of r, i.e.,

$$Y_k^1 = 0 \quad \text{as} \quad k = 3, 4, 7, 8, 10, 12. \tag{3.358}$$

As a result, there exists connections between the components of the vector X^1 in the first layer that can be represented in the form

$$X^1 = M Z^1\,, \quad Z^i = P_1 X^i\,, \quad i = 1, 2, ..., N+1. \tag{3.359}$$

Here Z^i is the vector-column with components $\{X_1^i, X_2^i, X_6^i, X_9^i, X_{11}^i\}$, and the matrices $M(12 \times 6)$ and $P_1(6 \times 12)$ are defined by the equations

$$M = m_1 \otimes m_1 \otimes m_2 \otimes m_2\,, \quad P_1 = m_3 \otimes m_3 \otimes m_4 \otimes m_4\,, \tag{3.360}$$

$$m_1 = \begin{bmatrix} 1 & 0 \\ 0 & 1 \\ 0 & 2 \\ 0 & 4 \end{bmatrix}, \quad m_2 = \begin{bmatrix} 1 \\ 0 \end{bmatrix}, \quad m_3 = \begin{bmatrix} 1 & 0 & 0 & 0 \\ 0 & 1 & 0 & 0 \end{bmatrix}, \quad m_4 = [1 \;\; 0] \,. \tag{3.361}$$

Since the functions $\alpha_1, \alpha_3, \beta$ and β_T tend to zero at infinity, solution (3.327) inside the $(N+1)$-th layer (in the matrix) should contain only negative degrees of r, i.e.,

$$Y_i^{N+1} = 0 \quad \text{as} \quad i = 1, 2, 5, 6, 9, 11. \tag{3.362}$$

It follows from this equation that there is connection between the components of vector X^{N+1}. This connection is represented in the form

$$P_2 X^{N+1} = L P_1 X^{N+1}. \tag{3.363}$$

Here the matrices $L(6 \times 6)$ and $P_2(6 \times 12)$ are

$$L = l_1 \otimes l_1 \otimes l_2 \,, \quad P_2 = m_5 \otimes m_5 \otimes m_6 \otimes m_6 \,, \tag{3.364}$$

$$l_1 = \begin{bmatrix} -15 & -8 \\ 120 & 49 \end{bmatrix}, \quad l_2 = \begin{bmatrix} -3 & 0 \\ 0 & -3 \end{bmatrix}, \quad m_5 = \begin{bmatrix} 0 & 0 & 1 & 0 \\ 0 & 0 & 0 & 1 \end{bmatrix}, \quad m_6 = \begin{bmatrix} 0 & 1 \end{bmatrix} . \tag{3.365}$$

In order to find the vector Z^1, let us substitute $X^{N+1}(a_N)$ from (3.355) into (3.363). Taking into account (3.359) we get a linear algebraic equation for the vector $Z^1 = P_1 X^1$:

$$B Z^1 = f, \;\; B = (P_2 - L P_1) G^N M \,, \;\; f = (L P_1 - P_2) g^N, \tag{3.366}$$

where the matrices G^N and g^N are defined in (3.355). After solving this equation, we obtain the vector Z^1, and can find the vector X^1 from (3.359). Then, all the vectors $X^{i+1}(a_i)$ $(i = 1, 2, ..., N)$ may be found from (3.355) and (3.359). Finally, from (3.350) we calculate the vectors Y^i that define the solutions inside all the layers according to (3.327)–(3.329). Tensor $A(r, n)$ inside the i-th layer takes the form

$$A(r, n) = \sum_{k=1}^{6} A^k(r) E^k(n), \tag{3.367}$$

$$\begin{aligned} A^1(r) = {} & 3Y_1^i + 2Y_5^i + (5Y_2^i + 2Y_6^i) r^2 \\ & + 2Y_7^i r^{-3} + 2(-Y_4^i + Y_8^i) r^{-5}, \end{aligned} \tag{3.368}$$

$$\begin{aligned} A^2(r) = {} & \frac{1}{3}((-3Y_1^i - 2Y_5^i + Y_9^i) - (3Y_2^i + 4Y_6^i) r^2 \\ & + (Y_{10}^i - 3Y_3^i + Y_7^i) r^{-3} - 3(Y_4^i - Y_8^i) r^{-5}), \end{aligned} \tag{3.369}$$

$$A^3(r) = -2(Y_2^i - Y_6^i) r^2 + 3(Y_3^i - Y_7^i) r^{-3} + 5(Y_4^i - Y_8^i) r^{-5}, \tag{3.370}$$

$$A^4(r) = -\frac{2}{3}(3Y_2^i + 4Y_6^i)r^2 - (Y_{10}^i - 3Y_3^i + Y_7^i)r^{-3} + 5(Y_4^i - Y_8^i)r^{-5}, \tag{3.371}$$

$$A^5(r) = (6Y_2^i + 8Y_6^i)r^2 + 6(Y_3^i - 2Y_7^i)r^{-3} + 20(Y_4^i - Y_8^i)r^{-5}, \tag{3.372}$$

$$A^6(r) = -15(Y_3^i - Y_7^i)r^{-3} - 35(Y_4^i - Y_8^i)r^{-5}. \tag{3.373}$$

The strain tensor $\varepsilon(r, n)$ in the medium and the inclusion is calculated from (3.257) for an arbitrary external field ε_0.

As an example of the calculations, we consider an inclusion of unit radius, containing N spherical layers of the thickness $1/N$. Young modulus E_i inside the i-th layer is given by the equation

$$E_i = E(a_i)\,, \quad i = 1, 2, ..., N+1\,,$$
$$E(r) = E_0\left[1 + \delta\,\exp\left(\frac{\lambda r^2}{r^2-1}\right)\right] \quad \text{as} \quad r < 1\,, \tag{3.374}$$
$$E(r) = E_0 \quad \text{as} \quad r \geq 1\,,$$

where a_i is external radius of i-th layer. The Poisson ratio for the medium and all the layers was taken 0.4. External loading is uniaxial tension along the x_3-axis.

Figures 3.8 and 3.9 show the distribution of the stress component σ_{33} along the axis orthogonal to x_3, with the origin at the center of the inclusion. The calculations were performed for $\delta = -1$, $\lambda = 0, 0.1, 1, 10$ (soft inclusion, Fig. 3.8) and $\delta = 100$ with the same value of λ (hard inclusion, Fig. 3.9), and $N = 60$ for both cases.

It is clear that when $N \to \infty$, the Young modulus inside the inclusion tends to the continuous distribution (3.374). In order to check the capacity of the algorithm, the calculations were performed for an increasing number of the layers N until a stable picture of the stress distribution inside the inclusion is obtained. It turns out that for $N > 40$ this distribution did not change as N grew.

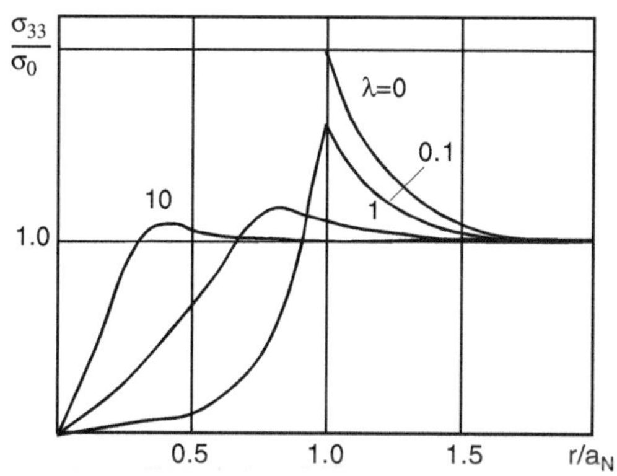

Fig. 3.8. Stress distribution in the medium with a soft spherically layered inclusion.

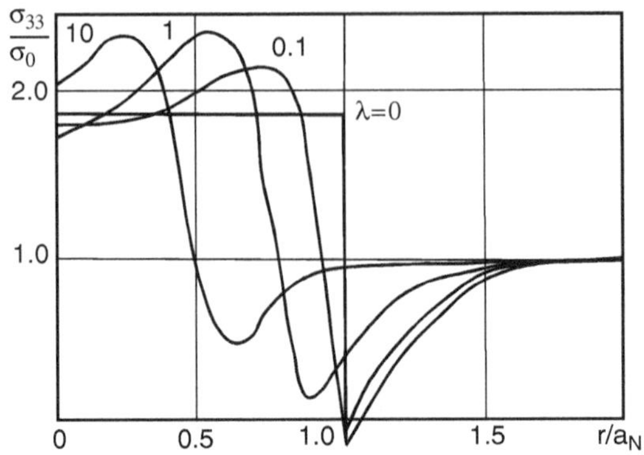

Fig. 3.9. Stress distribution in the medium with a hard spherically layered inclusion.

3.11 Axially symmetric inhomogeneity in an elastic medium

Let us consider a homogeneous medium with an inclusion in the form of an infinite cylinder. Suppose that the tensors of elastic moduli and thermal expansion coefficients of the inclusion are functions of the distance r from its central axis. The Cartesian (x_1, x_2, x_3) and cylindrical (r, n, z) coordinate systems are introduced to describe the position of an arbitrary point x in the medium with the cylindrical inclusion.

The axes x_3 and z coincide with the inclusion axis, n is the unit vector of the r-axis in the plane orthogonal to x_3 (Fig. 3.10). If the temperature field T of the medium and the external strain field are constant, the strain and stress tensors in the medium with the inclusion are independent of the coordinate x_3 (or z) and are functions of only x_1, x_2.

We start with the pure elastic problem ($T = 0$) and write the integral equation for the function $\varepsilon(x)$ (3.11) in the form

$$\varepsilon(\bar{x}) + \int \hat{\mathrm{K}}(\bar{x} - \bar{x}')C^1(\bar{x}')\varepsilon(\bar{x}')d\bar{x}' = \varepsilon^0, \quad \bar{x} = \bar{x}\,(x_1, x_2)\,, \tag{3.375}$$

where the kernel $\widehat{K}(\bar{x})$ is connected with the function $\mathrm{K}(x)$ in (3.12) by the equation

$$\hat{\mathrm{K}}(x_1, x_2) = \int_{-\infty}^{\infty} \mathrm{K}(x_1, x_2, x_3)dx_3\,. \tag{3.376}$$

The symbol of the operator $\hat{\mathrm{K}}$ (the Fourier transform $\hat{\mathrm{K}}^*(\bar{k})$ of the function $\hat{\mathrm{K}}(\bar{x})$ in (3.375)) is determined by the equation

$$\hat{\mathrm{K}}^*(\bar{k}) = \mathrm{K}^*(k_1, k_2, k_3)|_{k_3=0}\,, \quad \bar{k} = \bar{k}(k_1, k_2)\,, \tag{3.377}$$

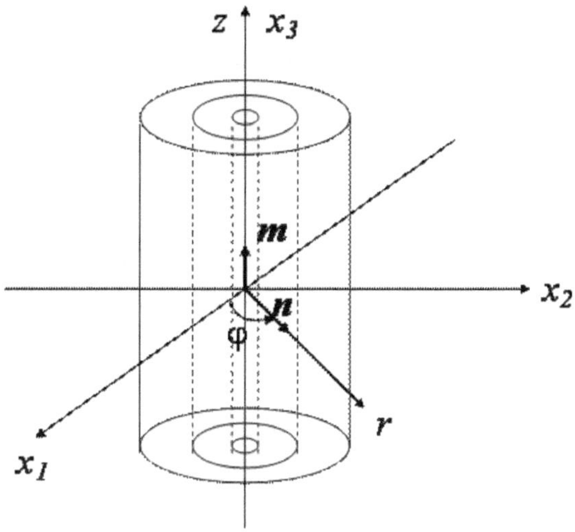

Fig. 3.10. The coordinate systems in the problem for a cylindrical multilayered inclusion.

where function $\mathrm{K}^*(k)$ is defined in (2.58).

Due to linearity of the problem, the solution of (3.375) has the form

$$\varepsilon(\bar{x}) = \varepsilon^0 + A(\bar{x})\varepsilon^0, \tag{3.378}$$

where the tensor $A(\bar{x})$ which vanishes at infinity satisfies the equation that follows from (3.375) and (3.378)

$$A(\bar{x}) + \int \hat{\mathrm{K}}(\bar{x}-\bar{x}')C^1(\bar{x}')A(\bar{x}')d\bar{x}' = -\int \hat{\mathrm{K}}(\bar{x}-\bar{x}')C^1(\bar{x}')d\bar{x}'. \tag{3.379}$$

For presentation of the tensor $A(\bar{x})$ we introduce three tensor bases built of the unit vectors m and n of the z and r-axes of the cylindrical coordinate system, respectively, and the rank two tensor $\theta_{\alpha\beta} = \delta_{\alpha\beta} - m_\alpha m_\beta$ (the projector on the normal plane to the z-axis).

The first of these bases (P-basis) was introduced in (2.11)–(2.13) (see also Appendix A.2). The tensors of the P-basis are similar to the tensors of the E-basis (3.262). The projector $\theta_{\alpha\beta} = \delta_{\alpha\beta} - m_\alpha m_\beta$ plays the role of the unit tensor $\delta_{\alpha\beta}$ for the P-tensors. The tensors $P^i(m)$, as well as $E^i(m)$, form a closed algebra with respect to the product operation defined in (3.264). The multiplication table of the P-tensors and the equation for the inverse tensor with respect to the tensor that belongs to the linear shell of the P-basis are presented in Appendix A.2.

The P-basis is convenient for presentation of the elastic moduli tensor C^0 of a transversally isotropic body with symmetry axis defined by the unit vector m. This representation has the form (2.10). The connection between the five moduli $(k_0, m_0, l_0, \mu_0, n_0)$ with "technical" elastic constants of an transversally isotropic body is given by the equalities

$$m_0 = \frac{E_{01}}{1+\nu_{012}}, \quad k_0 = (2\Delta_0 E_{03})^{-1}, \quad l_0 = \frac{\nu_{013}}{\Delta_0 E_{03}}, \tag{3.380}$$

$$n_0 = \frac{1-\nu_{012}}{\Delta_0 E_{01}}, \quad \mu_0 = \frac{E_{03}}{2(1+\nu_{03})}, \quad \Delta_0 = \frac{1-\nu_{012}}{E_{01}E_{03}} - 2\left(\frac{\nu_{013}}{E_{03}}\right)^2. \tag{3.381}$$

Here E_{01} is the Young modulus of the medium in the θ-plane that is normal to z-axis, E_{03} the same modulus in the z-axis direction, ν_{012}, ν_{013} are the Poisson coefficients. For an isotropic medium, we have:

$$m_0 = \mu_0, \; k_0 = \lambda_0 + \mu_0, \; l_0 = \lambda_0, \; n_0 = \lambda_0 + 2\mu_0. \tag{3.382}$$

In what follows, the material of the inclusion is supposed to be also transversally isotropic, and its elastic moduli tensor C has the form

$$C = kP^2 + 2m\left(P^1 - \frac{1}{2}P^2\right) + l(P^3+P^4) + 4\mu P^5 + nP^6, \tag{3.383}$$

where the parameters m, k, l, μ, n are functions of the r-coordinate.

We introduce the θ-basis that consist of six tensors θ^i belonging to the θ-plane:

$$\theta^1_{\alpha\beta\lambda\mu} = \theta_{\alpha(\lambda}\theta_{\mu)\beta}, \quad \theta^2_{\alpha\beta\lambda\mu} = \theta_{\alpha\beta}\theta_{\lambda\mu}, \quad \theta^3_{\alpha\beta\lambda\mu} = \theta_{\alpha\beta}n_\lambda n_\mu, \tag{3.384}$$

$$\theta^4_{\alpha\beta\lambda\mu} = n_\alpha n_\beta \theta_{\lambda\mu}, \quad \theta^5_{\alpha\beta\lambda\mu} = n_{(\alpha}\theta_{\beta)(\lambda}n_{\mu)}, \quad \theta^6_{\alpha\beta\lambda\mu} = n_\alpha n_\beta n_\lambda n_\mu, \tag{3.385}$$

$$\theta_{\alpha\beta} = \delta_{\alpha\beta} - m_\alpha m_\beta. \tag{3.386}$$

The set of these tensors is closed with respect to the product operation (3.264), the corresponding multiplication table is presented in Appendix A.3. Finally, we introduce the R-basis that consist of the follow tensors:

$$R^1_{\alpha\beta\lambda\mu} = n_\alpha n_\beta n_\lambda n_\mu, \quad R^2_{\alpha\beta\lambda\mu} = m_\alpha m_\beta m_\lambda m_\mu, \tag{3.387}$$

$$R^3_{\alpha\beta\lambda\mu} = m_\alpha m_\beta n_\lambda n_\mu, \quad R^4_{\alpha\beta\lambda\mu} = n_\alpha n_\beta m_\lambda m_\mu, \tag{3.388}$$

$$R^5_{\alpha\beta\lambda\mu} = n_{(\alpha}m_{\beta)}n_{(\lambda}m_{\mu)}. \tag{3.389}$$

The sets of elements of each of the P, θ and R-bases are closed with respect to the product operation (3.264).

Using (3.377) and (2.58), we present the symbol $\hat{\mathrm{K}}^*(\bar{k})$ of the operator $\hat{\mathrm{K}}$ in (3.375) in the form

$$\hat{\mathrm{K}}^*(m,l) = \frac{1}{\mu_0}\left[\theta^5(m,l) - \kappa_0\theta^6(m,l) + \frac{\mu_0}{m_0}R^5(m,l)\right], \quad l = \frac{\bar{k}}{|\bar{k}|}. \tag{3.390}$$

The medium is transversally isotropic with the elastic moduli tensor (2.10), $\kappa_0 = k_0/n_0$.

Let apply the Mellin transform with respect to the coordinate r to (3.379). Because $\hat{\mathrm{K}}$ is a singular integral operator with symbol which is a homogeneous function of the zero degree (3.390), its action on piece-wise smooth functions is determined by an equation similar to (3.266) and (3.267) when $d = 2$. Thus, after application of the Mellin transform, (3.379) takes the form

$$A^*(s,n) + \hat{\mathrm{K}}_s(C^1A)^*(s,n) = -(\mathrm{K}_sC^{1*})(s,n)\,, \tag{3.391}$$

where the action of the operator $\hat{\mathrm{K}}_s$ on the function $f^*(s,n)$ is determined by the equation

$$(\hat{\mathrm{K}}_sf^*)(s,n) = -\frac{1}{(2\pi)^2}\Gamma(2-s)\Gamma(s)\int_{l_1}(-n\cdot l)^{-s}dl \times \int_{l_1}\hat{\mathrm{K}}^*(m,l)f^*(s,e)(e\cdot l)^{s-2}de\,. \tag{3.392}$$

Here n, l, e are vectors on the unit circle l_1 in the θ-plane, and the tensor $\hat{\mathrm{K}}^*(m,l)$ is defined in (3.390).

Consider the action of the operator $\hat{\mathrm{K}}_s$ on the elements of P, θ and R-bases. To calculate the integrals in (3.392) we use the following equation that are obtained by direct integration over the unit circle l_1 ($\mathrm{Re}\, s < 1$):

$$\int_{l_1}(n\cdot l)^{-s}dl = \jmath(s)\,,\quad \int_{l_1}(n\cdot l)^{-s}\,l_\alpha l_\beta dl = \frac{\jmath\,(s)}{2-s}\left(\theta_{\alpha\beta} - sn_\alpha n_\beta\right)\,, \tag{3.393}$$

$$\int_{l_1}(n\cdot l)^{-s}\theta^6(l)dl = \frac{\jmath\,(s)}{(2-s)\,(4-s)}\left[\theta^2\,(n) + 2\theta^1\,(n) - s(\theta^3\,(n) + \theta^4(n) + 4\theta^5(n) + s(s+2)\theta^6(n)\right]\,, \tag{3.394}$$

$$\jmath(s) = \frac{\Gamma\left(\frac{1}{2}\right)\Gamma\left(\frac{1-s}{2}\right)}{\Gamma\left(\frac{2-s}{2}\right)}(1+e^{-i\pi s})\,. \tag{3.395}$$

These equations lead to the following results for the action of the operator $\hat{\mathrm{K}}_s$ on the elements of the P-basis:

$$\hat{\mathrm{K}}_sP^1 = \frac{1}{m_0(2-s)(4-s)}\left[\hat{T}_1^* + (1-\kappa_0)\hat{T}_3^*\right]\,, \tag{3.396}$$

$$\hat{\mathrm{K}}_sP^2 = \frac{1-\kappa_0}{m_0(2-s)}\hat{T}_2^*\,,\quad \hat{\mathrm{K}}_sP^3 = \frac{1-\kappa_0}{m_0(2-s)}\hat{T}_4^*\,, \tag{3.397}$$

$$\hat{\mathrm{K}}_sP^5 = \frac{1}{2m_0(2-s)}\hat{T}_5^*\,,\quad \hat{\mathrm{K}}_sP^4 = \hat{\mathrm{K}}_sP^6 = 0\,. \tag{3.398}$$

Here the five tensors $\hat{T}_i^*$ have the forms

$$\hat{T}_1^* = (4-s)(\theta^1 - s\theta^5) - \hat{T}_3^* \ , \ \ \hat{T}_2^* = \theta^2 - s\theta^4, \tag{3.399}$$

$$\hat{T}_3^* = \theta^2 + 2\theta^1 - s(\theta^3 + \theta^4 + 4\theta^5) + s(2+s)\theta^6, \tag{3.400}$$

$$\hat{T}_4^* = P^3 - sP^4, \ \ \hat{T}_5^* = P^5 - sR^5, \tag{3.401}$$

where the tensors $P^i(m), \theta^i(m)$ and $R^i(n,m)$ are determined in (2.11)–(2.14) and (3.384)–(3.389).

The actions of the operator $\hat{\mathrm{K}}_s$ on the elements of the θ-basis take the forms:

$$\hat{\mathrm{K}}_s\theta^1 = \hat{\mathrm{K}}_s P^1, \ \ \hat{\mathrm{K}}_s\theta^2 = \hat{\mathrm{K}}_s P^2, \tag{3.402}$$

$$\hat{\mathrm{K}}_s\theta^3 = \frac{1-\kappa_0}{m_0 s(2-s)(4-s)}\left[(4-s)\hat{T}_2^* - (2-s)\hat{T}_3^*\right], \tag{3.403}$$

$$\hat{\mathrm{K}}_s\theta^4 = \frac{(1-\kappa_0)(s-1)}{m_0 s(2-s)}\hat{T}_2^*, \tag{3.404}$$

$$\hat{\mathrm{K}}_s\theta^5 = \frac{1-\kappa_0}{2m_0 s(2-s)(4-s)}\left[2(s-1)\hat{T}_3^* + \frac{s}{1-\kappa_0}\hat{T}_1^*\right], \tag{3.405}$$

$$\mathrm{K}_s\theta^6 = \frac{(1-\kappa_0)(s-1)}{m_0 s(4-s^2)(4-s)}\left[(4-s)\hat{T}_2^* + \frac{2}{1-\kappa_0}\hat{T}_1^* + s\hat{T}_3^*\right], \tag{3.406}$$

and only three tensors $\hat{T}_i^*$ in (3.399), (3.400) are present in these equations. The action of the operator $\hat{\mathrm{K}}_s$ on the elements of the R-basis is defined by equations

$$\hat{\mathrm{K}}_s R^1 = \hat{\mathrm{K}}_s\theta^6, \ \ \hat{\mathrm{K}}_s R^2 = \hat{\mathrm{K}}_s R^3 = 0\,, \tag{3.407}$$

$$\hat{\mathrm{K}}_s R^4 = \frac{(1-\kappa_0)(s-1)}{m_0 s(2-s)}\hat{T}_4^*, \ \ \hat{\mathrm{K}}_s R^5 = \frac{(s-1)}{2\mu_0 s(2-s)}\hat{T}_5^*. \tag{3.408}$$

Let us turn to the solution of (3.391). For a transversally isotropic fiber, the tensor C^{1*} on its right-hand side has the form similar to (3.383)

$$\begin{aligned} C^{1*}(s) = {} & k_1^*(s)P^2 + 2m_1^*(s)\left(P^1 - \frac{1}{2}P^2\right) \\ & + l_1^*(s)(P^3 + P^4) + 4\mu_1^*(s)P^5 + n_1^*(s)P^6, \end{aligned} \tag{3.409}$$

where $m_1^*(s), k_1^*(s), l_1^*(s), \mu_1^*(s)$ and $n_1^*(s)$ are the Mellin transforms of the perturbations of the elastic moduli in the inclusion. Equation (3.398) implies that the right-hand side of (3.391) is a linear combination of the tensors $\hat{T}_i^*$ defined in (3.399)–(3.401).

If $A(r,n)$ in (3.391) is taken in the form of a linear combination of the elements of the P, θ and R-bases with coefficients dependent on r, the product

$C^1 A$ is a similar linear combination. Therefore, the tensor $\hat{K}_s(C^1 A)^*$ is a linear combination of the same five tensors $\hat{T}_i^*$. It is natural to find the tensor $A^*(s,n)$ in the form of a linear combination of not all the elements of the P, θ and R-bases but only five tensors $\hat{T}_i^*$:

$$A^*(s,n) = \sum_{i=1}^{5} \alpha_i^*(s)\hat{T}_i^*(n,s)\,. \tag{3.410}$$

Here $\alpha_i^*(s)$ are the s-representation of five scalar functions $\alpha_i(r)$. Because the multiplier $(-s)$ in the Mellin transforms corresponds to the operator $D = rd/dr$ in the r-space, we obtain from (3.399)–(3.401) and (3.410) the following equation for the tensor $A(r,n)$:

$$A(r,n) = \sum_{i=1}^{5} \hat{T}_i(n,D)\alpha_i(r)\,. \tag{3.411}$$

Here the differential operators $\hat{T}_i(n,D)$ are defined by the right-hand sides of (3.399)–(3.401) if the parameter $(-s)$ is replaced by the operator D.

The detailed equation for the tensor $A(r,n)$ has the form

$$\begin{aligned} A(r,n) = {} & \left[\theta^1 + \theta^5(n)D\right](4+D)\alpha_1(r) + \left[\theta^2 + \theta^4(n)D\right]\alpha_2(r) \\ & + \left\{\theta^2 + 2\theta^1 + \left[\theta^3(n) + \theta^4(n) + 4\theta^5(n)\right]D + \theta^6(n)D(D-2)\right\} \\ & \times(\alpha_3(r) - \alpha_1(r)) + \left[P^3 + R^4(n,m)\right]\alpha_4(r) + \left[P^5 + R^5(n,m)D\right]\alpha_5(r). \end{aligned} \tag{3.412}$$

Here $P^i = P^i(m)$, $\theta^i(n) = \theta^i(m,n)$. This equation contains five unknown scalar functions $\alpha_i(r)$. Multiplying (3.391) with the tensor θ^1 from the right we get an equation that includes only the functions $\alpha_1, \alpha_2, \alpha_3$:

$$A_\theta^* + \hat{K}_s(C^1 A_\theta)^* = -\hat{K}_s C_\theta^{1*}, \tag{3.413}$$

$$A_\theta^* = A^*\theta^1 = \alpha_1^*\hat{T}_1^* + \alpha_2^*\hat{T}_2^* + \alpha_3^*\hat{T}_3^*, \tag{3.414}$$

$$C_\theta^{1*} = C^{1*}\theta^1 = k_1^* P^2 + 2m_1^*\left(P^1 - \frac{1}{2}P^2\right) + l_1^* P^4. \tag{3.415}$$

Taking into account these equations and (3.399), we find the tensor $A_\theta(r,n)$ in the form

$$\begin{aligned} A_\theta(r,n) = {} & (\theta^1 + \theta^5(n)D)(4+D)\alpha_1(r) + (\theta^2 + \theta^4(n)D)\alpha_2(r) \\ & + \left\{\theta^2 + 2\theta^1 + \left[\theta^3(n) + \theta^4(n) + 4\theta^5(n)\right]D + \theta^6(n)D(D-2)\right\} \\ & \times(\alpha_3(r) - \alpha_1(r))\,, \end{aligned} \tag{3.416}$$

and the product $C^1 A_\theta$ in (3.413) is presented as follows:

$$C^1 A_\theta = \sum_{i=1}^{6} \hat{S}_i(r)\theta^i + \hat{S}_7 P^4 + \hat{S}_8 R^3, \tag{3.417}$$

$$\hat{S}_1 = 2m_1[(2+D)\alpha_1 + 2\alpha_3]\,, \tag{3.418}$$

$$\hat{S}_2 = l_1[(2+D)\alpha_2 + (4+D)\alpha_3] + 2m_1 D(\alpha_3 - \alpha_1)\,, \tag{3.419}$$

$$\hat{S}_3 = l_1 D(4+D)\alpha_3 + 2\mu_1 D(\alpha_3 - \alpha_1)\,, \tag{3.420}$$

$$\hat{S}_4 = 2m_1 D(\alpha_2 + \alpha_3 - \alpha_1)\,, \quad \hat{S}_5 = 2m_1 D(D\alpha_1 + 4\alpha_3)\,, \tag{3.421}$$

$$\hat{S}_6 = 2m_1 D(D-2)(\alpha_3 - \alpha_1)\,. \tag{3.422}$$

Explicit equations for the functions $\hat{S}_7$ and $\hat{S}_8$ are not essential for the future.

Substituting (3.415) and (3.417) into (3.413) and taking into account (3.398), (3.402)-(3.408) we get a tensor equation which left and right parts are linear combinations of the tensors $\hat{T}_1^*, \hat{T}_2^*, \hat{T}_3^*$. Equating the coefficients at the tensors $\hat{T}_1^*, \hat{T}_3^*$ in the left and right-hand sides, we obtain the equations connecting the functions α_1^* and α_3^*:

$$m_0 s(4-s^2)(4-s)\alpha_1^* + \hat{\Phi}_1^* = -2s(2+s)m_1^*(s)\,, \tag{3.423}$$

$$m_0 s(4-s^2)[\alpha_3^* - (1-\kappa_0)\alpha_1^*] - (1-\kappa_0)\hat{\Phi}_2^* = 0\,, \tag{3.424}$$

$$\hat{\Phi}_1^* = \frac{1}{2}\left[s(2+s)(2\hat{S}_1^* + \hat{S}_5^*) - 4(1-s)\hat{S}_6^*\right], \tag{3.425}$$

$$\hat{\Phi}_2^* = \frac{2-s}{2}\left[(2+s)(2\hat{S}_3^* + \hat{S}_5^*) - 2(1-s)\hat{S}_6^*\right]. \tag{3.426}$$

Here $\hat{S}_i^*(s)$ are the Mellin transforms of the functions $\hat{S}_i(r)$ in (3.418)–(3.422). Replacing the Mellin transforms by the original functions and the multiplier $(-s)$ by the operator D we obtain the system of ordinary differential equations of the fourth order for $\alpha_1(r)$ and $\alpha_3(r)$. If the elastic moduli are piece-wise smooth functions of r with derivatives that are zero when $r = 0$, the boundary conditions for the system has a form similar to (3.304):

$$D\alpha_j = D^2\alpha_j = 0\,, \quad j = 1, 3 \quad \text{as} \quad r = 0\,, \tag{3.427}$$

$$\alpha_1, \alpha_3 \to 0 \quad \text{as} \quad r \to \infty\,, \tag{3.428}$$

To obtain the equation for the function $\alpha_2(r)$ we multiply both parts of (3.413) with the tensor θ^2 and take into account the following equations:

$$C_\theta^{1*}\theta^2 = 2k_1^* P^2 + 2l_1^* P^4, \quad A_\theta^*\theta^2 = (\theta^2 - s\theta^4)\beta^*\,, \tag{3.429}$$

$$(C^1 A_\theta)^*\theta^2 = \hat{S}_9^*\theta^2 + \hat{S}_{10}^*\theta^4 + l_1^*(2-s)\beta^* P^4, \tag{3.430}$$

$$\hat{S}_9(r) = [2k_1(r) + (k_1(r) - m_1(r))D]\beta(r)\,, \tag{3.431}$$

$$\hat{S}_{10}(r) = 2m_1(r)D\beta(r)\,, \quad \beta^* = 2\alpha_2^* + (4-s)\alpha_3^*\,. \tag{3.432}$$

From here and (3.398)–(3.408) it follows that the left and right parts of the equation

$$A_\theta^*\theta^2 + \hat{\mathrm{K}}_s(C^1A_\theta)^*\theta^2 = -\mathrm{K}_sC_\theta^{1*}\theta^2 \tag{3.433}$$

are proportional to the tensor $\hat{T}_2^*$. Equating coefficients of this tensor on the left and right we obtain

$$m_0s(2-s)\beta^*(s) + (1-\kappa_0)\left[s\hat{S}_9^*(s) + (s-1)\hat{S}_{10}^*(s)\right]$$

$$= -2(1-\kappa_0)sk_1^*(s)\,. \tag{3.434}$$

Transition to the r-representation and taking account of the equation

$$\beta(r) = 2\alpha_2(r) + (4+D)\alpha_3(r) \tag{3.435}$$

that follows from (3.432), give us an ordinary differential equation of the second order for the function $\beta(r)$ with the following boundary conditions:

$$D\beta = 0 \quad \text{as} \quad r = 0\,; \quad \beta(r) \to 0 \quad \text{as} \quad r \to \infty. \tag{3.436}$$

To derive the differential equation for the functions $\alpha_4(r)$ and $\alpha_5(r)$ in (3.412), we introduce the tensor $A_\pi(r,n)$

$$A_\pi(r,n) = A(r,n) - A_\theta(r,n) = (P^3 + R^4D)\alpha_4 + (P^5 + R^5D)\alpha_5\,. \tag{3.437}$$

The difference of (3.391) and (3.413) leads to an equation with Mellin transform of the form

$$A_\pi^*(s,n) + \hat{\mathrm{K}}_s(C^1A_\pi)^*(s,n) = -\hat{\mathrm{K}}_sC_\pi^{1*}(s,n)\,. \tag{3.438}$$

where

$$C^1A_\pi = \hat{S}_{11}P^3 + \hat{S}_{12}P^5 + \hat{S}_{13}P^6 + \hat{S}_{14}R^4 + \hat{S}_{15}R^5, \tag{3.439}$$

$$C_\pi^1 = l_1P^3 + 4\mu_1P^5 + n_1P^6, \quad \hat{S}_{11} = [2k_1 + (k_1 - m_1)D]\alpha_4\,, \tag{3.440}$$

$$\hat{S}_{12} = 2\mu_1\alpha_5\,, \quad \hat{S}_{13} = l_1(2+D)\alpha_4\,, \tag{3.441}$$

$$\hat{S}_{14} = 2m_1D\alpha_4\,, \quad \hat{S}_{15} = 2\mu_1D\alpha_5\,. \tag{3.442}$$

From these equations and (3.398)–(3.408) we deduce that the right and left parts of (3.438) are linear combinations of the tensors $\hat{T}_4^*$ and $\hat{T}_5^*$. Equating the coefficients of these tensors we obtain

$$m_0s(2-s)\alpha_4^* + (1-\kappa_0)\left[s\hat{S}_{11}^* + (s-1)\hat{S}_{14}^*\right] = -(1-\kappa_0)sl_1^*(s)\,, \tag{3.443}$$

$$4\mu_0s(2-s)\alpha_5^* + 2\left[s\hat{S}_{12}^* + (s-1)\hat{S}_{15}^*\right] = -8s\mu_1^*(s)\,. \tag{3.444}$$

Transition to the r-representations gives two differential equations for the functions $\alpha_4(r)$ and $\alpha_5(r)$. The boundary conditions for these equations are similar to (3.436)

$$D\alpha_4 = D\alpha_5 = 0 \quad \text{if} \quad r = 0, \quad \alpha_4, \alpha_5 \to 0 \quad \text{as} \quad r \to \infty. \tag{3.445}$$

In conclusion, we consider a problem of temperature stresses in a medium with a cylindrically symmetric inhomogeneity caused by constant temperature field T. The thermal expansion coefficients are piece-wise smooth functions of r, and for a transversally isotropic material, the tensor α is presented in the form

$$\alpha_{\alpha\beta}(r) = \alpha_\theta(r)\theta_{\alpha\beta} + \alpha_m(r)m_\alpha m_\beta\,, \tag{3.446}$$

where α_θ is the coefficient of the thermal expansion in the plane orthogonal to the inclusion axis, and α_m is that for the axial direction.

The tensor of thermal stresses $\sigma(x)$ in the medium satisfies Equation (3.21) when $\sigma^0(x) = 0$. We introduce the tensors of elastic deformations ε_T^e, total deformations ε_T, and perturbation of deformations ε_T^1 due to the inhomogeneity.

$$\varepsilon_T^e = C^{-1}\sigma\,, \quad \varepsilon_T = \varepsilon_T^e + \alpha T, \quad \varepsilon_T^1 = \varepsilon_T - \alpha^0 T. \tag{3.447}$$

As in Section 3.9 we obtain the equation for the tensor ε_T^1 in the form

$$\varepsilon_T^1(\bar{x}) + \int \hat{\mathrm{K}}(\bar{x}-\bar{x}')C^1(\bar{x}')\varepsilon_T^1(\bar{x}')d\bar{x}'$$

$$= \int \hat{\mathrm{K}}(\bar{x}-\bar{x}')C(\bar{x})\alpha^1(\bar{x}')d\bar{x}'T, \tag{3.448}$$

where $\alpha^1(\bar{x}) = \alpha(\bar{x}) - \alpha^0$, and the kernel $\hat{\mathrm{K}}(\bar{x})$ is defined in (3.376). We take $T = 1$.

Applying the Mellin transform to both sides of (3.448) we find

$$\varepsilon_T^{1*}(s,n) + \hat{\mathrm{K}}_s(C^1\varepsilon_T^1)^*(s,n) = \hat{\mathrm{K}}_s(C\alpha^1)^*(s,n)\,, \tag{3.449}$$

$$C\alpha^1 = (2k\alpha_{1\theta} + l\alpha_{1m})\theta + (2l\alpha_{1\theta} + n\alpha_{1m})m\otimes m\,. \tag{3.450}$$

The operator $\hat{\mathrm{K}}_s$ was defined in (3.392). The action of this operator on the tensors $\theta, m\otimes m$ and $n\otimes n$ is determined by the equations

$$\hat{\mathrm{K}}_s\theta = \frac{1-\kappa_0}{m_0(2-s)}h(n,s)\,, \quad h(n,s) = \theta - sn\otimes n\,, \tag{3.451}$$

$$\hat{\mathrm{K}}_s(n\otimes n) = \frac{(1-\kappa_0)(s-1)}{m_0 s(2-s)}h(n,s)\,, \quad \hat{\mathrm{K}}_s(m\otimes m) = 0\,. \tag{3.452}$$

These equations show that the solutions of (3.448) and (3.449) have the form

$$\varepsilon_T^{1*}(s,n) = (\theta - sn\otimes n)\beta_T^*(s)\,, \text{ or } \varepsilon_T^1(r,n) = (\theta + n\otimes nD)\beta_T(r)\,, \quad (3.453)$$

where $\beta_T(r)$ is a scalar function with Mellin transform $\beta_T^*(s)$.

Substituting (3.453) into (3.449) and taking into account (3.451) we go to the equation for the function $\beta_T^*(s)$:

$$(\lambda_0 + 2\mu_0)s(2-s)\beta_T^*(s) + S_T^*(s) = s\gamma^*(s)\,, \quad (3.454)$$

$$S_T(r) = \{D[2(2+D)(k_1(r) - m_1(r)) + 2m_1(r)] \\ -2(D+1)m_1(r)D\}\beta_T(r)\,, \quad (3.455)$$

$$\gamma(r) = 2\alpha_{1\theta}(r)k(r) + \alpha_{1m}(r)l(r)\,.$$

Transition to the r-representation gives us an ordinary differential equation for the function $\beta_T(r)$. The boundary conditions for this equation have the forms

$$D\beta_T = 0 \text{ if } r = 0, \quad \beta_T \to 0 \text{ if } r \to \infty. \quad (3.456)$$

3.12 Multilayered cylindrical inclusion

Let the elastic moduli and thermal expansion coefficients of a cylindrical inclusion be piece-wise constant functions with jumps at the points $r = a_i$, $i = 1, 2, ..., N$; $0 < a_1 < a_2 < ... < a_N$. In this case, the inclusion consists of a cylindrical kernel and $(N-1)$ cylindrical layers with constant thermoelastic parameters..

In the regions with constant properties (inside the layers), the functions $\alpha_j(r)$ $(i = 1, 3, 4, 5)$, $\beta(r)$ and $\beta_T(r)$ satisfy the differential equations

$$D(2-D)(2+D)(4+D)\alpha_j(r) = 0\,, \quad j = 1, 3\,, \quad (3.457)$$

$$D(2+D)\alpha_j(r) = 0\,, \quad j = 4, 5\,, \quad (3.458)$$

$$D(2+D)\beta(r) = 0\,, \quad D(2+D)\beta_T(r) = 0\,, \quad (3.459)$$

that follow from (3.423), (3.424), (3.434), (3.443), (3.444), and (3.455).

The general solutions of these equations in the intervals $a_{i-1} < r < a_i$, $i = 1, 2, ..., N+1$ ($a_0 = 0$, $a_{N+1} = \infty$) have the forms

$$\alpha_1 = Y_1^i + Y_2^i r^2 + Y_3^i r^{-2} Y_4^i r^{-4}, \quad (3.460)$$

$$\alpha_3 = Y_5^i + Y_6^i r^2 + Y_7^i r^{-2} + Y_8^i r^{-4}, \quad (3.461)$$

$$\alpha_4 = Y_{11}^i + Y_{12}^i r^{-2}, \quad \beta = Y_9^i + Y_{10}^i r^{-2}, \quad (3.462)$$

$$\alpha_5 = Y_{13}^i + Y_{14}^i r^{-2}, \quad \beta_T = Y_{15}^i + Y_{16}^i r^{-2}, \quad (3.463)$$

where Y_j^i are arbitrary constants. Hence, inside the i-th layer, the solution is defined by 16 arbitrary constants.

Consider the discontinuity of the functions $\alpha_j(r)$, $j = 1,3,4,5$, $\beta(r)$ and $\beta_T(r)$ and their derivatives on the layer interfaces $r = a_i$. As in Section 3.9 it is possible to show that the following equations for the jumps of the unknown functions hold ($[\varphi]_i = \varphi(a_i + 0) - \varphi(a_i - 0)$):

$$[\alpha_j]_i = 0\,,\ \ j = 1,3,4,5\,;\ \ [\beta]_i = [\beta_T]_i = 0\,,\ \ [D\alpha_j]_i = 0\,,\ \ j = 1,3\,, \tag{3.464}$$

$$[mD^2\alpha_1]_i = -2[m]_i - 4[m\alpha_1]_i - 2[mD\alpha_1]_i - 4[m\alpha_3]_i - 4[mD\alpha_3]_i\,, \tag{3.465}$$

$$[mD^3\alpha_1]_i = 12[m]_i + 24[m\alpha_1]_i + 16[mD\alpha_1]_i + 24[m\alpha_3]_i + 12[mD\alpha_3]_i\,, \tag{3.466}$$

$$\begin{aligned}[(k+m)D^2\alpha_3]_i = -2[m]_i - 4[m\alpha_1]_i - 4[mD\alpha_1]_i \\ -4[m\alpha_3]_i - 2[(2k+m)D\alpha_3]_i\,,\end{aligned} \tag{3.467}$$

$$[(k+m)D^3\alpha_3]_i = 12[m]_i + 24[mD\alpha_1]_i + 24[m\alpha_3]_i + 16[(k+m)D\alpha_3]_i\,, \tag{3.468}$$

$$[(k+m)D\beta]_i = -2[k(1+\beta)]_i\,,\ \ [(k+m)D\alpha_4]_i = -[l]_i - 2[k\alpha_4]_i\,, \tag{3.469}$$

$$[\mu D\alpha_5]_i = -[\mu(2+\alpha_5)]_i\,,\ \ [(k+m)D\beta_T]_i = [\gamma]_i - 2[k\beta_T]_i\,. \tag{3.470}$$

These equations lead to an algorithm for calculating all the constants Y_j^i in the solution (3.460)–(3.463). Let us introduce $(N+1)$ 16D-vectors Y^i with components that are the constants Y_j^i, and $(N+1)$ vectors $X^i(r)$ with the following components:

$$X_1^i = \alpha_1\,,\ \ X_2^i = D\alpha_1\,,\ \ X_3^i = D^2\alpha_1\,, \tag{3.471}$$

$$X_4^i = D^3\alpha_1\,,\ \ X_5^i = \alpha_3\,,\ \ X_6^i = D\alpha_3\,, \tag{3.472}$$

$$X_7^i = D^2\alpha_3\,,\ \ X_8^i = D^3\alpha_3\,,\ \ X_9^i = \beta\,, \tag{3.473}$$

$$X_{10}^i = D\beta\,,\ \ X_{11}^i = \alpha_4\,,\ \ X_{12}^i = D\alpha_4\,, \tag{3.474}$$

$$X_{13}^i = \alpha_5\,,\ \ X_{14}^i = D\alpha_5\,,\ \ X_{15}^i = \beta_T\,, X_{16}^i = D\beta_T\,. \tag{3.475}$$

Here $\alpha_j = \alpha_j(r)\,,\ \beta = \beta(r)\,,\ \beta_T = \beta_T(r)\,,\ a_{i-1} < r < a_i,\ i = 1,2,...,N+1$.

Because of (3.460)–(3.463), the vectors Y^i and X^i are connected by the equations

$$X^i(r) = H(r)Y^i,\ \ Y^i = H^{-1}(r)X^i(r)\,, \tag{3.476}$$

$$H(r) = h_1(r) \otimes h_1(r) \otimes h_2(r) \otimes h_2(r) \otimes h_2(r) \otimes h_2(r)\,, \tag{3.477}$$

$$h_1 = \begin{bmatrix} 1 & r^2 & r^{-2} & r^{-4} \\ 0 & 2r^2 & -2r^{-2} & -4r^{-4} \\ 0 & 4r^2 & 4r^{-2} & 16r^{-4} \\ 0 & 8r^2 & -8r^{-2} & -64r^{-4} \end{bmatrix}\,, \qquad h_2 = \begin{bmatrix} 1 & r^{-2} \\ 0 & -2r^{-2} \end{bmatrix}\,. \tag{3.478}$$

As a result, the values of the vectors $X^i(r)$ on the left ($r = a_{i-1} + 0$) and right ($r = a_i - 0$) ends of the i-th interval are connected by the equations similar to (3.353)

$$X^i(a_i) = R^i X(a_{i-1}) \,, \quad R^i = H(a_i) H^{-1}(a_{i-1}) \,, \tag{3.479}$$

where the matrix H (16×16) is defined in (3.477).

Because of (3.464)–(3.470), on the interface of the i-th and $(i+1)$-th layers $(r = a_i)$ the following equation holds:

$$X^{i+1}(a_i) = F^i + \Gamma^i X^i(a_i) \,, \tag{3.480}$$

where the vector F^i and matrices Γ^i are reconstructed from (3.464)–(3.470). The explicit equations for these objects are presented in Appendix D.2.

From (3.479) and (3.480) we deduce that the vector of the solution $X^{i+1}(a_i)$ in the $(i+1)$-th layer is expressed via such a vector for the first layer $X^1(a_1)$ in the form similar to (3.355)

$$X^{i+1}(a_i) = g^i + G^i X^1(a_1), \quad i = 1, 2, ..., N, \tag{3.481}$$

$$G^i = \prod_{k=1}^{i} Q^k, \; g^1 = F^1, \tag{3.482}$$

$$g^i = F^i + \sum_{j=1}^{i} \left(\prod_{k=1}^{j+1} Q^k \right) F^j, \quad i = 2, 3, ..., N, \quad Q^k = \Gamma^k R^k, \tag{3.483}$$

where R^1 is the unit matrix, R^k $(k = 2, 3, ..., N)$ are defined in (3.479).

Because the solution is bounded when $r = 0$ and tends to zero when $r \to \infty$, the vectors Y^1_j and Y^{N+1}_j that define the solutions in the first and in the $(N+1)$-th layers have to satisfy the conditions:

$$Y^1_j = 0 \,, \quad j = 3, 4, 7, 8, 10, 12, 14, 16 \,, \tag{3.484}$$

$$Y^{N+1}_j = 0 \,, \quad j = 1, 2, 5, 6, 9, 11, 13, 15 \,. \tag{3.485}$$

The equation for the vector $X^1(a_1)$ follows from these conditions in the form

$$X^1 = MZ \,, \quad BZ = f \,, \tag{3.486}$$

where Z is an unknown 8D-vector, the matrices M (16×8), $B (8 \times 8)$ and the vector f are defined by the equations

$$M = m_1 \otimes m_1 \otimes m_2 \otimes m_2 \,, \quad L = l_1 \otimes l_1 \otimes l_2 \otimes l_2 \otimes l_2 \otimes l_2 \,, \tag{3.487}$$

$$B = LG^N M \,, \quad f = -Lg^N \,, \tag{3.488}$$

$$m_1 = \begin{bmatrix} 1 & 0 \\ 0 & 1 \\ 0 & 2 \\ 0 & 4 \end{bmatrix} \,, \quad m_2 = \begin{bmatrix} 1 & 0 \\ 0 & 0 \\ 0 & 1 \\ 0 & 0 \end{bmatrix} \,, \quad l_1 = \begin{bmatrix} 8 & 6 & 1 & 0 \\ -48 & -28 & 0 & 1 \end{bmatrix} \,, \quad l_2 = \begin{bmatrix} 2 & 1 \end{bmatrix} \,, \tag{3.489}$$

where matrix G^N and vector g^N are defined in (3.481).

After calculating the vector $X^1(a_1)$ from the solution of the system (3.486) we can find all the vectors $X^{i+1}(a_i)$ $(i = 1, 2, 3, ..., N)$ from (3.481); Equation (3.476) gives all the constants Y^i_j in the solution inside each layer.

3.13 Notes

Integral equations for strains and stresses in an inhomogeneous medium were used in the work of many authors. The general theory of such equations was constructed at the end of the 1960s. Historical surveys and references to the original works may be found in [149], [48].

The conditions for the strain and stress fields on the boundary of two elastic media were obtained in [110], [108].

A homogeneous isotropic elastic medium with an ellipsoidal isotropic inclusion was considered in the works of many authors (e.g., [179], [176]); remarkable results were obtained in [45], [46]. The polynomial conservation theorem for anisotropic medium and inclusion and an arbitrary polynomial external field was proved in [110].

Derivation of the integral equation for a crack from the integral equation for a cavity was proposed in [108]. A regularization of the integral operator in the integral equation for a crack in an anisotropic medium was obtained in [74], [79].

The analogy of the polynomial conservation theorem for an elliptical crack was proved in [212]. It is shown in [108] that this theorem is a consequence of the polynomial conservation theorem for an ellipsoid.

A spherical inclusion with a layer in a homogeneous elastic medium was considered in [95], [29], [140]. The solution of this problem for a cylindrical inclusion with a layer was obtained in [59]. The method of the solution of the problem of thermoelasticity for spherical or cylindrical inclusions with any number of layers was proposed in [88], [89], [91]. The contents of Sections 3.8, 3.9 is based on these works.

4. Thin inclusion in a homogeneous elastic medium

Thin hard inclusions are often used as a filler in composite materials. For the same volume concentration of fillers, rigid laminas increase the elastic moduli by 1.5–2 times more than spherical inclusions or fibers with the same elastic properties. On the other hand, a medium containing thin soft inclusions or cracks is an appropriate model of damaged materials. In both cases, the elastic moduli of thin inclusions differ essentially from the moduli of the matrix material. The properties of such inclusions are characterized by two dimensionless parameters. The "geometrical" parameter δ_1 is the ratio of the minimal and maximal sizes of the inclusion, and the "physical" parameter δ_2 is the ratio of the characteristic elastic modulus of the inclusion and the medium. The parameter δ_1 is always small, and the parameter δ_2 is either small (soft inclusion), or large (hard inclusion). The main terms of the asymptotic expansions of the elastic fields in the vicinity of thin inclusions with respect to these parameters are the objectives of this chapter. The problem of construction of these terms is reduced to integral equations on the middle surface of a thin inclusion. For thin ellipsoidal inclusions, the solutions of these equations are obtained in explicit forms.

4.1 External expansions of elastic fields

Let an infinite homogeneous elastic medium contain an inclusion that occupies a finite region V. We assume that one characteristic size of the region V is much smaller that two other sizes, and the middle surface Ω of V is smooth and bounded by a close contour Γ. At each point x on Ω we introduce a local coordinate system y_1, y_2, z with the z-axis directed along the normal $n(x)$ to Ω, and we denote by $h(x)$ the transverse size of V along the z-axis (Fig. 4.1).

We assume that $h(x)$ is a smooth function presented in the form

$$h(x) = \delta_1 l(x)\ , \quad \delta_1 << 1\ , \quad x \in \Omega\ , \tag{4.1}$$

where δ_1 is a small dimensionless parameter, $l(x)$ has the order of the maximal size of V, and satisfies the condition

$$|\partial_\alpha h(x)| << 1 \tag{4.2}$$

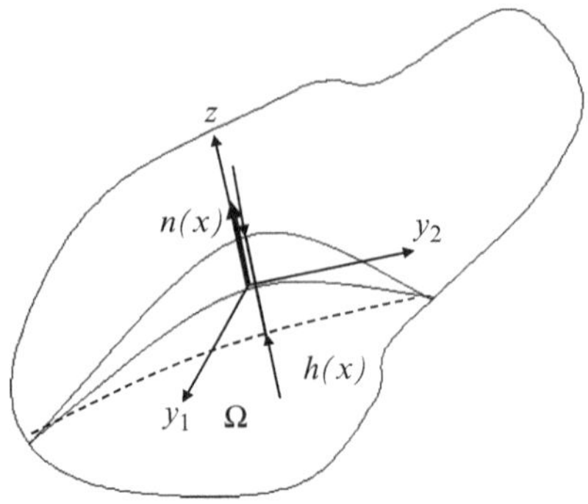

Fig. 4.1. Thin inclusion in a homogeneous medium.

everywhere on Ω except maybe a small vicinity of the boundary contour Γ of Ω. Here ∂ is the gradient along the surface Ω

$$\partial_\alpha = \nabla_\alpha - n_\alpha(x)n_\beta(x)\nabla_\beta \,. \tag{4.3}$$

Supposing that the elastic moduli of the inclusion are constant, we consider Equations (3.26), (3.27) for the stress and strain fields far from the inclusion (external solutions [204]). For a fixed point $x \notin V$, the kernels $\mathrm{K}(x-x')$ and $\mathrm{S}(x-x')$ in (3.26), (3.27) are smooth and bounded functions as $x' \in V$. Therefore, the principal terms of the elastic fields far from the inclusion take the forms

$$\varepsilon(x) = \varepsilon^0(x) - \int_\Omega \mathrm{K}(x-x')C^1\bar{\varepsilon}^+(x',h)d\Omega', \tag{4.4}$$

$$\sigma(x) = \sigma^0(x) + \int_\Omega \mathrm{S}(x-x')B^1\bar{\sigma}^+(x',h)d\Omega', \tag{4.5}$$

$$\bar{\sigma}^+(x,h) = \int_{-h(x)/2}^{h(x)/2} \sigma^+(x+n(x)z)dz\,, \tag{4.6}$$

$$\bar{\varepsilon}^+(x,h) = \int_{-h(x)/2}^{h(x)/2} \varepsilon^+(x+n(x)z)dz\,. \tag{4.7}$$

As a rule, these terms have the greatest interest in applications. In order to construct the principal terms of the asymptotic expansion of the functions $\bar{\sigma}^+(x,h)$ and $\bar{\varepsilon}^+(x,h)$ for small parameters of the problem, we consider the limiting values of the potentials in (4.4), (4.5) on the surface Ω (internal limits of the external solutions [204]).

4.2 Properties of potentials (4.4) and (4.5)

Let us introduce the following notation:

$$m(x) = B^1\bar{\sigma}^+(x,h) = -B^0C^1\bar{\varepsilon}^+(x,h)\ , \tag{4.8}$$

where $\bar{\sigma}^+(x,h)$ and $\bar{\varepsilon}^+(x,h)$ have the forms (4.6), and rewrite (4.4), (4.5) in the forms

$$\varepsilon(x) = \varepsilon^0(x) + \varepsilon^1(x)\ ,\ \sigma(x) = \sigma^0(x) + \sigma^1(x)\ , \tag{4.9}$$

$$\varepsilon^1(x) = \int_{\Omega} \mathrm{K}(x-x')C^0m(x')d\Omega', \tag{4.10}$$

$$\sigma^1(x) = \int_{\Omega} \mathrm{S}(x-x')m(x')d\Omega'. \tag{4.11}$$

Supposing that $m(x)$ is an arbitrary smooth tensor field on Ω we consider limiting values of the potentials $\sigma^1(x)$ and $\varepsilon^1(x)$ when $x \to \Omega$. To that end, we use projection operators (3.48), (3.49) on the normal $n(x)$ ($\Pi(n)$) and tangent plane($\Theta(n)$) to Ω at the point $x \in \Omega$. Representations of these operators in the E-basis (3.262)–(3.263) have the forms

$$\Pi(n) = 2E^5(n) - E^6(n)\,,\quad \Theta(n) = E^1 - 2E^5(n) + E^6(n)\,. \tag{4.12}$$

The operators Π and Θ split any symmetric rank two tensor $m(x)$ on Ω into "normal" $m^\pi(x)$ and "tangential" $m^\theta(x)$ components

$$m(x) = m^\pi(x) + m^\theta(x)\,, \tag{4.13}$$

$$m^\pi_{\alpha\beta} = \Pi_{\alpha\beta\lambda\mu}(n)m_{\lambda\mu}(x)\,,\quad m^\theta_{\alpha\beta}(x) = \Theta_{\alpha\beta\lambda\mu}(n)m_{\lambda\mu}(x)\,. \tag{4.14}$$

Note, that the tensor $m^\pi(x)$ is presented in the form

$$m^\pi_{\alpha\beta}(x) = n_{(\alpha}(x)b_{\beta)}(x), \tag{4.15}$$

where $b(x)$ is a vector field on Ω, and the tensor $m^\theta(x)$ satisfies the condition

$$n_\alpha(x)m^\theta_{\alpha\beta}(x) = 0, \tag{4.16}$$

and is the tensor of the surface Ω.

Let us present the density $C^0m(x)$ of the potential $\varepsilon^1(x)$ in (4.10) in the form of the following sum:

$$C^0_{\alpha\beta\lambda\mu}m_{\lambda\mu}(x) = q_{\alpha\beta}(x) + C^0_{\alpha\beta\lambda\mu}n_\lambda(x)b_\mu(x), \tag{4.17}$$

where the vector $b(x)$ satisfies the equation

$$n_\alpha(x)C^0_{\alpha\beta\lambda\mu}n_\lambda(x)b_\mu(x) = n_\alpha(x)C^0_{\alpha\beta\lambda\mu}m_{\lambda\mu}(x). \tag{4.18}$$

As a result, the tensor $q(x)$ in (4.17) satisfies the equations

$$n_\alpha(x)q_{\alpha\beta}(x) = 0\,, \quad \Theta_{\alpha\beta\lambda\mu}(n)q_{\lambda\mu}(x) = q_{\alpha\beta}(x), \tag{4.19}$$

and is the tensor of the surface Ω. We consider a potential $\varepsilon_1(x)$ with density that is either the tensor q or the tensor $C^0_{\alpha\beta\lambda\mu}n_\lambda(x)b_\mu(x)$ in (4.17).

1. Let $C^0 m = q$, and q be a tensor of the surface Ω. Using Gauss' theorem for surfaces (Appendix C.1), we present the potential $\varepsilon^1(x)$ with such a density in the form

$$\varepsilon^1_{\alpha\beta}(x) = \int_\Omega \mathrm{K}_{\alpha\beta\lambda\mu}(x-x')\Theta_{\lambda\mu\rho\nu}(x')q_{\rho\nu}(x')d\Omega' =$$
$$= -\nabla_{(\alpha}\int_\Omega G_{\beta)\mu}(x-x')\partial'_\lambda q_{\lambda\mu}(x')d\Omega' +$$
$$+ p.v. \int \nabla_{(\alpha}G_{\beta)\mu}(x-x')q_{\lambda\mu}(x')e_\lambda(x')d\Gamma', \tag{4.20}$$

where $e(x)$ is the external normal to the contour Γ, located in the tangent plane to Ω at the point $x \in \Gamma$. The function $\mathrm{K}(x)$ is defined in (2.53). Here the equation

$$\nabla_\lambda G_{\beta\mu}(x-x')\Theta_{\lambda\mu\nu\rho}(x')q_{\nu\rho}(x') =$$
$$= -\partial'_\lambda[G_{\beta\mu}(x-x')q_{\lambda\mu}(x')] + G_{\beta\mu}(x-x')\partial'_\lambda q_{\lambda\mu}(x')\,, \tag{4.21}$$

was taken into account. This equation follows from (4.3) and the definition (4.12) of the projector Θ, ∂'_λ means the derivative with respect to x'.

The integral over the border contour Γ in (4.20) is a continuous function for $x \in \Omega$ ($x \notin \Gamma$), and the integral over Ω is the gradient of the potential of a simple layer of static elasticity with the density $\partial_\lambda q_{\lambda\mu}(x)$. Thus, the latter integral is a continuous function in the whole space except the surface Ω. Let ε^{1+} and ε^{1-} be the limit values of the potential ε^1 when the point x tends to Ω from the side of the normal and from the opposed side, consequently. The jump $[\varepsilon^1]$ of the considered potential on Ω is determined by the equation that follows from (2.132), (2.131)

$$[\varepsilon^1_{\alpha\beta}(x)] = \varepsilon^{1+}_{\alpha\beta}(x) - \varepsilon^{1-}_{\alpha\beta}(x) = n_{(\alpha}(x)G^*_{\beta)\mu}(n)\partial_\lambda q_{\lambda\mu}(x)\,, \tag{4.22}$$

where $G^*(k)$ is defined in (2.58).

Direct verification shows that the tangential component $\Theta(n)\varepsilon^1(x)$ of the potential $\varepsilon^1(x)$ is continuous on the surface Ω.

$$\left[\Theta_{\alpha\beta\lambda\mu}(n)\varepsilon^1_{\lambda\mu}(x)\right] = 0. \tag{4.23}$$

Note that the integral that presents the value of the function $\Theta(n)\varepsilon^1(x)$ on Ω

$$\Theta_{\alpha\beta\lambda\mu}(n)\varepsilon^1_{\lambda\mu}(x) = \int\limits_{\Omega} U_{\alpha\beta\lambda\mu}(x,x')q_{\lambda\mu}(x')d\Omega', \tag{4.24}$$

$$U_{\alpha\beta\lambda\mu}(x,x') = \Theta_{\alpha\beta\nu\rho}(x)\mathrm{K}_{\nu\rho\tau\delta}(x-x')\Theta_{\tau\delta\lambda\mu}(x') \tag{4.25}$$

diverges formally since $\mathrm{K}(x-x') \sim |x-x'|^{-3}$ when $x \to x'$. The regularization of this integral is obtained in Appendix C.3 in the form

$$\Theta_{\alpha\beta\lambda\mu}(x)\varepsilon^1_{\lambda\mu}(x) = p.v.\int_{\Omega} U_{\alpha\beta\lambda\mu}(x,x')\left[q_{\lambda\mu}(x') - q^0_{\lambda\mu}(x,x')\right]d\Omega' -$$

$$-\Theta_{\alpha\beta\lambda\mu}(x)p.v.\int \nabla_\lambda G_{\mu\nu}(x-x')q^0_{\nu\rho}(x,x')e_\rho(x')d\Gamma'. \tag{4.26}$$

Here symbol $p.v.\int ...d\Omega$ means the Cauchy principal value of the integral, $q^0(x,x')$ is a constant tensor field on Ω. The function $q^0(x,x')$ satisfies the equation $\partial' q(x,x') = 0$ and coincides with $q(x)$ at $x' = x$. For a plane surface Ω, the regularization (4.26) is transformed into the following one:

$$\Theta_{\alpha\beta\lambda\mu}(x)\varepsilon^1_{\lambda\mu}(x) = p.v.\int_{\bar{\Omega}} U_{\alpha\beta\lambda\mu}(x,x')\left[q_{\lambda\mu}(x') - q_{\lambda\mu}(x)\right]d\Omega', \tag{4.27}$$

where $\bar{\Omega}$ is the plane which the surface Ω belongs to. The function $q(x)$ is continued by zero outside Ω.

Let us go to the potential $\sigma^1(x)$ in (4.10) with the density $m(x) = B^0 q(x)$, where $q(x)$ is a tensor of the surface Ω. The definition (2.72) of the function $\mathrm{S}(x)$ implies

$$\sigma^1(x) = \int\limits_{\Omega} \mathrm{S}(x-x')B^0 q(x')d\Omega' = C^0\varepsilon^1(x) - q(x)\Omega(x), \tag{4.28}$$

where $\varepsilon^1(x)$ is the potential (4.20), $\Omega(x)$ is the delta-function concentrated on the surface Ω. Thus, the limiting values of the potential $\sigma^1(x)$ are the limiting values of $\varepsilon^1(x)$ multiplied with the tensor C^0. Equations (4.22) and (2.58) imply that the jump of the normal component $n(x)\sigma^1(x)$ of the potential $\sigma^1(x)$ on Ω is determined by the equation

$$\left[n_\beta(x)\sigma^1_{\alpha\beta}(x)\right] = n_\beta(x)\sigma^{1+}_{\alpha\beta}(x) - n_\beta(x)\sigma^{1-}_{\alpha\beta}(x) = \partial_\beta q_{\beta\alpha}(x)\,. \tag{4.29}$$

2. Let the density of the potential $\varepsilon^1(x)$ (4.10) be the second term in the right-hand side of (4.17):

$$\varepsilon^1_{\alpha\beta}(x) = \int\limits_{\Omega} \mathrm{K}_{\alpha\beta\lambda\mu}(x-x')C^0_{\lambda\mu\rho\tau}n_\rho(x')b_\tau(x')d\Omega'. \tag{4.30}$$

From definition (2.53) of $\mathrm{K}(x)$ it follows that $\varepsilon^1(x)$ is the symmetrized gradient of the double layer potential of static elasticity. If $b(x)$ in (4.30) is a smooth vector field on Ω, the potential $\varepsilon^1(x)$ can be represented in the form

$$\varepsilon^1_{\alpha\beta}(x) = \varepsilon^{1r}_{\alpha\beta}(x) + n_{(\alpha}(x)b_{\beta)}(x)\Omega(x) \,, \tag{4.31}$$

where the function $\varepsilon^{1r}(x)$ is discontinuous on Ω. The limiting values of the potential $\varepsilon^1(x)$ on Ω are determined by the equation (Appendix C.2)

$$\varepsilon^{1\pm}_{\alpha\beta}(x) = p.v.\int_\Omega \mathrm{K}_{\alpha\beta\lambda\mu}(x-x')\, C^0_{\lambda\mu\tau\rho} n_\tau(x')[b_\rho(x') - b_\rho(x)]d\Omega' +$$

$$+R_{\alpha\beta\lambda}(x)b_\lambda(x) \mp \frac{1}{2}\Lambda_{\alpha\beta\lambda\mu}(n)\partial_\lambda b_\mu(x) \,, \tag{4.32}$$

$$R_{\alpha\beta\lambda}(x) = B^0_{\alpha\beta\tau\rho}\oint_\Gamma \mathrm{rot}_{\nu\gamma} Z_{\tau\rho\nu\lambda}(x-x')d\Gamma'_\gamma, \tag{4.33}$$

$$\Lambda_{\alpha\beta\lambda\mu}(n) = B^0_{\alpha\beta\nu\rho} S^*_{\nu\rho\lambda\mu}(n) \,, \tag{4.34}$$

where $Z(x)$ is the Green tensor for internal stresses connected with the tensor $\mathrm{S}(x)$ by Equation (2.67): $\mathrm{S}(x) = Rot Z(x)$, $d\Gamma_\gamma$ is the infinitesimal vector element on the contour Γ, the function $\mathrm{S}^*(n)$ is defined in (2.75).

For the potential $\sigma^1(x)$ with the density $m(x) = n(x)b(x)$ in (4.10), there is a representation similar to (4.28):

$$\sigma^1_{\alpha\beta}(x) = C^0_{\alpha\beta\lambda\mu}\varepsilon^1_{\lambda\mu}(x) - C^0_{\alpha\beta\lambda\mu}n_\lambda(x)b_\mu(x)\Omega(x) \,. \tag{4.35}$$

Equation (4.30) implies that the function $\sigma^1(x)$ has no singular components, and its jump on Ω is determined by the equation

$$[\sigma^1_{\alpha\beta}(x)] = -S^*_{\alpha\beta\lambda\mu}(n)\partial_\lambda b_\mu(x). \tag{4.36}$$

Using definition (2.75) of the tensor $\mathrm{S}^*(n)$ one can verify that the vector $n_\beta(x)\sigma^1_{\alpha\beta}(x)$ is continuous on Ω. The value of the vector $n_\beta(x)\sigma^1_{\alpha\beta}(x)$ on Ω is determined by the following regularized equation (Appendix C.2):

$$-n_\beta(x)\sigma^1_{\beta\alpha}(x) = \int_\Omega T_{\alpha\beta}(x,x')b_\beta(x')d\Omega' =$$

$$= p.v.\int_\Omega T_{\alpha\beta}(x,x')\,[b_\beta(x') - b_\beta(x)]d\Omega' +$$

$$+n_\beta(x)\oint_\Gamma \mathrm{rot}_{\nu\mu} Z_{\alpha\beta\nu\lambda}(x-x')d\Gamma'_\mu b_\lambda(x) \,, \tag{4.37}$$

$$T_{\alpha\beta}(x,x') = -n_\lambda(x)S_{\lambda\alpha\beta\mu}(x-x')n_\mu(x') \,. \tag{4.38}$$

If Ω is a part of a plane, the function $n_\beta(x)\sigma^1_{\alpha\beta}(x)$ on Ω is determined also by the following regularization:

$$-n_\beta(x)\sigma^1_{\beta\alpha}(x) = \int_\Omega T_{\alpha\beta}(x-x')b_\beta(x')d\Omega' =$$

$$= p.v.\int_{\bar\Omega} T_{\alpha\beta}(x-x')\,[b_\beta(x') - b_\beta(x)]d\Omega'. \tag{4.39}$$

Here $\bar\Omega$ is the plane which the surface Ω belongs to; the function $b(x)$ is continued by zero outside Ω.

4.3 External limit problems for a thin inclusion

In this section, we consider the limiting fields outside a thin inclusion when its transverse dimension h (parameter δ_1 in (4.1)) tends to zero. If the elastic moduli of the inclusion do not degenerate in the limit $\delta_1 \to 0$ $(C \neq 0, C \neq \infty)$, the elastic fields inside V remain finite. As a result, the functions $\bar{\varepsilon}^+(x,h)$ and $\bar{\sigma}^+(x,h)$ in (4.6), (4.7) vanish when $\delta_1 \to 0$. It follows from (4.5), (4.6) that in this limit the stress and strain fields outside the inclusion coincide with the external fields $\sigma^0(x)$, $\varepsilon^0(x)$ applied to the medium.

This result will change if the elastic moduli of a thin inclusion tend to zero or to infinity when δ_1 tends to zero. For an inclusion that is much softer than the medium, the elastic moduli tensor C is represented in the form

$$C = \delta_2 \bar{C}, \quad \delta_2 << 1, \quad \bar{C} = O(C^0). \tag{4.40}$$

If the inclusion is much harder than the medium, a similar representation holds for the elastic compliance tensor B of the inclusion

$$B = \delta_2 \bar{B}, \ \delta_2 << 1, \quad \bar{B} = O(B^0). \tag{4.41}$$

Let us consider the limit problems for thin inclusions when the "geometrical" δ_1 and "physical" δ_2 parameters simultaneously tend to zero, while their ratio has the order 1.

4.3.1 Thin soft inclusion

We start with the case of a thin soft inclusion when the parameter $\delta_1(h)$ and the tensor of elastic moduli C tend to zero. It follows from the equation

$$B^1 \bar{\sigma}^+ = (C^{-1} - B^0) C \bar{\varepsilon}^+ = -B^0 C^1 \bar{\varepsilon}^+ \tag{4.42}$$

that $B^1 \bar{\sigma}^+ \to \bar{\varepsilon}^+$ when $C \to 0$ $(C^1 \to -C^0)$. From this equation and (4.5), (4.6) it follows that if $h, C \to 0$ $(\delta_1, \ \delta_2 \to 0)$, the external solution of the problem for a thin inclusion takes the form

$$\varepsilon(x) = \varepsilon^0(x) + \int_{\Omega} \mathrm{K}(x - x') C^0 \bar{\varepsilon}^+(x') d\Omega', \tag{4.43}$$

$$\sigma(x) = \sigma^0(x) + \int_{\Omega} S(x - x') \bar{\varepsilon}^+(x') d\Omega', \tag{4.44}$$

$$\bar{\varepsilon}^+(x) = \lim \bar{\varepsilon}^+(x,h), \qquad h, \ C \to 0. \tag{4.45}$$

On the other hand, if $h, \ C \to 0$ the stress vector $n(x)\sigma(x)$ should be continuous on the surface Ω. At the same time, continuity of the displacement vector on Ω may be violated due to vanishing of the tensor C. Hence, the limiting values of the displacement $u(x)$ and stress $n(x)\sigma(x)$ vectors on the surface Ω have to satisfy the equations

$$[u(x)] = b(x), \quad [n(x)\sigma(x)] = 0, \quad x \in \Omega, \tag{4.46}$$

where $b(x)$ is an unknown jump of the displacement vector on the surface Ω.

It follows from (4.46) that the limiting strain field in the medium with the inclusion contains a singular component $\varepsilon^{1s}(x)$ concentrated on Ω that has the form

$$\varepsilon^{1s}_{\alpha\beta}(x) = n_{(\alpha}(x)b_{\beta)}(x)\Omega(x). \tag{4.47}$$

In this case, the function $\bar{\varepsilon}^{+}(x)$ in (4.43) and (4.45) is determined by an equation that follows from (4.7)

$$\bar{\varepsilon}^{+}_{\alpha\beta}(x) = n_{(\alpha}(x)b_{\beta)}(x), \tag{4.48}$$

and (4.43) and (4.44) take the forms

$$\varepsilon_{\alpha\beta}(x) = \varepsilon^{0}_{\alpha\beta}(x) + \int_{\Omega} \mathrm{K}_{\alpha\beta\lambda\mu}(x - x')C^{0}_{\lambda\mu\nu\rho}n_{\nu}(x')b_{\rho}(x')d\Omega', \tag{4.49}$$

$$\sigma_{\alpha\beta}(x) = \sigma^{0}_{\alpha\beta}(x) + \int_{\Omega} \mathrm{S}_{\alpha\beta\lambda\mu}(x - x')n_{\lambda}(x')b_{\mu}(x')d\Omega'. \tag{4.50}$$

The potentials on the right-hand sides of these equations coincide with the potentials (4.10), (4.11) when $m(x) = n(x)b(x)$. From the properties of such potentials indicated in Section 4.2 it follows that the condition (4.46) of continuity of the vector $n(x)\sigma(x)$ on Ω for the field $\sigma(x)$ in (4.50) is fulfilled automatically. In this case, with the accuracy of the terms of the order δ_1, the following equation holds ($h(x)$, $C \neq 0$)

$$n(x)\sigma(x) = h^{-1}(x)n(x)\bar{\sigma}^{+}(x,h), \; x \in \Omega, \tag{4.51}$$

where the function $\bar{\sigma}^{+}(x,h)$ is defined in (4.6).

Using Hooke's law for the material of the inclusion we can write

$$\bar{\sigma}^{+}(x,h) = C\bar{\varepsilon}^{+}(x,h). \tag{4.52}$$

From this equation and (4.48), (4.51) we find the connection between the vector $b(x)$ and the stress vector on the surface Ω

$$n_{\beta}(x)\sigma_{\beta\alpha}(x) = \lambda_{\alpha\beta}(x)b_{\beta}(x)\,, \quad \lambda_{\alpha\beta}(x) = \lim_{h,C\to 0} \frac{n_{\lambda}(x)C_{\lambda\alpha\beta\mu}n_{\mu}(x)}{h(x)}. \tag{4.53}$$

This equation and (4.50) give us the equation for the vector field $b(x)$ on Ω in the form

$$\lambda_{\alpha\beta}(x)b_\beta(x)+\int_\Omega T_{\alpha\beta}(x,x')b_\beta(x')d\Omega' = n_\beta(x)\,\sigma^0_{\beta\alpha}(x)\ , \tag{4.54}$$

$$T_{\alpha\beta}(x,x') = -n_\lambda(x)S_{\lambda\alpha\beta\mu}(x-x')n_\mu(x). \tag{4.55}$$

Here the action of the integral operator T with the kernel $T(x,x')$ is determined by the right-hand side of (4.37) or (4.39).

Hence, the solution of the limiting problem for a thin soft inclusion ($\delta_1,\delta_2\to 0$) has the form (4.49), (4.50), where the vector $b(x)$ is to be found from equation (4.54). If $\lambda(x)=0$, (4.54) is the equation for a crack with the surface Ω in a homogeneous medium.

4.3.2 Thin hard inclusion

Let us consider the case of a thin hard inclusion when the transverse dimension h and the tensor of the elastic compliances of the inclusion B both tend to zero. In this case,

$$C^1\bar{\varepsilon}^+ = (B^{-1}-C^0)B\bar{\sigma}^+ = -C^0B^1\bar{\sigma}^+, \tag{4.56}$$

and $C^1\bar{\varepsilon}^+\to\bar{\sigma}^+$ when $B\to 0$ $(B^1\to -B^0)$.

From this equation and (4.4), (4.5) we obtain that the limiting solution of the problem for thin hard inclusion takes the form

$$\varepsilon(x) = \varepsilon^0(x) - \int_\Omega \mathrm{K}(x-x')\bar{\sigma}(x')d\Omega', \tag{4.57}$$

$$\sigma(x) = \sigma^0(x) - \int_\Omega \mathrm{S}(x-x')B^0\bar{\sigma}(x')d\Omega', \tag{4.58}$$

$$\bar{\sigma}(x) = \lim\ \bar{\sigma}^+(x,h),\quad h,B\to 0. \tag{4.59}$$

Because the elastic moduli of the inclusion tend to infinity, the displacement vector and the tangential component of the strain tensor are continuous on the surface Ω, but the stress vector may have a jump on this surface. Thus, the limiting values of the displacement vector and the strain tensor satisfy the conditions

$$[u(x)]|_\Omega = 0,\quad [\Theta(x)\varepsilon(x)]|_\Omega = 0. \tag{4.60}$$

The first of these conditions shows that the strain field does not contain singular components concentrated on $\Omega(x)$. From the properties (4.23) of the potential $\varepsilon^1(x)$ it follows that the density $\bar{\sigma}(x)$ of the potentials in (4.57), (4.58) has to be a tensor of the surface Ω, i.e., $\bar{\sigma}(x)$ satisfies the equations

$$n_\beta(x)\bar{\sigma}_{\alpha\beta}(x) = 0,\quad \Theta_{\alpha\beta\lambda\mu}(x)\bar{\sigma}_{\lambda\mu}(x) = \bar{\sigma}_{\alpha\beta}(x). \tag{4.61}$$

For such a tensor $\bar{\sigma}_{\alpha\beta}(x)$, the field $\varepsilon(x)$ in (4.57) satisfies the second condition (4.60) automatically.

Because of continuity of the tangential components of the limiting strain tensor on Ω, up to terms of the order δ_1 we have the equation

$$\Theta(x)\varepsilon(x) = h^{-1}(x)\Theta(x)\bar{\varepsilon}^{+}(x,h), \quad x \in \Omega, \tag{4.62}$$

where the integral strain $\bar{\varepsilon}^{+}(x,h)$ has form (4.7).

Since the integral strain $\bar{\varepsilon}^{+}(x,h)$ and stress $\bar{\sigma}^{+}(x,h)$ are connected by Hooke's law (4.52), we obtain from (4.61) that in the limit $h, B \to 0$, the following equations hold:

$$\Theta(x)\varepsilon(x) = \mu(x)\bar{\sigma}(x)\,, \quad \mu_{\alpha\beta\lambda\mu}(x) = \lim_{h,B\to 0} \frac{\Theta_{\alpha\beta\nu\rho}(x)B_{\nu\rho\tau\delta}\Theta_{\tau\delta\lambda\mu}(x)}{h(x)}\,. \tag{4.63}$$

Substituting $\varepsilon(x)$ from (4.57) we obtain the equation for the density $\bar{\sigma}(x)$ in the form

$$\mu_{\alpha\beta\lambda\mu}(x)\bar{\sigma}_{\lambda\mu}(x) + \int_{\Omega} U_{\alpha\beta\lambda\mu}(x,x')\bar{\sigma}_{\lambda\mu}(x')d\Omega' = \Theta_{\alpha\beta\lambda\mu}(x)\varepsilon^{0}_{\lambda\mu}(x)\,. \tag{4.64}$$

$$U_{\alpha\beta\lambda\mu}(x,x') = \Theta_{\alpha\beta\nu\rho}(x)\mathrm{K}_{\nu\rho\tau\delta}(x-x')\Theta_{\tau\delta\lambda\mu}(x'). \tag{4.65}$$

The action of the operator U with the kernel $U(x,x')$ in this equation on a smooth functions $\bar{\sigma}(x)$ is determined by the right-hand sides of (4.26) or (4.27). If $\mu = 0$, (4.64) is the equation for a nonstretched membrane embedded into a homogeneous elastic medium.

The operators T and U in (4.54), (4.64) are elliptic pseudodifferential operators with symbols which are homogeneous functions of degree 1 (Section 3.6). The symbols of the operators in (4.54), (4.64) differ from the symbols of T and U by the positively defined summands $\lambda(x)$ and $\mu(x)$, and therefore, they are also elliptic. The existence and uniqueness of the solution of equation (4.54) take place in the class of functions that are equal to zero on the contour Γ [48]

$$b(x) = 0 \quad \text{as} \quad x \in \Gamma\,, \tag{4.66}$$

and equation (4.64) - in the class of functions satisfying the condition

$$\bar{\sigma}_{\alpha\beta}(x)e_{\beta}(x) = 0 \quad \text{as} \quad x \in \Gamma\,. \tag{4.67}$$

The previous consideration shows that the limiting external solutions of the problems for thin inclusions are not defined uniquely but depend on the limiting value of the ratio δ_1/δ_2. The value of this ratio determines the functions $\lambda(x)$ and $\mu(x)$ in equations (4.54) and (4.64). For unique definitions of these functions we have to match the external and internal asymptotic expansions of the elastic fields over the small parameters of the problem [204], [64].

4.4 Internal limiting problems and the matching procedure

Let us consider the limit fields inside a thin inclusion when its transverse dimension tends to zero. We choose an arbitrary point x on the surface Ω ($x \notin \Gamma$) as the center of the local coordinate system y_1, y_2, y_3. The natural internal variables of the problem for a thin inclusion are defined by equation [204]

$$\xi_i = h^{-1}(x) y_i \ , \quad i = 1, 2, 3 \ . \tag{4.68}$$

When the parameter $\delta_1(h)$ tends to zero, it leads to the internal limiting problem which (in the coordinates ξ_i) is the problem of the equilibrium of the medium containing an inclusion in the form of the plane layer of a unit thickness in the region $|\xi_3| \leq 1/2$ (Fig. 4.2)

Let $\sigma^e(x)$ and $\varepsilon^e(x)$ denote the stress and strain fields corresponding to the solution of the external limiting problem. In accordance with the method of matching external and internal asymptotic expansions [204], [64], we assume that on the interfaces between the inclusion and the medium, the solution of the internal limiting problem coincides with the limiting values of the solution of the external limiting problem on Ω

$$\lim_{\xi_3 \to \pm 1/2} \sigma(\xi) = \sigma^{e\pm}(x) \ , \quad \lim_{\xi_3 \to \pm 1/2} \varepsilon(\xi) = \varepsilon^{e\pm}(x) \ . \tag{4.69}$$

In these equations, the point $\xi(\xi_1, \xi_2, \xi_3)$ tends to the plane $\xi_3 = 1/2$ or $\xi_3 = -1/2$, from outside the layer.

Let the tensor of elastic moduli of inclusion C be small compared to the tensor of elastic moduli of the medium: $C(C^0)^{-1} = O(\delta_2)$, $\delta_2 << 1$. In this case, the external solutions σ^e, ε^e have the forms (4.49) and (4.50).

$$\sigma^e(x) = \sigma^0(x) + \int_\Omega \mathrm{S}(x - x') n(x') b^0(x') d\Omega' \ , \tag{4.70}$$

$$\varepsilon^e(x) = \varepsilon^0(x) + \int_\Omega \mathrm{K}(x - x') C^0 n(x') b^0(x') d\Omega' \ . \tag{4.71}$$

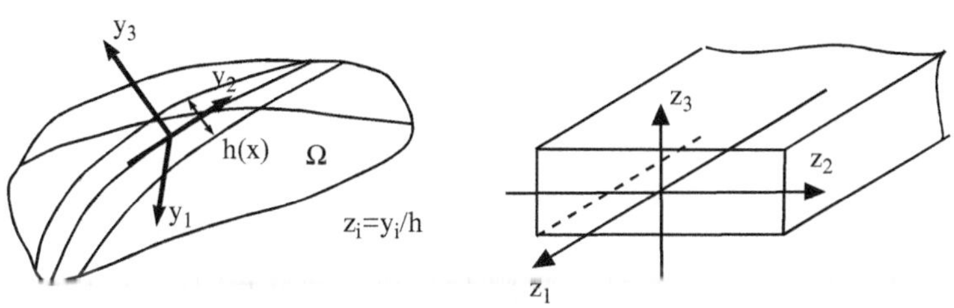

Fig. 4.2. Internal variables in the elasticity problem for a thin inclusion.

Here vector $b^0(x)$ is the solution of equation (4.54), where $\lambda(x)$ is a smooth unknown function of order C/h.

From the properties of the potential on the right-hand side of (4.70) it follows that the stress vector $n(x)\sigma^e(x)$ is continuous on Ω. Therefore, its limits $n(x)\sigma^{e+}(x)$ and $n(x)\sigma^{e-}(x)$ coincide at the point $x \in \Omega$ and have the form

$$n(x)\sigma^{e+}(x) = n(x)\sigma^{e-}(x) = \lambda(x)b^0(x) \,. \tag{4.72}$$

Here property (4.53) of the potential $\sigma^e(x)$ is taken into account.

Using the limiting values of potential (4.71) defined in (4.32) we can write the limiting values of the tangent components of the tensor $\varepsilon^e(x)$ on Ω:

$$\Theta(x)\varepsilon^{e\pm}(x) = \Theta(x)D(x) \pm \Theta(x)\partial b^0(x) \,, \tag{4.73}$$

$$D(x) = \varepsilon^0(x) + p.v.\int_\Omega \mathrm{K}\,(x - x')\, C^0 n\,(x')\left[b^0\,(x') - b^0\,(x)\right]d\Omega' + \\ + R\,(x)\, b^0\,(x) \,. \tag{4.74}$$

Let us turn to the internal problem, that is, the problem for the inclusion in the form of a plane layer (Section 3.5). The stress and strain fields that satisfy conditions (4.72) and (4.73) on the layer borders. Taking into account the continuity of the stress vector and the tangential component of the strain tensor on the layer boundary, we can find the fields $\sigma^+(x)$ and $\varepsilon^+(x)$ inside the layer in the form

$$n\sigma^+(\xi) = \lambda b^0, \tag{4.75}$$

$$\Theta\varepsilon^+(\xi) = \Theta D + 2\xi_3\Theta\partial b^0. \tag{4.76}$$

Because of Hooke's law for the inclusion material, the following equation holds:

$$n\sigma^+ = nC\Pi\varepsilon^+ + nC\Theta\varepsilon^+ = \lambda b^0. \tag{4.77}$$

Passing in (4.77) to the external variables y_i and substituting the result in (4.6) give us the equation for the integral strain field $\bar{\varepsilon}^+(x,h)$ inside the inclusion:

$$n(x)C\Pi(x)\bar{\varepsilon}^+(x) = h(x)\lambda(x)b^0(x) - n(x)C\Theta(x)\bar{\varepsilon}^+(x), \tag{4.78}$$

$$\Theta(x)\bar{\varepsilon}^+(x) = O(h). \tag{4.79}$$

Taking into account that tensor $\Pi\bar{\varepsilon}^+$ is represented in the form

$$\Pi_{\alpha\beta\lambda\mu}(x)\bar{\varepsilon}^+_{\lambda\mu}(x) = n_{(\alpha}(x)b_{\beta)}(x), \tag{4.80}$$

from (4.78), (4.79) we obtain the following equation for the vector b:

$$b_\alpha(x) = g_{\alpha\beta}(x)b^0_\beta(x) + O(h)\,, \tag{4.81}$$

$$g_{\alpha\beta}(x) = h(x)d^{-1}_{\alpha\mu}(x)\lambda_{\mu\beta}(x)\,, \tag{4.82}$$

$$d_{\alpha\beta}(x) = n_\lambda(x)C_{\lambda\alpha\beta\mu}n_\mu(x)\,. \tag{4.83}$$

Substituting $\bar{\varepsilon}^+(x)$ from (4.77)–(4.81) into the right-hand side of (4.4), we find the equation for the external limit of the internal solution $\varepsilon^i(x)$:

$$\varepsilon^i(x) = \varepsilon^0(x) - \int_\Omega \mathrm{K}(x-x')(C-C^0)\left(\Pi\bar{\varepsilon}^+(x) + \Theta\bar{\varepsilon}^+(x)\right)d\Omega' =$$
$$= \varepsilon^0(x) + \int_\Omega \mathrm{K}(x-x')\left[C^0 n\left(x'\right) g\left(x'\right) b^0\left(x'\right)\right] d\Omega' + O\left(\delta_1, \delta_2\right)\,. \tag{4.84}$$

Comparing this equation with the right-hand side of (4.71) we see that the principal terms of the external limit of the internal solution $\varepsilon^i(x)$ "match" the external solution $\varepsilon^e(x)$ if the tensor $g\left(x'\right)$ is the rank two unit tensor, i.e.,

$$g_{\alpha\beta}(x) = \delta_{\alpha\beta}. \tag{4.85}$$

From (4.81)–(4.83) we obtain that the parameter $\lambda(x)$ in (4.54) should have the form

$$\lambda(x) = h^{-1}(x)n(x)Cn(x), \tag{4.86}$$

where $h(x)$ and C are actual values of the transverse dimension and elastic moduli tensor of the inclusion.

Hence, the external expansions of the elastic fields outside a thin soft inclusion are determined by the equation

$$\sigma(x) = \sigma^e(x) + O(\delta_1, \delta_2)\,, \quad \varepsilon(x) = \varepsilon^e(x) + O(\delta_1, \delta_2)\,, \tag{4.87}$$

where the functions σ^e and ε^e are defined in (4.69), (4.70), and the vector b^0 is the solution of equation (4.54) when $\lambda(x)$ has the form (4.86).

Let the material of a thin inclusion be harder than the matrix material: $B(B^0)^{-1} = O(\delta_2)$, $\delta_2 << 1$. Matching the external and internal asymptotic expansions we choose the external solution in the form

$$\sigma^e(x) = \sigma^0(x) - \int_\Omega S(x-x')B^0\bar{\sigma}(x')d\Omega', \tag{4.88}$$

$$\varepsilon^e(x) = \varepsilon^0(x) - \int_\Omega \mathrm{K}(x-x')\bar{\sigma}(x')d\Omega', \tag{4.89}$$

where the density $\sigma(x)$ is the tensor of the surface Ω and satisfies equation (4.64), and $\mu(x)$ is an arbitrary smooth function of order B/h.

The internal limit problem has the same sense as in the previous case. For the solution of this problem, it is necessary to substitute (4.88) into condition (4.69) for the external solution. It is possible to show that the external limit of the internal solution matches the external solution (4.88) if the density $\bar{\sigma}(x)$ of the potentials in (4.88) satisfies equation (4.64), where the function $\mu(x)$ has the form

$$\mu_{\alpha\beta\lambda\mu}(x) = h^{-1}(x)\Theta_{\alpha\beta\nu\rho}(x)B_{\nu\rho\tau\delta}\Theta_{\tau\delta\lambda\mu}(x). \tag{4.90}$$

Here h and B are the actual values of the transverse dimension and the tensor of compliances of the inclusion.

4.5 Singular models of thin inclusions

To calculate the elastic field near a thin inclusion in a homogeneous medium, it is natural to replace the 3D-problem of elasticity by an equivalent 2D-problem. In order to perform this change, heuristic models of thin inclusions have been proposed in many works. For instance in [28], a thin inclusion was replaced by its middle surface Ω with the following boundary conditions:

$$[n\sigma]|_{\Omega} = 0\,, \quad [u]|_{\Omega} = b\,, \quad n\sigma|_{\Omega} = \lambda b\,, \tag{4.91}$$

where tensor $\lambda(x)$ has the form

$$\lambda_{\alpha\beta}(x) = h^{-1}(x)n_{\lambda}(x)C_{\lambda\alpha\beta\mu}n_{\mu}(x)\,. \tag{4.92}$$

In [28] the first of conditions (4.91) is called the equation of equilibrium, and the last one is Hooke's law for the inclusion. The boundary conditions (4.91) coincide with (4.46), (4.53). Therefore, the solution of this problem has the form (4.49), (4.50), where the vector $b(x)$ is to be found from Equation (4.54). Strictly speaking, condition (4.66) should be added to (4.91).

It follows from the results of Section 4.3 that the thinner the inclusion (the smaller is δ_1) and the smaller the ratio δ_2 of the elastic moduli of the inclusion and the medium, the better the model problem with condition (4.91)and (4.92) approximates the exact solution. It is natural to call such an inclusion a crack-like one. By using the singular model with condition (4.91) on Ω, we neglect the effect of strain constraints on the surface of the inclusion. As a result, the tangential component of the strain tensor can have a jump on the surface Ω.

For a thin hard inclusion, it is natural to replace the latter by its middle-surface with the distribution of the sources of internal stress that generate the strain and stress fields in the medium determined by (4.57), (4.58). The density $\bar{\sigma}(x)$ of the potentials in these equations is defined in (4.64), where the function $\mu(x)$ has the form

$$\mu_{\alpha\beta\lambda\mu}(x) = h^{-1}(x)\Theta_{\alpha\beta\nu\rho}(x)B_{\nu\rho\tau\delta}\Theta_{\tau\delta\lambda\mu}(x)\,. \tag{4.93}$$

The boundary conditions on the middle surface of the inclusion that are equivalent to the problem (4.64), (4.57), (4.58) have the forms (4.60), (4.63):

$$[u(x)]|_{\Omega} = 0\,, \quad [\Theta(x)\varepsilon(x)]|_{\Omega} = 0\,, \tag{4.94}$$

$$\Theta(x)\varepsilon(x) = \mu(x)\bar{\sigma}(x)\,, \quad \Theta(x)\bar{\sigma}(x) = \bar{\sigma}(x)\,, \quad x \in \Omega\,, \tag{4.95}$$

Equations (4.94) are the compatibility conditions for the deformation of the inclusion and the medium (the second condition follows from the first.) The first condition (4.95) is Hooke's law for the inclusion. Note that the tensor $\bar{\sigma}(x)$ in (4.95), (4.59) is the integral stress tensor (forces) in the cross section of the inclusion. The equation of equilibrium for these forces is a consequence of (4.28) and (4.58), and has the form

$$\partial_\alpha \bar{\sigma}_{\alpha\beta}(x) = -[n_\alpha(x)\sigma_{\alpha\beta}(x)]\,, \quad x \in \Omega. \tag{4.96}$$

Note that the forces in a thin momentless shell satisfy a similar equation [159]. Condition (4.67) on the boundary contour Γ of Ω should be added to the boundary conditions (4.94)–(4.96). Hence, conditions (4.94)–(4.96) and (4.67) correspond to the problem of a momentless elastic shell embedded in a homogeneous elastic medium. The model problem with the indicated conditions on Ω describes the fields in the vicinity of a thin hard inclusion as better when the parameters δ_1 and δ_2 are small.

4.6 Thin ellipsoidal inclusions

Let a thin inclusion occupy an ellipsoidal region V_h with the semiaxes a_1, a_2, h. In this case, the middle surface of the inclusion is an ellipse with semiaxes a_1, a_2, and the function $h(x)$ is represented in the form

$$h(x) = 2hz(x)\,, \quad z(x_1, x_2) = \sqrt{1 - \left(\frac{x_1}{a_1}\right)^2 - \left(\frac{x_2}{a_2}\right)^2}\,, \tag{4.97}$$

where x_1, x_2 are the Cartesian coordinates connected with the principal axes of the ellipse Ω, and a_1, a_2 are its semiaxes. In what follows, $V_h(x)$ is the characteristic function of the region V_h, and $P_m(x)$ is a polynomial of degree m in the coordinates x_1, x_2, x_3.

Because of the polynomial conservation theorem for an ellipsoid, the integral operator K with the kernel $\mathrm{K}(x)$ (3.12) transforms the function

$$P_{m,h}(x) = P_m(x)V_h(x)(2h)^{-1} \tag{4.98}$$

into a polynomial of degree m inside the ellipsoidal domain V_h. As the result, the operators $\bar{U}$, $\bar{T}$ with the kernels

$$\bar{U}(x) = \Theta(n)\mathrm{K}(x)\Theta(n)\,, \quad \bar{T}(x) = n\mathrm{S}(x)n \tag{4.99}$$

also transform polynomials of degree m into polynomials of the same degree. Here n is the normal to the ellipsoid middle surface, $\Theta(n)$ is the projector (3.49). Operators T and U in (4.54), (4.64) for thin inclusions are connected with the operators $\bar{T}$ and $\bar{U}$ by the equations

$$(Tb)(x) = (\bar{T}b\Omega)(x) = \int \bar{T}(x-x')b(x')\Omega(x')dx', \tag{4.100}$$

$$(U\bar{\sigma})(x) = (\bar{U}\bar{\sigma}\Omega)(x) = \int \bar{U}(x-x')\bar{\sigma}(x')\Omega(x')dx'. \tag{4.101}$$

Here $\Omega(x)$ is the delta-function concentrated on the surface Ω.

The symbols of the operators $\bar{U}$ and $\bar{T}$ have the forms

$$\bar{U}^*(k) = \Theta(n)\mathrm{K}^*(k)\Theta(n)\,, \quad \bar{T}^*(k) = -n\mathrm{S}^*(k)n\,, \tag{4.102}$$

and are homogeneous functions of the degree zero. Therefore, $\bar{U}$ and $\bar{T}$ are bounded and, consequently, closed in the Sobolev space of generalized functions $H(R^3)$ [48]. Thus, the following equation holds:

$$\lim_{h\to 0}\bar{U}P_{m,h} = \bar{U}(\lim_{h\to 0}P_{m,h})\,, \quad \lim_{h\to 0}\bar{T}P_{m,h} = \bar{T}(\lim_{h\to 0}P_{m,h})\,, \tag{4.103}$$

and the limit on the right-hand sides has the form

$$\lim_{h\to 0}P_{m,h}(x) = P_m(x)z(x)\Omega(x), \tag{4.104}$$

where the function $z(x)$ is defined in (4.97).

Since the left-hand sides of (4.103) are polynomial of degree m inside the region V_h, the right-hand sides of the limit function $P_{m,h}$ are polynomial on Ω. From here and (4.100), (4.101) we obtain the following properties of the operators T and U in Equations (4.54), (4.64).

If Ω is the elliptical surface, the operators T and U transform a polynomial of degree m multiplied by the function $z(x)$ in (4.97) into a polynomial of the same degree on Ω. Thus, for a thin ellipsoidal inclusion and a polynomial external field, the solutions of (4.54), (4.64) have the following forms:

$$b(x) = \bar{B}(x)z(x),\ \bar{\sigma}(x) = \bar{Q}(x)z(x), \tag{4.105}$$

where $\bar{B}(x)$ and $\bar{Q}(x)$ are polynomial of the same degree m as the external field. Particularly, if σ^0 (or ε^0) is a constant tensor, $\bar{B}$ and $\bar{Q}$ in (4.105) are constant tensors. The equation for these tensors follow from (4.54), (4.64) in the forms

$$\bar{B} = (\lambda^0 + T^0)^{-1}n\sigma^0, \quad \bar{Q} = (\mu^0 + U^0)^{-1}\Theta(n)\varepsilon^0, \tag{4.106}$$

$$\lambda^0 = (2h)^{-1}nCn\,, \quad \mu^0 = (2h)^{-1}\Theta(n)B\Theta(n)\,, \tag{4.107}$$

$$T^0 = \int T(x_1,x_2)[z(x_1,x_2)-1]dx_1dx_2\,, \tag{4.108}$$

$$U^0 = \int U(x_1, x_2)[z(x_1, x_2) - 1]dx_1 dx_2 . \tag{4.109}$$

Here regularizations (4.26), (4.37) were used, the integrals are calculated over the whole plane and converge in the ordinary sense, the function $z(x)$ is continued by zero outside Ω.

For an ellipsoidal inclusion and a polynomial external field of degree m, the solution of (4.54), (4.61) can be constructed as follows. First, the result of the operators $T_\lambda = \lambda(x)+T$ and $U_\mu = \mu(x)+U$ on a homogenous polynomial $x^{(m)}$ multiplied with $z(x)$ should be found, $x^{(m)} = x_1^{q_1} x_2^{q_2}$, q_1, q_2 are integer positive number and $q_1 + q_2 = m$. The result is a polynomial $Q_m(x)$; coefficients may be found by calculating the sequential derivatives of $T_\lambda x^{(m)} z$ and $U_\mu x^{(m)} z$ with respect to x at the point $x = 0$. Then, solving the system of linear equations, one can find the coefficients of the polynomial $P_m(x)$ such that the product $P_m(x)\ z(x)$ is transformed by the operator $T_\lambda(U_\mu)$ into the homogeneous polynomial $x^{(m)}$. The solution for the polynomial external field of degree n is a linear combination of the functions $P_m(x)z(x)$, $m = 0, 1, 2, ..., n$. Hence, the problem is reduced to the calculation of standard integrals, and the inversion of nonsingular matrices.

We present here the equation for the tensors $B = (\lambda^0 + T^0)^{-1}$ and $\Lambda = (\mu^0 + U^0)^{-1}$ in the right-hand sides of (4.106) for an isotropic inclusion and matrix. Using Equations (4.38), (3.204) for the kernel $T(x)$ and calculating the integral T^0, we find

$$B_{\alpha\beta} = (\lambda^0 + T^0)^{-1}_{\alpha\beta} = B_1 e^{(1)}_\alpha e^{(1)}_\beta + B_2 e^{(2)}_\alpha e^{(2)}_\beta + B_3 n_\alpha n_\beta , \tag{4.110}$$

$$B_1 = \left(\frac{\mu}{h} + T_1^0\right)^{-1}, \quad B_2 = \left(\frac{\mu}{h} + T_2^0\right)^{-1}, \quad B_3 = \left(\frac{\lambda + 2\mu}{h} + T_3^0\right)^{-1}, \tag{4.111}$$

where $e^{(1)}$, $e^{(2)}$ are the unit vectors of the principal axes of the ellipse Ω; the coefficients T_1^0 , T_2^0 , T_3^0 are defined in (3.233)–(3.235).

For a plane circular soft inclusions, these equations become simpler, and take the forms ($a_1 = a_2 = a$)

$$B_1 = B_2 = \left[\frac{\mu}{h} + \frac{\pi}{8a} \cdot \frac{\mu_0(2 - \nu_0)}{(1 - \nu_0)}\right]^{-1}, \tag{4.112}$$

$$B_3 = \left[\frac{\lambda + 2\mu}{h} + \frac{\pi}{4a} \cdot \frac{\mu_0}{(1 - \nu_0)}\right]^{-1}. \tag{4.113}$$

Let us consider a thin hard ellipsoidal inclusion. In this case, the tensor

$$\Lambda = (\mu^0 + U^0)^{-1} \tag{4.114}$$

in (4.106) may be calculated in explicit form. The equation for the tensor U^0 has the form (4.109) and is presented as follows:

$$U^0_{\alpha\beta\lambda\mu} = \Theta_{\alpha\beta\nu\rho}(n)\mathrm{K}^0_{\nu\rho\tau\delta}\Theta_{\tau\delta\lambda\mu}(n)\,, \quad \mathrm{K}^0 = \int \mathrm{K}\,(x)\,[z\,(x) - 1]d\Omega\,, \quad (4.115)$$

where the integration is performed over the whole plane x_1, x_2, the function $z(x)$ is defined in (4.97), and is continued by zero outside Ω. The tensor $\mathrm{K}(x)$ for the isotropic medium is

$$\mathrm{K}(x) = \frac{1}{4\pi\mu_0|x|^3}\Big\{E^1 - 3E^5(n) - \frac{\kappa_0}{2}\big[E^2{+}2E^1$$
$$-3\left(E^3(n){+}E^4(n) + 4E^5(n)\right) + 15E^6(n)\big]\Big\}\,, \quad n = x/|x|, \quad (4.116)$$

where $E^i(n)$ are the elements of the tensor basis defined in Appendix A.1.

Substituting this equation into (4.115) and introducing the coordinates r, φ in the plane x_1, x_2,

$$x_1 = ra_1 \cos\,\varphi\,, \quad x_2 = ra_2 \sin\,\varphi, \quad (4.117)$$

where a_1, a_2 are the semiaxes of the ellipse Ω , we obtain

$$\mathrm{K}^0_{\alpha\beta\lambda\mu} = -\frac{a_1a_2}{8\pi\mu_0}\int\limits_0^\infty \frac{z(r)-1}{r^2}dr \int\limits_0^{2\pi} \Big\{2\left(\frac{3}{t^2\,(\varphi)}\delta_{\alpha)(\lambda}m_{\mu)(\beta}\,(\varphi) - E^1_{\alpha\beta\lambda\mu}\right)$$
$$-\kappa_0\left[\frac{3}{t^2(\varphi)}\left(\delta_{\alpha\beta}m_{\lambda\mu}\,(\varphi) + \delta_{\lambda\mu}m_{\alpha\beta}\,(\varphi) + 4\delta_{\alpha)(\lambda}m_{\mu)(\beta}\,(\varphi)\right)\right.$$
$$\left.-2E^1_{\alpha\beta\lambda\mu} - E^2_{\alpha\beta\lambda\mu} - \frac{15}{t^4\,(\varphi)}m_{\alpha\beta\lambda\mu}(\varphi)\right]\Big\}\frac{d\varphi}{t^3\,(\varphi)}\,, \quad (4.118)$$

$$t^2\,(\varphi) = a_1^2\cos^2\varphi + a_2^2\sin^2\varphi\,, \quad (4.119)$$

$$z\,(r) = \sqrt{1-r^2} \quad (r \le 1), \quad z\,(r) = 0 \quad (r > 1), \quad (4.120)$$

$$m_{11} = a_1^2\cos^2\varphi, \quad m_{22} = a_2^2\sin^2\varphi\,, \quad (4.121)$$

$$m_{12} = m_{21} = a_1a_2\sin\varphi\cos\varphi\,,$$

$$m_{\alpha 3} = m_{3\alpha} = 0, \quad m_{1111} = (m_{11})^2, \quad m_{2222} = (m_{22})^2, \quad (4.122)$$

$$m_{1122} = m_{2211} = m_{11}m_{22}\,, \quad (4.123)$$

$$m_{\alpha 3\lambda\mu} = m_{3\alpha\lambda\mu} = m_{\alpha\mu 3\lambda} = m_{\alpha\mu\lambda 3} = 0\,. \quad (4.124)$$

Calculating the integrals in these equations we find that the nonzero components of the tensor U^0 in the Cartesian basis connected with the principal axes of the ellipse Ω have the forms

$$U^0_{1111} = \frac{a_2}{4a_1^2\mu_0}[6\,(1+3\kappa_0)\,J_1 - (2+3\kappa_0)\,J_0 - 15\kappa_0\,J_{11}], \quad (4.125)$$

$$U^0_{1122} = \frac{a_2}{4a_1^2\mu_\circ}\kappa_0[3(J_1+J_0) - J_0 - 15J_{12}], \tag{4.126}$$

$$U^0_{2222} = \frac{a_2}{4a_1^2\mu_0}[6\,(1+3\kappa_0)\,J_2 - (2+3\kappa_0)\,J_0 - 15\kappa_0 J_{22}], \tag{4.127}$$

$$U^0_{1212} = \frac{a_2}{4a_1^2\mu_0}\left[\frac{3}{2}\,(1+2\kappa_0)\,(J_1+J_2) - (1+\kappa_\circ)\,J_0 - 15\kappa_0 J_{12}\right]. \tag{4.128}$$

Here the coefficients J_i, J_{ij} $(i,j=1,2)$ are

$$J_0 = \frac{\mathrm{E}(k)}{1-k}, \quad J_1 = \frac{1}{3}\left[J_0 - \frac{\mathrm{E}(k)-\mathrm{K}(k)}{k}\right], \tag{4.129}$$

$$J_2 = \frac{1}{2}\left[2J_0 + \frac{\mathrm{E}(k)-\mathrm{K}(k)}{k}\right], \tag{4.130}$$

$$J_{11} = \frac{3k^2-3k-2}{15k^2(1-k)}\mathrm{E}(k) - \frac{4k^2-2k-2}{15k^2(1-k)}\mathrm{K}(k), \tag{4.131}$$

$$J_{22} = \frac{1}{15k^2}\left[\frac{3k^2+7k-2}{1-k}E\,(k) - (6k-2)\,\mathrm{K}\,(k)\right], \tag{4.132}$$

$$J_{12} = \frac{1}{15k^2}\left[\frac{2\,(k^2+5k-5)}{1-k}E\,(k) - 5\,(k-2)\,\mathrm{K}(k)\right], \tag{4.133}$$

where the notation is the same as in (3.236). For a circular thin ellipsoidal inclusion, the equation for the tensor Λ (4.114) takes the form $(a_1 = a_2 = a)$.

$$\Lambda = \Lambda_1 P^2(n) + \Lambda_2\left(P^1(n) - \frac{1}{2}P^2(n)\right), \tag{4.134}$$

$$\Lambda_1 = \mu_0 a\left[\xi\frac{1-\nu}{1+\nu} + \frac{\pi}{8}(2-\kappa_0)\right]^{-1}, \quad \Lambda_2 = 2\mu_0 a\left[\xi + \frac{\pi}{16}(4-\kappa_0)\right]^{-1}, \tag{4.135}$$

where μ, ν are the shear modulus and Poisson's ratio of the inclusion; $\xi = a\mu_0/(2h\mu)$, P^1, P^2 are the elements of the P-basis (Appendix A.2).

Let Ω be a plane strip of width a directed along the x_1-axis. In this case $a_1 \to \infty$, $a_2 = a$, and the essential components of the tensor Λ in (4.134) are determined by the equations

$$\Lambda_{1111} = \frac{2a\mu_0}{\xi}\cdot\frac{2\xi + (2-\kappa_0)(1+\nu)}{2\xi(1-\nu)+2-\kappa_0}, \quad \Lambda_{1122} = \Lambda_{2211} = \nu\Lambda_{2222}, \tag{4.136}$$

$$\Lambda_{2222} = 4a\mu_0[2\xi(1-\nu)+(2-\kappa_0)]^{-1}, \quad \Lambda_{1212} = \frac{2a\mu}{1+2\xi}, \quad \xi = \frac{a\mu_0}{2h\mu}. \tag{4.137}$$

4.7 Notes

Heuristic models of thin inclusions were proposed in [4], [193], [96]. General methods for the construction of asymptotic expansions for the solutions of the partial differential equations in the vicinity of thin regions were developed in [204], [64], [142]. Thin inclusions with properties that differ essentially from the properties of the medium are considered in [180], [187], [188], [170], [104].

The contents of this chapter is based on the works [79], [80].

5. Hard fiber in a homogeneous elastic medium

In this chapter, we consider equilibrium of a homogeneous elastic medium with an inclusion in the form of a long hard fiber. Hard fibers are widely used as reinforcing elements for composite materials. The solution of this problem depends on two small parameters: the ratio of a characteristic transverse dimension of the fiber to its length, and the ratio of the elastic moduli of the medium and the fiber. This chapter is devoted to the construction of the principal terms of the elastic fields in the medium with a hard fiber, for small values of the parameters.

5.1 External and internal limiting solutions

Consider an infinite homogeneous elastic medium with an inclusion in the form of a long curvilinear fiber. The middle line Γ of the fiber is a smooth curve, and its cross section is a circle with the radius $a(z)$, where z is a point on Γ. We suppose that $a(z)$ is a piece-wise smooth continuous function that satisfies the condition

$$\left|\frac{da}{dz}\right| \ll 1 \tag{5.1}$$

everywhere on Γ except, may be, the ends of the fiber. For the characteristic transversal size of the fiber is much smaller than its length, the function $a(z)$ is presented in the form

$$a(z) = \delta_1 l(z), \quad \delta_1 << 1, \tag{5.2}$$

where $\delta_1$is the small dimensionless parameter, and the function $l(z)$ has the order of the fiber length. The medium is subjected to an external stress (strain) field $\sigma^0(x)$ $(\varepsilon^0(x))$.

The principal terms of the stress and strain tensor in the medium outside the fiber have the form of the following integrals:

$$\sigma(x) = \sigma^0(x) + \int_\Gamma S(x-z)B^1\bar{\sigma}(z)\mathrm{d}\Gamma_z \ , \tag{5.3}$$

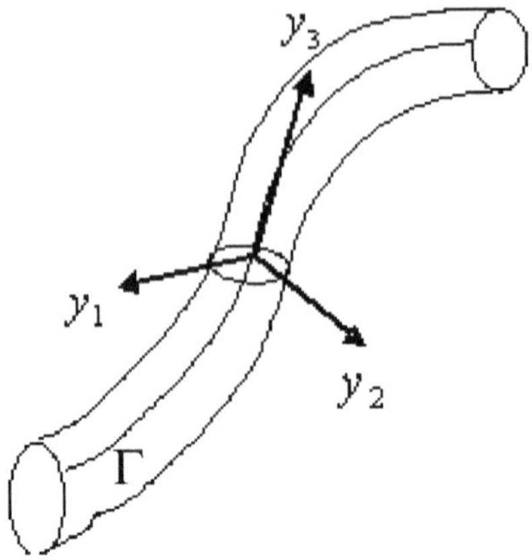

Fig. 5.1. Local coordinate system in the elasticity problem for a hard fiber.

$$\varepsilon(x) = \varepsilon^0(x) - \int_\Gamma \mathrm{K}(x-z)C^1\bar{\varepsilon}(z)\mathrm{d}\Gamma_z \,, \tag{5.4}$$

$$\bar{\sigma}(z) = \int_{\Omega(z)} \sigma^+(x)\mathrm{d}\Omega \,, \quad \bar{\varepsilon}(z) = \int_{\Omega(z)} \varepsilon^+(x)\mathrm{d}\Omega \,, \tag{5.5}$$

where $\Omega(z)$ is the cross section of the region V occupied by the fiber.

Let us consider the internal limit solution of the fiber problem when δ_1 tends to zero, and take a point $z \in \Gamma$ ($z \neq z_0$, z_l, where z_0, z_l are the the ends of the fiber) as the origin of the local Cartesian coordinate system y_1, y_2, y_3 with the y_3-axis directed along the tangent line to Γ (see Fig. 5.1).

We introduce the dimensionless variables $\xi_i = a^{-1}(z)y_i$ ($i = 1,2,3$) and let the parameter δ_1 (the radius $a(z)$ of the fiber) tend to zero. The region V in the coordinates ξ_i is transformed into the region V_0 of the circular cylinder of unit radius with the generatrix parallel to ξ_3. Equations (3.26), (3.27) in the coordinates ξ_i take the forms

$$\sigma^{(0)}(\xi, z) - \int_{V_0} \mathrm{S}(\xi-\xi')B^1\sigma^{(0)}(\xi', z)\mathrm{d}\xi' = \sigma^0(z) \,, \tag{5.6}$$

$$\varepsilon^{(0)}(\xi, z) + \int_{V_0} \mathrm{K}(\xi-\xi')C^1\varepsilon^{(0)}(\xi', z)\mathrm{d}\xi' = \varepsilon^0(z) \,. \tag{5.7}$$

Here we took into account that $\mathrm{S}(x)$ and $\mathrm{K}(x)$ are homogeneous functions of degree -3

$$\sigma^{(0)}(\xi, z) = \lim_{\delta_1 \to 0} \sigma^+(y) \,, \quad \varepsilon^{(0)}(\xi, z) = \lim_{\delta_1 \to 0} \varepsilon^+(y) \,. \tag{5.8}$$

Since $\sigma^0(z)$, $\varepsilon^0(z)$ are independent of the coordinates ξ_i, the solutions of equations (5.6), (5.7) inside the circular cylinder V_0 are also independent on ξ and are presented in the forms

$$\sigma^{(0)}(z) = \Lambda^\sigma \sigma^0(z), \quad \varepsilon^{(0)}(z) = \Lambda^\varepsilon \varepsilon^0(z), \tag{5.9}$$

$$\Lambda^\sigma = (E^1 - D^0 B^1)^{-1}, \quad \Lambda^\varepsilon = (E^1 + A^0 C^1)^{-1}. \tag{5.10}$$

Here the tensors A^0 and D^0 are determined by the relations

$$A^0 = \lim_{\delta_1 \to 0} \frac{1}{4\pi} \int_{\Omega_1} \mathrm{K}^*(k_1, k_2, \delta_1 k_3) d\Omega\,, \quad D^0 = C^0 A^0 C^0 - C^0, \tag{5.11}$$

where Ω_1 is the surface of the unit sphere in the k-space of the Fourier transform, the function $\mathrm{K}^*(k_1, k_2, k_3)$ is the Fourier image of the kernel $\mathrm{K}(x)$ (2.53), the system k_1, k_2, k_3 is coupled with y_1, y_2, y_3. For an isotropic medium, the tensor A^0 has the form (3.127)

$$A^0 = \frac{1}{4\mu_0}\left[(2 - \kappa_0)P^1(m) - \frac{1}{2}\kappa_0 P^2(m) + 2P^5(m)\right], \tag{5.12}$$

where $P^i(m)$ are the elements of the P-basis (Appendix A.2), m is the unit vector of the y_3-axis.

Let us introduce the projectors $\Theta'(m)$ and $\Pi'(m)$

$$\Theta'(m) = P^1(m) + 2P^5(m)\,, \quad \Pi'(m) = P^6(m)\,, \quad \Theta' + \Pi' = E^1, \tag{5.13}$$

$$\Theta'\Theta' = \Theta', \quad \Pi'\Pi' = \Pi', \quad \Theta'\Pi' = \Pi'\Theta' = 0\,. \tag{5.14}$$

These projectors split the linear shell of the P-basis into two orthogonal subspaces (Θ' and Π'). Using the algebra of the tensors P^i (Appendix A.2) we can show that the tensor A^0 in (5.12) belongs to the Θ' - subspace and has a nonsingular inverse tensor $(A^0)^{-1}$in this subspace

$$\Theta'(m)A^0(m) = A^0(m)\Theta'(m) = A^0(m)\,, \tag{5.15}$$

$$\Pi'(m)A^0(m) = A^0(m)\Pi'(m) = 0\,, \tag{5.16}$$

$$A^0(m)(A^0(m))^{-1} = (A^0(m))^{-1}A^0(m) = \Theta'(m)\,. \tag{5.17}$$

Let us turn to the densities $\bar{\sigma}(z)$ and $\bar{\varepsilon}(z)$ of the potentials in the external limiting solutions (5.3), (5.4). If the elastic moduli of the inclusion are bounded, the solutions (5.9) of the internal limiting problem are also bounded. It follows from (5.5), that if $\delta_1 \to 0$, the densities $\bar{\sigma}(z)$ and $\bar{\varepsilon}(z)$ vanish, and the external limiting solutions (5.3), (5.4) coincide with the unperturbed external fields $\sigma^0(x)$ and $\varepsilon^0(x)$.

We reach another result if the tensor $B = C^{-1}$ tends to zero together with δ_1. In this case, from (5.9) we have

$$B^1\sigma^+ = B^1\sigma^{(0)} \to -(C^0)^{-1}(C^{-1}+A^0)^{-1}\varepsilon^0, \tag{5.18}$$

$$C^1\varepsilon^+ = C^1\varepsilon^{(0)} \to (C^{-1}+A^0)^{-1}\varepsilon^0. \tag{5.19}$$

Because the tensor A^0 (5.12) is singular (it belongs to the Θ'- subspace), in the limit $C^{-1} = B \to 0$, some components of the tensors $B^1\sigma^+$ and $C^1\varepsilon^+$ may tend to infinity. Thus, in the limit $\delta_1, B \to 0$, the densities $B^1\bar{\sigma}$ and $C^1\bar{\varepsilon}$ of the potentials (5.3), (5.4) may be bounded.

Let us go to the detailed analysis of the stress field inside a hard fiber, and suppose that the tensor of elastic compliances B of the fiber is represented in the form

$$B = \delta_2 \bar{B}\ , \quad \bar{B}C^0 = O(1)\ , \tag{5.20}$$

where $\delta_2 << 1$ is a small dimensionless parameter.

5.2 Principal terms of the stress field inside a hard fiber

For simplicity, we consider a fiber in the form of a body of revolution, and place the origin of the Cartesian coordinate system y_1, y_2, y_3 in the middle of the fiber. The y_3-axis is directed along the fiber axis Γ. Using Hooke's law we rewrite the equation (3.26) in terms of the tensor $\sigma(x)$ inside the fiber in the form

$$C^{-1}\sigma(y) + \int_V \mathrm{K}(y-y')C^1C^{-1}\sigma(y')\mathrm{d}y' = \varepsilon^0(y). \tag{5.21}$$

In the limit $\delta_1 \to 0$, the solution of this equation is constant across a cross section of the fiber. Therefore, the principal terms of the solution of (5.21) with respect to δ_1, δ_2 are constant in a fiber cross section (at least far from the fiber ends). Let us substitute $\sigma(y) = \sigma(y_3)$ in the left-hand side of (5.21) and consider this equation at points on the fiber axis Γ. For each point $y \in V$ a unique representation holds

$$y = \bar{y} + y_3 m, \tag{5.22}$$

where $\bar{y} = \bar{y}(y_1, y_2)$ is the vector in the plane of the fiber cross-section $\Omega(y_3)$.

Let us introduce the dimensionless coordinates ($2l$ is the fiber length)

$$\xi = y_3/l, \quad \eta = \bar{y}/l, \tag{5.23}$$

From (5.21) the equation for $\sigma(\xi)$ follows in the form

$$C^{-1}\sigma(\xi) + \int_{-1}^{1} \bar{\mathrm{K}}(\xi,\xi')C^1C^{-1}\sigma(\xi')\mathrm{d}\xi' = \varepsilon^0\,(\xi)\ , \tag{5.24}$$

$$\bar{\mathrm{K}}(\xi,\xi') = \int\limits_{\Omega(\xi')} \mathrm{K}[(\xi-\xi')m - \eta']\mathrm{d}\eta' \,. \tag{5.25}$$

In what follows, it is convenient to consider C^0 and C as dimensionless tensors and

$$C^0 = O(1), \quad C^{-1} = O(\delta_2). \tag{5.26}$$

To that end, we multiply both sides of (5.24) with a characteristic elastic modulus of the medium.

The expression for the kernel $\bar{\mathrm{K}}(\xi,\xi')$ of the integral operator in Equation (5.24) is presented in the form ($k = k(k_1,k_2,k_3)$)

$$\bar{\mathrm{K}}(\xi,\xi') = \frac{1}{(2\pi)^3} \int\limits_{\Omega(\xi')} \mathrm{d}\,\eta' \int \mathrm{K}^*(k) \exp\{\, -ik\cdot[(\xi-\xi')m-\eta']\}\mathrm{d}k\,, \tag{5.27}$$

where $\mathrm{K}^*(k)$ is the Fourier transform of the function $\mathrm{K}(x)$ (2.58). Changing the order of integration in this equation (firstly over $\Omega(\xi')$, and then over k_1, k_2) leads to the equation

$$\bar{\mathrm{K}}(\xi,\xi') = \frac{1}{2\pi} \int\limits_{-\infty}^{\infty} \bar{\mathrm{K}}^*(\delta_1(\xi'),k_3) e^{-ik_3(\xi-\xi')}\mathrm{d}k_3\,, \;\; \delta_1(\xi) = \frac{a(\xi)}{l}\,. \tag{5.28}$$

Here the function $\bar{\mathrm{K}}^*(\delta_1(\xi'),k_3)$ for an isotropic medium is represented in the following expansion in the P-basis

$$\bar{\mathrm{K}}^*(\delta_1,k_3) = \sum_{i=1}^{6} \mathrm{K}_i^*(\delta_1,k_3)P^i(m)\,, \tag{5.29}$$

$$\mathrm{K}_1^* = \frac{1}{16\mu_0}[4(2-\kappa_0)M_1^* + \kappa_0 M_2^*]\,, \;\; \mathrm{K}_2^* = \frac{\kappa_0}{32\mu_0}(M_2^* - 4M_1^*)\,, \tag{5.30}$$

$$\mathrm{K}_3^* = \mathrm{K}_4^* = -\frac{\kappa_0}{4\mu_0}M_2^*\,, \;\; \mathrm{K}_5^* = \frac{1}{2\mu_0}(2 - M_1^* - 2\kappa_0 M_2^*)\,, \tag{5.31}$$

$$\mathrm{K}_6^* = \frac{1}{2\mu_0}[2(1-\kappa_0)(1-M_1^*) + \kappa_0 M_2^*]\,. \tag{5.32}$$

Functions $M_1^*(\delta_1,k)$, $M_2^*(\delta_1,k)$ are determined by the equations (henceforth, the index 3 of the variable k_3 is omitted)

$$M_1^*(\delta_1,k) = \delta_1|k|\mathrm{K}_1(\delta_1|k|)\,, \;\; M_2^*(\delta_1,k) = \delta_1^2k^2\mathrm{K}_0(\delta_1|k|)\,, \tag{5.33}$$

where K_0, K_1 are modified Bessel functions. From here and (5.28)–(5.32) it follows that the kernel $\bar{\mathrm{K}}(\xi,\xi')$ of the integral operator $\bar{\mathrm{K}}$ is determined by an expression similar to (5.29) in which M_1^*, M_2^* have to be replaced by

$$M_1(\xi,\xi') = \frac{\delta_1^2(\xi')}{2[(\xi-\xi')^2+\delta_1^2(\xi')]^{3/2}}, \tag{5.34}$$

$$M_2(\xi,\xi') = -\frac{1}{2}\frac{\mathrm{d}^2}{\mathrm{d}\xi^2}\left\{\frac{\delta_1^2(\xi')}{[(\xi-\xi')^2+\delta_1^2(\xi')]^{1/2}}\right\}. \tag{5.35}$$

These functions are the inverse Fourier transforms of the functions $M_1^*(\delta_1,k)$ and $M_2^*(\delta_1,k)$ (5.33) with respect to the variable k. $M_1(\xi,\xi')$ and $M_2(\xi,\xi')$ are the kernels of the integral operators M_1, M_2 that act on $\sigma(\xi)$ according to the equation

$$(M_i\sigma)(\xi) = \int_{-1}^{1} M_i(\xi,\xi')\sigma(\xi')\mathrm{d}\xi', \quad i=1,2. \tag{5.36}$$

In order to construct the principal term of the solution of (5.24) with respect to the small parameters δ_1,δ_2 we represent the function $\delta_1(\xi)$ in (5.28)–(5.35) in the form

$$\delta_1(\xi) = \delta_1\alpha(\xi), \quad \delta_1 << 1, \quad \alpha(\xi) = O(1), \tag{5.37}$$

and expand the Bessel function in (5.33) in a series with respect to the small parameter δ_1. Taking into account only two terms of these series [2]

$$\mathrm{K}_1(z) = \frac{1}{z} + \frac{z}{2}\ln z + O(z), \tag{5.38}$$

$$\mathrm{K}_0(z) = -\ln z + O(1), \tag{5.39}$$

we obtain the equation for the symbol $\bar{\mathrm{K}}^*(\delta_1,k)$ in (5.29) in the form

$$\bar{\mathrm{K}}^*(\delta_1,k) = A^0 + (\delta_1^2\ln\delta_1)\alpha^2(\xi)k^2A^1 + O(\delta_1^2), \tag{5.40}$$

where the constant tensor A^0 is defined in (5.12), and the tensor A^1 in the $P^i(m)$-basis is presented in the form

$$A^1 = \frac{1}{32\mu_0}\left[2(4-3\kappa_0)P^1 - 3\kappa_0P^2 + 8\kappa_0(P^3+P^4)+\right.$$
$$\left.+8(4\kappa_0-1)P^5 - 16P^6\right]. \tag{5.41}$$

Taking into account only two terms in the expansion (5.40) for the symbol $\bar{\mathrm{K}}^*$ of the operator K we transform (5.21) into the following equation:

$$C^{-1}\sigma(\xi) + A^0C^1C^{-1}\sigma(\xi) - (\delta_1^2\ln\delta_1)A^1\frac{\mathrm{d}^2}{\mathrm{d}\xi^2}\left[\alpha^2(\xi)\sigma(\xi)\right] = \varepsilon^0(\xi). \tag{5.42}$$

The solution of this equation may be found in the form of an expansion similar to (5.40)

$$\sigma(\xi) = \sigma^{(0)}(\xi) + (\delta_1^2 \ln \delta_1)\sigma^{(1)}(\xi) + \cdots \ . \tag{5.43}$$

We obtain the equation for the principal term $\sigma^{(0)}(\xi)$ of this expansion if we substitute (5.43) into (5.42) and keep the principal terms with respect to δ_1. Depending on the relation between the small parameters δ_1 and δ_2 there are two cases:

1. $\boldsymbol{\delta_2^{-1}\delta_1^2 ln\delta_1 = o(1)}$**.** In this case, the equation for $\sigma^{(0)}(\xi)$ takes the form

$$C^{-1}\sigma^{(0)}(\xi) + A^0 C^1 C^{-1}\sigma^{(0)}(\xi) = \varepsilon^0(\xi) \ . \tag{5.44}$$

Applying the projectors Θ' and Π' (5.13) to both sides of this equation and taking into account property (5.16) of the tensor A^0 we obtain the following two equations:

$$A^0\sigma_\theta^{(0)}(\xi) + (\Theta' - A^0C^{-1})C^{-1}\sigma^{(0)}(\xi) = \varepsilon_\theta^0(\xi) \ , \tag{5.45}$$

$$\Pi' C^{-1}\sigma^{(0)}(\xi) = \varepsilon_\pi^0(\xi) \ ,$$

$$\sigma^0 - \sigma_\theta^0 + \sigma_\pi^0 , \quad \sigma_\theta^0 = \Theta'\sigma^0 , \quad \sigma_\pi^0 = \Pi'\sigma^0 , \quad \varepsilon_\pi^0 = \Pi'\varepsilon^0 . \tag{5.46}$$

Since in the Θ'-subspace A^0 is a nonsingular tensor with components which have the order of 1, the following evaluations hold:

$$\sigma_\theta^0(\xi) = O(1), \quad \sigma_\pi^0(\xi) = O(\delta_2^{-1}). \tag{5.47}$$

Here we accept that $\varepsilon^0, \sigma^0 = O(1)$. Taking into account the structure of the tensor σ_π^0

$$\sigma_\pi^{(0)}(\xi) = \sigma_m^0(\xi) m \otimes m, \tag{5.48}$$

where $\sigma_m^0(\xi)$ is the scalar function, we reach the evaluation for $\sigma^{(0)}(\xi)$ in the form

$$\sigma^{(0)}(\xi) = \sigma_m^{(0)}(\xi) m \otimes m + O(1) \ , \quad \sigma_m^{(0)}(\xi) = O(\delta_2^{-1}) \ . \tag{5.49}$$

The expression for the axial component $\sigma_m^{(0)}(\xi)$ of the tensor $\sigma^{(0)}(\xi)$ follows from (5.45) and (5.46) and takes the form

$$\sigma_m^{(0)}(\xi) = \mathrm{E}_m \varepsilon_m^{(0)}(\xi) \ , \quad \varepsilon_m^0(\xi) = \varepsilon_{\alpha\beta}^0(\xi) m_\alpha m_\beta \ , \tag{5.50}$$

where E_m is Young's modulus along the fiber axis.

2. $\boldsymbol{\delta_2^{-1}\delta_1^2 \ln \delta_1 = O(1), \ C^0C^{-1} = O(\delta_1^2 \ln \delta_1)}$**.** In this case, the equation for $\sigma^{(0)}(\xi)$ in (5.43) coincides with (5.42). Applying the operators Θ' and Π' to both sides of (5.42) we see that the stress tensor $\sigma^{(0)}(\xi)$ is evaluated as in

(5.49), where the axial component $\sigma_m^{(0)}(\xi)$ of this tensor is the solution of the following differential equation:

$$\frac{d^2}{d\,\xi^2}\left[\alpha^2(\xi)\sigma_m^{(0)}(\xi)\right] - q^2\sigma_m^{(0)}(\xi) = -q^2 \mathrm{E}_m \varepsilon_m^0(\xi)\,, \tag{5.51}$$

$$q^2 = -\frac{\mathrm{E}_0}{\mathrm{E}_m}\cdot\frac{(1+\nu_0)^{-1}}{\delta_1^2\ln\delta_1} = -\frac{2\mu_0}{\mathrm{E}_m\delta_1^2\ln\delta_1}\,,\quad q = O(1)\,. \tag{5.52}$$

Hence, the principal term of the expansion (5.43) of the stress field inside the fiber is determined apart from two arbitrary constants appearing in the general solution of differential Equation (5.51).

Let us compare the formal expressions for $\sigma^{(0)}(\xi)$ with the exact solution of integral Equation (5.24). To that end, we substitute the function $\sigma^{(0)}(\xi)$ into the left-hand side of (5.24) and compare it with the right-hand side. If the evaluation (5.49) for $\sigma^{(0)}(\xi)$ holds, the Π'-component of the discrepancy presents its essential term. For the function $\sigma^{(0)}(\xi)$ in (5.48) we obtain

$$\alpha_m\sigma_m^{(0)}(\xi) - (M\sigma_m^{(0)})(\xi) = f_0(\xi) + R(\sigma_m^{(0)},\xi)\,, \tag{5.53}$$

$$\alpha_m = 1 + \frac{\mathrm{E}_0}{2\mathrm{E}_m}\cdot\frac{(1-\nu_0)}{(1-\kappa_0)}\,,\quad f_0(\xi) = \frac{1+\nu_0}{1-\kappa_0}\mathrm{E}_0\varepsilon_m^0(\xi)\,, \tag{5.54}$$

$$M = M_1 - \frac{\kappa_0}{2(1-\kappa_0)}M_2\,, \tag{5.55}$$

where $R(\sigma_m^{(0)},\xi)$ is the discrepancy, and the operators M_1 and M_2 are defined in (5.34)–(5.36).

To compensate for the discrepancy R in (5.53) we have to add the term $\sigma_m^{(1)}(\xi)$ to $\sigma_m^{(0)}(\xi)$, so that the sum

$$\sigma_m(\xi) = \sigma_m^{(0)}(\xi) + \sigma_m^{(1)}(\xi)\,, \tag{5.56}$$

satisfies the equation

$$\alpha_m\sigma_m(\xi) - (M\sigma_m)(\xi) = f_0(\xi)\,. \tag{5.57}$$

If the term $\sigma_m^{(1)}(\xi)$ in (5.56) can be neglected in comparison with $\sigma_m^{(0)}(\xi)$, the function $\sigma_m^{(0)}(\xi)m\otimes m$ is the principal term of the stress field inside the fiber. If $\sigma_m^{(1)}$ and $\sigma_m^{(0)}$ have the same order, the principal term of $\sigma_m^{(1)}$ should be added to the function $\sigma_m^{(0)}$.

Note that the Θ'-component of the discrepancy R is small compared with the axial component $\sigma_m^{(0)}(\xi)$ that has the order δ_2^{-1}. Let us go to the analysis of the discrepancy R in (5.53) for fibers of specific forms.

5.3 Stress fields inside fibers of various forms

5.3.1 Cylindrical fiber

For a cylindrical fiber of radius a, we have

$$\delta_1 = a/l\,, \quad \alpha(\xi) = 1. \tag{5.58}$$

In order to obtain the discrepancy R in (5.53) for this case, we have to evaluate the result of the action of the operators M_1 and M_2 (5.36) on a smooth bounded function $\sigma(\xi)$ of the order 1. The function $(M_1\sigma)(\xi)$ is presented in the form

$$(M_1\sigma)(\xi) = M_1^0(\xi)\sigma(\xi) + M_1^1(\xi)D_\xi\sigma(\xi) + \frac{1}{2}M_1^2D_\xi^2\sigma(\xi) + \\ + \int_{-1}^{1} M_1\left(\xi,\xi'\right)Q\left(\xi,\xi'\right)\mathrm{d}\xi', \tag{5.59}$$

$$Q\left(\xi,\xi'\right) = \sigma\left(\xi'\right) - \sigma\left(\xi\right) - D_\xi\sigma\left(\xi\right)\left(\xi'-\xi\right) - \frac{1}{2}D_\xi^2\sigma\left(\xi\right)\left(\xi'-\xi\right)^2, \tag{5.60}$$

$$M_1^k(\xi) = \int_{-1}^{1} M_1(\xi,\xi')(\xi'-\xi)^k\mathrm{d}\,\xi', \quad k=0,1,2, \;\; D_\xi = \frac{\mathrm{d}}{\mathrm{d}\,\xi}, \tag{5.61}$$

where the kernel $M_1(\xi,\xi')$ is defined in (5.34) with $\delta_1(\xi') = \delta_1$.

Calculating integrals (5.61) and substituting the results in (5.59), we obtain

$$(M_1\sigma)\left(\xi\right) = \sigma(\xi) - \left[\varPhi_0\left(\frac{1-\xi}{\delta_1}\right) + \varPhi_0\left(\frac{1+\xi}{\delta_1}\right)\right]\sigma(\xi) - \\ -\delta_1\left[\varPhi_1\left(\frac{1-\xi}{\delta_1}\right) - \varPhi_1\left(\frac{1+\xi}{\delta_1}\right)\right]D_\xi\sigma(\xi) - \frac{1}{2}\delta_1^2\ln\delta_1 D_\xi^2\sigma(\xi) + \\ + \frac{1}{2}\delta_1^2\ln\delta_1\left[\varPhi_2\left(1-\xi,\delta_1\right) + \varPhi_2\left(1+\xi,\delta_2\right)\right]D_\xi^2\sigma(\xi) + O(\delta_1^2)\,. \tag{5.62}$$

It is taken into account that the integral item in (5.59) has the order δ_1^2, $\varPhi_0, \varPhi_1, \varPhi_2$ are the boundary layer functions localized in the vicinity of the fiber ends $\xi = \pm 1$:

$$\varPhi_0(t) = \frac{1}{2}\mathrm{sign}\,t\left(1 - \frac{|t|}{\sqrt{1+t^2}}\right)\,, \quad \varPhi_1(t) = \frac{1}{2\sqrt{1+t^2}}\,, \tag{5.63}$$

$$\varPhi_2(t) = 1 - \frac{1}{2\ln\delta_1}\ln\left(\sqrt{t^2+\delta_1^2} - |t|\right)\,. \tag{5.64}$$

In the same way it may be shown that the following evaluation holds:

$$(M_2\sigma)(\xi) = -\delta_1\left[\Phi_1\left(\frac{1-\xi}{\delta_1}\right) - \Phi_1\left(\frac{1+\xi}{\delta_1}\right)\right] D_\xi\sigma(\xi) - \delta_1^2 \ln\delta_1 D_\xi^2\sigma(\xi) - \\ -\frac{1}{2}\delta_1^2 \ln\delta_1 [\Phi_2(1-\xi,\delta_1) + \Phi_2(1+\xi,\delta_1)] D_\xi^2\sigma(\xi) + O(\delta_1^2)\ . \quad (5.65)$$

Consider the case of a fiber with "geometrical" δ_1 and "physical" δ_2 parameters that satisfy the relation

$$\delta_2^{-1}\delta_1^2 \ln\delta_1 = o(1)\ . \quad (5.66)$$

In this case, the function $\sigma_m^{(0)}(\xi)$ has the form (5.50). Substituting (5.50) into (5.53) and taking into account (5.62), (5.65), we obtain an expression for the discrepancy R

$$R = \left[\Phi_0\left(\frac{1-\xi}{\delta_1}\right) + \Phi_0\left(\frac{1+\xi}{\delta_1}\right)\right]\sigma_m^{(0)}(\xi) + O(\delta_1\sigma_m(0))\ . \quad (5.67)$$

To compensate for this discrepancy that has the order $\sigma_m^{(0)}$ near the fiber ends, we have to add to the function $\sigma_m^{(0)}$ the items σ_m^+ and σ_m^-, depending on the "fast" variables τ_+, τ_-

$$\tau_+ = \frac{(1-\xi)}{\delta_1}\ , \quad \tau_- = \frac{(1+\xi)}{\delta_1}. \quad (5.68)$$

Substituting the sum

$$\sigma_m(\xi) = \sigma_m^{(0)}(\xi) + \sigma_m^+(\tau_+) + \sigma_m^-(\tau_-)\ , \quad (5.69)$$

into (5.53), introducing the new variable $\tau_+(\tau_-)$ and letting δ_1 tend to zero, we obtain the equation for the functions σ_m^+ and σ_m^-

$$\alpha_m\sigma_m^\pm(\tau_\pm) - (M_\tau\sigma_m^\pm)(\tau_\pm) = -\Phi_0(\tau_\pm)\sigma_m^{(0)}(\pm 1)\ , \quad (5.70)$$

$$(M_\tau\sigma_m)(\tau) = \int_0^\infty M_\tau(\tau-\tau')\sigma_m(\tau')\mathrm{d}\tau', \quad (5.71)$$

$$M_\tau(\tau) = \frac{1}{2}\left(1+\tau^2\right)^{-3/2}\left[1 - \frac{\kappa_0}{1-\kappa_0}\left(1 - \frac{3\tau^2}{1+\tau^2}\right)\right]. \quad (5.72)$$

Here, (5.70) is a Wiener–Hopf equation [52]. The symbol $M_\tau^*(k)$ of the operator M_τ is defined by the expression

$$M_\tau^*(k) = \alpha_m - k\mathrm{K}_1(k) + \frac{\kappa_0}{1-\kappa_0}k^2\mathrm{K}_0(k)\ , \quad (5.73)$$

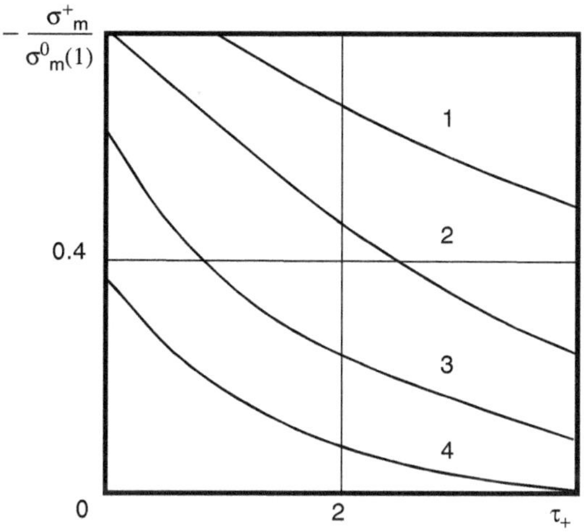

Fig. 5.2. The results of the numerical solution of equation (5.70).

where $\alpha_m > 1$ is determined in (5.54); $\mathrm{K}_0, \mathrm{K}_1$ are the modified Bessel functions. Since the function $M^*_\tau(k)$ does not equal to zero on the real axis, (5.70) has a unique solution in the class of continuous bounded functions [52]. The numerical solution of this equation is presented in Fig. 5.2 by curves 1–4 corresponding to the values of the parameter $\alpha_m = 1.01; 1.05; 1.2; 2$.

Analysis of the numerical solution leads to the following conclusions:

1. If $\alpha_m - 1 = O(1)$, then $\sigma^\pm_m(\tau_\pm)$ are the boundary layer functions localized in the vicinity of the end $\tau = 0$.
2. If $\alpha_m \to 1$, the functions $\sigma^\pm_m(\tau_\pm)$ are not functions of boundary layer types. The changing of the solution behavior is connected with the singularity of the symbol $M^*_\tau(k)$ at the point $k = 0$ as $\alpha_m = 1$.

Remark. Qualitatively, the solution of Equation (5.70) is similar to the solution of the following Wiener–Hopf equation:

$$(L_\alpha v)(\tau) = \alpha v(\tau) - \frac{1}{2}\int\limits_0^\infty e^{-|\tau-\tau'|} v(\tau')\mathrm{d}\tau' = e^{-\tau}. \tag{5.74}$$

Here L_α is the Wiener–Hopf operator with the symbol

$$L^*_\alpha(k) = \alpha - \frac{1}{1+k^2} \, . \tag{5.75}$$

When $\alpha > 1$ the solution of Equation (5.74) has the form

$$v(\tau) = \frac{1}{2\left[\sqrt{\alpha(\alpha-1)}+\alpha\right]} \exp\left[-\sqrt{\frac{\alpha-1}{\alpha}}\,\tau\right], \tag{5.76}$$

and properties 1) and 2) are clearly seen in this case.

Let τ be the fast variable similar to (5.68). Returning in (5.76) to the "slow" variable $\xi = \delta_1 \tau$ gives

$$v(\xi) = \frac{1}{2\left[\sqrt{\alpha(\alpha-1)}+\alpha\right]} \exp\left(-\frac{\xi}{\gamma}\right), \quad \gamma = \sqrt{\frac{\alpha}{\alpha-1}}\delta_1 . \tag{5.77}$$

Thus, if $\gamma = O(1)$ ($\sqrt{\alpha-1} = O(\delta_2)$), the function $v(\xi)$ is not localized at the end $\xi = 0$ for any δ_1. If $\gamma = o(1)$, i.e., $\gamma \to 0$ when $\delta_1 \to 0$, $v(\xi)$ is a bounder layer function localized at the vicinity of the end $\xi = 0$ as $\delta_1 \to 0$.

Let us return to Equation (5.70). Numerical analysis of this equation shows that when $\delta_2^{-1}\delta_1^2 \ln \delta_1 = o(1)$, the boundary layer functions $\sigma_m^+(\frac{1-\xi}{\delta_1})$ and $\sigma_m^-(\frac{1+\xi}{\delta_1})$ compensate the principal term in the discrepancy R in (5.53) in the vicinity of the fiber ends $\xi = \pm 1$. The remaining part of the discrepancy is compensated by terms having the order $\delta_1 \sigma_m^{(0)}$ in the expression for σ_m. These terms may be neglected in comparison with the principal term that has form (5.69).

Consider the case $\delta_2^{-1}\delta_1^2 \ln \delta_1 = O(1)$ when the function $\sigma_m^{(0)}(\xi)$ satisfies the equation (5.51) with $\alpha(\xi) = 1$ (cylindrical fiber)

$$\frac{d^2}{d\xi^2}\sigma_m^{(0)}(\xi) - q^2 \sigma_m^{(0)}(\xi) = -q^2 \mathrm{E}_m \varepsilon_m^0(\xi) . \tag{5.78}$$

If we substitute the general solution of this equation into the left-hand side of (5.53) and take into account (5.62)–(5.65), the discrepancy R still has form (5.67). Thus, the principal term of R is small if the function $\sigma_m^{(0)}(\xi)$, being the solution of Equation (5.78), satisfies the boundary conditions

$$\sigma_m^{(0)}(-1) = \sigma_m^{(0)}(1) = 0 . \tag{5.79}$$

These conditions allow us to find the values of the arbitrary constants in the general solution of (5.78).

Note that if $q >> 1$ ($\delta_2^{-1}\delta_1^2 \ln \delta_1 = o(1)$), the solutions of (5.78), (5.79) and (5.50) become close except for the δ_1-vicinities of the fiber ends, where the difference of these solutions is described by the boundary layer functions. Thus, the solution of (5.78) with conditions (5.79) correctly describes the slowly varying part of the field $\sigma(\xi)$ inside the fiber for all relations between the small parameters δ_1 and δ_2.

To visualize the character of the approximation of the solution of (5.57) by the function $\sigma_m^{(0)}(\xi)$ in (5.78), (5.79), we consider a model equation similar to (5.57):

$$\alpha_m \sigma - M_1 \sigma = -\frac{1}{2}\delta_1^2 \ln \delta_1 , \quad \alpha_m = 1 - \frac{1}{2} q^2 \delta_1^2 \ln \delta_1 , \tag{5.80}$$

where the operator M_1 is defined in (5.34), (5.36) with $\delta_1(\xi) = \delta_1 \ll 1$.

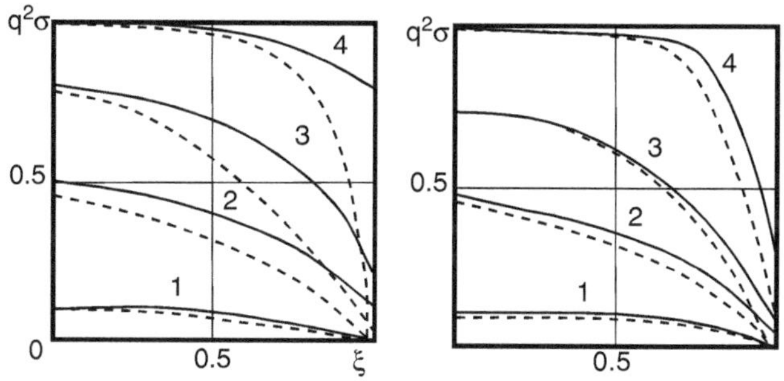

Fig. 5.3. The results of the numerical solution of equation (5.80) (left part: $\delta_1 = 0.1$, right part: $\delta_1 = 0.01$).

The principal term of the solution of this equation satisfies the ordinary differential equation

$$\frac{d^2}{d\xi^2}\sigma^{(0)}(\xi) - q^2\sigma^{(0)}(\xi) = -1 , \tag{5.81}$$

with boundary conditions (5.79). The solution of this equation has the form

$$\sigma^{(0)}(\xi) = \frac{1}{q^2}\left(1 - \frac{\cosh(q\xi)}{\cosh q}\right) . \tag{5.82}$$

A comparison of this function with the results of the numerical solution of equation (5.80) is presented in Fig. 5.3 (left part $\delta_1 = 0.1$, right part $\delta_1 = 0.01$) by the solid lines; the dashed lines correspond to the function $\sigma^{(0)}(\xi)$ in (5.82). Curves 1–4 correspond to the values of the parameter $q = 0.4; 1.2; 2; 10$. When δ_1 decreases, the difference between $\sigma^{(0)}(\xi)$ and $\sigma(\xi)$ becomes important only in the δ_1-vicinity of the fiber ends.

5.3.2 Prolate ellipsoidal fiber

In this case, the function $\delta_1(\xi)$ has the form

$$\delta_1(\xi) = \delta_1\alpha(\xi) , \quad \delta_1 = a/l , \quad \alpha(\xi) = \sqrt{1-\xi^2}, \tag{5.83}$$

where a and l are the ellipsoid semiaxes, and the action of the operators M_1 and M_2 in (5.36) on a smooth function $\sigma(\xi)$ is estimated as follows:

$$(M_1\sigma)(\xi) = \sigma(\xi) - \frac{1}{2}\delta_1^2 \ln\delta_1 D_\xi^2\left[\left(1-\xi^2\right)\sigma(\xi)\right] + O\left(\delta_1^2\right) , \tag{5.84}$$

$$(M_2\sigma)(\xi) = \delta_1^2 \ln\delta_1 D_\xi^2\left[\left(1-\xi^2\right)\sigma(\xi)\right] + O\left(\delta_1^2\right) . \tag{5.85}$$

Hence, in contrast to a cylindrical fiber (see (5.58), (5.65)), the principal terms of $(M_1\sigma)(\xi)$ and $(M_2\sigma)(\xi)$ in the expansion in series in δ_1 do not contain the boundary layer functions. From (5.84), (5.85) it follows that the principal term of the stress field inside the fiber is $\sigma_m^{(0)}(\xi) m \otimes m$, where the scalar function $\sigma_m^{(0)}(\xi)$ satisfies Equation (5.51) with $\alpha(\xi) = \sqrt{1-\xi^2}$

$$D_\xi^2\left[\left(1-\xi^2\right)\sigma_m^{(0)}(\xi)\right] - q^2\sigma_m^{(0)}(\xi) = -q^2 \mathrm{E}_m \varepsilon_m^0(\xi)\,. \tag{5.86}$$

The discrepancy R in (5.53) has the order $\delta_1^2\sigma_m^{(0)}$

$$R = O(\delta_1^2\sigma_m^{(0)})\,. \tag{5.87}$$

Since each of two linearly independent solutions of homogeneous Equation (5.86) has a singularity of the type $(1-\xi)^{-1}$ or $(1+\xi)^{-1}$ near the ends $\xi = \pm 1$ of the fiber, the arbitrary constants in the general solution of (5.86) should be taken equal to zero. The corresponding function $\sigma_m^{(0)}(\xi)$ is bounded and presents the principal term of the stress field inside the fiber. The terms in the components of the exact stress tensor $\sigma(\xi)$ that compensate for the discrepancy R in (5.53) have the order $(\ln\delta_1)^{-1}\sigma_m^{(0)}$ and may be neglected in comparison with $\sigma_m^{(0)}$.

If ε_m^0 is constant, the bounded solution of (5.86) is also constant and has the form

$$\sigma_m^{(0)} = \frac{q^2}{2+q^2} E_m \varepsilon_m^0 \tag{5.88}$$

Note that for a constant external field, the integral equation (5.21) for an ellipsoidal region V has a known exact solution (see Section 3.4). It follows from Equations (3.104), (3.125) for the field inside a prolate spheroid that the principal term of the asymptote to the exact solution when $\delta_1 \to 0$ has the form

$$\sigma^+(x) = \sigma_m^{(0)} m_\alpha m_\beta, \tag{5.89}$$

and the scalar $\sigma_m^{(0)}$ coincides with (5.88).

5.3.3 Fiber in the form of a double cone

In this case, the function $\delta_1(\xi)$ has the form

$$\delta_1(\xi) = \delta_1\alpha(\xi)\,, \quad \alpha(\xi) = 1 - |\xi|, \tag{5.90}$$

and the action of the operators M_1 on a smooth bounded function $\sigma(\xi)$ is represented in the form

$$M_1 = M_1^- + M_1^+, \tag{5.91}$$

$$\left(M_1^- \sigma\right)(\xi) = \int_{-1}^{0} M_1^- \left(\xi, \xi'\right) \sigma\left(\xi'\right) \mathrm{d}\xi', \tag{5.92}$$

$$\left(M_1^+ \sigma\right)(\xi) = \int_{0}^{1} M_1^+ \left(\xi, \xi'\right) \sigma\left(\xi'\right) \mathrm{d}\xi', \tag{5.93}$$

$$M_1^\pm \left(\xi, \xi'\right) = \frac{1}{2} \frac{\delta_1^2 \left(1 \mp \xi'\right)^2}{\left[\left(\xi - \xi'\right)^2 + \delta_1^2 \left(1 \mp \xi'\right)^2\right]^{3/2}}. \tag{5.94}$$

Let $H_+(\xi)$ be the Heaviside function ($H_+(\xi) = 1, \xi > 0; H_+(\xi) = 0, \xi < 0), H_-(\xi) = H_+(-\xi)$) and $\sigma_+(\xi)$ ($\sigma_-(\xi)$) be a smooth bounded function on the positive (negative) semi-axis R_+ (R_-) of the ξ-axis, $\sigma(\xi) = \sigma_-(\xi) H_-(\xi) + \sigma_+(\xi) H_+(\xi)$.

Presenting the function $(M_1^\mp \sigma_\pm)(\xi)$ in the form similar to (5.59) and calculating the integrals in this equation we obtain

$$\begin{aligned} \left(M_1 \sigma\right)(\xi) = \sigma(\xi) - \frac{1}{2}\delta_1^2 \ln \delta_1 D_\xi^2 \left[(1 - |\xi|)^2 \sigma(\xi)\right] + \\ + \Phi_0 \left(\frac{\xi}{\delta_1}\right) \left(\sigma_-(\xi) - \sigma_+(\xi)\right) - \\ - \delta_1 \Phi_1 \left(\frac{\xi}{\delta_1}\right) \left[\sigma_-(\xi) + \sigma_+(\xi) + D_\xi \sigma_-(\xi) - D_\xi \sigma_+(\xi)\right] - \\ - \delta_1^2 \ln \delta_1 \Phi_2\left(\xi, \delta_1\right) D_\xi^2 \left[(1 + \xi)^2 \sigma_-(\xi) + (1 - \xi)^2 \sigma_+(\xi)\right] - \\ - \frac{1}{2\delta_1^2} \left[\sigma_-(\xi) \ln(1 + \xi) + \sigma_+(\xi) \ln(1 - \xi)\right] + O\left(\delta_1^2\right). \end{aligned} \tag{5.95}$$

Here the functions Φ_0, Φ_1, Φ_2 are defined in (5.63)–(5.64). If $\sigma_\pm(0) \neq 0$, singular terms arising from the differentiation of $|\xi|$ have to be omitted. With the same restrictions it is possible to write for the operator M_2

$$\begin{aligned} \left(M_2 \sigma\right)(\xi) = \delta_1^2 \ln \delta_1 D_\xi^2 \left[(1 - |\xi|)^2 \sigma(\xi)\right] + \\ + \frac{1}{2}\delta_1^2 D_\xi^2 \left[(1 - |\xi|)^2 \sigma(\xi) \ln(1 - |\xi|)\right] + O\left(\delta_1^2\right). \end{aligned} \tag{5.96}$$

Equation (5.51) for the principal term of the stress field inside the double-cone takes the form:

$$D_\xi^2 \left[(1 - |\xi|)^2 \sigma_m^{(0)}(\xi)\right] - 2\bar{q}\sigma_m^{(0)}(\xi) = -2\bar{q}\mathrm{E}_m \varepsilon_m^0(\xi), \quad \bar{q} = (1/2)q^2. \tag{5.97}$$

Here, as well (5.95), (5.96), a singular term arising by differentiation of $|\xi|$ should be omitted.

The general solution of equation (5.97) has the form

$$\sigma_m^{(0)}(\xi) = \bar{\sigma}_m(\xi) + c_1 (1 - |\xi|)^{\beta_-} + c_2 (1 - |\xi|)^{\beta_+} , \tag{5.98}$$

$$\beta_\pm = -\frac{1}{2}\left(3 \pm \sqrt{1 + 8\bar{q}}\right) , \quad \bar{q} > 0 , \tag{5.99}$$

where $\bar{\sigma}_m(\xi)$ is a bounded particular solution of (5.97); c_1, c_2 are arbitrary constants.

Since the elastic field near a conical point on the inclusion–medium interface is square-integrable [98], the constant c_2 in front of the nonintegrable in the interval [-1,1] function $(1 - |\xi|)^{\beta_+}$ should be equal to zero. Substituting (5.98) with $c_2 = 0$ into the left-hand side of (5.53) and using (5.95), (5.96), we obtain the following evaluation of the discrepancy R near the middle point ($\xi = 0$) of the fiber:

$$\begin{aligned} R = 2\delta_1 \left[\Phi_1\left(\frac{\xi}{\delta_1}\right) - \frac{\xi}{\delta_1}\Phi_0\left(\frac{\xi}{\delta_1}\right)\right] c\beta + 2\delta_1 \Phi_1\left(\frac{\xi}{\delta_1}\right)[\bar{\sigma}_m(0) + c] + \\ + O(\delta_1^2 \ln \delta_1 \sigma_m^{(0)}) , \quad \beta = \beta_- , \quad c = c_1 . \end{aligned} \tag{5.100}$$

From here and (5.98) we see that the principal term of the discrepancy R decays as $|\xi/\delta_1|^{-1}$ in moving away from the middle of the fiber. To eliminate this term, the constant c should be taken in the form

$$c = -\frac{2}{2 + \beta}\sigma_m^{(0)}(0) . \tag{5.101}$$

In this case, the asymptotic of the principal term of the discrepancy R decays as $|\xi/\delta_1|^{-3}$, and the expression for $\sigma_m^{(0)}(\xi)$ takes the form

$$\sigma_m^{(0)}(\xi) = \bar{\sigma}_m(\xi) - \frac{2}{2 + \beta}\bar{\sigma}_m(0)(1 - |\xi|)^{\beta} . \tag{5.102}$$

In particular, if the field ε^0 is constant, we obtain

$$\sigma_m^{(0)}(\xi) = \frac{\bar{q}}{\bar{q} - 1}\left[1 - \frac{2}{2 + \beta}(1 - |\xi|)^{\beta}\right] E_m \varepsilon_m^0 , \quad \bar{q} \neq 1 , \tag{5.103}$$

$$\sigma_m^{(0)}(\xi) = -\frac{1}{3}[1 - 2\ln(1 - |\xi|)] E_m \varepsilon_m^0 , \quad \bar{q} = 1 . \tag{5.104}$$

Equations (5.102), (5.95), (5.96) imply that near the fiber ends, the discrepancy R in (5.53) has as a logarithmic singularity $R \sim \delta_1^2 \ln(1 - |\xi|)\sigma_m^{(o)}(\xi)$, and to compensate for this discrepancy it is necessary to add boundary layer functions to $\sigma_m^{(0)}(\xi)$. The equations for these functions are similar to (5.70).

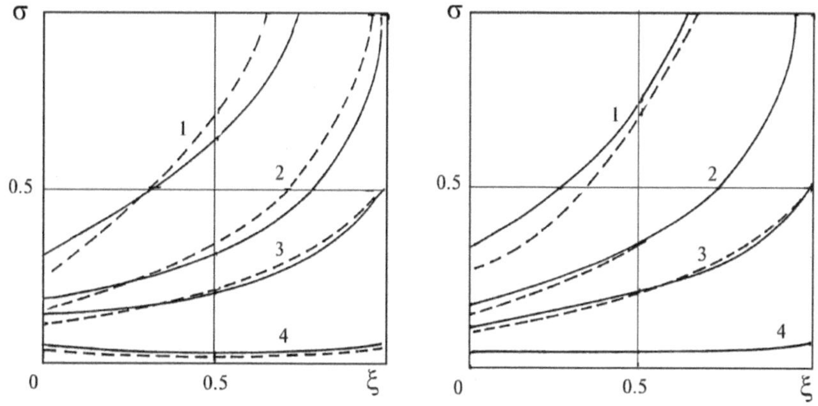

Fig. 5.4. The results of the numerical solution of equation (5.105) (left part: $\delta_1 = 0.1$, right part: $\delta_1 = 0.01$).

In order to visualize the difference between the asymptotic solution (5.103), (5.104) and the solution of integral Equation (5.57) we consider a model integral equation similar to (5.57).

$$\alpha_m \sigma - M_1 \sigma = -\frac{1}{2}\delta_1 \ln \delta_1 \,, \quad \alpha_m = 1 - \bar{q}\delta_1^2 \ln \delta_1 \,, \tag{5.105}$$

where the operator M_1 is defined in (5.34), (5.36) with $\delta_1(\xi) = \delta_1(1 - |\xi|)$. The principal term $\sigma^{(0)}(\xi)$ of the asymptotic solution of this equation has the form (5.103) when $E_m \varepsilon_m^0 = 1/2$.

The results of the numerical solution of Equation (5.105) are presented in Fig. 5.4 (left part: $\delta_1 = 0.1$, right part: $\delta_1 = 0.01$) by solid lines, the dashed line are the functions $\sigma^{(0)}(\xi)$ in (5.103). Lines 1–4 correspond to the values of the parameter $\bar{q} = 0.4, 1.2, 2, 10$. The discrepancies between the solid and dashed lines may be reduced by adding to $\sigma^{(0)}(\xi)$ boundary layer functions localized in the middle and the ends of the fiber.

At the end of this section, we consider the "transverse" component $\sigma_\theta^{(0)}$ of the stress field inside a hard fiber. Taking into account the evaluations (5.49)

$$\sigma^{(0)} = \sigma_\theta^{(0)} + \sigma_\pi^{(0)} \,, \quad \sigma_\theta^{(0)} = O(1) \,, \quad \sigma_\pi^{(0)} = O\left(\delta_2^{-1}\right) \,, \tag{5.106}$$

we apply the projector Θ' to (5.42) and, as a result, obtain the equation

$$A^0 \sigma_\theta^{(0)}(\xi) + \left(\Theta' - A^0 C^0\right) C^{-1} \sigma_\pi^{(0)}(\xi) - \left(\delta_1^2 \ln \delta_1\right) \times$$

$$\times \Theta' A^1 D_\xi^2 \left[\alpha^2(\xi)\, \sigma_\pi^0(\xi)\right] = \Theta' \varepsilon^0(\xi) \,. \tag{5.107}$$

From Equation (5.51) for the Π'-component of the function $\sigma^{(0)}(\xi)$ we have

$$\delta_1^2 \ln \delta_1 D_\xi^2 \left[\alpha^2(\xi)\, \sigma_\pi^{(0)}(\xi)\right] = 2\mu_0 \left[\varepsilon_\pi^0(\xi) - E_m^{-1} \sigma_\pi^{(0)}(\xi)\right] . \tag{5.108}$$

Substituting this equation into (5.107) and solving the resulting equation with respect to $\sigma_\theta^{(0)}$ we find

$$\begin{aligned}\sigma_\theta^{(0)} = \left(A^0\right)^{-1} \varepsilon^0 - \left[\left(A^0\right)^{-1} - \Theta' C^0\right] C^{-1} \sigma_\pi^{(0)} + \\ + 2\mu_0 \left(A^0\right)^{-1} A^1 \left(\varepsilon_\pi^0 - E_m^{-1} \sigma_\pi^{(0)}\right) .\end{aligned} \tag{5.109}$$

Here the inversion of the tensor A^0 is to be performed in the Θ'-subspace.

For a constant external field, the axial component $\sigma^{\pi(0)}(\xi)$ of the stress tensor has the form

$$\sigma_{\alpha\beta}^{\pi(0)}(\xi) = f(\xi)\, E_m P^6_{\alpha\beta\lambda\mu}(m) \varepsilon^0_{\lambda\mu} , \tag{5.110}$$

where the function $f(\xi)$ depends on the shape of the fiber. For the fibers considered above these functions are defined by the expressions (Fig. 5.5)

Cylinder:

$$f(\xi) = 1 - \frac{\cosh q\,\xi}{\cosh q} , \tag{5.111}$$

Ellipsoid:

$$f(\xi) = \frac{q^2}{2 + q^2} , \tag{5.112}$$

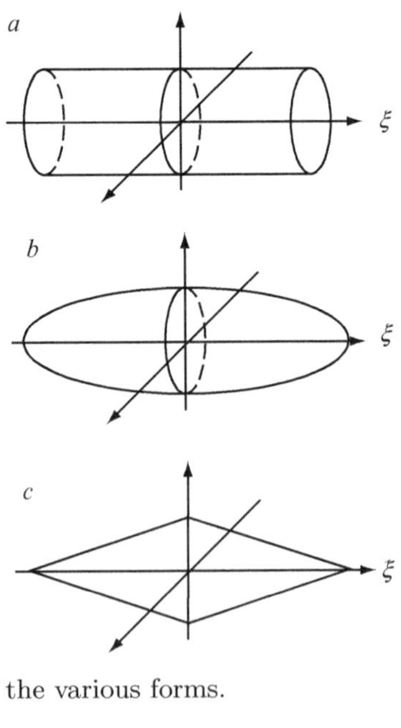

Fig. 5.5. Hard fibers of the various forms.

Double cone:

$$f(\xi) = \frac{q^2}{2-q^2}\left[1 - \frac{2}{2+\beta}(1-|\xi|)^{\beta}\right], \tag{5.113}$$

$$\beta = \frac{1}{2}\left(\sqrt{1+4q^2} - 3\right). \tag{5.114}$$

Substituting (5.110) into (5.109) we obtain the expression for the tensor $\sigma_{\theta}^{(0)}(\xi)$

$$\sigma^{(0)}(\xi) = \sigma_{\theta}^{(0)}(\xi) + \sigma_{\pi}^{(0)}(\xi) = \\ = \left[\left(A^0\right)^{-1} + \gamma(\xi)\left(P^3(m) + P^4(m)\right) + E_m f(\xi) P^6(m)\right]\varepsilon_0, \tag{5.115}$$

$$\left(A^0\right)^{-1} = \frac{\mu_0}{1-\kappa_0}P^2(m) + \frac{4\mu_0}{2-\kappa_0}\left(P^1(m) - \frac{1}{2}P^2(m)\right) + 8\mu_0 P^5(m), \tag{5.116}$$

$$\gamma(\xi) = \lambda_0 + \mu_0\left(1 - \frac{\mu}{\lambda+\mu}f(\xi)\right). \tag{5.117}$$

Here λ, μ are the Lamé parameters of the fiber material.

Equation (5.115) determines the principal terms of the stress field inside a hard fiber except near its ends, and the break points on the fiber surface.

5.4 Curvilinear fiber

Let the fiber axis Γ be a smooth curve and its transverse section $a(z)$ be a continuous piece-wise smooth function satisfying conditions (5.1), (5.2). The formal scheme of the construction of the principal terms of the stress field inside the hard fiber over the parameters δ_1, δ_2 is similar to that for an axisymmetric fiber. Supposing that the stress field is constant in the fiber cross sections $\Omega(z)$, $z \in \Gamma$, we integrate in (5.21) over the cross sections and consider this equation at the points of the curve Γ. In the dimensionless coordinates $\xi_i = x_i/l$, where $2l$ is the length of Γ, we obtain an equation for $\sigma(\xi)$ that is similar to (5.24)

$$C^{-1}\sigma(\xi) + \int_{\Gamma} \bar{\mathrm{K}}(\xi,\xi') C^1 C^{-1}\sigma(\xi')\, d\Gamma' = \varepsilon^0(\xi), \quad \xi \in \Gamma, \tag{5.118}$$

$$\bar{\mathrm{K}}(\xi,\xi') = \int_{\Omega(\xi')} \mathrm{K}(\xi - \xi' - \eta')\, d\eta', \tag{5.119}$$

where $\xi, \xi' \in \Gamma$, $\xi' + \eta'$ is the vector of the point in the fiber section $\Omega(\xi')$.

Omitting cumbersome calculations, we present the principal terms of the kernel $\bar{\mathrm{K}}(\xi, \xi')$ expansion in the series over the parameter δ_1:

$$\bar{\mathrm{K}}(\xi, \xi') = A^0(m)\delta(\xi - \xi') + \\ + \left(\delta_1^2 \ln \delta_1\right) A^1(m)\alpha^2(\xi') \frac{\mathrm{d}^2}{\mathrm{d}\xi^2}\delta(\xi - \xi') + O\left(\delta_1^2\right). \tag{5.120}$$

Here $m = m(\xi)$ is the unit vector tangential to Γ at the point ξ, the tensors $A^0(m)$ and $A^1(m)$ have forms (5.12), (5.41)

$$\delta_1 \alpha(\xi) = a(\xi)/l\,, \quad \delta_1 << 1\,, \quad \alpha(\xi) = O(1)\,. \tag{5.121}$$

Substituting (5.3) in (5.118), we obtain the following equation for the function $\sigma(\xi)$:

$$C^{-1}\sigma(\xi) + A^0(m)C^1C^{-1}\sigma(\xi) + \\ + \left(\delta_1^2 \ln \delta_1\right) A^1(m)\frac{\mathrm{d}^2}{\mathrm{d}\xi^2}\left[\alpha^2(\xi)\sigma(\xi)\right] = \varepsilon^0(\xi)\,. \tag{5.122}$$

For a hard fiber ($C^0C^{-1} = O(\delta_2)$, $\delta_2 << 1$), the function $\sigma(\xi)$ is evaluated as follows:

$$\sigma_{\alpha\beta}(\xi) = \sigma_m^{(0)}(\xi)\, m_\alpha(\xi)\, m_\beta(\xi) + O(1)\,, \quad \sigma_m^{(0)}(\xi) = O\left(\delta_2^{-1}\right). \tag{5.123}$$

The equation for the "axial" component $\sigma_m^{(0)}(\xi)$ of the stress field inside the hard fiber follows from (5.122) and has a form similar to (5.51)

$$\frac{\mathrm{d}^2}{\mathrm{d}\xi^2}\left[\alpha^2(\xi)\,\sigma_m^{(0)}(\xi)\right] - q^2\sigma_m^{(0)}(\xi) = -q^2\mathrm{E}_m\varepsilon_m^0(\xi)\,, \tag{5.124}$$

$$q^2 = -\frac{2\mu_0}{\mathrm{E}_m\delta_1^2 \ln \delta_1}\,, \quad \varepsilon_m^0(\xi) = \varepsilon_{\alpha\beta}^0(\xi)\, m_\alpha(\xi)\, m_\beta(\xi)\,. \tag{5.125}$$

Here the differentiation is performed with respect to the coordinate ξ along the middle line Γ of the curvilinear fiber.

The boundary conditions for Equation (5.124) follow from minimization of the discrepancy with the right-hand side of the initial integral Equation (5.1) after substituting the function $\sigma(\xi)$ from (5.123), (5.124) into its left-hand side . In particular, if $a(z)$ satisfies the condition (5.1) on Γ and $a(\xi_0) \neq 0, a(\xi_l) \neq 0$ (ξ_0, ξ_l are the points corresponding to the fiber ends), the conditions at the ends of the fiber take the form (5.79)

$$\sigma_m^{(0)}(\xi_0) = \sigma_m^{(0)}(\xi_l) = 0\,. \tag{5.126}$$

The external expansions of the stress and strain fields in the medium with a hard fiber follow from (5.3), (5.4). Taking into account the relations

$$C^1\varepsilon^+ = C^1C^{-1}\sigma^+ = \sigma^+ + O(\delta_2)\,, \tag{5.127}$$

we can write

$$\varepsilon(x) = \varepsilon^0(x) - \int_\Gamma \mathrm{K}(x-z)\bar{\sigma}(z)\mathrm{d}\Gamma_z\,, \tag{5.128}$$

$$\sigma(x) = \sigma^0(x) - \int_\Gamma S(x-z)(C^0)^{-1}\bar{\sigma}(z)\mathrm{d}\Gamma_z\,, \tag{5.129}$$

where the function $\bar{\sigma}(z)$ has the form

$$\bar{\sigma}(z) = s(z)\sigma^{(0)}(z)\,,\quad s(z) = \pi a^2(z)\,, \tag{5.130}$$

$$\sigma^{(0)}(z) = \sigma_\pi^{(0)}(z) + \sigma_\theta^{(0)}(z)\,, \tag{5.131}$$

$$\sigma_\pi^{(0)} = O(\delta_2^{-1})\,,\quad \sigma_\theta^{(0)} = O(1)\,. \tag{5.132}$$

Note that if the distance of the point x from the fiber ends exceeds its characteristic radius a, the contribution of the boundary layer functions appearing in the stress field are small compared to the component $\sigma^{(0)}(z)$ that changes slowly along Γ. Therefore, to calculate $\bar{\sigma}(z)$ in (5.128), (5.129) it is possible to use only the main term $\sigma^{(0)}(z)$ of the stress field inside the fiber obtained in the previous sections.

For a straight fiber and a constant external field, the stress and strain fields outside the fiber are determined by the same relations (5.128), (5.129), where the function $\bar{\sigma}(z)$ has the form

$$\bar{\sigma}(z) = \Lambda(z)\varepsilon^0,\quad \Lambda(z) = \pi a^2(z)B(z,m)\,, \tag{5.133}$$

$$B(z,m) = \left(A^0\right)^{-1} + \gamma(\xi)\left(P^3(m) + P^4(m)\right) + \mathrm{E}_m f\left(\xi\right)P^6(m)\,. \tag{5.134}$$

The tensor $(A^0)^{-1}$, and functions $\gamma(\xi)\,, f(\xi)$ are defined in (5.116), (5.117), (5.111)–(5.113).

5.5 Notes

The first solution for the medium reinforced with a hard cylindrical fiber was obtained in [32], where series of simplifications typical for the technical theory of elastic systems were accepted. A more exact approach to the solution of this problem was proposed in [28], [154]. Some mistakes of these works were corrected in [47].

This chapter is based on the works [81], [82].

6. Termal and electric fields in a medium with an isolated inclusion

Temperature and electric fields in composite materials are of interest in many applications. This chapter concerns the problem of calculation of these fields in a medium with an isolated inclusion. Fields in a homogeneous medium with an ellipsoidal inclusion, a thin inclusion, and a slander fiber are obtained and discussed. The methods for solving these problems are similar to those used in elasticity theory.

6.1 Fields with scalar potentials in a homogeneous medium with an isolated inclusion

Consider a homogeneous medium with properties described by a rank two tensor $C^0_{\alpha\beta}$. The medium contains a region V (inhomogeneity) with another property tensor $C_{\alpha\beta}$. The intensity $E(x)$ and the flux $D(x)$ of the field in the medium are described by the following system of differential equations:

$$\nabla_\alpha D_\alpha(x) = -q(x)\,, \quad D_\alpha(x) = C_{\alpha\beta}(x)E_\beta(x)\,, \quad \mathrm{rot}_{\alpha\beta}E_\beta(x) = 0\,, \tag{6.1}$$

where $q(x)$ is the scalar density of the field sources, $\mathrm{rot}_{\alpha\beta} = \epsilon_{\alpha\beta\lambda}\nabla_\lambda$, and $\epsilon_{\alpha\beta\lambda}$ is the Levi–Civita symbol. In (6.1), the first equation may be called the equilibrium equation, the second one the linear law connecting the intensity and flux vectors of the field. The third equation is satisfied automatically if the field $E_\alpha(x)$ is the gradient of a scalar function $\varphi(x)$, the potential of the field

$$E_\alpha(x) = \nabla_\alpha\varphi(x)\,. \tag{6.2}$$

The first two equations (6.1) give the equation for the potential $\varphi(x)$ in the form

$$\nabla_\alpha C_{\alpha\beta}(x)\nabla_\beta\varphi(x) = -q(x)\,. \tag{6.3}$$

The physical meaning of the potential $\varphi(x)$ depends on the problem under consideration. For heat transfer problems, $\varphi(x) = -T(x)$, and $T(x)$ is the temperature field, $D_\alpha(x)$ is the heat flux, and $C_{\alpha\beta}(x)$ is the tensor of the coefficients of thermoconductivity. For electrostatic fields in dielectric materials, $\varphi(x)$ is the potential of the electric field, $E_\alpha(x)$ is the vector of the electric field, $D_\alpha(x)$ is the electric displacement, and $C_{\alpha\beta}(x)$ is the tensor of the dielectric permittivity. For the problems of static elecroconductivity, $D_\alpha(x)$ is the current, and $C_{\alpha\beta}(x)$ is the tensor of electroconductivity.

Consider a medium with a finite inhomgeneity for which the tensor $C(x)$ is defined by the expression

$$C(x) = C^0 + C^1(x)\,. \tag{6.4}$$

Here C^0 is a constant tensor, $C^1(x)$ is a bounded function. As in Section 2.1, the vectors E and D are the solutions of the following integral equations:

$$E_\alpha(x) + \int \mathrm{K}_{\alpha\beta}(x-x')C^1_{\beta\mu}(x')E_\mu(x')\mathrm{d}x' = E^0_\alpha(x)\,, \tag{6.5}$$

$$D_\alpha(x) - \int S_{\alpha\beta}(x-x')B^1_{\beta\mu}(x')D_\mu(x')\mathrm{d}x' = D^0_\alpha(x)\,, \tag{6.6}$$

$$B^1 = B - B^0,\ \ B = C^{-1},\ \ B^0 = (C^0)^{-1}. \tag{6.7}$$

Here E^0 and D^0 are the fields in the medium without the inclusion ($C^1(x) = 0$) by the action of the same sources of the fields. The kernels $\mathrm{K}(x)$ and $\mathrm{S}(x)$ are determined by the equations similar to (3.12) and (3.22):

$$\mathrm{K}_{\alpha\beta}(x) = -\nabla_\alpha\nabla_\beta G(x)\,,\ \ \mathrm{S}_{\alpha\beta}(x) = C^0_{\alpha\lambda}\mathrm{K}_{\lambda\mu}(x)C^0_{\mu\beta} - C^0_{\alpha\beta}\delta(x)\,, \tag{6.8}$$

where $G(x)$ is the Green function for the infinite homogeneous medium with the property tensor C^0. This function satisfies the equation

$$\nabla_\alpha C^0_{\alpha\beta}\nabla_\beta G(x) = -\delta(x)\,. \tag{6.9}$$

The solution of this equation for an arbitrary anisotropic medium takes the form

$$G(x) = \frac{1}{4\pi\bar{r}(x)}\,,\ \ \bar{r}(x) = \sqrt{\det C^0 x_\alpha B^0_{\alpha\beta}x_\beta}\,. \tag{6.10}$$

The properties of the integral operators K and S in (6.5) and (6.6) are similar to those of the kernels in equation (3.11) and (3.23) for the elastic fields in a medium with an inclusion. Equations (6.5), (6.6) are equivalent to the original system (6.1). These equations have lower tensor dimensions than the equations of elasticity theory, and this fact simplifies the construction of their solutions.

6.2 Ellipsoidal inhomogeneity

6.2.1 Constant external field

The polynomial conservation theorem for an ellipsoidal inclusion with constant properties holds for equations (6.5) and (6.7) of the "scalar" theory. Therefore, if the external field E^0_α is constant, the field inside the ellipsoidal inclusion is also constant and determined by the expressions

$$E^+_\alpha = \Lambda^E_{\alpha\beta} E^0_\beta \,, \quad \Lambda^E_{\alpha\beta} = (\delta_{\alpha\beta} + A^0_{\alpha\lambda} C^1_{\lambda\beta})^{-1}, \tag{6.11}$$

$$A^0_{\alpha\beta} = \frac{1}{4\pi} \int_\Omega \mathrm{K}^*_{\alpha\beta}(a^{-1}k) \mathrm{d}\,\Omega \,, \quad \mathrm{K}^*_{\alpha\beta}(k) = \frac{k_\alpha k_\beta}{k_\lambda C^0_{\lambda\mu} k_\mu} \,. \tag{6.12}$$

If the medium is isotropic ($C^0_{\alpha\beta} = c_0 \delta_{\alpha\beta}$, $C^1_{\alpha\beta} = c_1 \delta_{\alpha\beta}$), and the inclusion is an ellipsoid with the semiaxes a_1, a_2, a_3, the tensor $A^0_{\alpha\beta}$ takes the form

$$A^0_{\alpha\beta} = A^0_1 e^1_\alpha e^1_\beta + A^0_2 e^2_\alpha e^2_\beta + A^0_3 e^3_\alpha e^3_\beta \tag{6.13}$$

$$A^0_k = \frac{a_1 a_2 a_3}{2c_0} \int_0^\infty \frac{\mathrm{d}\sigma}{(a_k^2 + \sigma)\sqrt{(a_1^2+\sigma)(a_2^2+\sigma)(a_3^2+\sigma)}}, \tag{6.14}$$

where e^k_α are the unit vectors of the ellipsoid principal axes, $k = 1, 2, 3$.

For a spheroidal inclusion with semi-axes $a_1 = a_2 = a, a_3$, the tensor $A^0_{\alpha\beta}$ in (6.13) takes the form

$$A^0_{\alpha\beta} = A^0_1 \theta_{\alpha\beta} + A^0_3 m_\alpha m_\beta, \quad \theta_{\alpha\beta} = \delta_{\alpha\beta} - m_\alpha m_\beta, \tag{6.15}$$

$$A^0_1 = \frac{1}{c_0} f_0(\gamma), \quad A^0_3 = \frac{1}{c_0}(1 - 2f_0(\gamma)), \quad \gamma = \frac{a}{a_3}, \tag{6.16}$$

and the function $f_0(\gamma)$ is defined in (3.96). For a spherical inclusion, the equations for the fields inside the inclusion take the forms

$$E^+_\alpha = \frac{3c_0}{3c_0 + c_1} E^0_\alpha, \quad D^+_\alpha = \frac{3c}{3c_0 + c_1} D^0_\alpha, \quad D^0_\alpha = c_0 E^0_\alpha. \tag{6.17}$$

6.2.2 Linear external field

Suppose that a medium with an ellipsoidal inhomogeneity is subjected to a linear external field

$$E^0_\alpha(x) = b^0_{\alpha\beta} x_\beta \,, \tag{6.18}$$

where b^0 is a constant tensor. In this case, the solution of Equation (6.5)

$$E^+_\alpha(x) + \int_V \mathrm{K}_{\alpha\beta}(x - x') C^1_{\beta\lambda} E^+_\lambda(x') \mathrm{d}\, x' = b^0_{\alpha\beta} x_\beta \tag{6.19}$$

is linear

$$E^{+}_{\alpha} = b^{1}_{\alpha\beta}x_{\beta}\,. \tag{6.20}$$

If we substitute this expression into (6.19), differentiate both parts with respect to x, and put $x = 0$, we obtain a linear algebraic equation for the constant tensor $b^{1}_{\alpha\beta}$

$$b^{1}_{\alpha\beta} + A^{1}_{\alpha\beta\lambda\mu}C^{1}_{\lambda\nu}b^{1}_{\nu\mu} = b^{0}_{\alpha\beta}\,, \tag{6.21}$$

$$A^{1}_{\alpha\beta\lambda\mu} = -\int\limits_{V}[\nabla_{\beta}\mathrm{K}_{\alpha\lambda}(x)]x_{\mu}\mathrm{d}\,x\,. \tag{6.22}$$

Using Gauss' theorem and Parseval's formula (see Appendix C.1), the equation for the tensor A^{1} is transformed into

$$A^{1}_{\alpha\beta\lambda\mu} = 3\mathrm{K}^{1}_{\alpha\beta\nu\rho}(a^{-1})_{\nu\lambda}a_{\mu\rho}\,, \tag{6.23}$$

$$\mathrm{K}^{1}_{\alpha\beta\lambda\mu} = \frac{1}{4\pi}\int\limits_{\Omega_1}\mathrm{K}^{*}_{\alpha\lambda}(a^{-1}k)k_{\beta}k_{\mu}\mathrm{d}\,\Omega\,. \tag{6.24}$$

For an isotropic medium and a spherical inclusion, these equations give

$$A^{1} = \frac{1}{5c_0}(E^{2} + 2E^{1})\,, \tag{6.25}$$

where the tensors E^{1} and E^{2} are the elements of the E-basis (Appendix A.1). Substituting this expression for A^{1} into (6.21) and solving this equation with respect to b^{1} we go to the equations for the symmetric $b^{1}_{(\alpha\beta)}$ and skewsymmetric $b^{1}_{[\alpha\beta]}$ parts of the tensor b^{1}

$$b^{1}_{(\alpha\beta)} = \Lambda^{1}_{\alpha\beta\lambda\mu}b^{0}_{\lambda\mu}\,,\quad b^{1}_{[\alpha\beta]} = b^{0}_{[\alpha\beta]}\,, \tag{6.26}$$

$$\Lambda^{1} = \frac{c_0}{3c}E^{2} + \frac{5c_0}{2c+3c_0}\left(E^{1} - \frac{1}{3}E^{2}\right)\,,\quad c = c_0 + c_1\,. \tag{6.27}$$

These equations show that the external field $b^{0}_{\alpha\beta}x_{\beta} = E^{0}_{\alpha}(x)$ is not perturbed by a spherical inhomogeneity if $b^{0}_{\alpha\beta}$ is a skewsymmetric tensor.

If the medium is isotropic and there are no field sources, the external field $E^{0}_{\alpha}(x)$ satisfies equation $\nabla_{\alpha}E^{0}_{\alpha}(x) = 0$. As a result, the tensor b^{0} in (6.18) satisfies the additional condition $\operatorname{tr} b^{0} = b^{0}_{\alpha\alpha} = 0$. This implies that the equation for the symmetric part of the tensor b^{1} is simplified and takes the form

$$b^{1}_{\alpha\beta} = \frac{5c_0}{5c_0 + 2c_1}\,b^{0}_{(\alpha\beta)}\,. \tag{6.28}$$

6.2.3 Spheroidal inhomogeneity in a transversely isotropic medium

If the medium is transversally isotropic, its property tensor $C^0_{\alpha\beta}$ takes the form

$$C^0_{\alpha\beta} = c^0_{11}\theta_{\alpha\beta} + c^0_{33}m_\alpha m_\beta, \tag{6.29}$$

where m_α is the unit vector along the isotropy axis. Let the medium contain a spheroidal inhomogeneity ($a_1 = a_2 = a, a_3$, $a/a_3 = \gamma$), and the spheroid semiaxis a_3 be directed along the vector m (the x_3-axis). In this case, the integral $A^0_{\alpha\beta}$ in (6.11) is calculated in the explicit form

$$A^0_{\alpha\beta} = A^0_1\theta_{\alpha\beta} + A^0_3 m_\alpha m_\beta \tag{6.30}$$

$$A^0_1 = \frac{\lambda^2}{2c^0_{11}}\left[1 - \frac{\gamma^2 c^0_{33}}{2c^0_{11}}\lambda\ln\left(\frac{\lambda+1}{\lambda-1}\right)\right] \tag{6.31}$$

$$A^0_3 = \frac{(\gamma\lambda)^2}{c^0_{11}}\left[\frac{1}{2}\lambda\ln\left(\frac{\lambda+1}{\lambda-1}\right) - 1\right], \quad \lambda = \left(1 - \gamma^2\frac{c^0_{33}}{c^0_{11}}\right)^{-\frac{1}{2}} \tag{6.32}$$

Note that these expressions arc valid for a complex λ.

6.3 Multilayered spherical inclusion in a homogeneous medium

In this section, we consider an isotropic medium that contains a spherical isotropic inclusion for which conductivity c is an arbitrary function of radial distance from its center. If the external field E^0_i is constant, the field inside the inclusion is presented in the form

$$E_i(r,n) = E^0_i + A_{ij}(r,n)E^0_j, \quad r = |x|, \ n = \frac{x}{r}. \tag{6.33}$$

Here $(r, n(\theta,\varphi))$ are spherical coordinates of a point x in the system with the origin at the center of the inclusion, $r = |x|$, $n = x/|x|$, φ, θ are polar and azimuthal angles. The tensor A satisfies an equation that follows from (6.5):

$$A + KC^1A = -KC^1, \tag{6.34}$$

$$C^1_{ij}(x) = c_1(r)\delta_{ij}, \quad c_1(r) = c(r) - c_0, \tag{6.35}$$

and c_0 is the property of the medium, $c(r)$ is the property of the inclusion.

Action of the integral operator K in (6.34) on a finite function $\psi(x)$ is defined by the equation (see Appendix B.4)

$$K\psi(r,n) = \left(M_s^{-1}K_sM_s\psi\right)(r,n), \tag{6.36}$$

where M_s is the operator of the Mellin transform. We denote by $\psi^*(s,n)$ the Mellin transform of the function $\psi(r,n)$ with respect to the r-variable

$$\psi^*(s,n) = (M_s\psi)(s,n) = \int_0^\infty r^{s-1}\psi(r,n)\mathrm{d}r, \tag{6.37}$$

$$\psi(r,n) = \left(M_s^{-1}\psi^*\right)(r,n) = -\frac{i}{2\pi}\int_{\tau-i\infty}^{\tau+i\infty} r^{-s}\psi^*(s,n)\mathrm{d}r. \tag{6.38}$$

The operator K_s acts on the Mellin transform of the function ψ according the equation

$$K_s\psi^*(s,n) = -\frac{i}{(2\pi)^3}\Gamma(3-s)\Gamma(s)\int_{\Omega_1}(-l\cdot n)^{-s}K^*(n)\,dn \times \int_{\Omega_1}(n\cdot m)^{s-3}\psi^*(s,m)\,\mathrm{d}m, \tag{6.39}$$

where $K^*(n)$ is the symbol of the operator K

$$K^*(n) = \int K(x)e^{ik\cdot x}\mathrm{d}x = \frac{1}{c_0}n\otimes n,\ \ n = \frac{k}{|k|}. \tag{6.40}$$

Note that the tensor H

$$H = \frac{1}{3-s}\left(\mathbf{1} - sn\otimes n\right) \tag{6.41}$$

is the eigen-tensor of the operator K_s. Direct calculation using (6.39) gives

$$K_sH = \frac{1}{c_0}H. \tag{6.42}$$

Actions of the operator K_s on the unit tensor $\mathbf{1}$ and the tensor $n\otimes n$ are defined by the equations

$$K_s\mathbf{1} = \frac{1}{c_0}H,\quad K_s\left(n\otimes n\right) = \frac{s-2}{c_0 s}H. \tag{6.43}$$

It follows from these equation that the Mellin transform of the tensor $A(r,n)$ over the variable r may be found in the form

$$A^*(s,n) = \left(\mathbf{1} - sn\otimes n\right)\alpha^*(s), \tag{6.44}$$

where $\alpha^*(s)$ is the Mellin transform of the scalar function $\alpha(r)$. As a result, the r-rcpresentation of the tensor $A(r,n)$ takes the form

$$A(r,n) = [\mathbf{1} + (n \otimes n)\, D]\, \alpha(r), \qquad D = r\frac{\mathrm{d}}{\mathrm{d}r}. \tag{6.45}$$

After substituting the product $c_1(r)A(r,n)$

$$c_1(r)A(r,n) = \beta_1(r)1 + \beta_2(r) n \otimes n, \tag{6.46}$$

$$\beta_1(r) = c_1(r)\alpha(r), \quad \beta_2(r) = c_1(r)D\alpha(r), \tag{6.47}$$

into (6.34) and applying the Mellin transform to the resulting equation we obtain

$$\left[(3-s)\alpha(s) + \frac{\beta_1(s)}{c_0} + \frac{\beta_2(s)(s-2)}{c_0 s}\right] H = -\frac{c_1(s)}{c_0} H. \tag{6.48}$$

Equating the coefficients in front of the tensor H in the right and left-hand sides of this equation and returning to the r-representation, we go to the differential equation for the function $\alpha(r)$:

$$D(3+D)\alpha(r) + \frac{1}{c_0} D\left(c_1 \alpha(r)\right) + \frac{1}{c_0}(2+D)\left(c_1(r)D\alpha(r)\right) = -\frac{1}{c_0} Dc_1(r). \tag{6.49}$$

If $c_1(r)$ is a smooth function, the solution of this equation is found from the following boundary conditions:

$$\alpha(0) < \infty, \quad \frac{\mathrm{d}\alpha}{\mathrm{d}r}|_{r=0} = 0, \quad \lim_{r\to\infty} \alpha(r) = 0. \tag{6.50}$$

If $c_1(r)$ is a stepwise function (many layered inclusion), in the regions where $c_1(r)$ is constant, (6.49) takes the form

$$D(3+D)\alpha = 0, \tag{6.51}$$

and the general solution of this equation is

$$\alpha(r) = Y_1 + Y_2 r^{-3}. \tag{6.52}$$

Here Y_1, Y_2 are constants that are different inside different layers.

As in Section 2.9, we can show that on the interfaces between the layers the following conditions hold:

$$[\alpha]_j = 0, \quad [cD\alpha]_j = -[c]_j (1 + \alpha(a_j)). \tag{6.53}$$

Here $r = a_j$ is the external border of the j-th layer,

$$[f]_j = f(a_j + 0) - f(a_j - 0). \tag{6.54}$$

Let us introduce the vector $X^i(r)$ with the components $X_1^i(r)$, $X_2^i(r)$ for the i-th layer

$$X_1^i(r) = \alpha(r) = Y_1^i + Y_2^i r^{-3}, \tag{6.55}$$

$$X_2^i(r) = D\alpha(r) = -3Y_2^i r^{-3}, \tag{6.56}$$

and the vector Y^i with constant components Y_1^i, Y_2^i. Connection between these vectors follows from (6.52) in the form

$$Y^i = h^i(a_i)X^i(a_i), \tag{6.57}$$

$$h^i(a_i) = \begin{bmatrix} 1, & 1/3 \\ 0, & -r^3/3 \end{bmatrix}. \tag{6.58}$$

The values of the vector $X^i(r)$ at the ends a_i and a_{i-1} of the i-th interval are connected by the equation

$$X^i(a_i) = R^i X^i(a_{i-1}), \tag{6.59}$$

where the transition matrix R^i has the form

$$R^i = \begin{bmatrix} 1, \frac{1}{3}\left[1 - \left(\frac{a_{i-1}}{a_i}\right)^3\right] \\ 0, \qquad \left(\frac{a_{i-1}}{a_i}\right)^3 \end{bmatrix}. \tag{6.60}$$

The solution in the $(i+1)$-th and the i-th layers are connected at the border a_i by an equation that follows from conditions (6.53):

$$X^{i+1}(a_i) = F^i + \Gamma^i X^i(a_i), \tag{6.61}$$

$$F^i = \begin{bmatrix} 0 \\ \frac{c_i - c_{i+1}}{c_{i+1}} \end{bmatrix}, \quad \Gamma^i = \begin{bmatrix} 1, & 0 \\ \frac{c_i - c_{i+1}}{c_{i+1}}, & \frac{c_i}{c_{i+1}} \end{bmatrix}. \tag{6.62}$$

Using Equations (6.59) and (6.61) of transition of the solution through the layer and through the border between the layers we obtain the connection between the solution inside the first layer (kernel) and in the $(i+1)$-the layer in the form

$$X^{i+1}(a_i) = g^i + G^i X^1(a_1), \tag{6.63}$$

$$g^i = F^i + \sum_{j=1}^{i-1} G_f(i,j)F^j, \quad i > 1 \;\; G^i = Q^i Q^{i-1}...Q^1, \tag{6.64}$$

$$Q^i = \Gamma^i R^i, \quad G_f(i,j) = G_f(i,j-1)Q^{i-j+1}, \tag{6.65}$$

$$G^1 = \Gamma^1, \quad g^1 = F^1. \tag{6.66}$$

Because the field $E(x)$ should be bounded in the center of the inclusion, (6.55) implies the equation

$$Y_2^1 = 0, \tag{6.67}$$

and for $\alpha \to 0$ at infinity, we obtain

$$Y_1^{N+1} = 0. \tag{6.68}$$

Here N is the total number of layers, index $N+1$ refers to the medium. The last equation gives us the equation for the constant Y_1^1

$$Y_1^1 = -\frac{g_2^N + 3g_1^N}{G_{21}^N + 3G_{11}^N}, \quad X^1 = \begin{bmatrix} Y_1^1 \\ 0 \end{bmatrix}. \tag{6.69}$$

Equations (6.69) present the solution inside the first layer, and using (6.63), (6.57), and (6.55) we obtain the solution inside all the layers. Thus, the problem is reduced to the multiplication 2×2 matrices. If the constants Y_1^i, Y_2^i are known, the field inside the i-th layer takes the form that follows from (6.45), (6.33)

$$E(r,n) = \left[(1 + Y_1^i)\mathbf{1} + (\mathbf{1} - 3n \otimes n)Y_2^i r^{-3}\right] \cdot E^0, \quad a_{i-1} < r < a_i, \tag{6.70}$$

where E^0 is an arbitrary constant external field, $i = 1, 2, ..., N+1$.

6.4 Thin inclusion in a homogeneous medium

Let a homogeneous medium with the property tensor $C^0_{\alpha\beta}$ contain an isolated inclusion with the property $C_{\alpha\beta}$. Suppose that one of the characteristic sizes of the region V occupied by the inclusion is much smaller than the other two. We denote this smallest size by h; if l is the order of the other sizes, the ratio $\delta_1 = h/l$ is small. Let us introduce the second (physical) dimensionless parameter δ_2 of the problem that is the ratio of the characteristic properties of the medium and the inclusion. We consider the cases when the parameter δ_2 is either small or large.

As in elasticity, the prime interest for applications are the principal terms of the asymptotic expansions of the fields $E_\alpha(x)$ and $D_\alpha(x)$ over the parameters δ_1, δ_2. In order to construct these terms, we use the same procedure of matching asymptotic expansions of the external and internal fields in the medium with the inclusion described in detail in Chapter 4. We present only the final results, and omit details of the derivations that are similar to those in the elastic case.

Consider a thin inclusion with thermo- or electroconductivity much smaller than the conductivity of the medium. In this case, the fields $E_\alpha(x)$ and $D_\alpha(x)$ in the medium are presented in the forms

$$E_\alpha(x) = E^0_\alpha(x) + \int_\Omega K_{\alpha\beta}(x - x')C^0_{\beta\lambda}n_\lambda(x')b(x')\mathrm{d}\Omega', \tag{6.71}$$

$$D_\alpha(x) = D^0_\alpha(x) - \int_\Omega S_{\alpha\beta}(x - x')n_\beta(x')b(x')\mathrm{d}\Omega', \tag{6.72}$$

where Ω is the middle surface of the inclusion, $n_\alpha(x)$ is the normal to this surface, and the scalar function $b(x)$ satisfies the equation

$$\lambda(x)b(x) + \int_\Omega T(x, x')b(x')\mathrm{d}\Omega' = n_\alpha(x)D^0_\alpha(x), \tag{6.73}$$

$$T(x, x') = -n_\alpha(x)S_{\alpha\beta}(x - x')n_\beta(x'). \tag{6.74}$$

The coefficient $\lambda(x)$ depends on the properties $C_{\alpha\beta}$ and the transverse size $h(x)$ of the inclusion

$$\lambda(x) = \frac{1}{h(x)}n_\alpha(x)C_{\alpha\beta}n_\beta(x), \tag{6.75}$$

and $n(x)$ is the normal to Ω at the point x. The action of the integral operator with kernel $T(x, x')$ on smooth functions is defined by the same equation as for elasticity (Section 4.2).

Consider a thin inclusion with conductivity much greater than the conductivity of the medium. In this case, the principal terms of the fields in the medium have the forms

$$E_\alpha(x) = E^0_\alpha(x) + \int_\Omega K_{\alpha\beta}(x - x')\zeta_\beta(x')\mathrm{d}\Omega', \tag{6.76}$$

$$D_\alpha(x) = D^0_\alpha(x) + \int_\Omega S_{\alpha\beta}(x - x')B^0_{\beta\lambda}\zeta_\lambda(x')\mathrm{d}\Omega', \tag{6.77}$$

where $\zeta_\alpha(x)$ is the vector of the middle surface Ω of the inclusion that satisfies the equation

$$\theta_{\alpha\beta}(x)\zeta_\beta(x) = \zeta_\alpha(x), \quad \theta_{\alpha\beta}(x) = \delta_{\alpha\beta} - n_\alpha(x)n_\beta(x). \tag{6.78}$$

The vector $\zeta_\alpha(x)$ is the solution of the following integral equation

$$\mu_{\alpha\beta}(x)\zeta_\beta(x) + \int_\Omega U_{\alpha\beta}(x, x')\zeta_\beta(x')\mathrm{d}x' = \theta_{\alpha\beta}(x)E^0_\beta(x), \tag{6.79}$$

$$U_{\alpha\beta}(x, x') = \theta_{\alpha\lambda}(x)K_{\lambda\mu}(x - x')\theta_{\mu\beta}(x'), \tag{6.80}$$

where the function $\mu_{\alpha\beta}(x)$ is

$$\mu_{\alpha\beta}(x) = \frac{1}{h(x)}\theta_{\alpha\lambda}(x)B_{\lambda\mu}\theta_{\mu\beta}(x). \tag{6.81}$$

The regularization of the integral operator with the kernel $U_{\alpha\beta}(x,x')$ in (6.79) is similar to that given in Section 4.4.

For a thin ellipsoidal inclusion, exact solutions of equations (6.73) and (6.79) may be found for constant external fields E^0_α and D^0_α. Let the semiaxes of a thin ellipsoidal inclusion be a_1, a_2, h $(h/a_1, h/a_2 \ll 1)$. Then, Ω is a plane elliptical surface with a constant normal n. In the Cartesian coordinates connected with the ellipsoid axes, the function $h(x)$ is presented in the form

$$h(x) = 2hz(x), \quad z(x) = \sqrt{1 - \left(\frac{x_1}{a_1}\right)^2 - \left(\frac{x_2}{a_2}\right)^2}. \tag{6.82}$$

The kernel of the integral operator in (6.73) takes the form

$$T(x,x') = T(x-x') = -n_\alpha S_{\alpha\beta}(x-x')n_\beta \tag{6.83}$$

The integral operator with such a kernel transforms the function $z(x)$ into a constant on Ω (the analogy of the polynomial conservation theorem for an elliptical crack). It allows us to find the solution of (6.73) for $D^0_\alpha = \text{const}$ in the form of a constant b multiplied by the function $z(x)$:

$$b(x) = bz(x). \tag{6.84}$$

Substituting (6.84) into (6.73) we obtain

$$b = \frac{2a_1^2}{a_2} B n_\alpha D^0_\alpha, \tag{6.85}$$

$$B = \frac{a_2}{2a_1^2}\left(\lambda_0 + T_0\right)^{-1}, \quad \lambda_0 = \frac{1}{2h} n_\alpha C_{\alpha\beta} n_\beta, \tag{6.86}$$

and the constant T_0 is the absolutely convergent integral

$$T_0 = \int n_\alpha S_{\alpha\beta}(x) n_\beta \left[z(x) - 1\right] \mathrm{d}\Omega. \tag{6.87}$$

Integration is performed over all the infinite plane that includes the surface Ω. The function $z(x)$ is continued by zero outside Ω.

For an isotropic inclusion and the matrix ($C_{\alpha\beta} = c\delta_{\alpha\beta}$, $C^0_{\alpha\beta} = c_0\delta_{\alpha\beta}$), $\lambda_0 = c/2h$, and the scalar T_0 is determined by the equation

$$T_0 = c_0^2 m_\alpha K^0_{\alpha\beta} m_\beta, \quad K^0_{\alpha\beta}(x) = \int K_{\alpha\beta}(x)\left[z(x) - 1\right] \mathrm{d}\Omega. \tag{6.88}$$

The explicit expression for the tensor $K_{\alpha\beta}(x)$ defined in (6.8) is

$$K_{\alpha\beta}(x) = \frac{1}{4\pi c_0 \left|x\right|^3}\left(\delta_{\alpha\beta} - \frac{3x_\alpha x_\beta}{\left|x\right|^2}\right). \tag{6.89}$$

Substituting this expression into (6.88) and introducing the coordinates r, φ in the plane x_1x_2: $x_1 = ra_1\cos\varphi,\ x_2 = ra_2\sin\varphi$ (a_1, a_2 are the semiaxes of the ellipse Ω), we obtain

$$K^0_{\alpha\beta} = -\frac{a_1a_2}{4\pi c_0}\int_0^\infty \frac{z(x)-1}{r^2}\mathrm{d}r \int_0^{2\pi}\left[\frac{3m_{\alpha\beta}(\varphi)}{t^2(\varphi)} - \delta_{\alpha\beta}\right]\frac{\mathrm{d}\varphi}{t^3(\varphi)}, \tag{6.90}$$

$$t^2(\varphi) = a_1^2\cos^2\varphi + a_2^2\sin^2\varphi, \tag{6.91}$$

$$m_{11} = a_1^2\cos^2\varphi, \quad m_{22} = a_2^2\sin^2\varphi, \quad m_{12} = m_{21} = a_1a_2\sin\varphi\cos\varphi. \tag{6.92}$$

$$z(r) = 1 - r^2 \quad (r \le 1), \quad z(r) = 0 \quad (r > 1). \tag{6.93}$$

Calculating the integrals in (6.90), we find

$$K^0_{11} = \frac{a_2}{2a_1^2c_0}\frac{K(k) - E(k)}{k}, \quad k = 1 - \left(\frac{a_2}{a_1}\right)^2, \quad a_1 \ge a_2 \tag{6.94}$$

$$K^0_{12} = 0, \quad K^0_{22} = \frac{a_2}{2a_1^2c_0}\frac{E(k) - (1-k)K(k)}{k(1-k)} \tag{6.95}$$

$$K^0_{13} = K^0_{23} = 0, \quad K^0_{33} = \frac{a_2E(k)}{2a_1^2c_0\,(1-k)}. \tag{6.96}$$

Here $K(k)$ and $E(k)$ are the complete elliptic integrals of the first and the second kind, respectively.

Hence, for an inclusion with a small conductivity we have

$$T_0 = \frac{c_0a_2}{2a_1^2}\frac{E(k)}{1-k}, \tag{6.97}$$

$$B = \frac{1}{c_0}\left[\frac{c}{c_0}\frac{a_2}{h} + \frac{E(k)}{1-k}\right]^{-1}. \tag{6.98}$$

After determination of the scalar $b(x)$, the fields $E_\alpha(x)$ and $D_\alpha(x)$ outside a thin inclusion with small conductivity can be expressed as follows:

$$E_\alpha(x) = E^0_\alpha - \int_\Omega K_{\alpha\beta}(x-x')c_0\Lambda^0_{\beta\lambda}Z(x')D^0_\lambda\mathrm{d}x', \tag{6.99}$$

$$D_\alpha(x) = D^0_\alpha - \int_\Omega S_{\alpha\beta}(x-x')\Lambda^0_{\beta\lambda}Z(x')D^0_\lambda\mathrm{d}x', \tag{6.100}$$

$$\Lambda^0_{\alpha\beta} = Bn_\alpha n_\beta, \quad Z(x) = \frac{2a_1^2}{a_2}z(x). \tag{6.101}$$

Let us consider Equation (6.79) for a thin inclusion with high conductivity. If $E^0_\alpha = \text{const}$, we find the solution of this equation in the form

$$\zeta_\alpha(x) = \zeta_\alpha z(x), \tag{6.102}$$

where ζ_α is a constant vector. Substituting this expression in (6.79), we obtain

$$\zeta_\alpha = \frac{2a_1^2}{a_2} d_{\alpha\beta}\theta_{\beta\lambda}e^0_\lambda, \tag{6.103}$$

$$d_{\alpha\beta} = \frac{a_2}{2a_1^2}\left(\mu^0_{\alpha\beta} + U^0_{\alpha\beta}\right)^{-1}, \quad \mu^0_{\alpha\beta} = \frac{1}{2h}\theta_{\alpha\lambda}(n)B_{\lambda\mu}\theta_{\mu\beta}(n), \tag{6.104}$$

and the constant tensor $U^0_{\alpha\beta}$ is expressed via the convergent integral

$$U^0_{\alpha\beta} = \int \theta_{\alpha\lambda}(n)K_{\lambda\mu}(x)\theta_{\mu\beta}(n)\left[z(x) - 1\right]\mathrm{d}\Omega. \tag{6.105}$$

For the isotropic medium and inclusion ($B_{\alpha\beta} = c^{-1}\delta_{\alpha\beta},\ B^0_{\alpha\beta} = c_0^{-1}\delta_{\alpha\beta}$), we obtain

$$\mu^0_{\alpha\beta} = \frac{1}{2hc}\theta_{\alpha\beta}, \tag{6.106}$$

and the tensor $U^0_{\alpha\beta}$ is

$$U^0_{\alpha\beta} = \theta_{\alpha\lambda}(n)K^0_{\lambda\mu}\theta_{\mu\beta}(n). \tag{6.107}$$

Taking into account (6.94) we obtain the components of the tensor $U^0_{\alpha\beta}$ in the form

$$U^0_{11} = \frac{a_2}{2a_1^2 c_0}\frac{K(k) - E(k)}{k}, \quad U^0_{12} = 0, \tag{6.108}$$

$$U^0_{22} = \frac{a_2}{2a_1^2 c_0}\frac{E(k) - (1-k)K(k)}{k(1-k)}. \tag{6.109}$$

After determination of the vector $\zeta_\alpha(x)$ one can calculate the fields $E(x)$ and $D(x)$ in the medium from the equations

$$E_\alpha(x) = E^0_\alpha - \int_\Omega K_{\alpha\beta}(x - x')\Lambda^0_{\beta\lambda}Z(x')D^0_\lambda \mathrm{d}x', \tag{6.110}$$

$$D_\alpha(x) = D^0_\alpha - \int_\Omega S_{\alpha\beta}(x - x')c_0^{-1}\Lambda^0_{\beta\lambda}Z(x')D^0_\lambda \mathrm{d}x', \tag{6.111}$$

where the constant tensor $\Lambda^0_{\beta\lambda}$ has the form

$$\Lambda^0_{\alpha\beta} = \theta_{\alpha\lambda}(n)d_{\lambda\mu}\theta_{\mu\beta}(n). \tag{6.112}$$

6.5 Axisymmetric fiber in a homogeneous media

Consider a slender fiber with a property (e.g., the conductivity) much larger than the property of the matrix. The integral equation for the field flux inside the inclusion may be written in the form

$$C^{-1}D(x)+\int_V K(x-x')C^1C^{-1}D(x')\mathrm{d}x'=E^0(x). \tag{6.113}$$

Here V is the region occupied by the inclusion.

We suppose that the field does not change essentially across the cross section of the fiber and write

$$D(x_1,x_2,x_3)=D(x_3). \tag{6.114}$$

Here x_3 is the coordinate along the fiber axis, and the fiber occupies the region $-l < x_3 < l$. Integrating over the cross section of the fiber leads to the following one-dimension integral equation for the field $D(x_3)$

$$C^{-1}D(\xi)+\int_{-1}^{1}\overline{K}(\xi,\xi')C^1C^{-1}D(\xi')\mathrm{d}\xi'=E^0(\xi). \tag{6.115}$$

Here the following dimensionless coordinates are introduced

$$\eta_i=\frac{x_i}{l},\quad i=1,2;\ \xi=\frac{x_3}{l}, \tag{6.116}$$

and $\overline{K}(\xi,\xi')$ is the following integral

$$\overline{K}(\xi,\xi')=\frac{1}{(2\pi)^3}\int_{\Omega(\xi')}\mathrm{d}\eta\int K^*(k)\exp[-ik\cdot((\xi-\xi')m-\boldsymbol{\eta})]\mathrm{d}k. \tag{6.117}$$

After integrating over the circular cross section $\Omega(\xi)$ we reach the following equation for the kernel $\overline{K}(\xi,\xi')$:

$$\overline{K}(\xi,\xi')=\frac{\overline{a}^2(\xi')}{(2\pi)^2}\int\frac{k\otimes k}{c_0k^2}\exp[-ik\cdot((\xi-\xi')m)]\frac{J_1(ka(\xi'))}{ka(\xi')}\mathrm{d}k$$
$$=\frac{1}{4\pi c_0}\int_{-\infty}^{\infty}[\overline{a}|k_3|K_1(\overline{a}|k_3|)\theta(m)+$$
$$+(1-\overline{a}|k_3|K_1(\overline{a}|k_3|))m\otimes m]\exp[-ik_3(\xi-\xi')]\mathrm{d}k_3. \tag{6.118}$$

Here K_1 is the modified Bessel function, $\theta(m)=\mathbf{1}-m\otimes m$ is the projector on the plane orthogonal to the axis of the fiber. Expanding the function

$K_1(\overline{a}|k_3|)$ in a series with respect to the small parameters $\overline{a}(\xi) = a(\xi)/l$ ($a(\xi)$ is the radius of the cross section) we obtain

$$\overline{K}(\xi,\xi') = \frac{1}{4\pi c_0}\int_{-\infty}^{\infty}\left[\theta(\mathbf{m}) - \overline{a}^2\ln(\overline{a})\,k_3^2 m\otimes m\right]\exp[-ik_3(\xi-\xi')]\mathrm{d}k_3$$

$$= \frac{1}{2c_0}\delta(\xi-\xi')\theta(m) + \frac{1}{2c_0}\overline{a}^2\ln(\overline{a})\frac{\mathrm{d}^2}{\mathrm{d}\xi}\delta(\xi-\xi')m\otimes m + \cdots . \tag{6.119}$$

Keeping only the two first terms of this expansion we find that the integral Equation (6.115) for the tensor $D(\xi)$ splits into two equations for the projection θD of D on the cross section of the fiber, and the flux along the fiber $m \cdot D = D^m$

$$\theta D + 2\frac{c_1}{c_0}\theta D = \theta D^0, \tag{6.120}$$

$$\frac{\mathrm{d}^2}{\mathrm{d}\xi^2}\left[\alpha^2(\xi)D^m(\xi)\right] - q^2 D^m(\xi) = -q^2\frac{c}{c_0}D^{0m}(\xi). \tag{6.121}$$

Here the function $\overline{a}(\xi)$ is presented in the form

$$\overline{a}(\xi) = \delta_1\alpha(\xi), \quad \delta_1 \ll 1, \ \alpha(\xi) = O(1), \tag{6.122}$$

$$q^2 = -\frac{2c_0}{c_1\delta_1^2\ln(\delta_1)}. \tag{6.123}$$

The field flux $D(\xi)$ inside the fiber is

$$D(\xi) = \frac{2c_0}{c_0+c}D^0\theta(m) + D^m(\xi)m\otimes m. \tag{6.124}$$

If $c_0/c = \delta_2 << 1$, the axial part $D^m(\xi)$ of $D(\xi)$ is much greater than the transverse part.

The boundary conditions for the differential Equation (6.121) are to be obtained from the minimization of the discrepancy in the solutions of the integral Equation (6.115) and (6.121). In particular, for the ends of finite radii, these conditions take the forms

$$D^m(1) = D^m(-1) = 0 \tag{6.125}$$

The solutions of (6.121) for a cylindrical fiber, a prolate ellipsoid, and a fiber in the form of a double cone are presented in Section 5.3.

7. Homogeneous elastic medium with a set of isolated inclusions

In this chapter, we consider the equilibrium of an infinite homogeneous elastic medium containing a set of isolated inclusions (matrix composite). The prime objective of the chapter is the solution of the homogenization problem, and the determination of effective elastic properties of the matrix composites with various types of inclusions. For this purpose, self-consistent schemes (versions of the effective medium and effective field methods) are developed. The predictions of the methods are compared with the exact solutions for regular composites, and with experimental data for composites with random microstructures. The areas of applications of various self-consistent methods are analyzed and discussed.

7.1 The homogenization problem

The heterogeneous materials that are considered in this chapter are called matrix composites. Such materials consist of a homogeneous background phase (matrix) and a set of isolated inclusions of other phases. As a rule, the sizes and properties of the inclusions, as well as their spatial distributions, are random. The problem of calculation of elastic, thermoelastic, dielectric, conductive, and other physical properties of heterogeneous materials has specific features, and differs from the classical problems of mathematical physics. First, the characteristic scale of fluctuations of physical properties in composite materials is usually much smaller than the total size of the macroscopic specimen. Random functions that describe these properties are homogeneous in the broad sense [172]. Therefore, neglecting specific boundary effects, the main information about physical fields in random heterogeneous materials can be obtained from consideration of an infinite medium. The physical properties of such a medium are statistically homogeneous random functions of coordinates.

It is known [55], [172] that complete information about a random function is contained in its characteristic functional, or in infinite set of its multipoint statistical moments. For instance, the complete solution of the problem of elasticity for a random (stochastic) composite is an infinite number of multipoint statistical moments of the random displacement vector $u(x)$, strain $\varepsilon(x)$ and stress $\sigma(x)$ tensors expressed via the statistical moments of the elastic moduli tensor $C(x)$ and given, nonrandom external loading (surface forces, body forces $q(x)$, temperature field $T(x)$, etc.). To find a complete solution the problem of elasticity for a stochastically inhomogeneous medium is possible only in very special cases. The main difficulty here is that in spite of linearity of the initial system of differential equations, the considered problem is stochastically nonlinear (see Introduction). For elasticity, this system has the form

$$(L_{\alpha\beta}u_{\beta})(x) = q_{\alpha}(x), \quad L_{\alpha\beta} = -\nabla_{\lambda}C_{\lambda\alpha\beta\mu}(x)\nabla_{\mu}. \tag{7.1}$$

On the left-hand side of this system,there is the product of the statistically dependent random functions $C(x)$ and $\nabla u(x)$. Due to this nonlinearity, every statistical moment of a finite order of the random functions $u(x)$ or $\varepsilon(x)$ is expressed via an infinite set of the moments of higher orders. Therefore, to make the problem solvable, one has to introduce some simplifying assumptions which specific forms depend on peculiarities of the composite microstructure. Note that complete information about random physical fields in a heterogenous medium is unnecessary. As a rule, only the first few statistical moments of these fields may have interest for engineering applications.

The most important characteristic of a physical field in a random heterogeneous material is its first statistical moment, the mathematical expectation. The first moment characterizes average response of the microheterogeneous material to external loading. If such a response is known, the actual composite material may be replaced by some homogeneous material with the equivalent reaction to external loading. Construction of a model of a homogeneous medium that is equivalent to the actual heterogeneous composite material is the central problem of the theory of composites (homogenization problem). The solution of this problem allows one to calculate macroscopic properties of the composite material in terms of the properties of the phases and peculiarities of spatial distributions of the inclusions in the matrix phase. Knowledge of such relations is the basis for directed synthesis of composite materials with prescribed macroproperties.

It should be emphasized that in the calculation of the first statistical moment of a physical field, the averaging is performed over the ensemble realizations of the random set of inclusions in the composite. Let every realization of the random set of inclusions be an element ω of a functional space with the given known probabilistic measure $\mu(\omega)$ [55]. Hence, the field $u(x)$ in the heterogeneous material depends on ω, i.e., on a specific realization of the random set (ensemble) of the inclusions: $u(x) = u(x, \omega)$. The ensemble averaging of the field $u(x, \omega)$ at an arbitrary point x of the medium is the mean value of the random function $u(x, \omega)$ over the ensemble realizations ω, i.e.,

$$\langle u(x) \rangle = \int u(x, \omega) \mathrm{d}\mu(\omega). \tag{7.2}$$

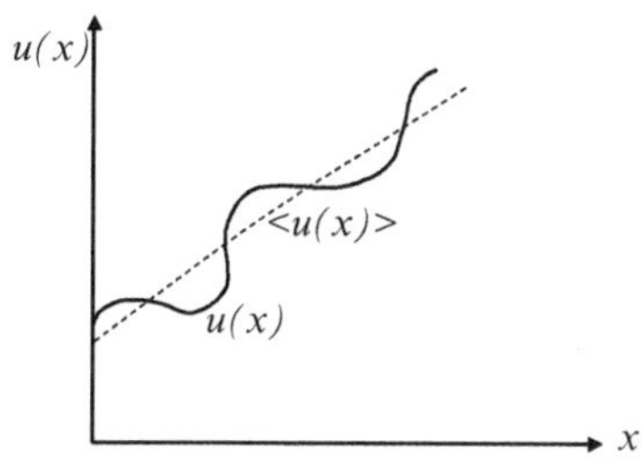

Fig. 7.1. Statistical averaging of a function $u(x)$.

Here the integral is understood as a continuous integral over the functional space of the realizations of the random set of inclusions (see [55], [172]). In Fig. 7.1, the solid line is a detailed function $u(x)$, the dashed line is the mean value of this function.

Let us introduce an operator L_* that allows one to calculate the mean value $\langle u(x) \rangle$ of a random field $u(x)$ from the equation

$$(L_* \langle u \rangle)(x) = q(x), \tag{7.3}$$

without the solution of the elasticity problem for every realization of the random field of inclusions and subsequent averaging of these solutions according to (7.2). If such an operator exists, it is called an effective operator of the microheterogeneous material. For statistically homogeneous composites, this operator describes a physical field in some homogeneous medium that is equivalent to the original composite material. The homogenization problem is in essence the construction of the operator L_*.

In some cases, the ensemble averaging may be replaced by averaging over a representative volume V_0 of the heterogeneous material. According to [62], the volume V_0 should contain a sufficiently large number of inclusions to be considered as macroscopically homogeneous, i.e., the linear dimensions of this volume should be much larger than the those of the inclusions and the distances between them. Then, the mean value $\langle u(x) \rangle$ of a random function $u(x)$ that describes a physical field in the composite is defined on the basis of a typical fixed realization of this function according to the equation

$$\langle u(x) \rangle = \frac{1}{V_0} \int_{V_0} u(x - y) \mathrm{d}y \,. \tag{7.4}$$

It is known [172] that the volume averaging (7.4) coincides with the averaging over the ensemble realizations, if the function $u(x)$ is ergodic, and its correlation radius is essentially smaller than the linear sizes of the volume V_0. Strictly speaking, the stress and strain fields in a statistically homogeneous material are ergodic only for a constant external field applied to the medium. If the external field changes in space, e.g., for wave propagation, only the ensemble averaging may be used for the calculation of the mean fields in the composite. For a nonhomogeneous external field, the result of spatial averaging (7.4) depends on the size of the representative volume V_0 that is not defined uniquely. This dependence becomes stronger if the external field varies considerably inside the representative volume.

The problem of the connection between the ensemble and spatial averages of physical fields in heterogeneous materials has been discussed by many authors (see, e.g., [130]). For instance, it was shown that the systems described by the Schrödinger equation have the so-called self-averaging properties. The latter means that spatial averaging of extensive physical fields similar to the energy density coincides with the ensemble averaging. The necessary conditions for the self-averaging are spatial homogeneity of the system, and rapid decay of the correlations between physical parameters in different points of the system will increase of the distance between these points. Such a property exists for elastic stress and strain fields in statistically homogeneous composite materials.

7.2 Integral equations for the elastic fields in a medium with isolated inclusions

Consider a homogeneous elastic medium, with tensor of elastic moduli C^0, containing a set of isolated inclusions that occupy regions v_k with the characteristic functions $v_k(x)$, $k = 1, 2, ...$ The elastic moduli tensor of the medium with the inclusions $C(x)$ is represented in the form

$$C(x) = C^0 + C^1(x)V(x) \ , \quad V(x) = \sum_k v_k(x) \ , \tag{7.5}$$

where the function $C^1(x)v_k(x)$ describes the perturbation of the elastic moduli inside the k-th inclusion. In particular, if the tensor of elastic moduli is constant in every region v_k, then

$$C(x) = C^0 + \sum_k C_k^1 v_k(x), \tag{7.6}$$

where $C^0 + C_k^1$ is the elastic moduli tensor for the k-th inclusion.

If the strain $\varepsilon^+(x)$ and stress $\sigma^+(x)$ fields inside the inclusions are known,

$$\varepsilon^+(x) = \varepsilon(x)V(x), \quad \sigma^+(x) = \sigma(x)V(x), \tag{7.7}$$

the tensors $\varepsilon(x)$ and $\sigma(x)$ in the medium are represented in the form (3.26), (3.28):

$$\varepsilon(x) = \varepsilon^0(x) - \int \mathrm{K}\,(x - x')\, C^1\,(x')\, \varepsilon^+\,(x')\, \mathrm{d}x', \tag{7.8}$$

$$\sigma(x) = \sigma^0(x) + \int \mathrm{S}(x-x')\, B^1(x')\, \sigma^+(x')\, \mathrm{d}x'. \tag{7.9}$$

Here $\varepsilon^0(x)$, $\sigma^0(x)$ are external strain and stress fields applied to the medium ($\sigma^0 = C^0\varepsilon^0$) that are bounded functions of oscillating type. If the inclusions are homogeneously distributed in the medium, the functions $\varepsilon^+(x)$ and $\sigma^+(x)$ are functions of the same type as $\varepsilon^0(x)$ and $\sigma^0(x)$. The integrals in (7.8), (7.9) that express the action of the operators K and S on such functions formally diverge at $x = 0$ and $|x| \to \infty$.

Consider the problem of regularization of these integrals, and suppose that with the accuracy to square integrated terms the functions

$$m(x) = B^1(x)\sigma(x)V(x), \quad q(x) = C^1(x)\varepsilon(x)V(x) \tag{7.10}$$

are represented in the form of exponential series with, generally speaking, incommensurable wave vectors k^j:

$$m(x) = m^0 + \sum_j m^j \exp\left[i\, k^j \cdot x\right], \quad q(x) = q^0 + \sum_j q^j \exp\left[i\, k^j \cdot x\right]. \tag{7.11}$$

Here m^0, q^0 are constant components of the functions $m(x)$ and $q(x)$, and for the given coefficients m^j and q^j, these series converge, maybe in a generalized sense.

Using the convolution property, one can present the action of the operators K and S on the function $\exp[i\, k^j \cdot x]$ as the product with the constant tensors $\mathrm{K}^*(k^j)$ and $\mathrm{S}^*(k^j)$:

$$\int \mathrm{K}(x-x')e^{ik^j\cdot x'}\mathrm{d}x' = \mathrm{K}^*(k^j)e^{ik^j\cdot x}, \tag{7.12}$$

$$\int \mathrm{S}(x-x')e^{ik^j\cdot x'}\mathrm{d}x' = \mathrm{S}^*(k^j)e^{ik^j\cdot x}, \tag{7.13}$$

where $\mathrm{K}^*(k)$ and $\mathrm{S}^*(k)$ are the Fourier transforms of the functions $K(x)$ and $S(x)$. Since the functions $\mathrm{K}^*(k)$ and $\mathrm{S}^*(k)$ defined in (2.58), (2.75) are homogeneous functions of degree zero with respect to k, they are uniquely defined and bounded for all $k^j\,(k^j \neq 0)$. Hence, by action of the operators K and S on the series in (7.11) the latter are transformed into analogous series with the coefficients $\mathrm{K}^*(k^j)q^j$, $\mathrm{S}^*(k^j)m^j$; they converge if the initial series converges.

The action of the operators K and S on a square integrable function $\varphi(x)$ can be determined by the equations

$$(\mathrm{K}\varphi)(x) = \frac{1}{(2\pi)^3}\int \mathrm{K}^*(k)\varphi^*(k)e^{-ik\cdot x}\mathrm{d}k, \tag{7.14}$$

$$(\mathrm{S}\varphi)(x) = \frac{1}{(2\pi)^3}\int \mathrm{S}^*(k)\varphi^*(k)e^{-ik\cdot x}\mathrm{d}k, \tag{7.15}$$

where the integrals converge absolutely.

It remains to define the action of the operators K and S on the constants q^0 and m^0 in (7.11). Let us consider a model problem in which the "inclusion" has constant elastic moduli $C^0 + C^1$ and occupies the whole space. As the result, the medium remains homogeneous with elastic moduli $C = C^0 + C^1$. If the external stress field σ^0 applied to the medium is a fixed constant tensor, the strain and stress fields in the medium are also constant and have the forms

$$\varepsilon = (C^0 + C^1)^{-1}\sigma^0, \quad \sigma = \sigma^0. \tag{7.16}$$

It is easy to see that the solutions of (7.8), (7.9) coincide with (7.16) if the following equations hold

$$\int \mathrm{K}(x - x')\mathrm{d}x' = (C^0)^{-1}, \quad \int \mathrm{S}(x - x')\mathrm{d}x' = 0\,. \tag{7.17}$$

Let the external strain field ε^0 be fixed. In this case, the strain and stress fields in the medium take the forms

$$\varepsilon = \varepsilon^0, \quad \sigma = (C^0 + C^1)\varepsilon^0, \tag{7.18}$$

and the solutions of Equations (7.8), (7.9) coincide with (7.18) if the actions of the operators K and S on the constant are defined by the equations

$$\int \mathrm{K}\,(x - x')\,\mathrm{d}x' = 0\,, \quad \int \mathrm{S}(x - x')\mathrm{d}x' = -C^0\,. \tag{7.19}$$

Thus, in what follows, we have to define what field (stress or strain) is fixed in the problem. Note that another way of deriving Equations (7.17)–(7.19) was considered in Section 2.2 (Equations (2.78)–(2.81)).

7.3 Tensor of the effective elastic moduli

Using the function $q(x)$ defined in (7.10), we can rewrite Equations (7.8), (7.9) in the forms

$$\varepsilon(x) = \varepsilon^0 - \int \mathrm{K}\,(x - x')\,q(x')\mathrm{d}\,x', \tag{7.20}$$

$$\sigma(x) = \sigma^0 - \int \mathrm{S}(x - x')\,B^0 q(x')\mathrm{d}\,x'. \tag{7.21}$$

In what follows, we assume that the external strain ε^0 and stress σ^0 fields are constant.

Because $\mathrm{K}(x)$ and $\mathrm{S}(x)$ are nonrandom functions, the mean values of the fields in (7.20), (7.21) are presented in the forms

$$\langle\varepsilon(x)\rangle = \varepsilon^0 - \int \mathrm{K}\,(x-x')\,\langle q(x')\rangle \mathrm{d}\,x'\,, \tag{7.22}$$

$$\langle\sigma(x)\rangle = \sigma^0 - \int \mathrm{S}(x-x')\,B^0\langle q(x')\rangle \mathrm{d}\,x'. \tag{7.23}$$

For a spatially homogeneous random set of inclusions, $\varepsilon(x)$ and, consequently, $q(x)$ are homogeneous random functions with ergodic property. Therefore, the mean $\langle q(x)\rangle$ in (7.22), (7.23) is a constant tensor whose value can be found on the basis of a fixed typical realization of the random function $q(x)$:

$$\langle q\rangle = \lim_{W\to\infty}\frac{1}{W}\int_W q(x)\mathrm{d}\,x\,. \tag{7.24}$$

Here W is a region in R^3 that occupies the whole space in the limit $W\to\infty$. Taking into account the linearity of the problem, the strain tensor $\varepsilon(x)$ in the medium is represented in the form

$$\varepsilon(x) = \Lambda(x)\varepsilon^0\,, \tag{7.25}$$

where $\Lambda(x)$ is a tensor function of rank four. Substituting this equation for $\varepsilon(x)$ in Equation (7.10) for $q(x)$, and averaging the latter function according to (7.24), we obtain

$$\langle q\rangle = pP\varepsilon^0,\quad P=\langle P_v\rangle\,,\quad P_v = \langle v\rangle^{-1}\int_v C^1(x)\Lambda(x)\mathrm{d}\,x\,. \tag{7.26}$$

Here p is the volume concentration of the inclusions, the integral P_v is taken over the volume v of each inclusion, the means $\langle P_v\rangle$ and $\langle v\rangle$ are calculated over the ensemble distributions of the tensors P_v and inclusion volumes v in the composite.

Suppose that the mean strain field $\langle\varepsilon\rangle$ is fixed in the problem. This value is independent of the properties and the volume concentrations of the inclusions, and coincides with the external field ε^0 applied to the medium. Note that for a composite, fixing the strain tensor means assignment of the displacement vector on the border Ω of this body. If the displacement field $u^0(x)$ given on Ω has the form $u^0_\alpha(x) = \varepsilon^0_{\alpha\beta}x_\beta$, where $\varepsilon^0_{\alpha\beta}$ is a constant symmetric tensor, the mean value of the strain field in the composite is equal to ε^0. It follows from the definition of the volume average and Gauss' theorem

$$\langle\varepsilon_{\alpha\beta}\rangle = \frac{1}{2V}\int_V\left(\frac{\partial\,u_\alpha}{\partial\,x_\beta}+\frac{\partial\,u_\beta}{\partial\,x_\alpha}\right)\mathrm{d}\,x = \frac{1}{2V}\int_\Omega\left(u^0_\alpha n_\beta + u^0_\beta n_\alpha\right)\mathrm{d}\,\Omega = \varepsilon^0_{\alpha\beta}\,. \tag{7.27}$$

This property holds for a region V occupying the whole space.

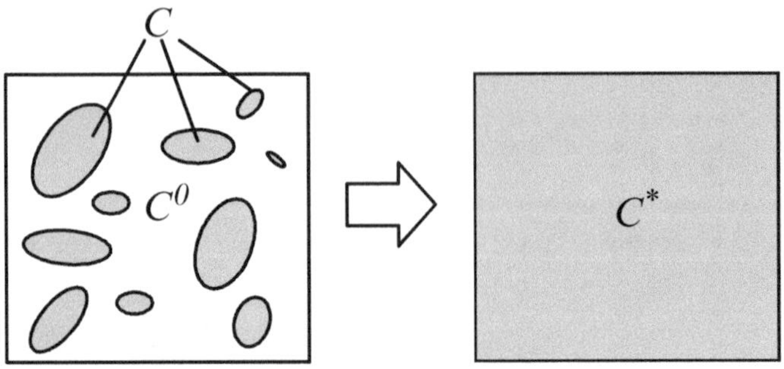

Fig. 7.2. Homogenization of a microheterogeneous medium.

When the mean strain field is fixed in the problem, the action of the operators K and S in (7.22) and (7.23) on constants is determined in (7.19). From these equations and (7.26) we obtain the means $\langle \varepsilon \rangle$ and $\langle \sigma \rangle$ in the forms

$$\langle \varepsilon \rangle = \varepsilon^0, \tag{7.28}$$

$$\langle \sigma \rangle = \sigma^0 + \langle q \rangle = (C^0 + pP)\varepsilon^0 = C^* \langle \varepsilon \rangle , \tag{7.29}$$

$$C^* = C^0 + pP, \tag{7.30}$$

where the tensor C^*connects the mean stress and strain tensors in the composite medium and is called the tensor of effective elastic moduli of the composite.

Hence, for a fixed external strain or stress tensor, the solution of the homogenization problem is equivalent to calculation of the tensor of the effective elastic moduli of the composite (Fig. 7.2). The problem is reduced to the construction of the tensor P in (7.26), and the main difficulty is calculating the tensor $\Lambda(x)$ in (7.25) that defines the detailed fields in the set of interacting inclusions. In what follows, self-consistent methods are used for the construction of the tensor P in Equation (7.30) for C^*.

7.4 The effective medium method and its versions

The first self-consistent method that was applied for the solution of the homogenization problem in the mechanics of heterogeneous media is based on the following hypothesis. It is assumed that each inclusion in the composite behaves as an isolated one in a homogeneous medium with properties which coincide with the overall properties of the composite material. The exciting field acting on every inclusion in the composite is the external field ε^0 applied to the heterogeneous medium. Self-consistent methods based on the hypotheses of this type are called *effective medium methods (*EMM*)* (Fig. 7.3). This hypothesis reduces the problem of interaction between many inclusions in the composite to the problem for one isolated inclusion embedded in a homogeneous medium (a one particle problem). Using this hypothesis, one can find

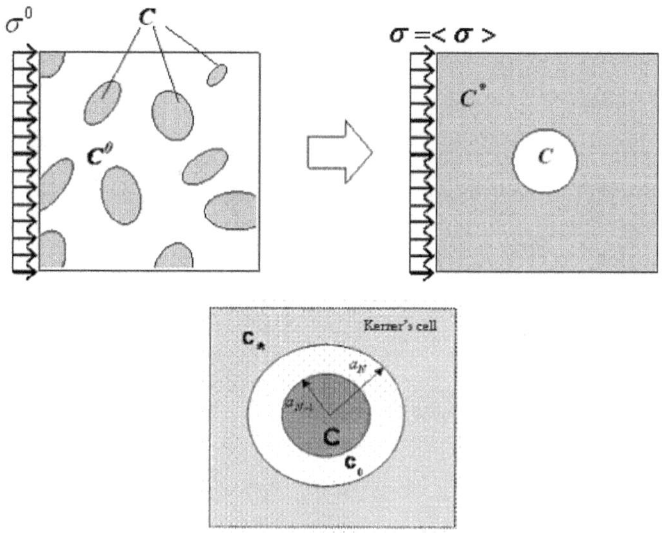

Fig. 7.3. The schemes of the effective medium method.

the dependence of the tensor P_v in (7.26) on the parameters a_i, $i = 1, 2, ...$ characterizing the form of the inclusion, its elastic moduli C, and the tensor of the effective elastic moduli of the composite C^*.

$$P_v = P_v(C^*, C, \{a\}). \tag{7.31}$$

Here $\{a\} = (a_1, a_2, ...)$ is the set of the form parameters. Averaging P_v over the ensemble distributions of the form parameters and the tensor C and substituting the result in (7.30), we obtain an algebraic equation for the components of the unknown tensor C^*

$$C^* = C^0 + p\langle P_v\left(C^*, C, \{a\}\right)\rangle. \tag{7.32}$$

Let isotropic spherical inclusions be homogeneously distributed in an isotropic matrix, and λ_0, μ_0 (λ, μ) be the Lamé coefficients of the matrix (the inclusions). Using the hypothesis of the EMM and the solution of the problem for a homogeneous spherical inclusion embedded in a medium with the effective Lamé coefficients λ^*, μ^* of the composite (3.136), we obtain the following equation for the tensor P_v:

$$P_v = \left(C - C^0\right)\left[I + A^*\left(C - C^*\right)\right]^{-1}. \tag{7.33}$$

Here the tensor A^* is defined in (3.133), where the parameters λ_0 and μ_0 should be replaced by λ^*, μ^*. From this equation and (7.32) we obtain the following system of algebraic equations for the effective bulk $K^* = \lambda^* + 2\mu^*/3$ and shear μ^* moduli of the composite

$$K_* = K_0 + p\,(K - K_0)\left[1 + \frac{3\,(K - K_*)}{3K_* + 4\mu_*}\right]^{-1}, \tag{7.34}$$

$$\mu_* = \mu_0 + p\,(\mu - \mu_0)\left[1 + \frac{6\,(\mu - \mu_*)\,(K_* + 2\mu_*)}{5\mu_*\,(3K_* + 4\mu_*)}\right]^{-1}. \tag{7.35}$$

Nonlinear algebraic Equations (7.34) and (7.35) can be solved numerically. To visualize the solution, consider a particular case when the inclusions are spherical pores ($K = 0$, $\mu = 0$). From (7.34) we find

$$K_* = \frac{4(1-p)K_0\mu_*}{3pK_0 + 4\mu_*}, \tag{7.36}$$

and substituting this equation into (7.35) we find that μ_* satisfies the quadratic equation

$$16\mu_*^2 + \left[K_0(3-p) - 4\mu_0(4-5p)\right]\mu_* - 3\mu_0 K_0(1-2p) = 0. \tag{7.37}$$

In the limit $K_0 \to \infty$ that corresponds to an incompressible matrix, the effective bulk and shear moduli of the composite take the forms

$$K_* = \frac{4(1-p)(1-2p)}{p(3-p)}\mu_0, \quad \mu_* = \frac{3(1-2p)}{3-p}\mu_0. \tag{7.38}$$

These equations predict that the effective elastic moduli vanish at $p = 0.5$; this rarely happens in actual composites. Thus, the EMM should be used with some precaution in application to composite materials with contrasting properties of components. For absolutely rigid spheres in an incompressible matrix, (7.35) gives for the effective shear modulus of the composite

$$\mu_* = \frac{\mu_0}{1 - \frac{5}{2}p} \tag{7.39}$$

In this case, the modulus μ_* becomes infinite when $p = 0.4$.

The shortcomings of this self-consistent scheme were pointed out in a number of works (see, e.g., [30]). In particular, the effective elastic moduli of the medium with relatively rigid spherical inclusions ($C^0C^{-1} = O(\delta)$, $\delta << 1$), or pores ($C = 0$), calculated from equations (7.34) differ essentially from those measured in experiments if the volume concentration of the inclusions (pores) p exceeds 0.3.

Better correspondence with experimental data can be obtained by using a modification of the effective medium method proposed in [95] and used in [29], [191], [203]. In these works, a layer of matrix material was introduced between the effective medium and the inclusion by the formulation of the one-particle problem (three phase model). Let us consider this variant of the effective medium method in detail for the composite reinforced by spherical multi-layered inclusions.

Let each inclusion in the composite consists of $N-1$ spherical layers with different elastic properties. Suppose for simplicity that all the inclusions are identical, and the material of the matrix and each of the layers is homogeneous and isotropic. We choose an arbitrary inclusion and put in its center the origin of a spherical coordinate system (r, n) $(r=|x|,\, n=x/|x|)$. The values $r=a_i$, $(i=1,2,...,N-1\,;\;\; a_0=0<a_1<a_2<...<a_{N-1})$ correspond to the layer boundaries. We will find the strain field inside every inclusion from the solution of following problem. Let us add to the layers of the inclusion a layer with the properties of the matrix material C^0 (the N-th layer), and put this multilayered inclusion in a homogeneous medium with the effective elastic properties of the composite C^*. The region inside the multilayered inclusion of radius a_N containing the inclusion and the mentioned layer of matrix material is called the Kerner cell V^k. Following [95], [29], we choose the external radius a_N of the N-th layer from the equation

$$\left(\frac{a_{N-1}}{a_N}\right)^3 = p\,, \tag{7.40}$$

where p is the volume concentration of the inclusions. It is implied in this equation that the volumes of the matrix phase and the inclusion inside Kerner's cell are related as it is for the whole material. The external strain field that acts on this inclusion coincides with the external field ε^0 applied to the medium. The strain field in the medium with the isolated multilayered inclusion can be represented in the form

$$\varepsilon(r,n) = \Lambda(r,n)\varepsilon^0, \quad \Lambda(r,n) = \mathrm{E}^1 + A(r,n)\,, \tag{7.41}$$

where the tensor $A(r,n)$ has form (3.288), (3.327)–(3.329) and depends on ten constants Y_j^i $(j=1,2,...,10\,;\; i=1,2,...,N+1)$. The algorithm for the construction of these constants is presented in Section 3.10. Note that the $(N+1)$-th layer $(a_N<r<a_{N+1},\quad a_{N+1}=\infty)$ is the effective medium outside the Kerner cell V^k.

Substituting $\Lambda(r,n)$ from (7.41) into (7.26) and taking into account equations (3.288), (3.327)–(3.329) for the tensor $A(r,n)$, we obtain the tensor P in (7.30) in the form

$$P = K_0\left(\pi_1 - q_1\right)E^2 + 2\mu_0\left(\pi_2 - q_2\right)\left(E^1 - \frac{1}{3}E^2\right), \tag{7.42}$$

$$q_j = \sum_{i=1}^{N} q_j^i\,,\; j=1,2\,;\; \pi_1 = \sum_{i=1}^{N} \overline{K}_i q_1^i\,,\; \pi_2 = \sum_{i=1}^{N} \bar{\mu}_i q_2^i\,, \tag{7.43}$$

$$\overline{K}_i = \frac{K_i}{K_0}\,,\; \bar{\mu}_i = \frac{\mu_i}{\mu_0}\,,\; \bar{a}_i = \frac{a_i}{a_N}\,, \tag{7.44}$$

$$q_1^i = \left(1 + Y_9^i\right)\left(\bar{a}_i^3 - \bar{a}_{i-1}^3\right), \tag{7.45}$$

$$q_2^i = \left(1 + 3Y_1^i + 2Y_5^i\right)\left(\bar{a}_i^3 - \bar{a}_{i-1}^3\right) + \frac{7}{5}a_N^2\left(3Y_2^i + Y_6^i\right)\left(\bar{a}_i^5 - \bar{a}_{i-1}^5\right), \quad (7.46)$$

where K_i, μ_i are the bulk and shear moduli of the i-th layer, K_0, μ_0 are the corresponding moduli for the matrix.

From (7.32), (7.42) we find the equations for the effective bulk K_* and shear μ_* moduli of the composite:

$$K_* = K_0[1 + p(\pi_1 - q_1)], \quad \mu_* = \mu_0[1 + p(\pi_2 - q_2)]. \quad (7.47)$$

Here the parameters q_1, q_2, π_1, π_2 depend on the properties of all the layers, their sizes, and the moduli K_*, μ_* of the effective medium via the solution of the one-particle problem. Therefore, Equation (7.47) is in essence a system of nonlinear equations for the effective parameters K_*, μ_*. Equations that correspond to the simplest version of the effective medium method can be obtained from (7.47) by letting the thickness of the N-th layer $(a_N \to a_{N-1})$ to zero.

Note that the equations obtained in [29] for the effective elastic moduli K_*, μ_* of composite reinforced by spherical inclusions do not follow from (7.46) because another condition of self-consistency was used in the derivation of the latter equations. It was assumed in [29] that the strain and stress field averaged over the Kerner cell V^k are connected by the tensor of effective elastic moduli for the composite

$$\langle\sigma\rangle_{V^k} = C^*\langle\varepsilon\rangle_{V^k}, \quad \langle f\rangle_{V^k} = \frac{1}{V^k}\int\limits_{V^k} f(x)\mathrm{d}\,x. \quad (7.48)$$

Calculating the integrals in (7.48) and taking into account (7.41) and the equation for the tensor $A(r,n)$ in (3.288), (3.327)–(3.329), we arrive at the following equations for the effective elastic moduli K_*, μ_*:

$$K_* = K_0\frac{\pi_1}{q_1}, \quad \mu_* = \mu_0\frac{\pi_2}{q_2}, \quad (7.49)$$

where the parameters q_i , π_i $(i = 1, 2)$ are the same as in (7.46).

The solution of (7.47),(7.49) can be found by an iterative procedure based on the relations

$$K_*^{(n)} = F_k\left(K_*^{(n-1)}, \mu_*^{(n-1)}\right), \quad \mu_*^{(n)} = F_\mu\left(K_*^{(n-1)}, \mu_*^{(n-1)}\right), \quad (7.50)$$

where $K_*^{(n)}, \mu_*^{(n)}$ are the effective moduli at the n-th step of the iteration process. The functions F_k and F_μ are defined by the right-hand sides of Equation (7.47) or (7.49). The calculations show that the values of the elastic moduli K_*, μ_*, found from (7.47) and (7.49) practically coincide. At the same time, the iterative process (7.50) turns out to be more stable if it is based on Equation (7.49).

Figure 7.4 presents the comparison of the theoretical and experimental values of the effective Young modulus E_* of a composite with spherical inclusions. The upper part of this figure corresponds to the case of hard spherical

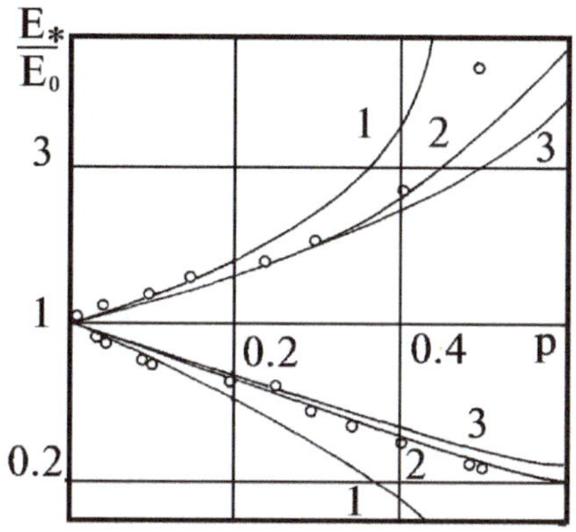

Fig. 7.4. Effective Young's modulus for the spherical stiff inclusions (upper part) and spherical pores (lower part).

inclusions ($E/E_0 = 28,7$, $\nu_0 = 0,394$, $\nu = 0,23$), and the lower part to spherical pores, p is the volume concentration of the inclusions, E, ν are Young's modulus and Poisson's ratio of the inclusions, E_0, ν_0 are the corresponding quantities for the matrix.

Curves 1 correspond to the simplest variant of the EMM, curves 2 to the modified version of this method, circles are experimental data presented in [192]. It is clear that the modified version of the EMM describes the experimental data better than the simplest version. However, if the volume concentration of hard inclusions is high ($p > 0.4$), the error of the modified version exceeds 20%.

If the distribution of the inclusions in space is anisotropic, application of the EMM faces some technical difficulties. Also, uncertainties appear in the formulation of the one-particle problem. Consider, for example, a composite material with an isotropic matrix and isotropic spheroidal inclusions. Suppose that the inclusions have the same sizes and properties, and their symmetry axes are parallel to the x_3-axis of the global Cartesian coordinate system. The overall properties of the composite material will be transversally isotropic with the symmetry axis x_3. The effective medium method leads to the following equation for the tensor of effective elastic moduli C^* [213]:

$$C^* = C^0 + p\left(C - C^0\right)\left[I + A^*\left(C - C^*\right)\right]^{-1}, \tag{7.51}$$

where the tensor A^* has to be calculated from the solution of the one-particle problem for a spheroidal inclusion embedded in a transversally isotropic medium with unknown elastic properties. This problem was solved in Section 3.6, and the equation for the tensor A^* takes the form

$$A^* = A_1^* P^2 + A_2^*\left(P^1 - \frac{1}{2}P^2\right) + A_3^*\left(P^3 + P^4\right) + A_5^* P^5 + A_6^* P^6, \tag{7.52}$$

where P^k are the elements of the P-basis (Appendix A.2), and the coefficients $A_1^*, ...A_6^*$ are defined in (3.168), where the elastic moduli of the matrix $k_0, m_0, l_0, \mu_0, n_0$ have to be replaced by the effective moduli $k_*, m_*, l_*, \mu_*, n_*$.

Using the algebra of the P-tensors we find

$$[I + A^* (C - C^*)]^{-1} =$$

$$= \frac{B_6}{2\Delta} P^2 + \frac{1}{B_2}\left(P^1 - \frac{1}{2}P^2\right) - \frac{B_3}{\Delta}P^3 - \frac{B_4}{\Delta}P^4 + \frac{4}{B_5}P^5 + \frac{2B_1}{\Delta}, \quad (7.53)$$

$$\Delta = 2\left(B_1 B_6 - B_3 B_4\right), \quad (7.54)$$

$$B_1 = \frac{1}{2} + 2A_1^*(\lambda + \mu - k_*) + A_3^*(\lambda - l_*), \quad (7.55)$$

$$B_2 = 1 + 2A_2^*(\mu - m_*), \quad (7.56)$$

$$B_3 = 2A_1^*(\lambda - l_*) + A_3^*(\lambda + 2\mu - n_*), \quad (7.57)$$

$$B_4 = 2A_3^*(\lambda + \mu - k_*) + A_6^*(\lambda - l_*), \quad (7.58)$$

$$B_5 = 2\left[1 + 2A_5^*\left(\mu - \mu_*\right)\right], \quad (7.59)$$

$$B_6 = 1 + A_6^*\left(\lambda + 2\mu - n_*\right) + 2A_3^*(\lambda - l_*). \quad (7.60)$$

From (7.51) we obtain the components C_{1133}^* and $C_{3311}^*$of the tensor C^* in the form

$$C_{1133}^* = l_* = \lambda_0 + \frac{2p}{\Delta}\left[(\lambda - \lambda_0)B_1 - (\lambda + \mu - \lambda_0 - \mu_0)\, B_3\right], \quad (7.61)$$

$$C_{3311}^* = l_*' = \lambda_0 + \frac{p}{\Delta}\left[(\lambda - \lambda_0)B_6 - (\lambda + 2\mu - \lambda_0 - 2\mu_0)\, B_4\right]. \quad (7.62)$$

Thus, the effective elastic moduli tensor C^* obtained in the framework of the effective medium method turns out to be nonsymmetric with respect to pairs of indices.

In [213] and [16], the so-called symmetric effective medium method (SEMM) was proposed. In this method, *each* component of the composite with the elastic properties C^r is considered as an ellipsoid embedded in the homogeneous medium with the effective properties of the composite. The symmetric EMM leads to the equation

$$\sum_{r=1}^{n} p_r \left[(C^r - C^*)^{-1} + A^*\right]^{-1} = 0. \quad (7.63)$$

The tensor of the effective moduli of the composite obtained from this equation turns out to be symmetric. But in this method, the matrix and inclusions are treated as equivalent phases. Meanwhile, they are topologically different in the matrix composites, and the latter is a continuous host phase that does not consist of a set of ellipsoidal regions. Thus, one cannot consider the field in the matrix phase as the field in ellipsoidal inclusions. It is more appropriately to apply this method to polycrystals or granular materials, where the matrix phase does not exist, then to matrix composites.

Note that for two-phase composites with spherical inclusions, Equation (7.51) of the EMM and Equation (7.63) of the SEMM coincide.

7.4.1 Differential effective medium method

The differential effective medium method (DEMM) is based on the concept that a matrix composite with a finite volume concentration of inclusions may be built up by sequential addition to a host material (matrix) a small number of particles of the second phase. After every such addition, the homogenization procedure is performed, and the new composite is considered as a homogeneous host material. The process is continued until the required concentration c of the inclusions is achieved (Fig. 7.5). As the result, the effective elastic moduli turn out to be functions of the total amount of inserted particles, $C^* = C^*(c)$. Note that at every step of the process, a small concentration of particles is added $\delta c << 1$, and therefore, the equations for the elastic moduli of the composites with small volume concentrations of inclusions may be used (the dilute approximation). For ellipsoidal inclusions in a homogeneous matrix with the moduli C^0, the dilute approximation has the form

$$C^* = C^0 + p(C - C^0)\,\langle \Lambda^{\varepsilon} \rangle\,. \tag{7.64}$$

Here the tensor Λ^{ε} for ellipsoidal inclusions is defined in (3.81), and averaging in (7.64) is performed over the orientations and size distribution of the inclusions. Hence, for the described process we can write

$$C^*(c + \delta c) = C^*(c) + \delta c(C - C^*)\,\langle \Lambda^{\varepsilon *} \rangle\,, \tag{7.65}$$

where the tensor $\Lambda^{\varepsilon *}$ is determined by the same equation (3.81) from the solution of the one-particle problem for a ellipsoidal inclusion in the medium with the effective elastic properties C^*.

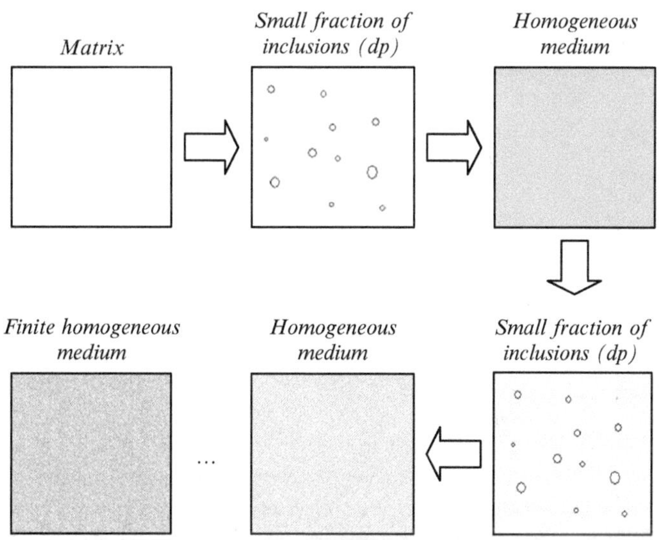

Fig. 7.5. The differential scheme of the effective medium method.

In the limit $\delta c \to 0$, (7.65) is transformed in the differential equation for the unknown function $C^*(c)$

$$\frac{\mathrm{d}C^*}{\mathrm{d}c} = (C - C^*)\,\langle \Lambda^{\varepsilon *} \rangle\,. \tag{7.66}$$

The natural initial condition for this equation is

$$C^*(0) = C^0. \tag{7.67}$$

It was noted in [145] that the total concentration c of particles introduced into the matrix does not coincide with the volume concentration p of the second phase. The reason is that when adding a new portion δc of particles, some of them will "fall" into the region occupied by the previously deposited ones. Only the part $(1-p)\delta c$, that "falls" within the matrix, will increase the second phase concentration p. Thus, in the limit $\delta c \to 0$ we have

$$\mathrm{d}p = (1-p)\mathrm{d}c. \tag{7.68}$$

Now consider a composite material with an isotropic matrix containing a random set of isotropic spherical inclusions homogeneously distributed in the matrix. In this case, the composite material is isotropic, and the tensor of the effective elastic moduli of this material satisfies the differential equation that follows from (7.66)

$$(1-p)\frac{\mathrm{d}C^*_{\alpha\beta\lambda\mu}}{\mathrm{d}p} = \left[(C_{\alpha\beta\lambda\mu} - C^*_{\alpha\beta\lambda\mu})^{-1} + A^*_{\alpha\beta\lambda\mu}\right]^{-1}, \tag{7.69}$$

$$A^*_{\alpha\beta\lambda\mu} = \frac{1}{3(3K_* + 4\mu_*)}E^2 + \frac{3(K_* + 2\mu_*)}{5\mu_*(3K_* + 4\mu_*)}\left(E^1 - \frac{1}{3}E^2\right). \tag{7.70}$$

Here K_* and μ_* are the bulk and shear effective elastic moduli of the composite. This tensor equation is equivalent to the system of two differential equations for K_* and μ_*

$$(1-p)\frac{\mathrm{d}K_*}{\mathrm{d}p} = \frac{(K - K_*)(3K_* + 4\mu_*)}{3K + 4\mu_*}, \tag{7.71}$$

$$(1-p)\frac{\mathrm{d}\mu_*}{\mathrm{d}p} = \frac{5\mu_*(\mu - \mu_*)(3K_* + 4\mu_*)}{\mu_*(9K_* + 8\mu_*) + 6\mu_*(K_* + 2\mu_*)}, \tag{7.72}$$

with the initial conditions: $K_*(0) = K_0$, $\mu_*(0) = \mu_0$. Numerical solutions of these equations are given in Fig. 7.6 by curves 2 for hard inclusions (upper part of the figure) and spherical pores (lower part of the figure). For comparison, the predictions of the EMM and the modified EMM are also presented in this figure by lines 1 and 3, correspondingly. The properties of the phase are the same as those used for the construction of the graphs in Fig. 7.4.

Consider a matrix composite material containing a random set of aligned spheroidal inclusions. In this case, the material has transversely isotropic

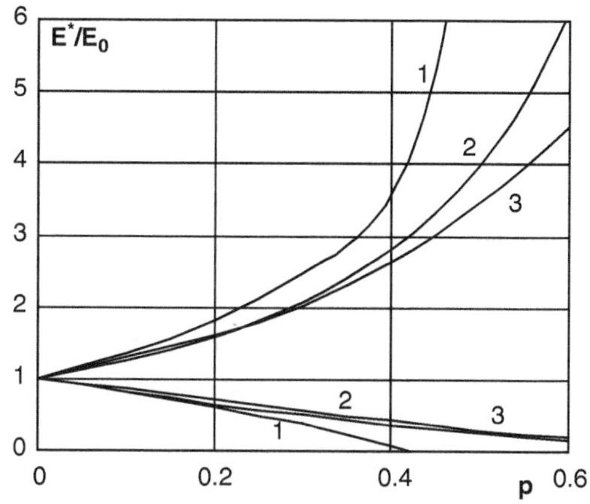

Fig. 7.6. Effective Young's modulus for the spherical stiff inclusions (upper part) and spherical pore (lower part).

symmetry and is characterized by five independent elastic moduli: $k_*, m_*, \lambda_*, \mu_*, n_*$. The DEMM scheme leads to the following system of differential equations for these moduli

$$(1-p)\frac{\mathrm{d}k_*}{\mathrm{d}p} = \frac{1}{2D}\left(\frac{2(k-k_*)}{\Delta} + A_6^*\right), \tag{7.73}$$

$$(1-p)\frac{\mathrm{d}m_*}{\mathrm{d}p} = \left(\frac{1}{m-m_*} + 2A_2^*\right), \tag{7.74}$$

$$(1-p)\frac{\mathrm{d}\lambda_*}{\mathrm{d}p} = \frac{1}{D}\left(\frac{2(\lambda-\lambda_*)}{\Delta} - A_3^*\right), \tag{7.75}$$

$$(1-p)\frac{\mathrm{d}\mu_*}{\mathrm{d}p} = \left(\frac{1}{\mu-\mu_*} + 2A_5^*\right), \tag{7.76}$$

$$(1-p)\frac{\mathrm{d}n_*}{\mathrm{d}p} = \frac{2}{D}\left(\frac{(n-n_*)}{2\Delta} + A_1^*\right). \tag{7.77}$$

The coefficients Δ and D in these equations are

$$\Delta = 2\left[(k-k_*)(n-n_*) - (\lambda-\lambda_*)^2\right], \tag{7.78}$$

$$D = 2\left\{\left[\frac{n-n_*}{2\Delta} + A_1^*\right]\left[\frac{2(k-k_*)}{\Delta} + A_6^*\right] - \left(\frac{\lambda-\lambda_*}{\Delta} - A_3^*\right)^2\right\}, \tag{7.79}$$

and the coefficients $A_1^*, ..., A_6^*$ are defined in (7.52).

Note that for multiphase composites, the result of application of the DEMM depends on the order of implantation of the particles of different components into the host phase. The detailed discussion of this problem is presented in [156].

7.5 The effective field method

An approximate solution of the homogenization problem may be obtained by a self-consistent method different from the EMM. In this method, every inclusion in the composite is considered as an isolated one embedded into the homogeneous medium with the elastic properties C^0 of the matrix. Existence of other inhomogeneities is taken into account by the effective exciting field acting on this inclusion. The effective exciting field of strain ε^* (stress σ^*) is the sum of the external field $\varepsilon^0(\sigma^0)$ applied to the medium and the fields induced by surrounding inclusions (see Fig. 7.7). In the simplest version of the method, the effective field is taken as constant and the same for all the inclusions (quasycrystalline approximation).

Self-consistent methods accounting for interactions between many inclusions by introducing a local exciting field acting on each inclusion are called effective field methods (EFM).

Consider Equations (7.6)–(7.8) for the strain field in a homogeneous medium with a set of isolated inclusions. The strain field inside the i-th inclusion satisfies the following equation:

$$\varepsilon(x) = \varepsilon^*(x) - \int_{v_i} K(x - x')C^1\varepsilon(x')\mathrm{d}x', \tag{7.80}$$

$$\varepsilon^*(x) = \varepsilon^0 - \sum_{k \neq i} \int_{v_k} K(x - x')C^1\varepsilon(x')\mathrm{d}x'. \tag{7.81}$$

The first of these equations shows that the i-th inclusion may be considered as an isolated one in the background matrix by the action of an exciting field $\varepsilon^*(x)$. The latter is the sum of the external field ε^0 applied to the medium

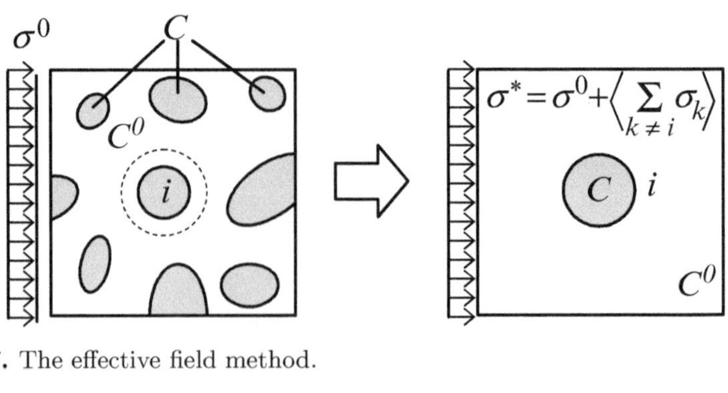

Fig. 7.7. The effective field method.

and the field induced in the region v_i by all the surrounding inclusions. The main hypothesis of the method is that the field ε^* is constant and the same for all the inclusions. For the constant ε^*, the solution of Equation (7.80) for the field inside the i-th inclusion is presented in the form

$$\varepsilon(x) = \Lambda^i(x)\varepsilon^*, \tag{7.82}$$

where the tensor $\Lambda^i(x)$ depends on the elastic properties and the form of the i-th inclusion. The function $q(x)$ (7.10) in Equations (7.20)–(7.23) takes the following form:

$$q(x) = C^1(x)\Lambda(x)\varepsilon^* V(x), \tag{7.83}$$

where the function $\Lambda(x)$ coincides with $\Lambda^i(x)$ inside the i-th inclusion.

Let us introduce the function $V(x; x')$ by the equation

$$V\left(x; x'\right) = \sum_{k \neq i} v_k\left(x'\right) \quad \text{when} \quad x \in v_i\ , \tag{7.84}$$

where $v_i(x)$ is the characteristic function of the region v_i, occupied by the i-th inclusion. With this function, the local exciting field at a point $x \in V$ ($V = \bigcup_i v_i$) is presented in the form

$$\varepsilon^*(x) = \varepsilon^0 - \int \mathrm{K}\left(x - x'\right) C^1\left(x'\right) \Lambda\left(x'\right) \varepsilon^* V\left(x; x'\right) \mathrm{d}x'. \tag{7.85}$$

If we average this equation under the condition that $x \in V$

$$\langle \varepsilon^*(x)|x\rangle = \varepsilon^0 - \int \mathrm{K}\left(x - x'\right)\left\langle C^1\left(x'\right) \Lambda\left(x'\right) V\left(x; x'\right) |x\right\rangle \mathrm{d}x' \varepsilon^*, \tag{7.86}$$

and identify the mean $\langle \varepsilon^*(x)|x\rangle$ with the effective exciting field ε^*

$$\langle \varepsilon^*(x)|x\rangle = \varepsilon^*, \tag{7.87}$$

we obtain from (7.86) a closed equation for the effective field ε^*. Here $\langle \cdot |x\rangle$ is averaging over the realizations of the random set of inclusions by the condition $x \in V$.

The conditional mean under the integral in (7.86) is presented in the form

$$\left\langle C^1\left(x'\right) \Lambda\left(x'\right) V\left(x; x'\right) |x\right\rangle = \left\langle C^1\left(x'\right) \Lambda\left(x'\right) V\left(x'\right)\right\rangle \Psi\left(x, x'\right)\ ,$$

$$\Psi\left(x, x'\right) = \frac{\left\langle V\left(x; x'\right) |x\right\rangle}{\langle V(x)\rangle}\ . \tag{7.88}$$

Here we assume statistical independence of the inclusion properties on their spatial locations, and use the definition of the condition means [172]

$$\langle f(x)|x\rangle = \frac{\langle f(x)V(x)\rangle}{\langle V(x)\rangle}.$$

Using the ergodic properties of the considering functions, we can write

$$\left\langle C^1(x)\Lambda(x)V(x)\right\rangle =$$
$$\lim_{W\to\infty}\frac{1}{W}\int\limits_W C^1(x)\Lambda(x)V(x)\mathrm{d}x = \lim_{W\to\infty}\frac{1}{W}\sum_{v_i\in W}\int\limits_{v_i} C^1(x)\Lambda(x)\mathrm{d}x\,. \quad (7.89)$$

After averaging both sides of this equation once more over the ensemble realizations of the random set of inclusions we obtain

$$\left\langle C^1(x)\Lambda(x)V(x)\right\rangle = \lim_{W\to\infty}\frac{N}{W}\langle v\rangle P = pP, \quad (7.90)$$

$$P = \frac{1}{\langle v\rangle}\left\langle \int\limits_v C^1(x)\Lambda(x)\mathrm{d}x\right\rangle, \quad (7.91)$$

where N is the number of inclusions in the region W that occupies the whole space in the limit $W\to\infty$, v is the volume of a typical inclusion. Accounting for (7.87), (7.90), Equation (7.86) takes the form

$$\varepsilon^* = \varepsilon^0 - p\int \mathrm{K}\left(x-x'\right)P\Psi\left(x-x'\right)\mathrm{d}x'\varepsilon^*. \quad (7.92)$$

Here it is taken into account that for a spatially homogeneous set of inclusions, the function $\Psi(x,x')$ in (7.88) depends on the difference $x-x'$. Let us consider this function in detail. From the definition of the conditional mean we have

$$\Psi(x) = \frac{\left\langle V\left(x';x'+x\right)|x'\right\rangle}{\langle V(x)\rangle} = \frac{\left\langle V\left(x';x'+x\right)V\left(x'\right)\right\rangle}{\langle V(x)\rangle^2}. \quad (7.93)$$

To obtain the numerator in this equation, we consider, firstly, the covariance $\langle V(x'+x)V(x')\rangle$, where $V(x)$ is the characteristic function of the region occupied by the inclusions. Because of the ergodic property of this function, we have

$$\left\langle V\left(x'\right)V\left(x'+x\right)\right\rangle =$$
$$= \lim_{W\to\infty}\frac{1}{W}\int\limits_W\left[\sum_i v_i\left(x'\right)v_i\left(x'+x\right) + \sum_{i,j\,(i\neq j)} v_i\left(x'\right)v_j\left(x'+x\right)\right]\mathrm{d}x'. \quad (7.94)$$

Note, that $\int v_i\left(x'\right)v_i\left(x'+x\right)\mathrm{d}\,x'$ is the volume of intersection of two identical regions moved on a vector x with respect to each other.

The mean $\langle V(x';x'+x)V(x')\rangle$ in the numerator of (7.93) coincides with the correlation function (7.94) if one does not take into account the contribution in (7.93) the realizations in which points x and x' are inside the same inclusion. The latter follows from the definition (7.84) of the function

$V(x;x')$. Therefore, taking away the first sum in (7.94) we find the equation in the numerator in (7.93), and can write the function $\Psi(x)$ in the form ($\langle V(x)\rangle = p$)

$$\Psi(x) = \frac{1}{p^2} \lim_{W\to\infty} \frac{1}{W} \int_W \sum_{i,j\,(i\neq j)} v_i\,(x')\,v_j\,(x'+x)\,\mathrm{d}x'. \tag{7.95}$$

One consequence of this equation is that $\Psi(x)$ is a continuous function with mean value equal to 1

$$\lim_{W\to\infty} \frac{1}{W} \int_W \Psi(x)\mathrm{d}x = 1. \tag{7.96}$$

Because the regions v_i $(i = 1, 2, 3, ...)$ do not overlap, we have

$$\Psi(x) = 0 \quad \text{when} \quad x = 0. \tag{7.97}$$

In the general case, the function $\Psi(x)$ is a sum of a constant part (equal to 1), an oscillating function with mean value equal to zero, and a finite function. The function $\Psi(x)$ depends on the geometrical properties of the inclusion distribution in the matrix phase. Symmetry (in a statistical sense) of the random field of inclusions defines the symmetry of the function $\Psi(x)$. In particular, if the random set of inclusions is isotropic in space, $\Psi(x)$ is spherically symmetric, i.e., $\Psi(x) = \Psi(|x|)$ (Fig. 7.8).

Deviation of the random set of inclusions from statistical isotropy leads to anisotropy of the macroproperties of the composite material (texture) in spite of isotropy of the matrix and inclusion phases. In many cases, the texture symmetry can be described by a rank two tensor $\alpha_{\alpha\beta}$. In this case (Fig. 7.9),

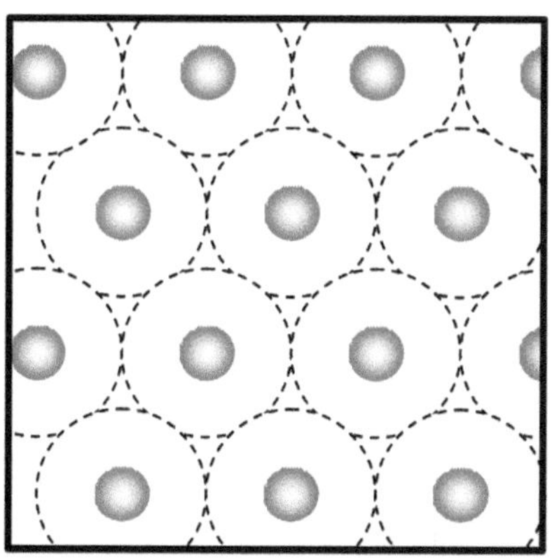

Fig. 7.8. Spacial distribution of inclusions for a spherically symmetric function $\Psi(|x|)$.

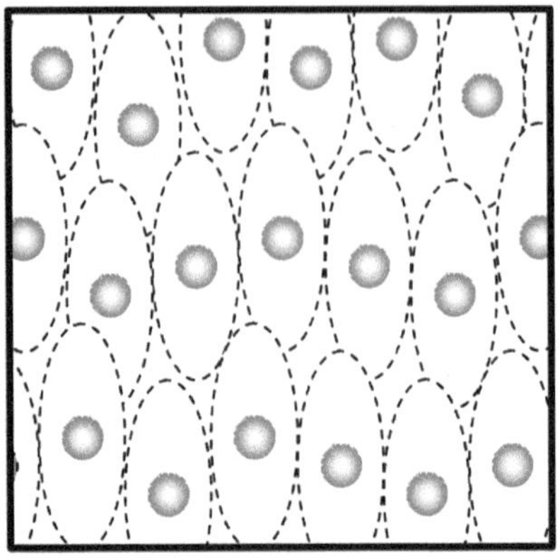

Fig. 7.9. Spacial distribution of inclusions in the case of the ellipsoidal symmetry of the function $\Psi(x)$.

the tensor α defines a linear space transformation that converts the function $\Psi(x)$ into a spherically symmetric one

$$\Psi(\alpha\, x) = \Psi(|x|). \tag{7.98}$$

In the general case, such a transform does not exist.

Using the mentioned properties of the function $\Psi(x)$, let us calculate the integral in (7.92). Taking into account the regularizations (2.81) and (7.19) we obtain

$$\int \mathrm{K}\,(x - x')\, \Psi\,(x - x')\, \mathrm{d}x' = \int \mathrm{K}(x)\,(\Psi(x) - 1)\, \mathrm{d}x =$$

$$= -A(\alpha) + (\det \alpha)\, p.v. \int \mathrm{K}(\alpha\, x)\,(\Psi(\alpha\, x) - 1)\, \mathrm{d}x\,, \tag{7.99}$$

where the tensor $A(\alpha)$ is defined in (3.82) when $a = \alpha$; the last integral is understood in the sense of a Cauchy principal value, and converges at zero and at infinity. When Equation (7.98) holds, the integral on the right-hand side of (7.99) vanishes, and this equation takes the form

$$\int \mathrm{K}\,(x - x')\, \Psi\,(x - x')\, \mathrm{d}x' = -A(\alpha), \tag{7.100}$$

$$A(\alpha) = \frac{1}{4\pi} \int_{\Omega_1} K^*(\alpha^{-1} k) \mathrm{d}\Omega. \tag{7.101}$$

Substituting this relation in (7.92) and solving the equation so obtained for the effective field ε^*, we find

$$\varepsilon^* = \Lambda^* \varepsilon^0, \; \Lambda^* = \left[I - pA(\alpha)P\right]^{-1}. \tag{7.102}$$

The equation for the mean value of the function $q(x)$ in (7.83) follows from (7.102) in the form

$$\langle q(x)\rangle = \left\langle C^1(x)\Lambda(x)\varepsilon^* V(x)\right\rangle = pP^*\varepsilon^0, \quad P^* = P\Lambda^*, \tag{7.103}$$

where the tensor P is defined in (7.91).

Finally, (7.26), (7.30) give us the following equation for the tensor of the effective elastic moduli C^* composite

$$C^* = C^0 + pP\Lambda^* = C^0 + pP[I - pA(\alpha)P]^{-1}. \tag{7.104}$$

In this equation, the symmetry of the texture defines the tensor $A(\alpha)$, and the forms and properties of the inclusions define the tensor P. Let us consider this equation in particular cases.

7.5.1 Homogeneous elastic medium with a set of ellipsoidal inclusions

For ellipsoidal inclusions, the tensor Λ in (7.82) is constant and has the form

$$\Lambda = \left(I + A(a)C^1\right)^{-1}, \tag{7.105}$$

where the tensor $A(a)$ is defined in (3.82), $a_{\alpha\beta} = a_\alpha\delta_{\alpha\beta}$ (no sum on α !), a_1, a_2, a_3 are the ellipsoid semiaxes. As a result, the tensor P in (7.90) takes the form

$$P = \frac{1}{\langle v(a)\rangle}\left\langle C^1\left[I + A(a)C^1\right]^{-1} v(a)\right\rangle, \quad v(a) = \frac{4}{3}\pi a_1 a_2 a_3\,, \tag{7.106}$$

where the averaging is performed over the ensemble distribution of the ellipsoid semiaxes, their orientations, and the elastic properties C^1.

Let the inclusions be spheres of a random radius a isotropically distributed in space. In this case,

$$A(\alpha) = A(a) = A^0, \tag{7.107}$$

where the tensor A^0 has form (3.82), $a_{\alpha\beta} = \delta_{\alpha\beta}$, and Equation (7.104) for the tensor C^* takes the form

$$C^* = C^0 + pC^1\left[I + (1-p)A^0C^1\right]^{-1}. \tag{7.108}$$

Note that for $p = 1$, this equation gives a physically correct result

$$C^* = C^0 + C^1. \tag{7.109}$$

Hence, the equation obtained for C^* is physically correct in the whole range of the volume concentrations of inclusions. Note that when $p \to 1$, the hypothesis of the effective field method are violated because isolated inhomogeneities disappear in this limit.

In the framework of the EFM, it is possible to evaluate not only the effective elastic moduli, but some other statistical characteristics of the elastic fields in matrix composites. According to the main hypothesis of the method, every inclusion in the composite behaves as an isolated one in the matrix phase C^0 by action of the constant external field ε^*. From the solution of the one particle problem (Section 3.3) it follows that the strain tensor inside the inclusion takes the form

$$\varepsilon^+ = \left(I + A(a)C^1\right)^{-1}\varepsilon^*. \tag{7.110}$$

Using the results of Section 3.4 one can find strain concentrations on the inclusions in the composite. Interaction between the inclusions is taken into account through the effective exciting field ε^* defined in (7.102).

Let us consider a composite material that consists of an isotropic matrix and a set of isotropic spheroidal inclusions. The random field of inclusions is statistically isotropic in space. It means that $\Psi(x) = \Psi(|x|)$, and the tensor $A(\alpha)$ in (7.104) is isotropic and can be presented in the form

$$A = A^0 = \frac{1}{9K_A}E^2 + \frac{1}{2\mu_A}\left(E^1 - \frac{1}{3}E^2\right), \tag{7.111}$$

$$K_A = K_0 + \frac{4}{3}\mu_0, \quad \mu_A = \frac{5\mu_0(3K_0 + 4\mu_0)}{6(K_0 + 2\mu_0)}. \tag{7.112}$$

The tensor $P = C^1 A$ in (7.104) may be presented in the P-basis as follows:

$$P = k_p P^2 + 2m_p\left(P^1 - \frac{1}{2}P^2\right) + \lambda_p\left(P^3 + P^4\right) + 4\mu_p P^5 + n_p P^6, \tag{7.113}$$

$$k_p = \frac{1}{\Delta}\left(K_1 + \frac{1}{3}\mu_1 + 3K_1\mu_1 A_6\right), \quad m_p = \mu_1\left(1 + 2\mu_1 A_2\right)^{-1}, \tag{7.114}$$

$$\lambda_p = \frac{1}{\Delta}\left(K_1 - \frac{2}{3}\mu_1 - 6K_1\mu_1 A_3\right), \quad \mu_p = \mu_1\left(1 + \mu_1 A_5\right)^{-1}, \tag{7.115}$$

$$n_p = \frac{1}{\Delta}\left(K_1 + \frac{4}{3}\mu_1 + 12K_1\mu_1 A_1\right), \quad \Delta = 1 + K_1(4A_1 + 4A_3 + A_6)$$
$$+\frac{4}{3}\mu_1\left(A_1 - 2A_3 + A_6\right) + 12K_1\mu_1(A_1A_6 - A_3^2) \tag{7.116}$$

where $K_1 = K - K_0$, $\mu_1 = \mu - \mu_0$, and the coefficients $A_1, ...A_6$ are determined in (3.93)–(3.96).

Let the spheroids be homogeneously distributed over the orientations. Using equations for averaging the P-tensors (Appendix A.5) we obtain that the composite material is macro isotropic with the following bulk K_* and shear μ_* effective elastic moduli

$$K_* = K_0 + pK_P \left(1 - p\frac{K_P}{K_A}\right)^{-1}, \tag{7.117}$$

$$\mu_* = \mu_0 + pG_P \left(1 - p\frac{G_P}{\mu_A}\right)^{-1}, \tag{7.118}$$

$$K_P = \frac{1}{9}\left(4k_p + 4\lambda_p + n_p\right), \tag{7.119}$$

$$G_P = \frac{1}{15}\left(k_p + 6m_p - 2\lambda_p + 6\mu_p + n_p\right). \tag{7.120}$$

When all the inclusions have the same orientation, Equation (7.104) takes the form

$$C^* = C^0 + p\left[(C^1)^{-1} + A(a) - pA(\alpha)\right]^{-1}. \tag{7.121}$$

If the symmetry of the correlation function $\Psi(x)$ coincides with the symmetry of a typical spheroid, the tensor $A(\alpha)$ has form (3.92) with the aspect ratio α different from the aspect ratio γ of a typical spheroid. If these aspect ratios coincide, (7.121) is simplified and takes the form

$$C^* = C^0 + p\left[(C^1)^{-1} + (1-p)A(\alpha)\right]^{-1}. \tag{7.122}$$

The medium has transversal symmetry, and its five independent effective elastic moduli are defined by the expressions

$$k_* = K_0 + \frac{1}{3}\mu_0 + \frac{p}{2\Delta}\left[\frac{1}{3}\left(\frac{1}{3K_1} + \frac{1}{\mu_1}\right) + (1-p)A_6\right], \tag{7.123}$$

$$m_* = \mu_0 + \frac{1}{2}p\left[\frac{1}{2\mu_1} + (1-p)A_2\right]^{-1}, \tag{7.124}$$

$$\lambda_* = K_0 - \frac{2}{3}\mu_0 - \frac{p}{\Delta}\left[\frac{1}{3}\left(\frac{1}{3K_1} - \frac{1}{2\mu_1}\right) + (1-p)A_3\right], \tag{7.125}$$

$$\mu_* = \mu_0 + p\left[\frac{1}{\mu_1} + (1-p)A_5\right]^{-1}, \tag{7.126}$$

$$n_* = K_0 + \frac{4}{3}\mu_0 + \frac{2p}{\Delta}\left[\frac{1}{3}\left(\frac{1}{3K_1} + \frac{1}{4\mu_1}\right) + (1-p)A_1\right], \tag{7.127}$$

$$\Delta = \frac{1}{6K_1\mu_1} + \frac{2}{3}(1-p)\left[\frac{1}{3K_1}\left(A_1 - 2A_3 + A_6\right) + \right.$$
$$\left. + \frac{1}{4\mu_1}\left(4A_1 + 4A_3 + A_6\right)\right] + 2(1-p)^2\left(A_1A_6 - A_3^2\right). \tag{7.128}$$

7.5.2 Elastic medium with a set of spherically layered inclusions

Let a set of multilayered spherical inclusions be homogeneously and isotropically distributed in an isotropic matrix. Suppose for simplicity that all the inclusions have layers of the same sizes and properties, and K_i and μ_i are the bulk and shear moduli of the i-th layer ($i = 1, 2, ..., N$). Using the solution of the problem for a spherically layered inclusion obtained in Section 3.10 it is possible to show, that tensor P in (7.91) takes the form

$$P = P_1 E^2 + P_2 \left(E^1 - \frac{1}{3} E^2 \right), \tag{7.129}$$

where the scalar coefficients P_1^0, P_2^0 are determined by the expressions

$$P_1 = K_0 \sum_{i=k}^{N} \left(\overline{K}_i - 1 \right) \left(1 + Y_9^i \right) \left(\bar{a}_i^3 - \bar{a}_{i-1}^3 \right), \tag{7.130}$$

$$\overline{K}_i = \frac{K_i}{K_0}, \quad \bar{\mu}_i = \frac{\mu_i}{\mu_0}, \quad \bar{a}_i = \frac{a_i}{a_N}, \tag{7.131}$$

$$P_2 = 2\mu_0 \sum_{i=1}^{N} \left(\bar{\mu}_i - 1 \right) \left[\left(1 + 3Y_1^i + 2Y_5^i \right) \left(\bar{a}_i^3 - \bar{a}_{i-1}^3 \right) + \right.$$

$$\left. + \frac{7}{5} a_N^2 \left(3Y_2^i + Y_6^i \right) \left(\bar{a}_i^5 - \bar{a}_{i-1}^5 \right) \right], \tag{7.132}$$

where a_i are the layer radii ($i = 1, 2, ..., N$). The algorithm for the calculation of the constants Y_j^i is presented in Section 3.8.

In this case, the tensor $A(\alpha)$ in (7.100), (7.104) has the form (3.133). The expression for the effective elastic moduli tensor follows from (7.104) in the form

$$C^* = K_* E^2 + 2\mu_* \left(E^1 - \frac{1}{3} E^2 \right), \tag{7.133}$$

$$K_* = K_0 + pP_1 \left(1 - ps_1 \bar{P}_1 \right)^{-1}, \quad \mu_* = \mu_0 + pP_2 \left(2 - ps_2 P_2 \right)^{-1}, \tag{7.134}$$

$$s_1 = \frac{3K_0}{3K_0 + 4\mu_0}, \quad s_2 = \frac{1}{5} (3 - s_1) = \frac{6 (K_0 + 2\mu_0)}{5 (3K_0 + 4\mu_0)},$$

$$\bar{P}_1 = \frac{1}{K_0} P_1, \quad \bar{P}_2 = \frac{1}{\mu_0} P_2, \tag{7.135}$$

where the coefficients P_1 and P_2 are defined in (7.104). For homogeneous inclusions with the bulk K and shear μ moduli, these coefficients take the form

$$P_1 = \frac{K - K_0}{K_0 + s_1 (K - K_0)}, \quad P_2 = \frac{2 (\mu - \mu_0)}{\mu_0 + s_2 (\mu - \mu_0)}. \tag{7.136}$$

The effective elastic moduli K_* and μ_* are determined by the expressions

$$K_* = K_0 + p\frac{K_0\,(K - K_0)}{K_0 + (1-p)\,s_1\,(K - K_0)}\,, \tag{7.137}$$

$$\mu_* = \mu + p\frac{\mu_0\,(\mu - \mu_0)}{\mu_0 + (1-p)\,s_2\,(\mu - \mu_0)}\,. \tag{7.138}$$

For the inclusions that are harder than the matrix, these expressions for K_*, μ_* coincide with the low bound of the Hashin–Strikman variational bounds [58], and for the inclusions softer than the matrix, they coincide with the similar upper bound.

The dependence of the effective Young's modulus E_* of the composite on the volume concentrations of the homogeneous spherical inclusions can be calculated on the base of (7.137), (7.138). These dependencies are presented in Fig. 7.4 by curves 3 for hard inclusions and for porous media. Comparison with experimental data that are also presented in Fig. 7.4 shows that for hard spherical inclusions, this version of the EFM predicts reliable values of the elastic moduli if the value of the volume concentration p does not exceed 0.4 ($p < 0.4$).

7.6 The Mori–Tanaka method

An approach known in the literature as the Mori–Tanaka method is a particular case of the EFM developed in the previous Section. The original derivation in [151] was expressed in terms of eigenstrains, equivalent inclusion, energy transformation, etc., and at the first glance, is totally different in spirit from self-consistent approaches in the mechanics of heterogeneous media. Clarification of the nature of this approximation was proposed in [10]. It was shown that the method is based in fact on the following assumption.

Every inclusion in a composite behaves as an isolated one in the matrix material under the action of a constant external strain (or stress) field that is the same for all the inclusions. This field coincides with the mean strain (or stress) field over the matrix.

Let us consider for simplicity a two-phase composite that consists of a homogeneous matrix and ellipsoidal inclusions randomly distributed in the matrix. We can write obvious relations for the mean stress and strain fields in the composite

$$\langle\sigma\rangle = p\langle\sigma\rangle_{inc} + (1-p)\langle\sigma\rangle_m, \quad \langle\varepsilon\rangle = p\langle\varepsilon\rangle_{inc} + (1-p)\langle\varepsilon\rangle_m, \tag{7.139}$$

where p is the volume concentration of the inclusions, $\langle\cdot\rangle_{inc}$ and $\langle\cdot\rangle_m$ are averaging over the regions occupied by the inclusions and the matrix.

Because Hooke's law holds for every phase, we can write

$$\langle\sigma\rangle = pC\langle\varepsilon\rangle_{inc} + (1-p)C^0\langle\varepsilon\rangle_m = C^0\langle\varepsilon\rangle + pC^1\langle\varepsilon\rangle_{inc}. \tag{7.140}$$

The hypothesis of the Mori–Tanaka method allows us to write for the fields inside the inclusions the following equation:

$$\varepsilon(x) = \Lambda(x)\langle\varepsilon\rangle_m, \quad \Lambda(x) = (I + A(x)C^1)^{-1}, \quad x \in V \tag{7.141}$$

where the function $A(x)$ coincides with the constant tensor $A(a_k)$ defined in (3.82) when $x \in v_k$ $(k = 1, 2, ...)$, $V = \sum_k v_k$. As a result, the mean strain field over the region occupied by the inclusions takes the form

$$\langle\varepsilon\rangle_{inc} = \langle\Lambda(x)\rangle \langle\varepsilon\rangle_m, \tag{7.142}$$

Substitution of this equation into the second equation in (7.139) gives for the mean strain field in the composite the value

$$\langle\varepsilon\rangle = [p\langle\Lambda(x)\rangle + (1 - p)I] \langle\varepsilon\rangle_m, \tag{7.143}$$

and for the mean strain field over the matrix, we obtain

$$\langle\varepsilon\rangle_m = \left[I - p\left\langle A(x)C^1\Lambda(x)\right\rangle\right]^{-1} \langle\varepsilon\rangle. \tag{7.144}$$

Here Equation (7.141) for $\Lambda(x)$ is taken into account.

Equation (7.142) gives the equation for the mean strain field over the inclusions in the form

$$\langle\varepsilon\rangle_{inc} = \langle\Lambda(x)\rangle \left(I - p\langle A(x)C^1\Lambda(x)\rangle\right)^{-1} \langle\varepsilon\rangle. \tag{7.145}$$

Substituting this equation in the right-hand side of (7.140), we obtain finally

$$\langle\sigma\rangle = C^*\langle\varepsilon\rangle, \quad C^* = C^0 + p\langle P(x)\rangle \left[I - p\langle A(x)P(x)\rangle\right]^{-1}, \tag{7.146}$$

$$P(x) = C^1\Lambda(x). \tag{7.147}$$

Equation (7.146) coincides in fact with equation (7.104) obtained in the framework of the EFM.

Equation (7.146) may be extended to multiphase composites as follows:

$$C^* = C^0 + n_0\langle vP(x)\rangle \left(I - n_0\langle vA(x)P(x)\rangle\right)^{-1}, \tag{7.148}$$

where n_0 is the numerical concentration of the inclusions, v is the volume of a typical inclusion, $P(x) = C^1(x)\Lambda(x)$, $C^1(x)$ is the constant tensor $C^1_{(k)}$ when $x \in v_k$.

Note that this formula can be also obtained in the framework of the EFM. Suppose that a set of ellipsoidal inclusions consists of a finite number n of subsets (families) which occupy the regions $V_1, V_2, ..., V_n$. The ellipsoids in each grope are supposed to have the same sizes, elastic property and orientation, but these parameters are different for different groups. As above, we assume that the effective field that acts on each inclusion is constant and the same for all the inclusions. In this case, Equation (7.85) can be written in the form

$$\varepsilon^*(x)=\varepsilon^0-\int K(x-x')P(x')V(x;x')\mathrm{d}x'\cdot\varepsilon^*,\ P(x)=C^1(x)\Lambda^0(x). \quad (7.149)$$

Averaging this equation by the condition $x \in V$ gives

$$\varepsilon^* = \varepsilon^0 + \sum_{m=1}^{n} p_m A^m P^m \varepsilon^*, \quad P^m = C^{1m}\Lambda(a_m) \quad (7.150)$$

$$A^m = -\int K(x-x')\Psi_m(x-x')\mathrm{d}x' \quad (7.151)$$

Here p_k is the volume concentration of the inclusions of the k-th family, $\Psi_m(x-x') = \langle V(x',x)|x,m\rangle$ and $\langle\cdot|x,m\rangle$ is the averaging by the condition that point x belongs to the inclusions of the same family (the orientation m). If the function $\Psi_m(x)$ has the symmetry of the ellipsoidal inclusion of the orientation m, we find from (7.150)

$$\varepsilon^* = \left(I - \sum_{m=1}^{n} p_m A^m P^m\right)^{-1} \varepsilon^0, \quad (7.152)$$

This equation may be written in the form

$$\varepsilon^* = \Lambda^*\varepsilon^0, \quad \Lambda^* = \left(I - n_0\left\langle v(x)A(x)P(x)\right\rangle\right)^{-1}, \quad (7.153)$$

where n_0 is the numerical concentration of inclusions, $A(x) = A^m$, $P(x) = P^m$ inside an inclusion of the orientation m.

Now we can write Equation (7.104) of the EFM for the tensor of the effective elastic moduli of the composite with different families of ellipsoidal inclusions in the form

$$C^* = C^0 + n_0\left\langle v(x)P(x)\right\rangle \Lambda^* =$$
$$= C^0 + n_0\left\langle v(x)P(x)\right\rangle \left(I - n_0\langle v(x)A(x)P(x)\rangle\right)^{-1}. \quad (7.154)$$

This is Equation (7.148) of the Mori–Tanaka method.

Hence, if the symmetry of the function $\Psi_m(x)$ coincides with the symmetry of the inclusion of the orientation m, (7.154) coincides with Equation (7.146) obtained by the Mori–Tanaka method. When the symmetry of the function $\Psi_m(x)$ is the same for all the inclusions ($A(x) = A(\alpha)$), (7.154) is transformed in (7.104).

Simplicity of Equations (7.146), makes the Mori–Tanaka method very attractive. This equation (as well as more general Equation (7.154)) serves for composite materials with ellipsoidal inclusions. The drawback of these equations were noted in [173], [11], [50]. To illustrate this drawback, let us consider a composite material containing two families of spheroidal inclusions of the same sizes. In the first family, the inclusions are homogeneously distributed over orientations and have the volume concentration p_1. The inclusions of the second family have the same orientation, and their volume concentration is p_2. In this case, Equation (7.146) takes the form

$$C^* = C^0 + (p_1\langle P\rangle + p_2 P)D^{-1}, \quad P = C^1\Lambda^0, \tag{7.155}$$

$$D = I - p_1\langle AP\rangle - p_2 AP. \tag{7.156}$$

Here averaging is carried out over the orientations of the inclusions of the first family, the tensor A is defined in (3.92). After averaging the P^i-tensors (Appendix A.5) we obtain

$$\langle P\rangle = K_P E^2 + 2G_P\left(E^1 - \frac{1}{3}E^2\right), \tag{7.157}$$

$$K_P = \frac{1}{9}(4k_P + 4l_P + n_P), \quad G_P = \frac{1}{15}(k_P + 6m_P - 2l_P + 6\mu_P + n_P), \tag{7.158}$$

where the coefficients k_P, m_P, l_P, μ_P and n_P are defined in (7.114)–(7.116). Then, we have for the product AP

$$AP = b_1 P^2 + b_2\left(P^1 - \frac{1}{2}P^2\right) + b_3 P^3 + b_4 P^4 + b_5 P^5 + b_6 P^6, \tag{7.159}$$

$$b_1 = 2A_1 k_P + A_3 l_P, \quad b_2 = 2A_2 m_P, \quad b_3 = 2A_1 l_P + A_3 n_P, \tag{7.160}$$

$$b_4 = 2A_3 k_P + A_6 l_P, \quad b_5 = 2A_5\mu_P, \quad b_6 = A_6 n_P + 2A_3 l_P, \tag{7.161}$$

and averaging (7.159) over the orientation gives

$$\langle AP\rangle = B_1 E^2 + B_2\left(E^1 - \frac{1}{3}E^2\right), \tag{7.162}$$

$$B_1 = \frac{1}{9}(4b_1 + 2b_3 + 2b_4 + b_6), \tag{7.163}$$

$$B_2 = \frac{1}{15}(2b_1 + 6b_2 - 2b_3 - 2b_4 + 3b_5 + 2b_6). \tag{7.164}$$

Thus, for the sum $p_1\langle P\rangle + p_2 P$ we obtain

$$p_1\langle P\rangle + p_2 P = a_1 P^2 + a_2\left(P^1 - \frac{1}{2}P^2\right) + a_3(P^3 + P^4) + a_5 P^5 + a_6 P^6, \tag{7.165}$$

$$a_1 = p_1\left(K_P + \frac{1}{3}G_P\right) + p_2 k_P, \quad a_2 = 2(p_1 G_P + p_2 m_P), \tag{7.166}$$

$$a_3 = p_1\left(K_P - \frac{2}{3}G_P\right) + p_2 l_P, \quad a_5 = 4(p_1 G_P + p_2\mu_P), \tag{7.167}$$

$$a_6 = p_1\left(K_P + \frac{4}{3}G_P\right) + p_2 n_P. \tag{7.168}$$

In the P-basis, the tensor D in (7.155) is

$$D = d_1 P^2 + d_2\left(P^1 - \frac{1}{2}P^2\right) + d_3 P^3 + d_4 P^4 + d_5 P^5 + d_6 P^6, \tag{7.169}$$

$$d_1 = \frac{1}{2} - p_1\left(B_1 + \frac{1}{6}B_2\right) - p_2 b_1, \quad d_2 = 1 - p_1 B_2 - p_2 b_2, \tag{7.170}$$

$$d_3 = -p_1\left(B_1 - \frac{1}{3}B_2\right) - p_2 b_3, \quad d_4 = -p_1\left(B_1 - \frac{1}{3}B_2\right) - p_2 b_4, \tag{7.171}$$

$$d_5 = 2 - 2p_1 B_2 - p_2 b_5, \quad d_6 = 1 - p_1\left(B_1 + \frac{2}{3}B_2\right) - p_2 b_6. \tag{7.172}$$

Using the algebra of the tensors of the P-basis (Appendix A.2) we obtain the final equation for the tensor of the effective moduli of the composite in the form

$$C^* = k_* P^2 + 2m_*\left(P^1 - \frac{1}{2}P^2\right) + l_* P^3 + l'_* P^4 + 4\mu_* P^5 + n_* P^6, \tag{7.173}$$

$$k_* = K_0 + \frac{1}{3}\mu_0 + \frac{1}{\Delta_d}(a_1 d_6 - a_3 d_4), \quad m_* = \mu_0 + \frac{a_2}{2d_2}, \tag{7.174}$$

$$\Delta_d = 2(d_1 d_6 - d_3 d_4),$$

$$l_* = K_0 - \frac{2}{3}\mu_0 + \frac{2}{\Delta_d}(a_3 d_1 - a_1 d_3), \tag{7.175}$$

$$l'_* = K_0 - \frac{2}{3}\mu_0 + \frac{1}{\Delta_d}(a_3 d_6 - a_6 d_4), \tag{7.176}$$

$$\mu_* = \mu_0 + \frac{a_5}{2d_5}, \quad n_* = K_0 + \frac{4}{3}\mu_0 + \frac{2}{\Delta_d}(a_6 d_1 - a_3 d_3). \tag{7.177}$$

Because, in general case, $l_* \neq l'_*$, the tensor C^*_{ijkl} in (7.173) is not symmetric with respect to pairs of its indices: $C^*_{ijkl} \neq C^*_{klij}$, This fact was noted in the mentioned publications devoted to the Mori–Tanaka method. At the same time, when $p_1 = 0$, $p_2 \neq 0$, (or $p_1 \neq 0$, $p_2 = 0$) tensor C^*_{ijkl} has the required symmetry

$$p_1 = p, \quad p_2 = 0, \quad l_* = l'_* = \frac{p}{\Delta_d}\left[K_P - \frac{2}{3}G_P + p(K_P B_2 - 2G_P B_1)\right],$$

$$p_2 = p, \quad p_1 = 0, \quad l_* = l'_* = \frac{p}{\Delta_d}\left[l_P + 2pA_3(k_P n_P - l_P^2)\right]. \tag{7.178}$$

The reason of such a drawback of Equations (7.146) and (7.154) is the assumption that the effective field is the same for all the inclusions. The version of the EFM that is based on other hypothesis and allows one to obtain physically correct equation for the tensor of the effective elastic moduli is considered in Section 7.17. Note that in the framework of Mori–Tanaka method, it is impossible to improve the equation for the tensor C^* and make it physically correct.

7.7 Regular lattices

In this section, the EFM is applied to the calculation of the effective elastic moduli of composites with regular lattices of identical inclusions. This problem has special interest because for regular composites, exact values of the effective elastic moduli may be found. Therefore, there appear a possibility to compare approximate methods with exact solutions and to estimate the area of application of the these approximations.

Let identical inclusions form a regular lattice in a homogeneous matrix. The effective elastic moduli tensor for regular composites is defined by the relation similar to (7.29)

$$\langle\sigma(x)\rangle = C^* \langle\varepsilon(x)\rangle \ , \tag{7.179}$$

where $\varepsilon(x)$, $\sigma(x)$ are the detailed strain and stress fields in the composite medium under the action of a constant external field $\varepsilon^0(\sigma^0)$. The averaging here is performed over the volume of the elementary cell of the periodical lattice, and such averaging is equivalent to averaging over the whole space. The mean values of $\varepsilon(x)$ and $\sigma(x)$ over the whole space coincide with the mean values of these functions at a fixed point x by all possible spatial translations of the regular lattice of inclusions. Hence, the averaging in (7.179) may be performed over the ensemble realizations of the random function $V(x)$ defined by the equation

$$V(x) = V_0(x + r) \ , \tag{7.180}$$

where $V_0(x)$ is the characteristic function of a fixed region occupied by a regular lattice of the inclusions, r is a stochastic vector homogeneously distributed in space. Note that for a constant external strain (stress) field, the ergodic properties of the fields $\varepsilon(x)$ and $\sigma(x)$ are fulfilled automatically, and the general scheme of the EFM developed in the previous sections can be applied to the case of regular structures without any changes.

We start with the construction of the mean $\Psi(x)$ in Equation (7.88) for regular lattices of inclusions. Let all the inclusions have the same ellipsoidal shapes. Thus, there exists a nondegenerate linear transformation of the x-space that converts these ellipsoids into spheres of unit radius. Let us consider Equation (7.94) for the mean $\langle V(x' + x)V(x')\rangle$ in the case of a regular lattice of inclusions. As was mentioned above, the integral $\int V_i(x' + x)V_i(x')dx'$ in this equation is the volume of intersection of two identical ellipsoids the centers of which are separated by the vector x. Thus, we can write

$$\int V_i(x')\, V_i(x'+x)\, dx' = v(a)J(x)\ , \tag{7.181}$$

$$J(x) = \begin{cases} \left(1-\frac{1}{2}|a^{-1}x|\right)^2\left(1+\frac{1}{4}|a^{-1}x|\right)\ , & |a^{-1}x| \le 2\ , \\ 0\ , & |a^{-1}x| > 2 \end{cases}\ , \tag{7.182}$$

where $v(a) = \frac{4}{3}\pi a_1 a_2 a_3$ is the volume of the ellipsoid with semiaxes a_1, a_2, a_3.

For a regular lattice of identical inclusions, the covariance $\langle V(x')V(x'+x)\rangle$ is a periodic function of the variable x, the periods of which coincide with the periods of the lattice composed of the centers of the inclusions. In the vicinity of the point $x = 0$, the form of this function is defined by the first term in (7.94). From here and (7.182) it follows that the covariance $\langle V(x')V(x'+x)\rangle$ has the form

$$\langle V(x')V(x'+x)\rangle = p\sum_m J(\mathbf{x}-\mathbf{m}), \tag{7.183}$$

where p is the volume concentration of the inclusions, $\mathbf{m}$ is the vector of the lattice composed by the centers of inclusions. Thus, the function $\Psi(x)$ in (7.95) is defined by the equation

$$\Psi(x) = \frac{1}{p}\sum_m{}' J(\mathbf{x}-\mathbf{m})\,, \tag{7.184}$$

where the prime over the sum sign means exclusion of the term with $\mathbf{m} = \mathbf{0}$, the function $J(x)$ has the form (7.182). It follows from this equation that the integral in Equation (7.92) for the effective strain field ε^* is presented in the form of the following converging series:

$$\int \mathrm{K}(x-x')\,\Psi(x-x')\,\mathrm{d}x' = -A\ , \tag{7.185}$$

$$A = A^0 - \sum_m{}' p.v. \int \mathrm{K}(x)\left[\frac{1}{p}J(\mathbf{x}-\mathbf{m}) - 1\right]\mathrm{d}x\,. \tag{7.186}$$

Here the tensor A^0 has form (3.82) with $a_{\alpha\beta} = \delta_{\alpha\beta}$, and relations (7.96) and (7.19) are taken into account. Substituting this result into (7.92) and solving the resulting equation with respect to ε^*, we obtain

$$\varepsilon^* = \Lambda^*\varepsilon^0,\ \ \Lambda^* = (I - pAP)^{-1}\,,\ \ P = \frac{1}{v(a)}\int_v C^1(x)\Lambda^0(x)\mathrm{d}x\,, \tag{7.187}$$

where v is the inclusion volume, and the function $\Lambda^0(x)$ is the same as in (7.82). From here and (7.104) it follows that the expression for the tensor of the effective elastic moduli for the composite with a regular lattice of the inclusions has the form

$$C^* = C^0 + pP\left(I - pAP\right)^{-1}. \tag{7.188}$$

Consider an example of an isotropic medium in which spherical isotropic inclusions form a cubic lattice. In this case, the tensor C^* takes the form

$$C^* = C^0 + pC^1\left[I + (1-p)\, A^0 C^1 + p\alpha(p)\Gamma C^1\right]^{-1}. \tag{7.189}$$

Here the tensor A^0 is defined in (3.133), and the components of the tensor Γ in the system of the basis vectors $e^{(1)}, e^{(2)}, e^{(3)}$ of the cubic lattice are defined by the expressions

$$\Gamma_{\alpha\beta\lambda\mu} = \frac{1}{\mu_0(1-\nu_0)}\left[E^2_{\alpha\beta\lambda\mu} + 2E^1_{\alpha\beta\lambda\mu} - 5\sum_{i=1}^{3}\delta_{i\alpha}\delta_{i\beta}\delta_{i\lambda}\delta_{i\mu}\right]. \tag{7.190}$$

The coefficient $\alpha(p)$ (7.189) is presented in the form of a convergent series of the integrals that is similar to (7.186)

$$\alpha(p) = \sum_m{}' \int k\,(x)\left[\frac{1}{p}J(\mathbf{x}-\mathbf{m}) - 1\right]dx\,, \tag{7.191}$$

$$k(x) = \frac{1}{32\pi|x|^3}\left[-3 + 5\left(\left(n\cdot e^{(1)}\right)^4 + \left(n\cdot e^{(2)}\right)^4 + \left(n\cdot e^{(3)}\right)^4\right)\right], \tag{7.192}$$

where $n = x/|x|$, the function $J(x)$ has form (7.182) with $a_{\alpha\beta} = \delta_{\alpha\beta}$. Here the explicit expression (B.14) for the function $\mathrm{K}(x)$ was used. The plot of the function $\alpha(p)$ versus p obtained by numerical summation of this series is presented in Fig. 7.10.

Let us go to 2D-problems. In this case, all the constructions are similar to the 3D-case. The expression for the tensor of the effective elastic moduli has the same form (7.189), where all the tensors are to be replaced by their

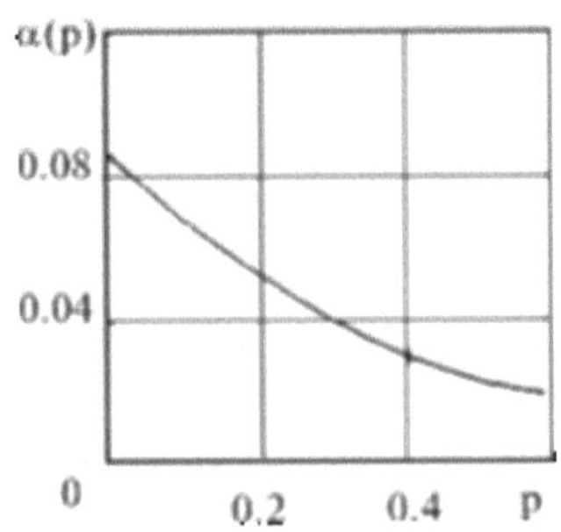

Fig. 7.10. Function $\alpha(p)$.

2D-analogs. For an isotropic medium, the expression for the function $\mathrm{K}(x)$ and the tensor A^0 in (7.186) take the forms ((B.19) in Appendix B.2)

$$\mathrm{K}(x) = \frac{1}{2\pi\mu_0|x|^2}\Big[E^1 - 2E^5 - \frac{\kappa_0}{2}\left(4E^1 - E^2 - 12E^5(n) + 8E^6(n)\right)\Big], \tag{7.193}$$

$$A^0 = \frac{1-\kappa_0}{4\mu_0}E^2 + \frac{2-\kappa_0}{4\mu_0}\left(E^1 - \frac{1}{2}E^2\right), \tag{7.194}$$

where $n = x/|x|$, $\kappa_0 = [2(1-\nu_0)]^{-1}$ for plane strain, and $\kappa_0 = (1+\nu_0)/2$ for plane stress; $E^i(n)$ are the elements of the E-basis for the plane case (Appendix A.6).

Let circular isotropic inclusions form a triangular lattice on a plane. In this case, the integral in (7.185) can be presented in the form

$$\int \mathrm{K}(x-x')\,\Psi(x-x')\,\mathrm{d}x' = -A^0 + p.v.\int \mathrm{K}(x)\,[\Psi(x)-1]\mathrm{d}x, \tag{7.195}$$

where integration is carried out over the whole plane. For the function $\Psi(x)$, properties (7.96), (7.97) hold, and the symmetry of this function coincides with the symmetry of the triangular lattice. The integral on the right-hand side of (7.195) is a tensor whose symmetry coincides with the symmetry of $\Psi(x)$. It is known [132] that the basis of rank four tensors that have the symmetry of a right triangular lattice contains only isotropic tensors. Therefore, the integral in (7.195) has the following tensor structure:

$$p.v.\int \mathrm{K}(x)\,[\Psi(x)-1]\mathrm{d}x = \alpha E^1 + \beta E^2, \tag{7.196}$$

where α and β are scalar coefficients.

Direct checking shows that convolution of the tensor $\mathrm{K}(x)$ (7.195) over the first and the second pairs of indices is zero. One obtains the same result by convolution of this tensor, firstly, over the external, and then, over the internal pairs of indices

$$\mathrm{K}_{\alpha\alpha\beta\beta}(x) = \mathrm{K}_{\alpha\beta\beta\alpha}(x) = 0. \tag{7.197}$$

This equation and (7.196) show that $\alpha = \beta = 0$, and for a triangular lattice of the circular inclusions, the integral in the right-hand side of (7.195) disappears.

Taking into account (7.188) and the solution of the problem for a circular inclusion in a constant external field, we obtain the expression for the tensor of the effective elastic moduli C^* of the medium with a triangular lattice of inclusions

$$C^* = C^0 + pC^1\left[I + (1-p)\,A^0C^1\right]^{-1}, \tag{7.198}$$

where the tensor A^0 has the form (7.194).

If the circular inclusions form a square lattice, the integral (7.196) is not zero. In the plane case, the analogue of the function $J(x)$ in (7.182) is the following function:

$$J(x) = \begin{cases} \frac{2}{\pi}\left[\arctan\left(\frac{\sqrt{1-b}}{b}\right) - b\sqrt{1-b}\right], & x < 2a \\ 0, & x > 2a \end{cases}, \tag{7.199}$$

where a is the inclusion radius, and $b = |x|/2a$. Function $\Psi(x)$ is determined in (7.184), and the integral (7.196) takes the form

$$p.v. \int \mathrm{K}(x)\left[\Psi(x) - 1\right]\mathrm{d}x = \alpha(p)\,\Gamma, \tag{7.200}$$

$$\Gamma_{\alpha\beta\lambda\mu} = \frac{2\kappa_0}{\mu_0}\left(2E^1_{\alpha\beta\lambda\mu} + E^2_{\alpha\beta\lambda\mu} - 4\sum_{i=1}^{2}\delta_{i\alpha}\delta_{i\beta}\delta_{i\lambda}\delta_{i\mu}\right), \tag{7.201}$$

where the components of the tensor Γ are presented in the basis of the vectors $e^{(1)}, e^{(2)}$ of the lattice cell, and the coefficient $\alpha(p)$ has the form of a convergent sum of integrals

$$\alpha(p) = \sum_m{}' \int k(x)\left[\frac{1}{p}J(\mathbf{x} - \mathbf{m}) - 1\right]\mathrm{d}x\,, \tag{7.202}$$

$$k(x) = \frac{1}{8\pi|x|^2}\left[1 - 8\left(n\cdot e^{(1)}\right)^2\left(n\cdot e^{(2)}\right)^2\right], \quad n = \frac{x}{|x|} \tag{7.203}$$

The plot of the function $\alpha(p)$ obtained by numerical summation of this series is presented in Fig. 7.11.

The tensor of the effective elastic moduli of a composite with a square lattice of circular inclusions is defined by the expression

$$C^* = C^0 + pC^1\left[I + (1-p)\,A^0C^1 + p\alpha(p)\Gamma C^1\right]^{-1}, \tag{7.204}$$

where the tensors A^0 and Γ are determined in (7.194) and (7.201).

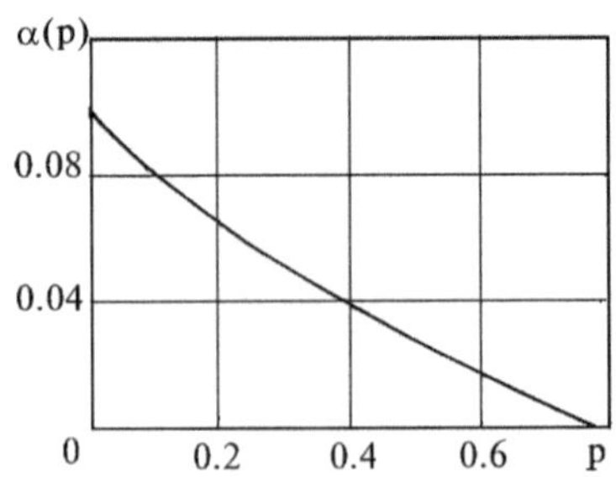

Fig. 7.11. Function $\alpha(p)$ in Equations (7.202), (7.204) for the effective elastic properties of a plane with a square lattice of circular inclusions.

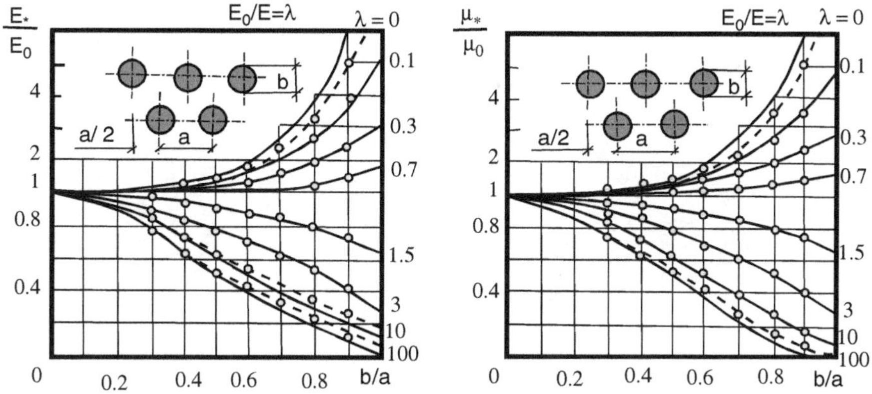

Fig. 7.12. Comparison of exact and approximate values of the effective Young's and shear moduli of composites with a triangle lattice of circular inclusions.

In [57], the stress state in plane periodical lattices of circular inclusions subjected to a constant external field was solved with the help of the expansion of elastic fields in series of doubly periodic functions. The values of the effective elastic moduli found in [57] may be considered as exact ones.

Figure 7.12 presents comparison of exact values of effective elastic moduli and obtained with the help of the EFM for a right-angled triangular lattice of circular inclusions (7.198). Here E_0, μ_0 and E, μ are Young's and shear elastic moduli of the matrix and inclusions; E_*, μ_* are the effective moduli of the composite. The solid lines correspond to the exact solutions, the dashed lines with light circles are predictions of the EFM.

In Figs. 7.13, 7.14, the effective elastic parameters of an elastic plane with a square lattice of the circular inclusions calculated from (7.204) are presented. In the same figures, exact values of these parameters given in [57] are given by solid lines.

Analysis of these figures shows that for the regular structures, that were considered the approximate solutions obtained by the EFM coincide practically with the exact ones when the ratio of the matrix and the inclusions elastic moduli is inside the bounds

$$0,1 \leq \frac{\mathrm{E}_0}{\mathrm{E}} \leq 10\,. \tag{7.205}$$

When $E_0/E > 10$ or $E_0/E < 0,1$, and for the concentration of inclusions $p \geq 0,4$, the error of the EFM solutions is significant and increases with the inclusion concentration.

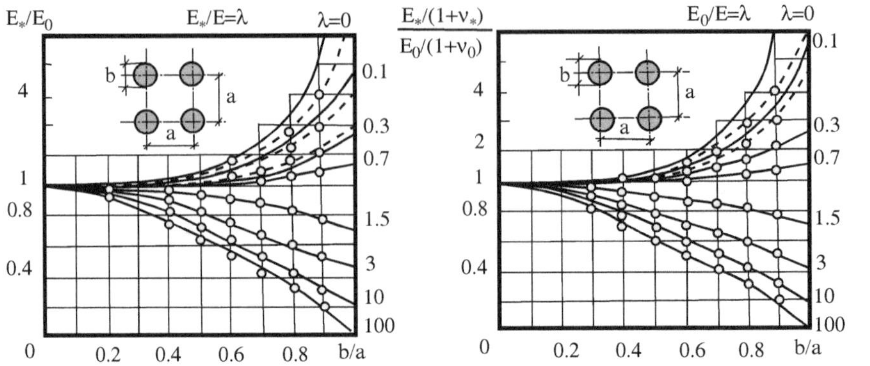

Fig. 7.13. Comparison of the exact and approximate effective Young's and shear moduli of a plane with circular inclusions.

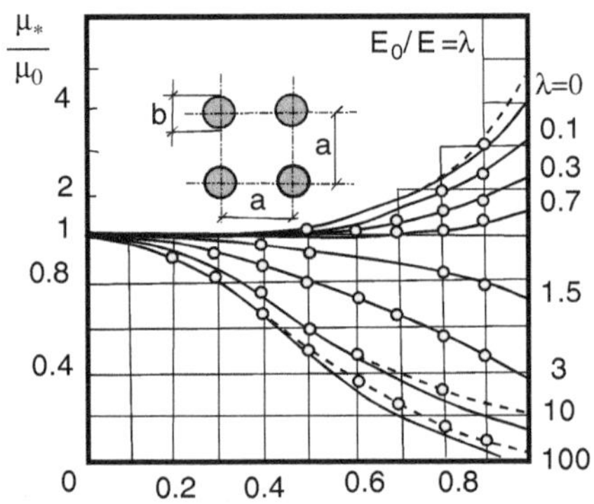

Fig. 7.14. Comparison of the exact and approximate effective shear moduli of composites with square lattices of circular inclusions.

7.8 Thin inclusions in a homogeneous elastic medium

Thin inclusions with elastic moduli that differ appreciatively from the moduli of the matrix are of prime interest for applications. Such inclusions are characterized by two dimensionless parameters. The first, "geometrical" parameter δ_1 is the ratio of the minimal and maximal linear dimension of the inclusions, and this parameter is always small. The second, "physical" parameter δ_2 is the ratio of the elastic moduli of the medium and the inclusions. This parameter is small for inclusions softer than the matrix, and large for inclusions harder than the matrix. The principal terms of the asymptotic expansions for the elastic fields in the vicinity of thin inclusions over the parameters δ_1, δ_2 (or δ_2^{-1}) were obtained in Chapter 4. By the solution of the homogenization problem, only the principal terms of the asymptotic expansions will be taken into account. In what follows, we consider thin inclusions with plane middle surfaces Ω_i ($i = 1, 2, ...$), and denote the unit vector normal to Ω_i by m_i.

We apply the EFM for the approximate solution of the homogenization problem and, as in Section 7.5, assume that the local exciting field ε^* acting on every inclusion in the composite is constant and the same for all the inclusions. As a result, the equations for the strain $\varepsilon(x)$ and stress $\sigma(x)$ fields in the composite with thin inclusions are presented in forms similar to (7.20), (7.21)

$$\varepsilon(x) = \varepsilon^0 - \int \mathrm{K}(x - x')q(x')\mathrm{d}x', \tag{7.206}$$

$$\sigma(x) = \sigma^0 - \int \mathrm{S}(x - x')q(x')\mathrm{d}x', \tag{7.207}$$

$$q(x) = \Lambda(x)\varepsilon^*\Omega(x), \quad \Omega(x) = \sum_i \Omega_i(x). \tag{7.208}$$

Here $\Omega_i(x)$ is the generalized function concentrated on the middle surface of the i-th inclusion Ω_i. The function $\Lambda(x)$ for every inclusion is to be found from the solution of the elasticity problem for an isolated thin inclusion subjected to a constant external field ε^*. In particular, for a thin ellipsoidal inclusion, the equation for the function $\Lambda(x)$ takes the form (see Section 4.6)

$$\Lambda(x) = \Lambda(m_i)z_i(x), \; x \in \Omega_i, \tag{7.209}$$

where the function $z_i(x)$ for the elliptical surface Ω_i with the semi-axes a_1^i, a_2^i is defined in (4.97)

$$z_i(x) = \sqrt{1 - \left(\frac{x_1}{a_1^i}\right)^2 - \left(\frac{x_2}{a_2^i}\right)^2}, \tag{7.210}$$

x_1, x_2 are the Cartesian axes directed along the principal axes of the ellipse Ω_i. The tensor $\Lambda(m_i)$ in (7.209) depends on the inclusion orientation m_i and the elastic moduli of the inclusion and the matrix (Section 5.7).

Let us introduce the generalized function

$$\Omega(x; x') = \sum_{i \neq k} \Omega_i(x') \quad \text{when} \quad x \in \Omega_k. \tag{7.211}$$

This function allows us to express the local exciting field at a point x on the middle surface of an arbitrary inclusion ($x \in \Omega$, $\Omega = \sum_i \Omega_i$) in the following form:

$$\varepsilon^*(x) = \varepsilon^0 - \int \mathrm{K}(x - x')\,\Lambda(x')\,\Omega(x; x')\,\mathrm{d}x'\varepsilon^*. \tag{7.212}$$

Let us average this equation under the condition that the point x is in any middle surface Ω_i, and denote this averaging by $\langle\cdot|x\rangle$. Identifying the mean $\langle\varepsilon^*(x)|x\rangle$ with the effective field ε^* we obtain a closed equation for the tensor ε^*

$$\langle \varepsilon^*(x)|x\rangle = \varepsilon^* = \varepsilon^0 - \int \mathrm{K}(x-x')\,\langle \Lambda(x')\,\Omega(x;x')\,|x\rangle \mathrm{d}x'\varepsilon^*. \tag{7.213}$$

For calculation of the mean under the integral in this equation we take into account (7.206) and assume that the elastic properties of the inclusions are statistically independent of their spatial position. As a result, we obtain

$$\langle \Lambda(x')\,\Omega(x;x')\,|x,m\rangle = \left\langle \bar{\Lambda}(m)\Psi_m(x-x')\right\rangle_m, \tag{7.214}$$

$$\bar{\Lambda}(m) = \langle z(x)\Omega(x)\rangle \Lambda(m), \quad \Psi_m(x-x') = \frac{\langle \Omega(x;x')\,|x,m\rangle}{\langle \Omega(x)\rangle}, \tag{7.215}$$

where $\langle\cdot|x,m\rangle$ denotes the averaging under the condition that the point x is on the middle surface of the inclusion of the orientation m; $\langle\cdot\rangle_m$ is the averaging over the orientations. The function $\Psi_m(x)$ depends on the geometrical peculiarities of the spatial distribution of inhomogeneities in the composite. The definition (7.211) of the function $\Omega(x,x')$ implies the properties of the function $\Psi_m(x)$ in (7.215)

$$\Psi_m(x) = 0 \quad \text{when} \quad x = 0, \tag{7.216}$$

$$\Psi_m(x) \to 1 \quad \text{when} \quad |x| \to \infty. \tag{7.217}$$

As with the function $\Psi(x)$ in (7.88), the function $\Psi_m(x)$ defines the density of distribution of the inclusions around an inclusion of orientation m whose center occupies the origin $x = 0$ of the chosen coordinate system. Let us assume that there exists a linear transformation of the x-space $\alpha(m)$ that converts the function $\Psi_m(x)$ into a spherically symmetric one, i.e.,

$$y = \alpha(m)x\,, \quad \Psi_m\left(\alpha^{-1}(m)y\right) = \Psi_m(|y|)\,. \tag{7.218}$$

In this case, the symmetry of the function $\Psi_m(x)$ coincides with the symmetry of the ellipsoid with the semiaxes are $\alpha_1, \alpha_2, \alpha_3$ defined by the equation

$$|\alpha(m)x| \leq 1. \tag{7.219}$$

Let us substitute (7.214) into (7.213) and calculate the integral in this equation taking into account (7.99), (7.100). As the result, we obtain

$$\varepsilon^* = \varepsilon^0 + \left\langle A(m)\bar{\Lambda}(m)\right\rangle \varepsilon^*, \tag{7.220}$$

$$A(m) = \int \mathrm{K}(x)[1-\Psi_m(x)]\mathrm{d}x, \tag{7.221}$$

where the mean in (7.220) is calculated over the ensemble distributions of the orientations and properties of the inclusions. Solving this equation with respect to ε^* we obtain the effective strain tensor in the form

$$\varepsilon^* = \left[E^1 - \left\langle A(m)\bar{\Lambda}(m)\right\rangle\right]^{-1}\varepsilon^0. \tag{7.222}$$

After averaging Equations (7.206),(7.207) for $\varepsilon(x)$ and $\sigma(x)$ over the ensemble realizations of the random set of inclusions we obtain the mean values of the strain and stress fields in the composite

$$\langle\varepsilon\rangle = \varepsilon^0 - \int \mathrm{K}(x-x')\left\langle\bar{\Lambda}(m)\right\rangle \varepsilon^* \mathrm{d}x', \tag{7.223}$$

$$\langle\sigma\rangle = \sigma^0 - \int \mathrm{S}(x-x')\, B^0 \left\langle\bar{\Lambda}(m)\right\rangle \varepsilon^* \mathrm{d}x'. \tag{7.224}$$

Here the relation

$$\langle\Lambda(x)\Omega(x)\rangle = \langle z(x)\Omega(x)\rangle\langle\Lambda(m)\rangle_m = \langle\bar{\Lambda}(m)\rangle. \tag{7.225}$$

is taken into account.

Suppose that the external strain field is fixed in the problem. Taking into account (7.19), we find from (7.223) and (7.224)

$$\langle\varepsilon\rangle = \varepsilon^0\,, \quad \langle\sigma\rangle = C^*\langle\varepsilon\rangle, \tag{7.226}$$

$$C^* = C^0 + \left\langle\bar{\Lambda}(m)\right\rangle\left[I - \left\langle A(m)\bar{\Lambda}(m)\right\rangle\right]^{-1}, \tag{7.227}$$

Here C^* is the tensor of the effective elastic moduli of the composites with a set of thin inclusions. Special forms of this equation are considered in the next section.

7.9 Elastic medium reinforced with hard thin flakes or bands

Let us apply (7.227) to the calculation of the effective elastic moduli tensor of a composite reinforced with thin hard inclusions. We assume that the inclusions are oblate ellipsoids with the semiaxes a_1, a_2, h $(h/a_1, h/a_2 << 1)$. In this case, the function $\Lambda(x)$ in Equations (7.206), (7.207) for the stress and strain tensors in the medium with inclusions is to be found from the problem for an isolated inclusion subjected to a constant external strain field ε^*. Using the results of Section 4.6, we can write

$$\Lambda(x) = \Lambda(m)z(x), \tag{7.228}$$

$$\Lambda(m) = \left[\frac{1}{2h}\Theta(m)C^{-1}\Theta(m) + U^0(m)\right]^{-1}, \tag{7.229}$$

where the function $z(x)$ is defined in (7.210), C is the tensor of the elastic moduli of the inclusion, the tensor $\Theta(m)$ is defined in (4.12), $U^0(m)$ has form

(4.125)–(4.128), and x_1, x_2 are the principal axes of the middle surface of the inclusion.

Suppose that material of inclusions is transversely isotropic with the isotropy plane coinciding with the middle surface Ω of the inclusion. Let E_{11} and ν_{12} be Young's modulus and Poisson's ratio of the inclusion in the plane x_1, x_2. Then, the tensor $\Theta(m)C^{-1}\Theta(m)$ takes the form

$$\Theta(m)C^{-1}\Theta(m) = \frac{1-\nu_{12}}{2E_{11}}P^2(m) + \frac{1}{2\mu_{12}}\left(P^1(m) - \frac{1}{2}P^2(m)\right), \quad (7.230)$$

where $\mu_{12} = E_{12}/2(1+\nu_{12})$ is the shear modulus of the inclusion in the middle plane, the tensors $P^i(m)$ have been defined in (2.11)–(2.13) and in Appendix A.2.

If Ω is a circle of a radius a, the tensor $\Lambda(m)$ in (7.228) has the form (4.134) ($\xi = a\mu_0/(2h\mu_{12})$)

$$\Lambda(m) = \Lambda_1 P^2(m) + \Lambda_2\left(P^1(m) - \frac{1}{2}P^2(m)\right), \quad (7.231)$$

$$\Lambda_1 = a\mu_0\left[\xi\frac{1-\nu_{12}}{1+\nu_{12}} + \frac{\pi}{8}(2-\kappa_0)\right]^{-1}, \quad (7.232)$$

$$\Lambda_2 = 2a\mu_0\left[\xi + \frac{\pi}{16}(4-\kappa_0)\right]^{-1}. \quad (7.233)$$

Let Ω be a plane band of width $2a$ directed along the x_1-axis ($a_1 \to \infty$, $a_2 = a$). In this case, the nonzero components of the tensor $\Lambda(m)$ have forms similar to (4.136)

$$\Lambda_{1111}(m) = \frac{1}{2\xi}[2\xi + (2-\kappa_0)(1+\nu_{12})]\Lambda_{2222}(m), \quad (7.234)$$

$$\Lambda_{1122}(m) = \nu_{12}\Lambda_{2222}(m), \quad (7.235)$$

$$\Lambda_{2222}(m) = 4a\mu_0[2\xi(1-\nu_{12}) + (2-\kappa_0)]^{-1}, \quad (7.236)$$

$$\Lambda_{1212}(m) = 2a\mu_0(1+2\xi)^{-1}, \quad \xi = a\mu_0(2h\mu_{12})^{-1}. \quad (7.237)$$

Let us calculate the tensor $A(m)$ (7.221) in Equation (7.227) for the tensor of the effective elastic moduli of the composite. We assume that the function $\Psi_m(x)$ has the symmetry of an ellipsoid coaxial with the inclusion of the orientation m. If this ellipsoid is an ellipsoid of revolution with the semiaxes $\alpha_1 = \alpha_2 = \alpha$ and the semiaxis α_3 directed along the normal m to the middle inclusion surface Ω, the tensor $A(m)$ takes the form ($\gamma = \alpha/\alpha_3 > 1$)

$$A(m) = A_1 P^2(m) + A_2\left(P^1(m) - \frac{1}{2}P^2(m)\right) + $$
$$+ A_3\left(P^3(m) + P^4(m)\right) + A_5 P^5(m) + A_6 P^6(m), \quad (7.238)$$

$$A_1 = \frac{1}{2\mu_0}[(1-\kappa_0)\, f_0 + f_1], \quad A_2 = \frac{1}{2\mu_0}[(2-\kappa_0)\, f_0 + f_1], \tag{7.239}$$

$$A_3 = -\frac{1}{\mu_0} f_1\,, \quad A_5 = \frac{1}{\mu_0}\left(1 - f_0 - 4f_1\right), \tag{7.240}$$

$$A_6 = \frac{1}{\mu_0}[(1-\kappa_0)\,(1 - 2f_0) + 2f_1]\,, \quad f_0 = \frac{1-g}{2\,(1-\gamma^2)}\,, \tag{7.241}$$

$$f_1 = \frac{\kappa_0}{4(1-\gamma^2)^2}\left[(2+\gamma^2)\, g - 3\gamma^2\right], \quad g = \frac{\gamma^2}{\sqrt{\gamma^2-1}} \mathrm{arctg}\sqrt{\gamma^2-1}. \tag{7.242}$$

If $\Psi_m(x)$ has the symmetry of an infinite cylinder with the generatrix parallel to x_1 ($\alpha_1 \to \infty$; $\alpha_2, \alpha_3 < \infty$), the essential components of the tensor $A(m)$ in the basis x_1, x_2, x_3 have the forms ($\alpha_2/\alpha_3 = \gamma > 1$)

$$A_{2222}(m) = \frac{\kappa_0}{2\mu_0} \cdot \frac{1}{(1+\gamma)} \left(\frac{3\gamma+2}{1+\gamma} - 4\nu_0\right), \tag{7.243}$$

$$A_{2233}(m) = \frac{\kappa_0 \gamma}{2\mu_0\,(1+\gamma)^2}\,, \tag{7.244}$$

$$A_{3333}(m) = \frac{\kappa_0 \gamma}{2\mu_0\,(1+\gamma)} \left(\frac{2\gamma+3}{\gamma+1} - 4\nu_0\right), \tag{7.245}$$

$$A_{1212}(m) = \frac{1}{4\mu_0\,(1+\gamma)}, \quad A_{1313}(m) = \frac{\gamma}{4\mu_0\,(1+\gamma)}, \tag{7.246}$$

$$A_{2323}(m) = \frac{\kappa_0}{4\mu_0}\left[\frac{1+\gamma^2}{(1+\gamma)^2} + 1 - 2\nu_0\right]. \tag{7.247}$$

In these cases, the symmetry of the function $\Psi_m(x)$ is defined by the scalar parameter γ. If the location of inclusions in the space is statistically independent, γ has the order a/h that is the aspect of inclusion of the orientation m.

7.9.1 Elastic medium with thin hard spheroids (flakes) of the same orientation

For thin hard spheroids (flakes), the tensor $\Lambda(m)$ in Equation (7.227) for C^* is defined in (7.214) and has the form

$$\bar{\Lambda}(m) = n^0 \left\langle \frac{2}{3}\pi a^2 \Lambda(m) \right\rangle, \tag{7.248}$$

where n^0 is the numerical concentration of the inclusions, the tensor $\Lambda(m)$ is defined in (7.231). For the calculation of this tensor the following equation was taken into account

$$\langle z(x)\Lambda(m)\Omega(x)\rangle =$$
$$= \lim_{W\to\infty}\frac{1}{W}\int_W \sum_i z_i(x)\Lambda(m)\Omega_i(x)\mathrm{d}x = n^0\left\langle \frac{2}{3}\pi a^2\Lambda(m)\right\rangle. \tag{7.249}$$

From (7.238) and (7.227) it follows that for the flakes of the same orientations, the composite material is transversely isotropic with the isotropy axis directed along the common normal m to the middle surfaces of the flakes. The tensor of the effective elastic moduli C^* is defined by the equation

$$C^* = C^0 + \frac{\bar{\Lambda}_1}{1-4A_1\bar{\Lambda}_1}P^2(m) + \frac{\bar{\Lambda}_2}{1-A_2\bar{\Lambda}_2}\left[P^1(m) - \frac{1}{2}P^2(m)\right], \tag{7.250}$$

$$\bar{\Lambda}_1 = \frac{2}{3}\pi n^0\langle a^2\Lambda_1\rangle, \quad \bar{\Lambda}_2 = \frac{2}{3}\pi n^0\langle a^2\Lambda_2\rangle, \tag{7.251}$$

where the coefficients Λ_1, Λ_2 and A_1, A_2 are determined in (7.231), (7.238). When $\gamma >> 1$, with the accuracy of the terms of the order γ^{-1}, the following equations hold:

$$A_1 = \frac{\pi\,(2-\kappa_0)}{16\mu_0\gamma}, \quad A_2 = \frac{\pi\,(4-\kappa_0)}{16\mu_0\gamma}. \tag{7.252}$$

7.9.2 Elastic medium with thin hard spheroids homogeneously distributed over the orientations

In this case, the composite is macroscopically isotropic and the equation for the tensor C^* in (7.227) takes the form

$$C^* = K_* E^2 + 2\mu_*\left(E^1 - \frac{1}{3}E^2\right), \tag{7.253}$$

$$K_* = K_0 + \frac{2\bar{\Lambda}_1}{3\left[3 - 4\bar{\Lambda}_1\,(2A_1 + A_3)\right]}, \tag{7.254}$$

$$\mu_* = \mu_0 + \frac{\bar{\Lambda}_1 + 3\bar{\Lambda}_2}{15 - 2\left[2\bar{\Lambda}_1\,(A_1 - A_3) + 3\bar{\Lambda}_2 A_2\right]}, \tag{7.255}$$

where the coefficients $\bar{\Lambda}_{1,2}$ and $A_{1,2,3}$ are defined in (7.251) and in (7.239)–(7.242).

Let us compare these theoretical results with experimental data for the elastic moduli of plastics reinforced with of mica thin flakes ($a/h >> 1$) [134], [161], [162]. In these works, the "flexural" modulus of elasticity of composites in the plane of the flakes was measured. The orientation of the flakes was approximately the same.

In Fig. 7.15, we present the dependence of Young's modulus E^*_{11} (in the plane of the flakes) on the parameter $\tau = \frac{4\pi}{3}\left\langle a^3\right\rangle n^0$ for the polyether resin

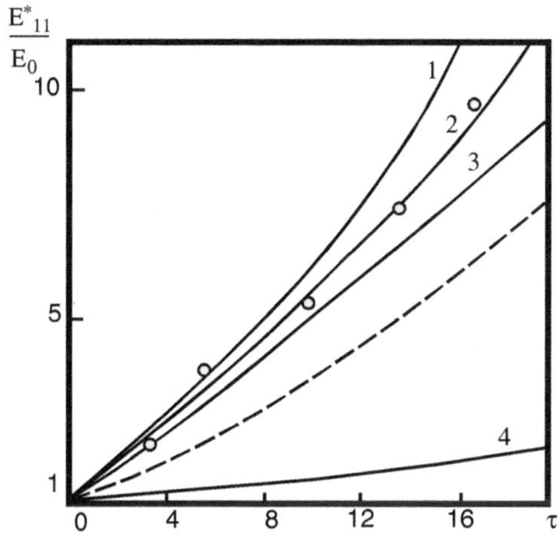

Fig. 7.15. Comparison of the EFM predictions with experimental data for the elastic modulus E^*_{11} of plastics reinforced with mica flakes of the same orientation.

($E_0 = 3.5\text{GPa}$, $\nu_0 = 0.4$) filled with mica flakes with average aspect ratio $\langle \frac{a}{h} \rangle = 39$, $E_{11} = 175GPa$, $\nu_{12} = 0.3$. The parameter τ is connected with the filler volume concentration p by the expression

$$\tau = \left\langle \frac{a}{h} \right\rangle p. \tag{7.256}$$

In Fig. 7.15, lines 1, 2, 3 are the theoretical predictions of E^*_{11} obtained from (7.250) when $\gamma = 10, 20, \infty$, correspondingly; line 4 is the dependence of $E^*_{33}(\tau)$ on τ when $\gamma = 20$, and light dots are the experimental values of E^*_{11} presented in [134]. The difference between the theory and experiments is minimal when $\gamma =< a/2h >$.

More than 100 experimental values of the moduli of elasticity of composites with different ratios of the moduli of the matrix and inclusions, different aspect ratios of the flakes, and different volume concentrations of the fillers [134], [161], [162] were used for analyzing the value of the relative difference Δ between the theoretical E^*_{11} and the experimental E^e_{11} values of the effective Young moduli

$$\Delta = \frac{|E^*_{11} - E^e_{11}|}{E^e_{11}}. \tag{7.257}$$

It turns out that the value of Δ does not depend strongly on the concentration of the fillers and is determined mainly by the values of the parameters γ and ξ

$$\xi = a\mu_0/(2h\mu_{12}) \approx aE_0/(2hE_{11}). \tag{7.258}$$

The minimal values of the relative error Δ are achieved when the parameter γ is taken equal to $< a/2h >$.

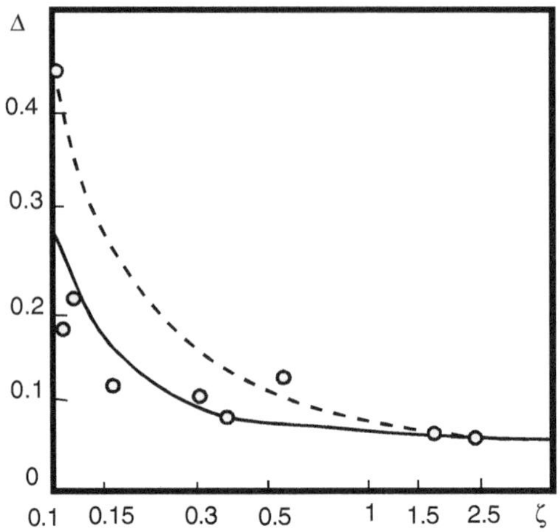

Fig. 7.16. The dependence of the relative difference $\Delta(\xi)$ between theoretical predictions and experimental data for Young's moduli of plastics reinforced with hard flakes on the parameter $\xi = a\mu_0/(2h\mu_{12})$. The solid line corresponds to (7.253), the dashed line corresponds to noninteracting inclusions ($\gamma^{-1} = 0$).

In Fig. 7.16, the solid line corresponds to the minimum of the mean-square deviation of $\Delta(\xi)$ from the "experimental" values presented in the figure by light dots. For the values of Δ, the maximum of the relative differences (7.257) for all the samples with the same ξ but different values of τ was taken. When $\xi > 0.2$ the deviation of E^*_{11} from E^e_{11} did not exceed 10% up to $\tau = 200$. The volume concentration of flakes attained $p = 0.7$ for all the samples [134].

7.9.3 Elastic medium with thin hard unidirected bands of the same orientation

In this case, the medium is orthotropic, and the tensor C^* takes the form

$$C^* = C^0 + \bar{P}. \tag{7.259}$$

The nonzero components of the tensor $\bar{P}$ in the basis of the axes x_1, x_2, x_3 (x_1-axis is directed along the bands, x_2 is in the plane parallel to their middle surfaces Ω, and x_3 is orthogonal to Ω) are determined by the equations

$$\bar{P}_{1111} = \langle p_{11}\rangle + \frac{\langle p_{12}^2 A_{22}\rangle}{1 - \langle p_{22}A_{22}\rangle}, \quad \bar{P}_{1122} = \frac{\langle p_{12}\rangle}{1 - \langle p_{22}A_{22}\rangle}, \tag{7.260}$$

$$\bar{P}_{2222} = \frac{\langle p_{22}\rangle}{1 - \langle p_{22}A_{22}\rangle}, \quad \bar{P}_{1212} = \frac{\langle p_{44}\rangle}{1 - \langle p_{44}A_{44}\rangle}, \tag{7.261}$$

$$p_{\alpha\beta} = \frac{1}{2}\pi a \Lambda_{\alpha\alpha\beta\beta}, \quad \alpha, \beta = 1, 2, \tag{7.262}$$

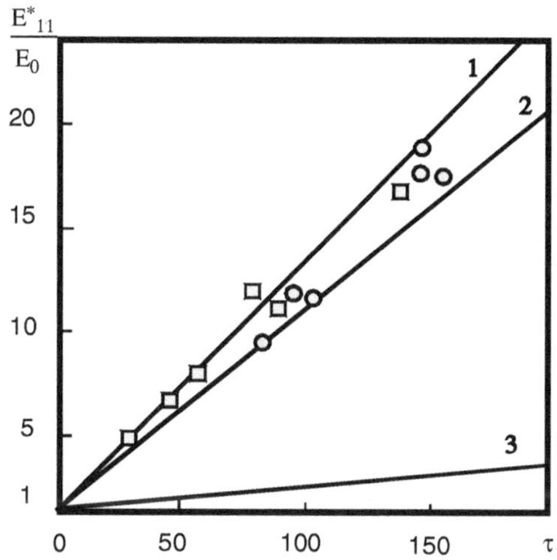

Fig. 7.17. Theoretical and experimental values of the Young moduli E_{11}^*, E_{22}^*, and E_{33}^* for the composites reinforced with thin hard bands.

$$p_{44} = \pi a \Lambda_{1212}\,, \quad A_{22} = A_{2222}(m)\,, \quad A_{44} = 2A_{1212}(m)\,, \tag{7.263}$$

where the components of the tensors $\Lambda_{\alpha\beta\lambda\mu}$ and $A_{\alpha\beta\lambda\mu}$ are defined in (7.209), (7.247).

Figure 7.17, presents theoretical dependences of the elastic moduli E_{11}^*, E_{22}^*, and E_{33}^* (lines 1–3, correspondingly) on the parameter $\tau = \pi n^0 \langle a^2 \rangle = p \langle a/h \rangle$ (p is the volume concentration of the bands), for a plastic reinforced by glass bands with the aspect $a/h = 230$ ($E_0 = 2$ GPa, $\nu_0 = 0.25$, $E_{11} = 59$ GPa, $\nu_{12} = 0.26$). When $\gamma = a/2h$, the theoretical dependences of $E_{ii}^*(\tau)$ turn out to be practically linear in the whole range of experimental values of τ and describe the experimental data with an accuracy of ~10%.

In conclusion, let us discuss possible errors in the predictions of the EFM for thin hard inclusions. These errors have two main sources. The first one is the use of only principal terms in the asymptotic solution of the one-particle problem with respect to the small parameters $\delta_1 = h/a$ and $\delta_2 = E_0/E_{11}$. The corresponding errors have the order $\max(\delta_1, \delta_2)$. Note that for relatively small aspect ratio ($a/h < 10$) and high stiffness of the inclusions ($E_0/E_{11} < 10^{-2}$) when the parameter ξ in (7.258) is small, the error caused by substitution of real flakes for thin ellipsoids increases. The elevation of the curve $\Delta(\xi)$ with small ξ on Fig. 7.16 is apparently related to this fact.

The second source of error is the use of the hypothesis of the quasicrystalline approximation for the description of interactions between many inclusions. But the analysis of experimental data shows that the overall effect of the interactions in these composites is relatively weak. As a result, the dependence of the effective elastic moduli on the volume concentration of the filler is quasilinear. In the framework of the EFM, the influence of the interactions correlates with the values of the parameter γ that describes the symmetry of the correlation function $\Psi_m(x)$. For noninteracting inclusions $\gamma^{-1} = 0$, and the relative error Δ increases in this case (the dashed line in Fig. 7.16 corresponds to $\gamma^{-1} = 0$).

Hence, comparison of theoretical predictions and experimental data leads to the following conclusion. For composites reinforced with thin hard inclusions, the area of applicability of the EFM may be described in terms of the dimensionless parameter

$$\xi = aE_0/hE_{11}. \tag{7.264}$$

When $\xi > 0.15$, the errors of the EFM in predictions of the elastic properties of the composites do not exceed 10–15%. This value of the error does not depend on the volume concentration of the filler. Better correspondence with experimental data is achieved if the value of the parameter γ is taken equal to half of the average aspect ratio of the reinforcing elements. Note that strictly speaking, the value of γ should be determined from the statistical analysis of the distribution of the filler in the actual composite material.

7.10 Elastic media with thin soft inclusions and cracks

In this section, we consider a homogeneous elastic medium with thin inclusions with elastic moduli considerably smaller than the moduli of the matrix ($C(C^0)^{-1} = O(\delta_2)$, $\delta_2 << 1$). For the solution of the homogenization problem, it is convenient to accept that the external stress field is fixed ($\langle\sigma\rangle = \sigma^0$), and the action of the operators K and S on constants is defined by Equations (7.17).

We start with consideration of a medium containing a homogeneous random set of thin soft ellipsoidal inclusions. Following the EFM (Section 7.8) we assume that every inclusion in the composite is subjected to a constant local effective stress field σ^*. The principal terms of the strain and stress fields in the composite with respect to the small parameters δ_2 and $\delta_1 = h/a$ (h and a are the values of the smallest and largest ellipsoid semiaxes) are presented in the form (see Chapter 4)

$$\varepsilon(x) = \varepsilon^0 + \int \mathrm{K}(x - x'^0) M(x') \Omega(x')\, dx' \sigma^*, \tag{7.265}$$

$$\sigma(x) = \sigma^0 + \int \mathrm{S}(x - x')\, M(x')\, \Omega(x')\, dx' \sigma^*, \tag{7.266}$$

$$\Omega(x) = \sum_i \Omega_i(x). \tag{7.267}$$

Here $\Omega_i(x)$ is the delta-function concentrated on the middle surface of the i-th inclusion. The function $M(x)$ for $x \in \Omega_i$ is to be found from the solution of the one-particle problem (Section 4.6). For a thin soft ellipsoid, this tensor has the form

$$M_{ijkl} = m_{(i}B_{j)(k}m_{l)}, \tag{7.268}$$

where the tensor B_{ij} is defined in (4.110), (4.111). For a flattened spheroid, the function $M(x)$ takes the form

$$M(x) = M(m)z(x), \tag{7.269}$$

$$M(m) = M_1 P^5(m) + M_2 P^6(m), \quad z(x) = \sqrt{1 - \frac{|x|^2}{a^2}}, \tag{7.270}$$

where a is the largest semiaxis of the spheroid,

$$M_1 = \frac{a}{\mu_0}\left[\frac{a\mu}{2h\mu_0} + \frac{\pi(2-\nu_0)}{8(1-\nu_0)}\right]^{-1}, \tag{7.271}$$

$$M_2 = \frac{a}{\mu_0}\left[\frac{a(\lambda+2\mu)}{2h\mu_0} + \frac{\pi}{4(1-\nu_0)}\right]^{-1}. \tag{7.272}$$

Here λ, μ are the Lamé coefficient of the inclusion, μ_0, ν_0 are the shear modulus and Poisson's ratio of the medium.

The equation for the effective field σ^* may be obtained as in (7.213), and takes the form

$$\sigma^* = \sigma^0 + \int \mathrm{S}(x-x')\,\langle M(x')\,\Omega(x;x')\,|x\rangle \mathrm{d}x'\sigma^*, \tag{7.273}$$

where the function $\Omega(x;x')$ is defined in (7.211). The mean value of the integrand in this equation is

$$\langle M(x')\,\Omega(x;x')\,|x\rangle = \left\langle \bar{M}(m)\Psi_m(x-x')\right\rangle_m, \tag{7.274}$$

$$\bar{M}(m) = \frac{2}{3}\pi a^2 n_0 M(m), \tag{7.275}$$

where tensor $M(m)$ is defined in (7.270)–(7.272) for a spheroid, and the function $\Psi_m(x)$ has the form (7.215) and satisfies conditions (7.216), (7.217). In what follows, this function is supposed to have the symmetry of an ellipsoid coaxial to the inclusion of the orientation m.

Calculating the integral in (7.273), taking into account (7.274), (7.17), and the properties of $\Psi_m(x)$, we obtain

$$\sigma^* = \sigma^0 - \left\langle D(m)\bar{M}(m)\right\rangle \sigma^*, \tag{7.276}$$

$$D(m) = \int \mathrm{S}(x)[1-\Psi_m(x)]\mathrm{d}x, \quad D(m) = C^0 A(m) C^0 - C^0, \tag{7.277}$$

where the tensor $A(m)$ is defined in (7.221).

After averaging equations (7.265), (7.267) for the strain and stress tensors, and taking into account (7.17) and the equation

$$\langle M(x)\Omega(x)\rangle = \langle \bar{M}(m)\rangle, \tag{7.278}$$

we obtain the equations for the mean strain and stress fields in the medium

$$\langle \varepsilon\rangle = \varepsilon^0 + \langle \bar{M}(m)\rangle \left[I + \langle D(m)\bar{M}(m)\rangle\right]^{-1}\sigma^0, \quad \langle \sigma\rangle = \sigma^0. \tag{7.279}$$

Since $\varepsilon^0 = B^0\sigma^0$, one can rewrite these equations in the forms

$$\langle \varepsilon\rangle = B^*\langle \sigma\rangle, \quad \langle \sigma\rangle = C^*\langle \varepsilon\rangle, \quad C^* = (B^*)^{-1}, \tag{7.280}$$

$$B^* = B_0 + \langle \bar{M}(m)\rangle \left[I + \langle D(m)\bar{M}(m)\rangle\right]^{-1}, \tag{7.281}$$

where $C^*$is the tensor of elastic moduli of the medium with thin soft inclusions.

Let the medium be isotropic and the function $\Psi_m(x)$ have the symmetry of a spheroid with the semiaxes $\alpha_1 = \alpha_2 = \alpha$, and α_3, and this spheroid be coaxial to the inclusion of the orientation m. In this case, the tensor $D(m)$ in the P-basis (2.11)–(2.13) takes the form

$$D(m) = d_1 P^2(m) + d_2\left(P^1(m) - \frac{1}{2}P^2(m)\right) + \\ + d_3\left(P^3(m) + P^4(m)\right) + d_5 P^5(m) + d_6 P^6(m), \tag{7.282}$$

$$d_1 = -\mu_0[4\kappa_0 - 1 - 2(3\kappa_0 - 1) f_0 - 2f_1], \tag{7.283}$$

$$d_3 = -2\mu_0[(2\kappa_0 - 1) f_0 + 2f_1], \tag{7.284}$$

$$d_2 = -2\mu_0[1 - (2-\kappa_0) f_0 - f_1], \tag{7.285}$$

$$d_5 = -4\mu_0 (f_0 + 4f_1), \quad d_6 = -8\mu_0 (\kappa_0 f_0 - f_1), \tag{7.286}$$

where the functions $f_0(\gamma)$, $f_1(\gamma)$ ($\gamma = \alpha/\alpha_3$) are defined in (7.238).

If $\gamma >> 1$, the coefficients d_i ($i = 1, 2, ..., 6$) in (7.282) are transformed into the following ones:

$$d_1 = -\mu_0 (4\kappa_0 - 1) + \frac{\pi\mu_0}{4\gamma}(7\kappa_0 - 2), \tag{7.287}$$

$$d_2 = -2\mu_0 + \frac{\pi\mu_0}{4\gamma}(4-\kappa_0), \quad d_3 = -\frac{\pi\mu_0}{2\gamma}(3\kappa_0 - 1), \tag{7.288}$$

$$d_5 = -\frac{\pi\mu_0}{\gamma}(1 + 2\kappa_0), \quad d_6 = -\frac{\pi\mu_0}{\gamma}\kappa_0. \tag{7.289}$$

In the limit $\gamma \to \infty$, we obtain

$$D(m) = -2\mu_0\left[P^1(m) + (2\kappa_0 - 1) P^2(m)\right]. \tag{7.290}$$

The case $\gamma = 1$ corresponds to a spatial distribution of the inclusion such that there is a spherical area around every inclusion, and the probability of other inclusions to appear in this area is small. In this case, the function $\Psi_m(x)$ has spherical symmetry $\Psi_m(x) = \Psi_m(|x|)$, and the tensor $D(m)$ is an isotropic tensor

$$D(m) = D^0 = \frac{4}{9}\mu_0 (1 - 4\kappa_0) E^2 - \frac{2}{15}\mu_0 (5 + 4\kappa_0) \left(E^1 - \frac{1}{3}E^2 \right) . \quad (7.291)$$

Let us consider Equation (7.280) for B^* in particular cases.

7.10.1 Thin soft inclusions of the same orientation

In this case, Equations (7.280), (7.282) imply that the tensor B^* has the form

$$B^* = B^0 + \frac{4\bar{M}_1}{4 + \bar{M}_1 d_5} P^5(m) + \frac{\bar{M}_2}{1 + \bar{M}_2 d_6} P^6(m) ,$$

$$\bar{M}_1 = \left\langle \frac{\tau}{2\mu_0} \left[\frac{a\mu}{2h\mu_0} + \frac{\pi (2 - \nu_0)}{8 (1 - \nu_0)} \right]^{-1} \right\rangle, \quad (7.292)$$

$$M_2 = \left\langle \frac{\tau}{2\mu_0} \left[\frac{a (\lambda + 2\mu)}{2h\mu_0} + \frac{\pi}{4 (1 - \nu_0)} \right]^{-1} \right\rangle, \quad (7.293)$$

where the inclusions are supposed to be thin spheroids of random radii a ($\tau = 4\pi a^3 n^0/3$). The parameter τ is called the crack density.

For cracks ($\lambda = \mu = 0$) and $\gamma >> 1$, the equation for B^* takes the form

$$B^* = B^0 + \frac{4\langle\tau\rangle}{\pi\mu_0 (1 + 2\kappa_0)} \left(1 - \frac{\tau}{\gamma} \right)^{-1} P^5(m) +$$
$$+ \frac{\langle\tau\rangle}{\pi\mu_0\kappa_0} \left[1 - \frac{\tau (1+\kappa_0)}{\gamma\, \kappa_0} \right]^{-1} P^6(m) . \quad (7.294)$$

The limit $\gamma \to \infty$ corresponds to the case of noninteracting cracks.

7.10.2 Homogeneous distribution of thin soft inclusions over the orientations

In this case, the tensor B^* takes the form

$$B^* = B^0 + \frac{\bar{M}_2}{9 (1 + j_2)} E^2 + \frac{3\bar{M}_1 + 2\bar{M}_2}{15 (1 + j_1)} \left(E^1 - \frac{1}{3}E^2 \right) , \quad (7.295)$$

$$j_1 = \frac{2}{15}\bar{M}_2 (d_6 - d_3) + \frac{1}{10}\bar{M}_1 d_5 , \quad j_2 = \frac{1}{3}\bar{M}_2 (2d_3 + d_6) . \quad (7.296)$$

The equations for the effective bulk K_* and shear μ_* moduli of the composite follow from these equations in the forms

$$K_* = K_0 \left[1 + \frac{K_0 \bar{M}_2}{1 + j_2} \right]^{-1} , \tag{7.297}$$

$$\mu_* = \mu_0 \left[1 + \frac{2\mu_0}{15\,(1 + j_1)} \left(3\bar{M}_1 + 2\bar{M}_2\right) \right]^{-1} . \tag{7.298}$$

If $\lambda = \mu = 0$ and $\gamma \to \infty$, we go to the case of a medium with a set of circular cracks

$$K_* = K_0 \left[1 + \frac{\langle\tau\rangle K_0\,(3K_0 + 4\mu_0)}{\pi\mu_0\,(3K_0 + \mu_0)} \right]^{-1} , \tag{7.299}$$

$$\mu_* = \mu_0 \left[1 + \frac{4\langle\tau\rangle\,(9K_0 + 4\mu_0)\,(3K_0 + 4\mu_0)}{15\pi\,(3K_0 + 2\mu_0)\,(3K_0 + \mu_0)} \right]^{-1} . \tag{7.300}$$

If $\gamma = 1$, these equations take the forms

$$K_* = K_0 \left[1 - M_k\,(1 + s_1 M_k)^{-1} \right] , \tag{7.301}$$

$$\mu_* = \mu_0 \left[1 - M_\mu\,(1 + s_2 M_\mu)^{-1} \right] , \tag{7.302}$$

$$M_k = K_0 \bar{M}_2 , \quad M_\mu = \frac{2\mu_0}{15} \left(3\bar{M}_1 + 2\bar{M}_2\right) , \tag{7.303}$$

$$s_1 = \frac{3K_0}{3K_0 + 4\mu_0} , \quad s_2 = \frac{1}{5}\,(3 - s_1) . \tag{7.304}$$

7.10.3 Elastic medium with regular lattices of thin inclusions

Let thin ellipsoidal inclusions have the same sizes and orientations, and their centers occupy the nodes of a regular lattice in space. Composites with regular lattices of inclusions were considered in Section 7.7. It was shown that the EFM may be applied to the solution of the homogenization problem in this case, and the tensor of the effective moduli C^* of the composite has the same structure as in the stochastic case. But all the averaging in the equation for C^* should be carried out over all possible translations of the regular lattice of inclusions.

We start with the construction of the mean under the integral in (7.273). For regular structure, the function $M(x) = M(x) z(x)$ is the same for all the inclusions, and the mean $\langle M(x) \Omega\,(x; x')\,|x\rangle$ takes the form

$$\langle M(x) \Omega\,(x; x')\,|x\rangle = \left\langle \bar{M}(m) \Psi_m\,(x - x') \right\rangle_m , \tag{7.305}$$

$$\bar{M}(m) = \frac{2}{3}\pi a_1 a_2 n_0 M(m) , \quad \Psi_m\,(x - x') = \frac{\langle z\,(x')\,\Omega\,(x; x')\,|x, m\rangle}{\langle z(x) \Omega(x)\rangle} , \tag{7.306}$$

where $\Omega(x)$ is the sum of delta-functions, concentrated on the regular set of plane elliptical surfaces Ω_i, the function $z(x)$ has form (7.210) on every such surface. Using the definition of a conditional mean [172], we can write the function $\Psi_m(x-x')$ in the form

$$\Psi_m(x-x') = \frac{\langle z(x')\,\Omega(x;x')\,\Omega(x)\rangle}{\langle z(x)\Omega(x)\rangle\langle\Omega(x)\rangle}. \tag{7.307}$$

Here we take into account that all the surfaces Ω_i have the same orientation m. The mean in the numerator of this equation is calculated as follows:

$$\begin{aligned}\langle z(x')\,\Omega(x')\,\Omega(x)\rangle &= \lim_{W\to\infty}\frac{1}{W}\int_W \sum_{i,j} z_i(x')\,\Omega_i(x')\,\Omega_j(x')\,\mathrm{d}\,x' = \\ &= \lim_{W\to\infty}\frac{1}{W}\left[\sum_i \int_W z_i(x')\,\Omega_i(x')\,\Omega_i(x)\,\mathrm{d}\,x' + \right. \\ &\left. + \sum_{i,j\,(i\neq j)}\int_W z_i(x')\,\Omega_i(x')\,\Omega_j(x')\,\mathrm{d}\,x'\right],\end{aligned} \tag{7.308}$$

where the first sum defines the value of this mean in the vicinity of the diagonal $x=x'$. To calculate the terms in the first sum, we define the Cartesian coordinate system on the surfaces Ω_i with the x_3-axis directed along the normal m to Ω_i, and the axes x_1, x_2 coincided with the main axes of the ellipse Ω_i.. Performing the integration first over x_3', and then over x_1', x_2', we find $(x\to x'+x)$

$$\int z_i(x')\,\Omega_i(x')\,\Omega(x'+x)\,\mathrm{d}\,x' = a_1a_2J(x_1,x_2)\,\delta(x_3), \tag{7.309}$$

$$J(x_1,x_2) = \begin{cases}\int\limits_{\eta/2}^{1} \frac{(\xi-\eta)\sqrt{\eta(2\xi-\eta)}+\xi^2[\arcsin(1-\eta/\xi)+\pi/2]}{\sqrt{1-\xi^2}}\xi d\xi\,, & \eta\le 2\\ 0, & \eta>2\end{cases},$$

$$\eta = \sqrt{(x_1/a_1)^2+(x_2/a_2)^2}. \tag{7.310}$$

For a regular lattice of domains Ω_i, the mean (7.308) is a periodic function, and in the vicinity of the diagonal $x=x'$, this function is defined by the right-hand side of (7.309). Taking into account that the denominator in (7.307) has the form

$$\langle z(x)\,\Omega(x)\rangle = \frac{2}{3}\pi a_1a_2n_0,\quad \langle\Omega(x)\rangle = \pi a_1a_2n_0, \tag{7.311}$$

we obtain the equation for the function $\Psi_m(x)$ in the form

$$\Psi_m(x) = \sum_n{}' \bar{J}(\mathbf{x}-\mathbf{n})\,\bar{\Omega}(\mathbf{x}-\mathbf{n}). \tag{7.312}$$

Here $\mathbf{n}$ is the vector of the lattice composed by the centers of the inclusions, $\bar{\Omega}(\mathbf{x}-\mathbf{n})$ is the delta-function concentrated on the plane with the normal $\mathbf{m}$ that goes through the lattice node with the vector $\mathbf{n}$. The prime over the sum sign means excluding the item with $\mathbf{n}=\mathbf{0}$, and the function $\bar{J}(x)$ is defined by the expression

$$\bar{J}(x)=\frac{3J(x)}{2\pi^2 a_1 a_2 n_0}, \tag{7.313}$$

where the function $J(x)$ has form (7.310) on the plane $\bar{\Omega}(x)$. Direct calculation allows one to find that the mean value of the function $\Psi_m(x)$ is equal to 1:

$$\langle\Psi_m(x)\rangle=\lim_{W\to\infty}\frac{1}{W}\int_W \Psi_m(x)\,\mathrm{d}\,x=1. \tag{7.314}$$

Hence, the expression for the tensor $D(m)$ in (7.277) takes the form

$$D(m)=D^0+\sum_n{}' p.v.\int_{v_n} \mathrm{S}(x)\left[1-\bar{J}(\mathbf{x}-\mathbf{n})\,\bar{\Omega}(\mathbf{x}-\mathbf{n})\right]\mathrm{d}x+$$
$$+p.v.\int_{v_0}\mathrm{S}(x)\,\mathrm{d}x\,, \tag{7.315}$$

where v_n is the region of the periodic cell corresponding to the lattice vector $\mathbf{n}$. For an isotropic medium, the tensor D^0 is defined in (7.291). The series in this equation converges absolutely.

The effective stress field σ^* that acts on every inclusion in the lattices, and the tensor of the effective elastic compliances B^* take the forms

$$\sigma^*=\left[I+D(m)\,\bar{M}(m)\right]^{-1}\sigma^0, \tag{7.316}$$

$$B^*=B^0+\bar{M}(m)\left[I+D(m)\,\bar{M}(m)\right]^{-1}, \tag{7.317}$$

where the tensor $\bar{M}(m)$ is defined in (7.275), (7.269)–(7.272).

Let us consider an isotropic medium containing circular cracks of the same orientation with centers that form a lattice with three orthogonal plane of symmetry. For the calculation of the tensor $D(m)$ (7.315) we use Equation (B.26) of Appendix B.2 for the function $\mathrm{S}(x)$. In this case, the effective field σ^* (7.316) takes the form

$$\sigma^*=\frac{1}{1-\alpha}\left[I+\frac{\beta}{1-\alpha-\beta}E^6(m)\right]\sigma^0, \tag{7.318}$$

where the scalar coefficients α and β are determined as follows:

$$\alpha=\frac{4}{2-\nu_0}\left[\gamma_2-(2-\nu_0)\,\gamma_1\right],\quad \beta=\frac{4\,(\nu_0-3)}{2-\nu_0}\gamma_2\,, \tag{7.319}$$

$$\gamma_i = -\sum_{k,l=-\infty}^{\infty} \Gamma_i(k,l,0) +$$

$$+2\sum_{k,l=-\infty}^{\infty}\sum_{m=1}^{\infty}\left[\Gamma_i(k,l,m) - \Gamma_i^0(k,l,m)\right], \quad i=1,2, \tag{7.320}$$

$$\Gamma_i(k,l,m) =$$

$$= \int_0^{2\pi} \mathrm{d}\,\varphi \int_0^2 g_i \bar{d}_1 k + d_0\eta\cos\varphi,\ \bar{d}_2 l + d_0\eta\sin\varphi,\ \bar{d}_3 m J(\eta)\,\eta d\eta, \tag{7.321}$$

$$\Gamma_i^0(k,l,m) =$$

$$= \frac{1}{12\lambda_1\lambda_2\lambda_3}\int_{-\lambda_1}^{\lambda_1}\int_{-\lambda_1}^{\lambda_1}\int_{-\lambda_1}^{\lambda_1} g_i\left(\bar{d}_1 k + \frac{\xi_1}{2},\ \bar{d}_2 l + \frac{\xi_2}{2},\ \bar{d}_3 l + \frac{\xi_3}{2}\right) \mathrm{d}\xi_1 \mathrm{d}\xi_2 \mathrm{d}\xi_3, \tag{7.322}$$

$$g_1(t_1,t_2,t_3) = \frac{t_3^4 - t_1^2\left(t_1^2 + t_2^2 - t_3^2\right)}{\left(t_1^2+t_2^2+t_3^2\right)^{7/2}}, \tag{7.323}$$

$$g_2(t_1,t_2,t_3) = \frac{t_3^2\left(t_3^2 - 3t_1^2\right)}{\left(t_1^2+t_2^2+t_3^2\right)^{7/2}}, \tag{7.324}$$

$$d_0^3 = n_0 a^3,\ \bar{d}_1^3 = n_0 d_1^3,\ \bar{d}_2^3 = n_0 d_2^3, \tag{7.325}$$

$$\bar{d}_3^3 = n^0 d_3^3,\ n_0 = \left(d_1 d_2 d_3\right)^{-1}. \tag{7.326}$$

Here d_1, d_2, d_3 are the dimensions of the elementary cell of the rectangular lattice formed by the centers of the cracks of the radius a, the function $J(\xi)$ is defined in (7.310).

Figure 7.18 presents, the coefficients γ_1, γ_2 as the functions of the parameter d_0 for a regular cubic lattice of cracks; the planes of cracks are parallel to a fixed face of the cube.

For a cubic lattice of cracks, the expression (7.317) for the tensor B^* takes the form

$$B^* = B^0 + \frac{8a^3 n^0 (1-\nu_0)}{3\mu_0(1-\alpha)}\left[\frac{2}{2-\nu_0}E^5(m) + \right.$$

$$\left. + \left(\frac{2\beta}{1-\alpha-\beta} - \frac{\nu}{2-\nu_0}\right)E^6(m)\right]. \tag{7.327}$$

Note that within the framework of the EFM, the stress intensity factor at the crack edges in the lattice may also be estimated. According to the main assumptions of the method, every crack behaves as an isolated one subjected to the effective stresses field σ^* in the form (7.318). For circular cracks subjected to tension along the normal to the crack planes $\sigma^0_{\alpha\beta} = \sigma^0 m_\alpha m_\beta$, the stress intensity factor is defined by the equation

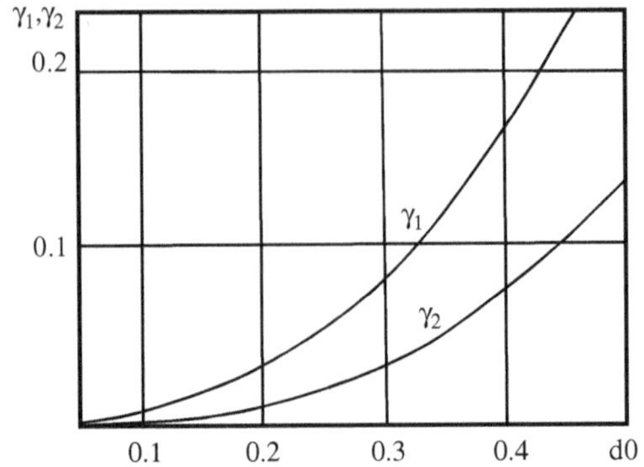

Fig. 7.18. Functions $\gamma_1(\alpha_0)$ and $\gamma_2(\alpha_0)$.

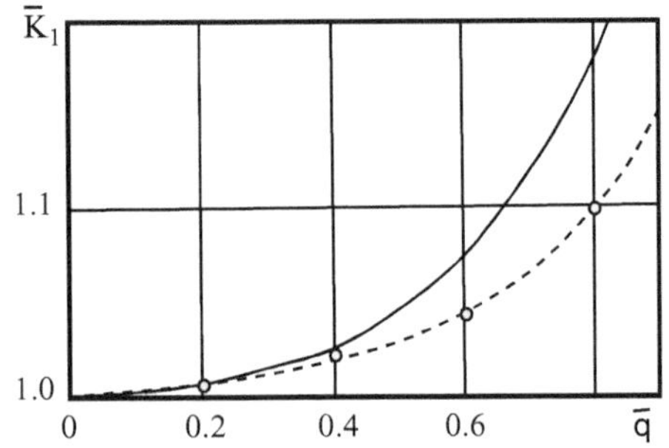

Fig. 7.19. Function $K_1(\overline{q})$.

$$\mathrm{K}_I = 2\sqrt{a/\pi}\sigma^*_{\alpha\beta}m_\alpha m_\beta = 2\sqrt{a/\pi}\sigma^0/(1-\alpha-\beta)\ . \tag{7.328}$$

Consider a set of circular cracks that compose a lattice on a plane. We go to this case from a 3D-lattice tending by letting one of its parameter (d_3) to infinity. Exact solution of the elasticity problem for a square lattice of circular cracks was obtained in [5], [6].

In Fig. 7.19, the stress intensity factor $K_1 = \sqrt{\pi/a}K_I/(2\sigma_0)$ obtained by the EFM (dashed line) is compared with the maximum value of K_1 on the crack edges obtained in [6] (continuous curve), $d_1 = d_2 = d, \quad \overline{q} = 2a/d$.

7.11 Plane problem for a medium with a set of thin inclusions

In this section, we consider the plane problem of elasticity for a medium with a set of thin soft elliptical inclusions or cracks . The EFM can be applied to the plane problem without any additional difficulties. For the plane problem, in the literature, there exists a number of exact solutions and experimental data. Thus, comparison of the EFM predictions with these data allows us to estimate the region of application of the method for the solution of the homogenization problem.

For a plane with a set of thin soft elliptical inclusions, the tensor of the elastic moduli is defined in (7.280), where the tensors $\bar{M}(m)$ and $D(m)$ have the forms

$$\bar{M}(m) = M_1 P^5(m) + M_2 P^6(m), \tag{7.329}$$

$$M_1 = \frac{\pi l^2 n^0}{2\mu_0} (\xi + \kappa_0)^{-1}, \tag{7.330}$$

$$M_2 = \frac{\pi l^2 n^0}{2\mu_0} \left(\xi \frac{\lambda + 2\mu}{\mu} + \kappa_0\right)^{-1}, \quad \xi = \frac{l\mu}{2h\mu_0}, \tag{7.331}$$

$$D(m) = d_1 P^1(m) + d_3 \left(P^3(m) + P^4(m)\right) + d_5 P^5(m) + d_6 P^6(m), \tag{7.332}$$

$$d_1 = -4\mu_0\kappa_0 (1 - 2f_0 + 3f_1), \quad d_3 = 4\mu_0\kappa_0 (f_0 - 3f_1), \quad d_5 = 4d_3, \tag{7.333}$$

$$d_6 = -12\mu_0\kappa_0 f_1, \quad f_0 = \frac{1}{1+\gamma}, \quad f_1 = \frac{2+\gamma}{6(1+\gamma)^2}, \quad \gamma = \frac{\alpha_2}{\alpha_1} > 1. \tag{7.334}$$

Here m is the normal to the middle line of an arbitrary inclusion, l and h are the semiaxes of the letter, λ_0, μ_0 are the Lamé constants for the medium; λ, μ are the same for the inclusions; γ is the aspect ration of the ellipse that defines the symmetry of the correlation function $\Psi_m(x)$. Equation (7.330) implies the equations

$$D(m) \to -4\mu_0\kappa_0 P^1(m) \quad \text{when} \quad \gamma \to \infty, \tag{7.335}$$

$$D(m) = D^0 = -\frac{\mu_0\kappa_0}{2}\left(E^2 + 2E^1\right) \quad \text{when} \quad \gamma = 1. \tag{7.336}$$

Consider Equation (7.280) for the effective compliance tensor B^* in particular cases.

7.11.1 A set of thin soft elliptical inclusions of the same orientation

In this case, the tensor B^* takes the form

$$B^* = B^0 + \frac{\langle M_1 \rangle}{1 + d_3 \langle M_1 \rangle} P^5(m) + \frac{\langle M_2 \rangle}{1 + d_6 \langle M_2 \rangle} P^6(m). \tag{7.337}$$

Here the averaging is performed over the random sizes and properties of inclusions. The coefficients M_1, M_2 and d_3, d_6 are defined in (7.329), (7.330).

7.11.2 Homogeneous distribution of thin inclusions over the orientations

In this case, the tensor B^* is isotropic

$$B^* = B^0 + \frac{\langle M_2\rangle}{4\,(1+j_2)} E^2 + \frac{\langle M_1+M_2\rangle}{4\,(1+j_1)} \left(E^1 - \frac{1}{2}E^2\right), \tag{7.338}$$

$$j_1 = \frac{1}{8}[M_1 d_5 + 2M_2(d_6 - d_3)], \quad j_2 = \frac{1}{2}M_2\,(d_3 + d_6)\,.$$

If $\gamma \to \infty$, then $j_1, j_2 \to 0$, and this is the case of noninteracting inclusions. For cracks ($\lambda = \mu = 0$), the equations for the effective Young's modulus E_* and Poisson's ratio ν_* of the cracked medium follow from (7.338) and take the forms

$$E_* = \frac{E_0}{1+\tau}, \quad \nu_* = \frac{\nu_0}{1+\tau}, \quad \tau = \pi n^0 \langle l^2\rangle\,. \tag{7.339}$$

If $\gamma = 1$, the equation for B^* is

$$B^* = B^0 + \frac{\tau}{4\mu_0\kappa_0\,(2-\tau)\,(4-\tau)}\left[8E^1 + \tau\left(E^2 - 4E^1\right)\right]. \tag{7.340}$$

where E^1, E^2 are the elements of the E-basis for the plane problem (Appendix A.5), and the effective Young's modulus and Poisson's ratio E_* and ν_* are defined by the equations

$$\frac{E_*}{E_0} = \left[1 + \frac{\tau\,(8-3\tau)}{(2-\tau)\,(4-\tau)}\right]^{-1}, \tag{7.341}$$

$$\frac{\nu_*}{\nu_0} = \frac{E_*}{E_0}\left[1 - \frac{\tau^2}{\nu_0\,(2-\tau)\,(4-\tau)}\right]. \tag{7.342}$$

In Fig. 7.20, the dependences (7.339) (line 1) , (7.341) (line 2) and (7.342) (line 3) are compared with experimental data [205]. The experiments were performed for rubber sheets containing random fields of rectilinear cuts. The experimental results are shown by the line with light dots (E_*/E_0) and by the dot-and-dash line (ν_*/ν_0). Note that for the crack field studied in [205], the best predictions of the EFM are obtained for $\gamma = 1$, corresponding to the model of a crack set with restrictions on the cracks intersections. Equations (7.341), (7.342) obtained by the EFM describe the experimental data with an accuracy that is sufficient for applications.

It is interesting to compare the predictions of various self-consistent methods for the plane problem for a medium with cracks. The EFM gives equation (7.341) for the effective Young's modulus of such a medium. The effective medium method (EMM) for the plane crack problem gives the equation

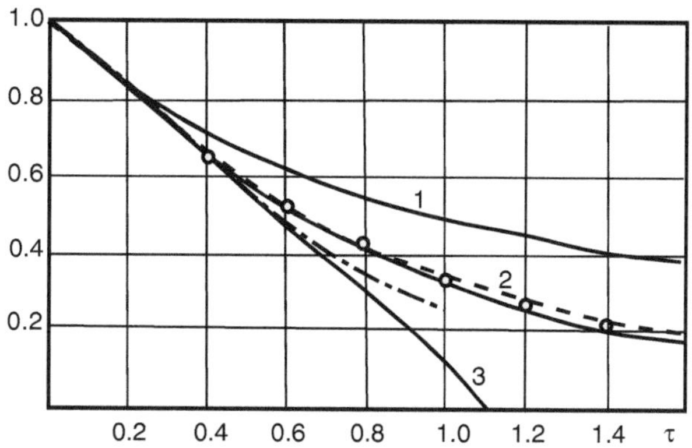

Fig. 7.20. Comparison of the experimental and predictions of the EFM for the elastic moduli of a cracked plane.

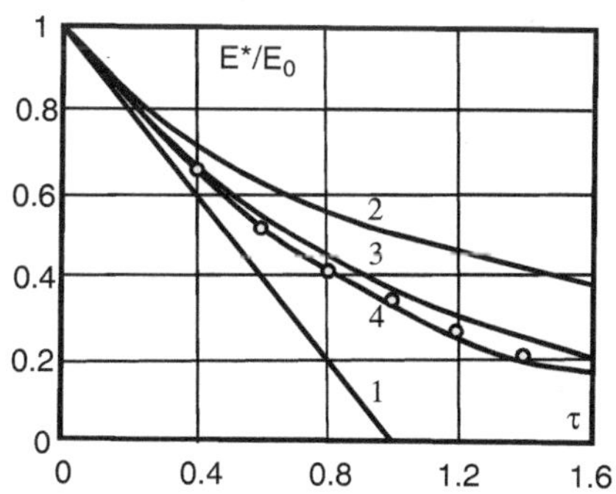

Fig. 7.21. Dependencies of Young's modulus on the crack density for a plane with direct cracks.

$$\frac{1}{E^*} = \frac{1}{E_0} + \frac{\tau}{E^*}, \quad \text{or} \quad \frac{E^*}{E_0} = 1 - \tau. \tag{7.343}$$

The differential effective medium method (DEMM) leads to the differential equation

$$\frac{\mathrm{d}E^*}{\mathrm{d}\tau} = -E^*. \tag{7.344}$$

with the initial condition $E^*(0) = E_0$. The solution of this equation is

$$\frac{E^*}{E_0} = \exp(-\tau). \tag{7.345}$$

In the Fig. 7.21, line 1 corresponds to the normalized effective young modulus obtained by the EMM, line 2 – to the model of noninteracting cracks (7.339), line 3 – to the DEMM, and line 4 – to the EFM. Black dots in this figure are experimental data presented in [205]. Thus, the differential effective medium method and the EFM with $\gamma = 1$ correspond best with the experimental data.

7.11.3 Regular lattices of thin inclusions in plane

For a regular lattice of rectilinear inclusions of the same orientations, the function $\Psi_m(x)$ in (7.312) takes the form

$$\Psi_m(x) = \sum_n{}' J(\mathbf{x}-\mathbf{n})\,\bar{L}(\mathbf{x}-\mathbf{n}), \tag{7.346}$$

where n is the vector of the lattice formed by the centers of the inclusions, the prime over the sum sign means exclusion of the term with $\mathbf{n} = \mathbf{0}$, $\bar{L}(\mathbf{x}-\mathbf{n})$ is the delta-function concentrated on the line with the normal $\mathbf{m}$ that goes through the lattice node with the vector $\mathbf{n}$. The function $J(x)$ is defined by the equation

$$J(x) =$$
$$= \frac{1}{2\pi n^0 l}\left(1-\frac{|x|}{l}\right)\sqrt{\left(2-\frac{|x|}{l}\right)\frac{|x|}{l}} + \arcsin(1-|x|/l) + \pi/2, \quad |x| \le 2l,$$
$$J(x) = 0, \quad |x| > 2l. \tag{7.347}$$

where x is the coordinate along the line. This function is an analogue of the function $J(x)$ in (7.310) in the plane case.

If $\Psi_m(x)$ has form (7.346), Equation (7.330) for the tensor $D(m)$ is presented in the form of the absolutely converging series

$$D(m) = D^0 + \sum_n{}' \left[\int_{v_n} \mathrm{S}(x)dx - \int \mathrm{S}(x)J(\mathbf{x}-\mathbf{n})\bar{L}(\mathbf{x}-\mathbf{n})dx\right]. \tag{7.348}$$

where v_n is the area of the cell that corresponds to the lattice vector $\mathbf{n}$, the tensor D^0 is defined in (7.336). For an isotropic medium, the formal equation for $\mathrm{S}(x)$ has the form (Appendix B.2)

$$\mathrm{S}(x) = \frac{\mu_0\kappa_0}{\pi|x|^2}\left(E^2 + 4E^5(n) - 8E^6(n)\right), \quad n = \frac{x}{|x|}. \tag{7.349}$$

For cracks of the same orientations that compose a lattice with two orthogonal axes of symmetry, the expressions for the effective field σ^* acting on every crack and for the tensor B^* are

$$\sigma^* = \frac{1}{1-\beta}\left[E^1 + \frac{\alpha}{1-\alpha-\beta}E^6(m)\right]\sigma^0, \tag{7.350}$$

$$B^* = B^0 + \frac{\tau}{2\mu_0 \kappa_0 (1-\beta)} \left[E^5(m) + \frac{\alpha}{1-\alpha-\beta} E^6(m) \right], \tag{7.351}$$

where the dimensionless coefficients α and β are the following double series:

$$\alpha = \lambda^2 \sum_{k=-\infty}^{\infty} \sum_{m=1}^{\infty} \left[\int_{-1}^{1} F_\alpha \left(\xi_1 \lambda + (k,m), m\right) j(\xi_1) \, d\xi - \right.$$

$$\left. - \int_{-1/2}^{1/2} d\,\xi_1 \int_{-1/2}^{1/2} F_\alpha \left(\xi_1 + (k,m), \xi_2 + m\right) d\xi_1 \right], \tag{7.352}$$

$$\beta = \frac{\lambda^2}{4} \left\{ \sum_{k=1}^{\infty} \int_{-1}^{1} \frac{j(\xi)}{\xi\lambda + k} d\xi + \sum_{k=-\infty}^{\infty} \sum_{m=1}^{\infty} \left[\int_{-1}^{1} F_\beta \left(\xi_1 \lambda + (k,m), m\right) j(\xi_1) \, d\xi_1 - \right.\right.$$

$$\left.\left. - \int_{-1/2}^{1/2} d\,\xi_1 \int_{-1/2}^{1/2} F_\beta \left(\xi_1 + (k,m), \xi_2 + m\right) d\xi_1 d\xi_2 \right] \right\}, \tag{7.353}$$

$$j(\xi) = \frac{1}{\pi} \left[2\left(1 - 2|\xi|\right) \sqrt{|\xi| \left(1 - |\xi|\right)} + \arcsin\left(1 - 2|\xi|\right) + \frac{\pi}{2} \right], \tag{7.354}$$

$$F_\alpha(\xi_1, \xi_2) = \frac{p_0^2 \xi_2 \left(3\xi_1 - p_0^2\right)}{\left[\xi_1^2 + (p_0 \xi_2)^2\right]^3}, \tag{7.355}$$

$$F_\beta(\xi_1, \xi_2) = \frac{\xi_1^4 + (p_0 \xi_2)^4 - 6(p_0 \xi_1 \xi_2)^2}{\left[\xi_1^2 + (p_0 \xi_2)^2\right]^3}. \tag{7.356}$$

It is supposed that the ξ_2-axis is directed along the normal to the crack lines, and ξ_1-axis is parallel to the cracks, $\lambda = 2l/d_1$, $p_0 = d_2/d_1$, where d_1, d_2 are the distances between the centers of the cracks along the axes ξ_1, ξ_2, correspondingly. The parameter (k,m) depends on the type of the lattice. Particularly, for a square lattice $(k,m) = k$, $p_0 = 1$, and for a triangular lattice $p_0 = \sqrt{3}/2$, $(k,m) = k + \frac{1}{2}[1 - (-1)^m]$.

For tension with the normal stress σ^0 along crack normals, the stress intensity factor for all the cracks is $k_{\rm I}$, and for pure shear with the shear stress τ^0, this factor is $k_{\rm II}$

$$k_{\rm I} = \sqrt{2\pi l}\, K_1 \sigma^0, \quad k_{\rm II} = \sqrt{2\pi l}\, K_2 \tau^0, \tag{7.357}$$

$$K_1 = (1 - \alpha - \beta)^{-1}, \quad K_2 = (1 - \beta)^{-1}. \tag{7.358}$$

Let us consider particular forms of these equations.

7.11.4 A triangular lattice of cracks

Figure 7.22 presents the dependencies of the effective Young's modulus E_*/E_0 (line 1) and shear modulus μ_*/μ_0 (line 2) on the parameter $\lambda = 2l/d$ of the lattice. Here d is the distance between the centers of the cracks. The dependences of the intensity factors K_1, K_2 (7.358) are shown in Fig. 7.23. The solid lines in these figures correspond to the exact solutions obtained in [51]

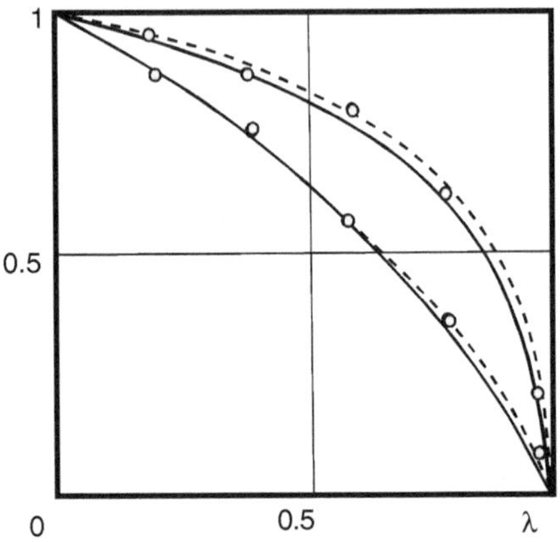

Fig. 7.22. Comparison of the exact and approximate values of the effective Young's E_*/E_0 and shear μ_*/μ_0 moduli of a plane with triangular lattice of direct cracks.

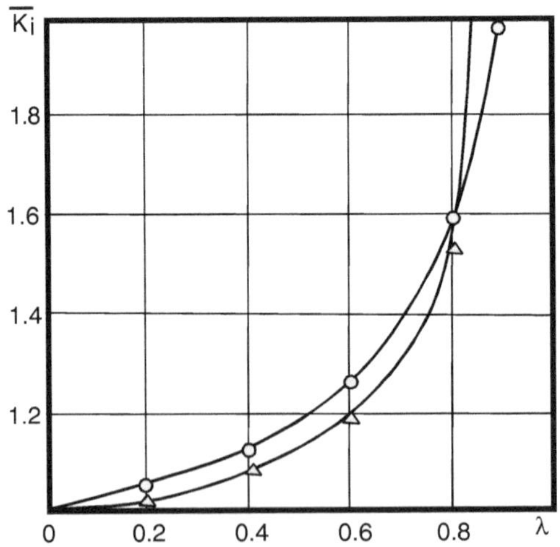

Fig. 7.23. Stress intensity factors for triangle lattice of cracks.

7.11.5 Collinear cracks

The values of the coefficients α and β in (7.352), (7.353) which correspond to this case can be obtained from the solution for a rectangular lattice when one of the characteristic sizes of the latter tends to infinity ($d_2 \to \infty$). In this $\alpha = 0$, and β has the form of the series

$$\beta = \frac{\lambda^2}{4}\sum_{k=1}^{\infty}\int_{-1}^{1}\frac{j(\xi)}{(\xi\lambda + k)^2}d\xi\,, \quad \lambda = 2l/d_1\,, \tag{7.359}$$

where the function $j(\xi)$ is defined in (7.353).

In the table, the value of the stress intensity factor K_1 is compared with the exact value K_1^0 of this factor obtained in [164]

$$K_1^0 = \frac{2\sin(\pi\lambda/2)}{\sqrt{\pi\lambda\sin\pi\lambda}}\,, \tag{7.360}$$

λ	0.2	0.4	0.6	0.8	0.99
K_1	1.02	1.07	1.20	1.50	5.10
K_1^0	1.02	1.07	1.20	1.55	6.20

7.11.6 Vertical row of parallel cracks

In Fig. 7.24, the values of the stress intensity factors k_1 and k_2 obtained by the EFM (line with light dots), are compared with the numerical solutions of

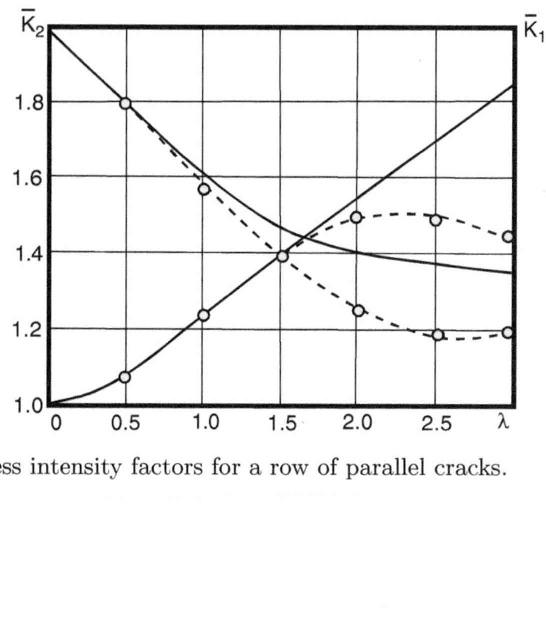

Fig. 7.24. Stress intensity factors for a row of parallel cracks.

the elasticity problem presented in [163] (solid lines). The coefficients α and β in (7.358) have the form

$$\alpha = \lambda^2 \sum_{m=1}^{\infty} \int_{-1}^{1} \frac{3(\xi\lambda)^4 - m^4}{\left[(\xi\lambda)^2 + m^2\right]^3} j(\xi)\, d\xi, \quad \lambda = \frac{2l}{d_2}, \tag{7.361}$$

$$\beta = \frac{\lambda^2}{4} \sum_{m=1}^{\infty} \int_{-1}^{1} \frac{(\xi\lambda)^4 + m^4 - 6m^2(\xi\lambda)^2}{\left[(\xi\lambda)^2 + m^2\right]^3} j(\xi)\, d\xi . \tag{7.362}$$

These examples show that in the plane case, the EFM and the quasicrystalline approximation are in good agreement with the results of more precise calculations of the elastic constants and of the stress intensity factors when the lengths of the cracks are smaller than the distances between their centers. This statement holds for other types of regular crack systems in a plane.

When the distance between the centers of the cracks in a lattice are smaller than the crack lengths, the local exciting field $\sigma^*(x)$ acting on every crack changes essentially, and the error of the EFM is connected with violation of the hypothesis that σ^* is constant.

7.12 Matrix composites reinforced by short axisymmetric fibers

Consider an isotropic elastic medium containing a random set of short axially symmetric fibers. Due to the fact that the fibers are oriented objects, we can use for the solution of the homogenization problem the same variant of the EFM as for the medium with thin inclusions. According to the method we suppose that the effective field ε^* is constant and the same for all the inclusions. The properties of the fibers are characterized by two small parameters: the ratio of the characteristic diameter to the fiber length (δ_1), and the ratio of the characteristic elastic moduli of the medium and the fibers (δ_2). Using the results of Chapter 5, and taking into account only the principal terms of expansion of the elastic fields with respect to these parameters, the equations for the strain and stress fields in the medium with fibers we can present in the forms (see (5.128)–(5.133))

$$\varepsilon(x) = \varepsilon^0 - \int \mathrm{K}(x - x')\, \Lambda(x', m')\, L(x')\, dx' \varepsilon^*, \tag{7.363}$$

$$\sigma(x) = \sigma^0 - \int S(x - x')\, B^0 \Lambda(x', m')\, L(x')\, dx' \varepsilon^*, \tag{7.364}$$

$$L(x) = \sum_i L_i(x), \quad m' = m(x'), \tag{7.365}$$

where $L_i(x)$ is a delta-function concentrated on the axis L_i of the i-th fiber. The function $\Lambda(x)$ on the axis of each fiber is defined by the relation that follows from (5.133)

$$\Lambda(x,m) = \pi a^2(x)\, B(x,m)\ , \quad x \in L\ , \tag{7.366}$$

$$B(x,m) = B^0(m) + b_1(x)\left(P^3(m) + P^4(m)\right) + E_m f(x)\, P^6(m)\ , \tag{7.367}$$

$$B^0(m) = \frac{E_0}{1-\nu_0^2}\left[(3-4\nu_0)\, P^1(m) - \frac{5}{4}P^2(m) + 4(1-\nu_0)\, P^5(m)\right], \tag{7.368}$$

$$b_1(x) = \frac{E_0}{2(1+\nu_0)}\left[\frac{1}{1-2\nu_0} - \left(1 - 2\nu_{1m}\frac{E_m}{E_1}\right) f(x)\right], \tag{7.369}$$

where E_0, ν_0 are Young's modulus and Poisson's ratio of the medium, E_m and E_1 are the Young's moduli of the fiber along its axis (m) and in the transverse direction, ν_{1m} is the Poisson's ratio in the transverse direction. The fiber is supposed to be transversely isotropic with the symmetry axis m. The function $f(x)$ in these relations depends on the fiber shape, and its explicit equation in some particular cases have forms (5.111)–(5.113), where $a(x)$ is the fiber radius.

Let us introduce the function $L(x;x')$

$$L(x;x') = \sum_{i\neq k} L_i(x') \quad \text{when } x \in L_k, \tag{7.370}$$

and present the local exciting field $\varepsilon^*(x)$ acting on the fiber $L_k (x \in L_k)$ in the form

$$\varepsilon^*(x) = \varepsilon^0 - \int \mathrm{K}(x-x')\, \Lambda(x',m')\, L(x;x')\, \mathrm{d}x' \varepsilon^*. \tag{7.371}$$

Averaging this equation under the condition that the point x is on the axis of any fiber, we obtain

$$\langle \varepsilon^*(x)|x\rangle = \varepsilon^0 - \int \mathrm{K}(x-x')\, \langle \Lambda(x',m')\, L(x;x')\, |x\rangle \mathrm{d}x' \varepsilon^*, \tag{7.372}$$

where $\langle \cdot\, |x\rangle$ is the conditional mean and $\langle \varepsilon^*(x)|x\rangle = \varepsilon^*$. If the properties of the fibers are statistically independent on their spatial locations, the mean in the integral in (7.372) is presented in the form

$$\langle \Lambda(x',m')\, L(x;x')\, |x\rangle = \langle \Lambda(m')\Psi_m(x-x')\rangle_m\ , \tag{7.373}$$

$$\Lambda(m) = 2n_0 l \int_{-l}^{l} \Lambda(z,m)\, \mathrm{d}z, \quad \Psi_m(x-x') = \frac{\langle L(x;x')\, |x, m\rangle}{\langle L(x)\rangle}, \tag{7.374}$$

where z is the coordinate along the axis of a typical fiber of length $2l$ with the origin in its center. As with thin inclusions (Section 7.8), the function $\Psi_m(x)$ has the properties

$$\Psi_m(0) = 0; \quad \Psi_m(x) \to 1 \text{ when } |x| \to \infty. \tag{7.375}$$

We assume hereafter that there is a linear transformation of the x-space that converts the function $\Psi_m(x)$ into a spherically symmetric one. Thus, the function $\Psi_m(x)$ has the symmetry of an ellipsoid of revolution with the axis m and the semiaxes equal to $\alpha_1 = \alpha_2 = \alpha$ and α_m.

Calculating the integral in (7.372) for the function $\Psi_m(x)$ with the mentioned properties, we obtain

$$\varepsilon^* = \varepsilon^0 + \langle A(m)\, \Lambda(m)\rangle\, \varepsilon^*, \tag{7.376}$$

$$A(m) = \int \mathrm{K}(x)\,[1 - \Psi_m(x)]\mathrm{d}\, x. \tag{7.377}$$

It was supposed here that the external strain field ε^0 is fixed in the problem, and Equations (7.19) hold. The tensor $A(m)$ for the mentioned function $\Psi_m(x)$ has the form (compare (3.92)–(3.125)) ($\gamma = \alpha/\alpha_m < 1$)

$$A(m) = a_1 P^2 + a_2\left(P^1 - \frac{1}{2}P^2\right) + a_3\left(P^3 + P^4\right) + \\ a_5 P^5 + a_6 P^6, \tag{7.378}$$

$$a_1 = \frac{1}{2\mu_0}[(1-\kappa_0)\, f_0 + f_1], \quad a_2 = \frac{1}{2\mu_0}[(2-\kappa_0)\, f_0 + f_1], \tag{7.379}$$

$$a_3 = -\frac{1}{\mu_0} f_1, \quad a_5 = \frac{1}{\mu_0}(1 - f_0 - 4f_1), \tag{7.380}$$

$$a_6 = \frac{1}{\mu_0}[(1-\kappa_0)(1-2f_0) + 2f_1], \tag{7.381}$$

$$f_0 = \frac{1-g}{2(1-\gamma^2)}, \quad f_1 = \frac{\kappa_0}{4(1-\gamma^2)^2}\left[(2+\gamma^2)\, g - 3\gamma^2\right], \tag{7.382}$$

$$g = \frac{\gamma^2}{2\sqrt{1-\gamma^2}} \ln \frac{1+\sqrt{1-\gamma^2}}{1-\sqrt{1-\gamma^2}}. \tag{7.383}$$

Solving Equation (7.376) with respect to the tensor ε^* we obtain

$$\varepsilon^* = \left[I - \langle A(m)\, \Lambda(m)\rangle\right]^{-1} \varepsilon^0. \tag{7.384}$$

Averaging Equation (7.364) for the stress tensor in the medium with fibers and introducing the notation

$$\langle \Lambda(x, m)\, L(x)\rangle = \langle \Lambda(m)\rangle, \tag{7.385}$$

we obtain

$$\langle \varepsilon \rangle = \varepsilon^0, \ \ \langle \sigma \rangle = C^* \langle \varepsilon \rangle \, , \tag{7.386}$$

$$C^* = C^0 + \langle \Lambda(m) \rangle \left[I - \langle A(m)\, \Lambda(m) \rangle \right]^{-1} . \tag{7.387}$$

Here $C^*$is the tensor of the effective elastic moduli of the composite reinforced by short hard fibers.

Let us consider Equation (7.387) for the tensor C^* in some particular cases. If all the fibers in a composite have the same sizes, properties, and orientations, the tensor $\Lambda(m)$ is defined by the equations $(P^i = P^i(m))$

$$\Lambda(m) = \lambda_1 P^2 + \lambda_2 \left(P^1 - \frac{1}{2} P^2 \right) + \lambda_3 \left(P^3 + P^4 \right) + \lambda_5 P^5 + \lambda_6 P^6, \tag{7.388}$$

$$\lambda_1 = p \frac{\mu_0 (1-\nu_0)}{1-2\nu_0}, \ \ \lambda_2 = 8 p \mu_0 \frac{1-\nu_0}{3-4\nu_0}, \tag{7.389}$$

$$\lambda_5 = 8 p \mu_0 \, , \ \ \lambda_6 - p \mathrm{E}_m \varphi(q) \, , \tag{7.390}$$

$$\lambda_3 = p \frac{\mu_0}{1-2\nu_0} \left\{ 1 - \varphi(q) \left[1 - 2(1-\nu_0)\, \nu_{1m} \frac{\mathrm{E}_m}{\mathrm{E}_1} \right] \right\} . \tag{7.391}$$

Here p is the volume concentration of the fibers, the function $\varphi(q)$ has the form

$$\varphi(q) = \left[\int_{-1}^{1} \alpha^2(\xi)\, \mathrm{d}\xi \right]^{-1} \int_{-1}^{1} \alpha^2(\xi)\, f(\xi, q)\, \mathrm{d}\xi \, , \tag{7.392}$$

$$q^2 = - \frac{\mathrm{E}_0}{\mathrm{E}_m} \frac{(1+\nu_0)^{-1}}{\delta_1^2 \ln \delta_1} = - \frac{2\mu_0}{\mathrm{E}_m \delta_1^2 \ln \delta_1}, \tag{7.393}$$

where $\xi = z/l$, z is the coordinate along the fiber axis, $\alpha(\xi)$ is the function of the fiber shape, $2l$ is the fiber length,

$$\frac{a(\xi)}{l} = \delta_1 \alpha(\xi) . \tag{7.394}$$

The function $f(\xi, q)$ is the same as in (5.111). For the types of the fibers considered in Chapter 5, the function $\varphi(q)$ is defined by the relations that follow from (7.392), (5.111)–(5.113).

a) Cylindrical fibers of the radius a $(\delta_1 = a/l)$

$$\varphi(q) = 1 - \frac{\operatorname{th} q}{q} . \tag{7.395}$$

b) Prolate spheroidal fibers with the semiaxes a and l $(\delta_1 = a/l)$

$$\varphi(q) = \frac{q^2}{2+q^2} . \tag{7.396}$$

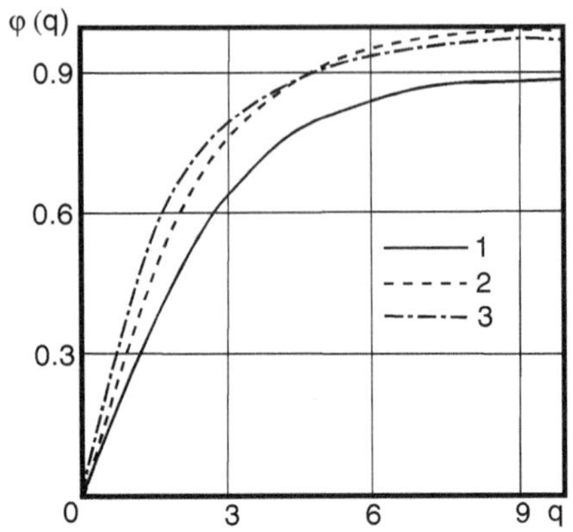

Fig. 7.25. Function $\varphi(q)$ for fibers of various forms.

c) Fibers in the form of a double cone $a(\xi) = a(1 - |\xi|)$, $\delta_1 = a/l$

$$\varphi(q) = \frac{q^2 \beta(5+\beta)}{(q^2-2)(2+\beta)(3+\beta)}, \quad \beta = \frac{1}{2}\left(\sqrt{1+8q} - 3\right). \tag{7.397}$$

The function $\varphi(q)$ characterizes the "underloading" of a fiber of finite length in comparison with the fiber of infinite length in the composite. The graphs of these functions are presented in Fig. 7.25 by the lines 1(a), 2(b), and 3(c).

For unidirected fibers, the equation for the tensor $I - \langle \Lambda(m) A(m) \rangle$ in (7.387)

$$H(m) = I - \langle \Lambda(m) A(m) \rangle \tag{7.398}$$

takes the form ($P^i = P^i(m)$)

$$H(m) = h_1 P^2 + h_2\left(P^1 - \frac{1}{2}P^2\right) + h_3 P^3 + h_4 P + h_5 P^5 + h_6 P^6, \tag{7.399}$$

$$h_1 = \frac{1}{2} - 2\lambda_1 a_1 - \lambda_3 a_3, \quad h_2 = 1 - \lambda_2 a_2, \tag{7.400}$$

$$h_3 = -(2\lambda_1 a_3 + \lambda_3 a_6), \quad h_4 = -(2\lambda_3 a_1 + \lambda_6 a_3), \tag{7.401}$$

$$h_5 = 2 - \frac{1}{2}\lambda_5 a_5, \quad h_6 = 1 - 2\lambda_3 a_3 - \lambda_6 a_6, \tag{7.402}$$

where the coefficients a_i and λ_i are defined in (7.378)–(7.383) and (7.388)–(7.391).

The tensor C^* in (7.387) is transversely isotropic and defined by the Equation ($P^i = P^i(m)$)

$$C^*(m) = k_* P^2 + 2m_*\left(P^1 - \frac{1}{2}P^2\right) + l_*\left(P^3 + P^4\right) + 4\mu_* P^5 + n_* P^6, \tag{7.403}$$

$$k_* = \frac{E_0}{2(1+\nu_0)(1-2\nu_0)} + \frac{1}{\Delta}(\lambda_1 h_6 - \lambda_3 h_4) , \tag{7.404}$$

$$m_* = \frac{E_0}{2(1+\nu_0)} + \frac{\lambda_2}{2h_2} , \tag{7.405}$$

$$l_* = \frac{E_0 \nu_0}{(1+\nu_0)(1-2\nu_0)} + \frac{1}{\Delta}(\lambda_3 h_6 - \lambda_6 h_4) , \tag{7.406}$$

$$\mu_* = \frac{1}{2}\left(\frac{E_0}{1+\nu_0} + \frac{\lambda_5}{h_5}\right) , \tag{7.407}$$

$$n_* = \frac{E_0(1-\nu_0)}{(1+\nu_0)(1-2\nu_0)} + \frac{2}{\Delta}(\lambda_6 h_1 - \lambda_3 h_3) , \quad \Delta = 2(h_1 h_6 - h_3 h_4) . \tag{7.408}$$

Connection of the constants $k_*, m_*, \mu_*, l_*, n_*$ with the "technical" elastic moduli of the transversely isotropic solid is given in (3.380).

Figure 7.26 presents the dependence of Young's modulus E_3^* (along the axis of reinforcement) on the volume concentration p of the fibers for the polyamide thermoplastic ($E_o = 2.5GPa, \nu_0 = 0.35$), reinforced by unidirected cylindrical glass fibers ($E_m = 70GPa, \nu_{1m} = 0.22, \delta_1 = 0.04$). The most satisfactory agreement of Equation (7.403) (the solid lines in Fig. 7.26) with experimental data [174] (light circles) is achieved if the parameter γ in (7.378) is taken equal to δ_1 (in Fig. 7.26, line 1 corresponds to $\gamma = \delta_1$, line 2 to $\gamma = 2\delta_1$).

Figure 7.27 presents the graphs of the Young moduli $E_1^*(p)$ and $E_3^*(p)$ for the composite reinforced with a periodic lattice of unidirected fibers in the form of parallelepipeds. For an isotropic medium and the inclusions ($E_0/E_m = 0.05, \nu_m = 0.3, \nu_0 = 0.35$), these moduli were calculated in [1] by the finite element method. Solid lines $1(E_3^*)$ and $1'(E_1^*)$ correspond to $\delta_1 = 0.1$; $2(E_3^*)$ and $2'(E_1^*)$ – $\delta_1 = 0.01$. When $\delta_1 = 0.1$, Equation (7.403) is in satisfactory agreement with the "exact" solution [1] if $\gamma = 3\delta_1$ (the

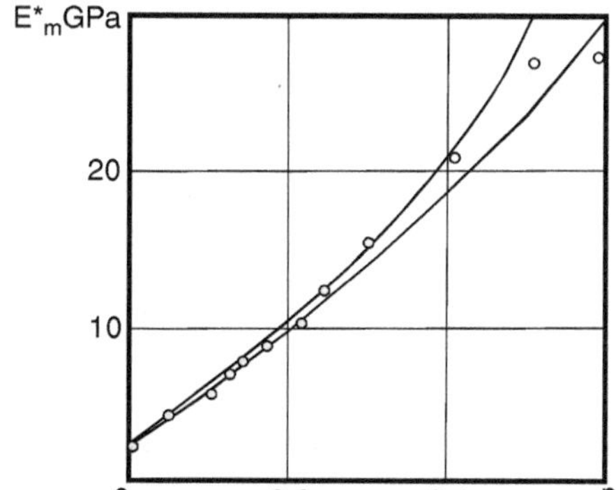

Fig. 7.26. The effective Young modulus E_3^* of the fiber reinforced composite versus the volume concentration p of the fibers.

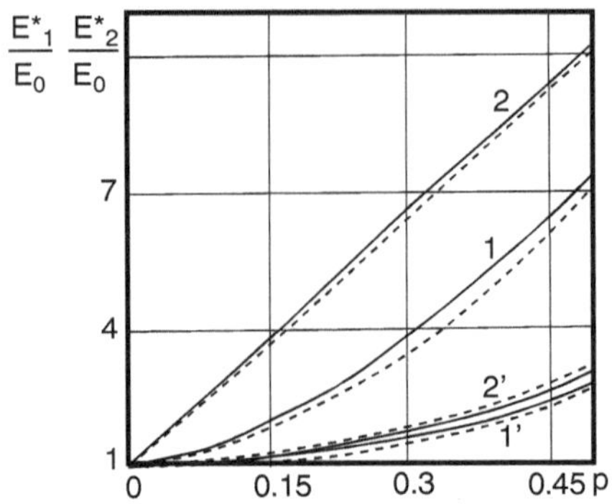

Fig. 7.27. Comparison of the approximate exact values of the effective Young modulus of the medium reinforced with a regular lattice of cylindrical fibers.

dashed lines in Fig. 7.20). When $\delta_1 = 0.01$ and $\gamma = O(\delta_1)$, Equation (7.403) practically coincided with the results obtained in [1].

Let the fibers be homogeneously distributed over the orientations. In this case, the composite is macroscopically isotropic, and its effective bulk K_* and shear μ_*, elastic moduli (7.386) have the forms

$$K_* = K_0 + \frac{\Lambda_1}{1 - j_1}, \quad \mu_* = \mu_0 + \frac{\Lambda_2}{1 - j_2}, \tag{7.409}$$

$$\Lambda_1 = \frac{1}{9}[4(\lambda_1 + \lambda_3) + \lambda_6], \tag{7.410}$$

$$\Lambda_2 = \frac{1}{15}\left[\lambda_1 + 3\lambda_2 - 2\left(\lambda_3 - \frac{3}{4}\lambda_5\right) + \lambda_6\right], \tag{7.411}$$

$$j_1 = \frac{1}{3}[8\lambda_1 a_1 + 6\lambda_3 a_3 + 4(\lambda_1 a_3 + \lambda_3 a_1) + 2(\lambda_3 a_6 + a_3\lambda_6) + \lambda_6 a_6], \tag{7.412}$$

$$j_2 = \frac{2}{15}\left[2\lambda_1 a_1 + 3(\lambda_3 a_3 + \lambda_2 a_2) - 2(\lambda_1 a_3 + \lambda_3 a_1) - (\lambda_3 a_6 + \lambda_6 a_3) + \right.$$
$$\left. + \frac{3}{4} a_5\lambda_5 + \lambda_6 a_6\right], \tag{7.413}$$

where the coefficients λ_i, a_i are defined in (7.378)–(7.383) and (7.388)–(7.391).

Figure 7.28 presents theoretical (solid lines) and experimental (light circles [135]) dependences of Young's modulus E_* and Poisson's ratio ν_* on the volume concentration p of the fibers for a composite with a thermoset plastic matrix ($E_0 = 2.25 GPa, \nu_0 = 0.4$) reinforced by randomly oriented cylindrical glass fibers ($E_m = 77\ GPa, \nu_m = 0.25, \delta_1 = 0.002$). The solid lines are obtained from (7.403)–(7.408) when $\gamma = 2\delta_1$.

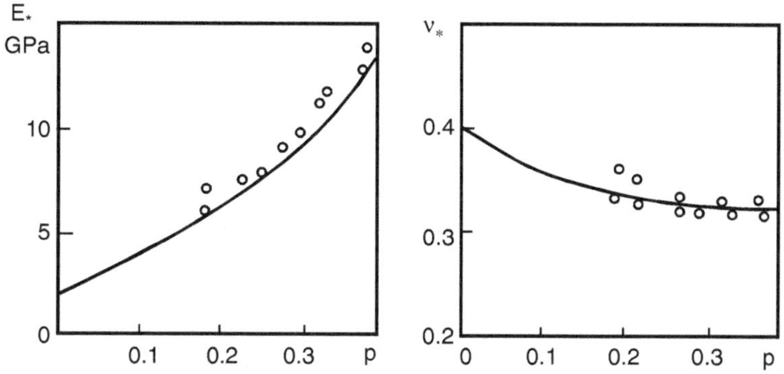

Fig. 7.28. Comparison of the theoretical predictions and experimental data for Young's modulus and Poisson's ratio of a composite reinforced with a random set of cylindrical fibers homogeneously distributed over orientations.

The results obtained in this section lead us to the following conclusions.

1. If the function $\alpha(\xi)$ does not change essentially along the fiber length, then for $q > 10$ (very long fibers) and for $q < 0.1$ (very hard fibers) the shape of the fibers practically does not influence the effective elastic moduli of the composite, and the tensor C^*depends only on the fiber volume concentration p $(q > 10)$, or on the dimensionless parameter $\chi = -n_0 l^3/\ln\delta_1, (q < 0.1)$.
2. In composites reinforced by fibers in the forms of ellipsoids or double cones, the material of the filler is used more rationally than in composites with cylindrical fibers. This follows from the fact that the graphs of the function $\varphi(q)$ corresponding to an ellipsoid or double cone (lines 2,3 in Fig.7.25) are above the graph corresponding to the cylinder (line 1).
3. The comparison of the predictions of the EFM with experimental data show that the parameter γ appearing in the theory should have the order of the average aspect δ_1 of the fibers. The parameter γ is maximal for a spatially correlated (regular) system of fibers $(\gamma \approx 3\delta_1)$, and it turns out to be smaller $(\gamma = (1 \div 2)\delta_1)$ for a random set of fibers. This corresponds to the geometrical meaning of the parameter γ which is connected with the probability of the intersection of fibers. For stochastic fiber sets, when the fiber axes can approach each other on the distance of the order of their diameters $(2a)$, the aspect γ has the order δ_1. For regular lattices of the fiber, the value of γ is defined by the ratio of the characteristic sizes of the elementary cell of the fiber lattice, and exceeds δ_1. The value of γ depends on the technological peculiarities of the composite formation and has to be determined from the stochastic analysis of the actual composite microstructure.

7.13 Elastic medium reinforced with unidirectional multilayered fibers

Consider a composite reinforced by a set of unidirectional infinite cylindrical fibers. Suppose that each fiber consists of N cylindrical layers with different elastic properties, the material of the matrix and the fibers is transversely isotropic with the symmetry axis parallel to the reinforcing direction. Distribution of the fibers is supposed to be statistically homogeneous and isotropic; the external field is a constant strain ε^0.

According to the EFM we assume that each fiber behaves as an isolated one subjected to a local exciting field ε^* that is constant and the same for all the inclusions. The field $\varepsilon(\bar{x})$ inside the fiber is presented in the form

$$\varepsilon(\bar{x}) = [I + A(\bar{x})]\varepsilon^*, \tag{7.414}$$

where $\bar{x} = \bar{x}(x_1, x_2)$, and x_1, x_2 are the Cartesian coordinates in the plane orthogonal to the fiber axis. The tensor $A(\bar{x})$ is defined by the relations (3.412). The algorithm of application of the EFM to the solution of the homogenization problem is similar to that developed in Section 7.3 for the medium with ellipsoidal inclusions. The equations for the tensor ε^* and the tensor of effective elastic moduli C^* are represented in a form similar to (7.102), (7.104) (n^0 is the numerical concentration of the fibers)

$$\varepsilon^* = (I - pAP)^{-1}\varepsilon^0, \quad p = \pi\langle a_N^2\rangle n^0, \tag{7.415}$$

$$C^* = C^0 + pP(I - pAP)^{-1}. \tag{7.416}$$

Here the tensors P and A are determined in (3.127))

$$P = \langle v\rangle^{-1}\left\langle \int_v C^1(\bar{x})[I + A(\bar{x})]d\bar{x}\right\rangle, \quad \langle v\rangle = \pi\langle a_N^2\rangle, \tag{7.417}$$

$$A(m) = \frac{1-\kappa_0}{4m_0}P^2(m) + \frac{2-\kappa_0}{4m_0}\left(P^1(m) - \frac{1}{2}P^2(m)\right) + \\ + \frac{2}{\mu_0}P^5(m), \; \kappa_0 = \frac{k_0}{k_0 + m_0}, \tag{7.418}$$

where v is the region in the plane (x_1, x_2) occupied by the inclusion with the external radius a_N, $\mathbf{m}$ is the unit vector along the fiber axis, k_0, m_0, μ_0 are the elastic moduli of the matrix. Note that the tensors C^0 and $C = C^0 + C^1$ in the $P^i(m)$-basis are presented in the forms ($r = |\bar{x}|$)

$$C^0 = k_0P^2 + 2m_0\left(P^1 - \frac{1}{2}P^2\right) + l_0\left(P^3 + P^4\right) + 4\mu_0P^5 + n_0P^6, \tag{7.419}$$

$$C(r) = k(r)P^2 + 2m(r)\left(P^1 - \frac{1}{2}P^2\right) + l(r)\left(P^3 + P^4\right) + \\ + 4\mu(r)P^5 + n(r)P^6. \qquad (7.420)$$

For a cylindrically layered inclusion, the elastic moduli of the i–th layer ($a_{i-1} < r < a_i$, $i = 1, 2, ...N$, $a_0 = 0$, a_i is the radius of the i-th layer) are $k_i, m_i, l_i, \mu_i, n_i$.

Equations (3.412), (3.460)–(3.463) transform the equation for the tensor P in (7.417) into

$$P = p_1P^2 + p_2\left(P^1 - \frac{1}{2}P^2\right) + p_3\left(P^3 + P^4\right) + p_5P^5 + p_6P^6, \qquad (7.421)$$

$$p_1 = \left\langle \sum_{i=1}^{N} (k_i - k_0)(1 + Y_9^i)(\bar{a}_i^2 - \bar{a}_{i-1}^2) \right\rangle, \quad \bar{a}_i = a_i/a_N, \qquad (7.422)$$

$$p_2 = \left\{ 2\sum_{i=1}^{N} (m_i - m_0)\left[\left(1 + 2Y_1^i + 2Y_5^i\right)\left(\bar{a}_i^2 - \bar{a}_{i-1}^2\right) + \right.\right. \\ \left.\left. +3a_N^2\left(Y_2^i + Y_6^i\right)\left(\bar{a}_i^4 - \bar{a}_{i-1}^4\right)\right]\right\}, \qquad (7.423)$$

$$p_3 = \left\langle \sum_{i=1}^{N} (l_i - l_0)\left(1 + Y_9^i\right)\left(\bar{a}_i^2 - \bar{a}_{i-1}^2\right) \right\rangle, \qquad (7.424)$$

$$p_5 = \left\langle \frac{1}{2}\sum_{i=1}^{N} \mu_1^i\left(2 + Y_{13}^i\right)\left(\bar{a}_i^2 - \bar{a}_{i-1}^2\right) \right\rangle, \qquad (7.425)$$

$$p_6 = \left\langle \sum_{i=1}^{N} \left[n_i - n_0 + 2\left(l_i - l_0\right)Y_{11}^i\right]\left(\bar{a}_i^2 - \bar{a}_{i-1}^2\right) \right\rangle, \qquad (7.426)$$

where the constants Y_j^i are determined from the solution of the one-particle problem for an isolated cylindrical multilayered inclusion subject to a constant external field. The algorithm of the calculation of these constants is described in Section 3.11.

The tensor C^* (7.416) in the $P^i(m)$-basis is presented in the form

$$C^* = k_*P^2 + 2m_*\left(P^1 - \frac{1}{2}P^2\right) + l_*\left(P^3 + P^4\right) + 4\mu_*P^5 + n_*P^6, \qquad (7.427)$$

$$k_* = k_0 + \frac{pp_1}{1 - 4pp_1b_1}, \quad m_* = m_0 + \frac{pp_2}{2\left(1 - pp_2b_2\right)}, \qquad (7.428)$$

$$l_* = l_0 + \frac{pp_3}{1 - 4pp_1b_1}, \quad \mu_* = \mu_0 + \frac{pp_5}{4 - pp_5b_5}, \qquad (7.429)$$

$$n_* = n_0 + pp_6 + p^2 \frac{4p_3^2 b_1}{1 - 4pp_1 b_1}, \tag{7.430}$$

$$b_1 = \frac{1 - \kappa_0}{4\mu_0}, \quad b_2 = \frac{2 - \kappa_0}{4\mu_0}, \quad b_5 = \frac{1}{2\mu_0}. \tag{7.431}$$

For homogeneous fibers ($N = 1$), the effective elastic constants of the composite take the forms

$$m_* = m_0 \left[1 + \frac{p\,(m - m_0)}{m_0 + (1-p)\,\beta_0\,(m - m_0)} \right], \tag{7.432}$$

$$\mu_* = \mu_0 \left[1 + \frac{2p\,(\mu - \mu_0)}{2\mu_0 + (1-p)\,(\mu - \mu_0)} \right], \tag{7.433}$$

$$l_* = l_0 + p\frac{(l - l_0)\,(k_0 + m_0)}{\Delta}, \quad k_* = k_0 + p\frac{(k - k_0)\,(k_0 + m_0)}{\Delta}, \tag{7.434}$$

$$n_* = n_0 + p\,(n - n_0) - p\,(1-p)\,\frac{(l - l_0)^2}{\Delta}, \tag{7.435}$$

$$\Delta = pk_0 + (1-p)\,k + m_0\,, \quad \beta_0 = \frac{k_0 + 2m_0}{2(k_0 + m_0)}. \tag{7.436}$$

Figure 7.29 compares the effective elastic parameters of the composite reinforced with unidirected cylindrical fibers calculated from (7.427)–(7.436) with the experimental data presented in [38].

In [38], the composites with the epoxy resin matrix ($E_0 = 5.27GPa$, $\nu_0 = 0.3$) and the transversely isotropic carbon fibers ($E_1 = 8GPa$, $E_3 = 410.6GPa$, $\nu_{12} = 0.568$, $\nu_{13} = 0.273$, $\mu_3 = 10.2GPa$) are studied. Line 1 in Fig. 7.29 corresponds to the theoretical dependence of the parameter $\lambda_* + 2m_*$ on the volume concentration of the fibers (light circles ($\circ$) are the experimental data), $2 - m_*(\bullet)$, $3 - \mu_*(\times)$, $4 - n_*(\blacksquare)$.

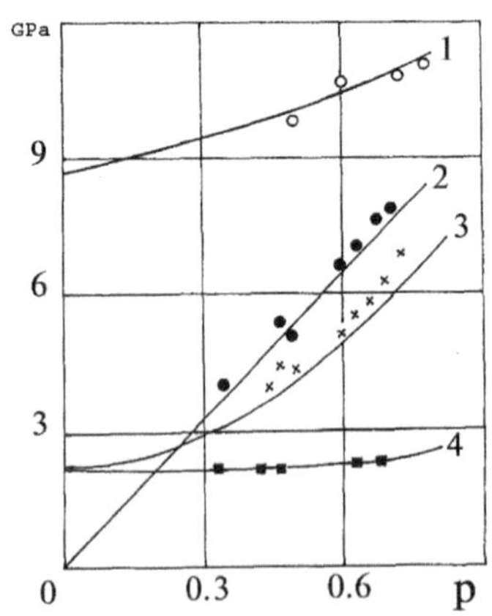

Fig. 7.29. Comparison of the EFM predictions for the effective elastic parameters of the composite reinforced with unidirectional cylindrical fibers with experimental data.

7.14 Thermoelastic deformation of composites with multilayered spherical or cylindrical inclusions

Consider thermal strains and stresses in a composite material reinforced with multilayered spherical or cylindrical inclusions, and subject to a constant temperature field. The medium is supposed to be unconstrained at infinity. We start with the calculation of the mean strain field in a composite material containing a random set of multilayered spherical inclusions. We use the EFM and suppose that every inclusion in a composite behaves as an isolated one in a homogeneous matrix with the tensor of elastic moduli C^0 and the tensor of temperature expansion coefficients α^0. A constant temperature field T and the effective exciting stress field σ^* induced by surrounding inhomogeneities act on every inclusion. In the framework of the method, the temperature stresses in the region v_i occupied by the i-the inclusion satisfy equation (3.21)

$$\sigma(x) - \int_{v_i} S(x - x') B_i^1(x') \sigma(x') \mathrm{d}x' =$$

$$= \int_{v_i} S(x - x') \alpha_i^1(x') \mathrm{d}x' + \overset{*}{\sigma} \tag{7.437}$$

$$B_i^1 = B_i - B^0, \quad \alpha_i^1 = \alpha_i - \alpha^0, \quad T = 1, \tag{7.438}$$

where B_i^1, α_i^1 are the perturbations of the tensors of elastic compliances and thermal expansion coefficients in the region v_i.

The solution of Equation (7.437) is presented in the form

$$\sigma(x) = \sigma^T(x) + \sigma^s(x), \tag{7.439}$$

where the functions $\sigma^T(x)$ and $\sigma^s(x)$ satisfy equation (7.437) with the right-hand side $\int_{v_i} \mathrm{S}(x - x')\alpha_i^1(x')\mathrm{d}x'$ and σ^*, correspondingly. The problem of the temperature stresses $\sigma^T(x)$ in the medium with an isolated multilayered spherical inclusion was solved in Section 3.9. In a spherical coordinate system (r, n) with the origin at the inclusion center, the equation for the tensor $\sigma^T(x)$ takes the form

$$\sigma^T_{\alpha\beta}(r, n) = 3k_i \left(Y_{11}^i - \alpha_i^1\right) \delta_{\alpha\beta} + 2\mu_i Y_{12}^i r^{-3} \left(\delta_{\alpha\beta} - 3n_\alpha n_\beta\right),$$

$$a_{i-1} < r < a_i, \quad i = 1, 2, ..., N+1, \quad a_0 = 0, \quad a_{N+1} = \infty. \tag{7.440}$$

Here we suppose that the inclusion consists of N layers with the external radii a_i. The bulk K_i, and shear μ_i moduli and thermal expansion coefficient α_i inside every layer are constants, the materials of the layers are isotropic. The algorithm of calculation of the constants Y_{11}^i and Y_{12}^i was developed in Section 3.9.

The tensor $\sigma^s(x)$ in (7.439) is the solutions of the problem for a multilayered spherical inclusion in the matrix phase subjected to a constant external field σ^*. It follows from the results of Chapter 3 that the solution of this problem is presented in the form

$$\sigma^s(x) = C(x)\left[I + A(x)\right]B^0\sigma^*, \tag{7.441}$$

where the expression for the tensor $A(x)$ is defined in (3.288), (3.327)–(3.329).

If $V(x)$ is the characteristic function of the region occupied by the inclusions, the strain and stress tensors in the composite in the temperature field $T = 1$ satisfy the equations

$$\sigma(x) = \int \mathrm{S}(x - x')\left[B^1(x')\sigma(x') + \alpha^1(x')\right]V(x')\,\mathrm{d}x', \tag{7.442}$$

$$\varepsilon(x) = \varepsilon^0 + \int \mathrm{K}(x - x')C^0\left[B^1(x')\sigma(x') + \alpha^1(x')\right]V(x')\,\mathrm{d}x'. \tag{7.443}$$

The expression for the effective stress field σ^* follows from (7.439), (7.441), and (7.442) after averaging the latter equation over the realizations of the random set of inclusions by the condition $x \in V$

$$\sigma^* = \int \mathrm{S}(x - x')\left\langle\left[\Lambda^T(x') + \Lambda^s(x')\sigma^*\right]V(x;x')\,|x\right\rangle \mathrm{d}x', \tag{7.444}$$

$$\Lambda^T(x) = B^1(x)\sigma^T(x) + \alpha^1(x), \tag{7.445}$$

$$\Lambda^s(x) = B^1(x)C(x)\left[I + A(x)\right]B^0. \tag{7.446}$$

If the properties of the inclusions are statistical independent of their spatial locations, the mean in (7.444) is presented as follows:

$$\left\langle\left[\Lambda^T(x') + \Lambda^s(x')\sigma^*\right]V(x;x')\,|x\right\rangle = \left(\bar{\Lambda}^T + \bar{\Lambda}^s\sigma^*\right)\Psi(x - x'), \tag{7.447}$$

$$\bar{\Lambda}^T = n_0\left\langle\int_v \Lambda^T(x)\,\mathrm{d}x\right\rangle, \quad \bar{\Lambda}^s = n_0\left\langle\int_v \Lambda^s(x)\,\mathrm{d}x\right\rangle, \tag{7.448}$$

$$\Psi(x - x') = \frac{\langle V(x;x')\,|x\rangle}{\langle V(x)\rangle}, \tag{7.449}$$

where v is the volume of a typical inclusion, n^0 is the numerical concentration of the inclusions.

Substituting (7.447) into (7.444) and calculating the integral which regularization is defined in (7.17), we obtain

$$\sigma^* = -D\left(\bar{\Lambda}^T + \bar{\Lambda}^s\sigma^*\right), \quad D = \int \mathrm{S}(x)\left[1 - \Psi(x)\right]\mathrm{d}x. \tag{7.450}$$

In what follows, we suppose that the function $\Psi(x)$ is spherically symmetric ($\Psi(x) = \Psi(|x|)$). For such a function, the tensor D^0 is defined in (7.291). Solving Equation (7.450) with respect to σ^*, we obtain

$$\sigma^* = -\left[I + D\bar{\Lambda}^s\right]^{-1} D\bar{\Lambda}^T. \tag{7.451}$$

Averaging Equation (7.443) for the strain tensor, we go the equation for the mean strain field in the form

$$\langle\varepsilon\rangle = \alpha^0 + \bar{\Lambda}^T + \bar{\Lambda}^s \sigma^*. \tag{7.452}$$

Because in this equation $\langle\varepsilon\rangle$ is the mean strain of the composite medium by the temperature change in one grade, $\langle\varepsilon\rangle$ coincides with the tensor $\alpha^*$of the temperature expansion coefficients. Thus, α^* is defined by the equation

$$\alpha^* = \alpha^0 + \left(I + D\bar{\Lambda}^s\right)^{-1} \bar{\Lambda}^T, \tag{7.453}$$

$$\bar{\Lambda}^s = \langle\Lambda_1\rangle E^2 + \langle\Lambda_2\rangle \left(E^1 - \frac{1}{3}E^2\right), \tag{7.454}$$

$$\Lambda_1 = -\frac{p}{3K_0^2}\sum_{i=1}^{N} (K_i - K_0)\left(1 + Y_9^i\right)\left(\bar{a}_i^3 - \bar{a}_{i-1}^3\right), \tag{7.455}$$

$$\Lambda_2 = -\frac{p}{2\mu_0^2}\sum_{i=1}^{N} (\mu_i - \mu_0)\left[\left(1 + 3Y_5^i + 2Y_6^i\right)\left(\bar{a}_i^3 - \bar{a}_{i-1}^3\right) + \right.$$
$$\left. + \frac{7}{5}a_N^2\left(3Y_2^i + Y_6^i\right)\left(\bar{a}_i^5 - \bar{a}_{i-1}^5\right)\right], \tag{7.456}$$

$$\bar{\Lambda}_{\alpha\beta}^T = \langle\Lambda_0^T\rangle\delta_{\alpha\beta}, \tag{7.457}$$

$$\Lambda_0^T = -\frac{p}{K_0}\sum_{i=1}^{N}\left[(K_i - K_0)\,Y_{11}^i - K_i\alpha_i^1\right]\left(\bar{a}_i^3 - \bar{a}_{i-1}^3\right). \tag{7.458}$$

Thus, the tensor α^* is isotropic $\alpha_{\alpha\beta}^* = \alpha_*\delta_{\alpha\beta}$, and the scalar coefficient α_* of temperature expansion takes the form

$$\alpha_* = \alpha_0 + \frac{\langle\Lambda_0^T\rangle}{1 + 9d_1\langle\Lambda_1\rangle}, \quad d_1 = -\frac{4\mu_0}{3\left(3k_0 + 4\mu_0\right)}. \tag{7.459}$$

The stress tensor $\sigma_{\alpha\beta}(r,n)$ in the vicinity of an arbitrary inclusion follows from (7.439), (7.441), and (7.451) in the form

$$\sigma_{\alpha\beta}(r,n) = 3k_i\left[Y_{11}^i - \alpha_i^1 + \varepsilon^*\left(1 + Y_9^i\right)\right]\delta_{\alpha\beta} +$$
$$+ 2\mu_i\left(Y_{12}^i + \varepsilon^* Y_{10}^i\right) r^{-3}\left(\delta_{\alpha\beta} - 3n_\alpha n_\beta\right), \tag{7.460}$$

$(a_{i-1} < r < a_i\,,\ i = 1, 2, ..., N+1)\,.$

Here interactions between the inclusions are taken into account via the terms proportional to the effective strain field ε^*

$$\varepsilon^* = -3d_1 \langle \Lambda_0^T \rangle \left(1 + 9d_1 \langle \Lambda_1 \rangle\right)^{-1}. \tag{7.461}$$

Note that the equation for $\sigma_{\alpha\beta}(r, n)$ has to be multiplied by the actual value of the temperature T.

For homogeneous inclusions, Equation (7.459) for α_* takes the form ($N = 1,\ \alpha_1 = \alpha,\ K_1 = K$)

$$\alpha_* = \alpha_0 + p\,(\alpha - \alpha_0) \left[1 - 4\,(1-p)\,\frac{\mu_0\,(K - K_0)}{K\,(3K_0 + 4\mu_0)}\right]^{-1}. \tag{7.462}$$

Particular forms of Equation (7.460) for some special cases were considered in [89].

In the conclusion of this section, we present the equations for the tensor of thermal expansion coefficients of a composite reinforced with many layered cylindrical fibers. Let the material of the medium and the inclusions be transversely isotropic with the unit vector m of the symmetry axis directed along the fibers. The tensors of the thermal expansion coefficients of the medium α^0 and of the i-th layer α_i in a fiber are defined by the expressions

$$\alpha^0_{\alpha\beta} = \alpha_{0\theta}\theta_{\alpha\beta}\,(m) + \alpha_{0m} m_\alpha m_\beta\,, \quad \alpha_{i\alpha\beta} = \alpha_{i\theta}\theta_{\alpha\beta}\,(m) + \alpha_{im} m_\alpha m_\beta\,, \tag{7.463}$$

where $\theta_{\alpha\beta}(m) = \delta_{\alpha\beta} - m_\alpha m_\beta\,;\ \alpha^0_\theta,\ \alpha_{i\theta}$ are the thermal expansion coefficients in the plane orthogonal to the reinforcing axis, $\alpha^0_m,\ \alpha_{im}$ are similar coefficients in the direction of the fiber axes. The tensor of the thermal expansion coefficients of the composite has the same structure as (7.453), where the tensor D is defined by the equation

$$D = C^0 A C^0 - C^0. \tag{7.464}$$

The tensor A^0 has the form (7.418), and the tensors $\bar{\Lambda}^s$ and $\bar{\Lambda}^T$ are ($P^i = P^i(m)$)

$$\bar{\Lambda}^s\,(m) = \langle \Lambda_1 \rangle P^2 + \langle \Lambda_2 \rangle \left(P^1 - \frac{1}{2}P^2\right) + \\ + \langle \Lambda_3 \rangle \left(P^3 + P^4\right) + \langle \Lambda_5 \rangle P^5 + \langle \Lambda_6 \rangle P^6, \tag{7.465}$$

$$\Lambda_1 = -p \sum_{i=1}^{N} (k_i - k_0)\left(1 + Y_9^i\right) \xi_i\,, \tag{7.466}$$

$$\Lambda_3 = -p \sum_{i=1}^{N} (l_i - l_0)\left(1 + Y_9^i\right) \xi_i\,, \tag{7.467}$$

$$\Lambda_2 = -2p\sum_{i=1}^{N}(m_i - m_0)\left[(1 + 2Y_1^i + 2Y_5^i)\xi_i + \right.$$
$$\left. +3\left(Y_2^i + Y_6^i\right)\left(a_i^2 + a_{i-1}^2\right)\xi_i\right] , \tag{7.468}$$

$$\Lambda_5 = -2p\sum_{i=1}^{N}(\mu_i - \mu_0)\left(2 + Y_{13}^i\right)\xi_i , \tag{7.469}$$

$$\Lambda_6 = -p\sum_{i=1}^{N}\left[(n_i - n_0) + 2(l_i - l_0)Y_{11}^i\right]\xi_i , \tag{7.470}$$

$$\Lambda_{\alpha\beta}^T = \Lambda_\theta^T\theta_{\alpha\beta} + \Lambda_m^T m_\alpha m_\beta , \tag{7.471}$$

$$\Lambda_\theta^T = -p\sum_{i=1}^{N}\left\langle t_\theta^i\xi_i\right\rangle , \quad \Lambda_m^T = -p\sum_{i=1}^{N}\left\langle t_\theta^i\xi_i\right\rangle , \tag{7.472}$$

$$t_\theta^i = \frac{l_0 n_0}{\Delta_0}\left[\left(\frac{l_i - l_0}{n_0} - \frac{k_i - k_0}{l_0}\right)Y_{15}^i + \right.$$
$$\left. +\left(\frac{k_i}{l_0} - \frac{l_i}{n_0}\right)\alpha_{i\theta}^1 + \frac{1}{2}\left(\frac{l_i}{l_0} - \frac{n_i}{n_0}\right)\alpha_{im}^1\right] , \tag{7.473}$$

$$t_e^i = \frac{l_0 k_0}{\Delta_0}\left[2\left(\frac{k_i - k_0}{k_0} - \frac{l_i - 4\mu_0}{l_0}\right)\left(Y_{15}^i - \alpha_{i\theta}^1\right) - \left(\frac{l_i}{k_0} - \frac{n_i}{l_0}\right)\alpha_{im}^1\right] , \tag{7.474}$$

$$\Delta_0 = k_0 n_0 - l_0^2 , \ \alpha_{i\theta}^1 = \alpha_{i\theta} - \alpha_{0\theta} , \ \alpha_{im}^1 = \alpha_{im} - \alpha_{0m} , \tag{7.475}$$

$$p = \pi n^0\left\langle a_N^2\right\rangle , \ \xi_i = \frac{\left(a_i^2 - a_{i-1}^2\right)}{a_N^2} . \tag{7.476}$$

For homogeneous fibers ($N = 1$) and isotropic matrix and inclusions, the nonzero components of the tensor α^* in the Cartesian basis with the x_3-axis directed along the fibers are

$$\alpha_{11}^* = \alpha_0 + p(\alpha - \alpha_0)A_1 , \quad \alpha_{33}^* = \alpha_0 + p(\alpha - \alpha_0)A_2 , \tag{7.477}$$

$$A_1 = \frac{1}{\Delta G_*}\left[E\Delta + 4(1-p)(1+\nu_0)(\nu - \nu_0)\right] , \tag{7.478}$$

$$A_2 = \frac{1}{\Delta G_*}\left\{\left[pE + (1+\nu)(1-p)E_0\right]\left(\frac{1}{k_0} + \frac{1}{\mu_0}\right) - \right.$$
$$\left. -(1-p)\nu_0 E\left(\frac{1}{k} + \frac{1}{\mu_0}\right) - 4(1-p)\nu_0(1+\nu)(\nu - \nu_0)\right\} , \tag{7.479}$$

$$\Delta = \frac{p}{k_0} + \frac{(1-p)}{k} + \frac{1}{\mu_0} , \quad G_* = n_* - \frac{(l_*)^2}{k_*} , \tag{7.480}$$

where $E, E_0; \nu, \nu_0$ are Young's moduli and Poisson's ratio of the inclusions and the matrix, the coefficients k_*, l_* and n_* are defined in (7.436).

The error in the equations for α^* was discussed in [97], [115]. It was noted that for the fiberglass plastics, the relative error incurred in calculating α^* from (7.453), (7.465) does not exceed 10–15% up to the fiber concentration corresponded to dense packing.

7.15 The point defect model in the theory of composite materials

In this section we consider the approximation based on the replacement of the inclusions of finite sizes by isolated point defects. For such a replacement, we use the condition that the perturbed field from an inclusion of a finite size should coincide at infinity with that for a point defect.

We consider an isolated inclusion with elastic moduli $C(x)$ in a homogeneous medium C^0; V is a finite region occupied by the inclusion. The stress and strain fields in the medium with the inclusion are presented in the forms

$$\sigma(x) = \sigma^0(x) + \int \mathrm{S}(x - x')m(x')\,dx', \quad m(x) = B^1(x)\,\sigma(x)\,V(x)\ , \tag{7.481}$$

$$\varepsilon(x) = \varepsilon^0(x) + \int \mathrm{K}(x - x')\,C^0 m(x')\,d\,x', \;\; B^1(x) = B(x) - B^0, \tag{7.482}$$

where $m(x)$ is the density of the dislocation moments induced in the region occupied by the inclusion and equivalent to the considered inhomogeneity. Transition to the point defect corresponds to replacement of the density $m(x)$ by the first term of the multipole expansion (Section 2.4)

$$m(x) = M\delta(x - \xi) + \ldots\ , \qquad M = \int m(x)\,d\,x, \tag{7.483}$$

where ξ is a point inside the inclusion that is taken as its center of weight. If the external field σ^0 is constant inside V, the density $m(x)$ and the tensor M are defined by the equations

$$m(x) = M, \;\; M = M^0\sigma^0, \;\; M^0 = \int M(x)\,d\,x\ , \tag{7.484}$$

where the function $M(x)$ is determined from the solution of the problem for an isolated inhomogeneity in a constant external field σ^0. In particular, for an ellipsoidal inclusion, the tensor M^0 has the following form:

$$M^0 = -vB^0C^1\left[I + A(a)\,C^1\right]^{-1}B^0, \tag{7.485}$$

where v is the inclusion volume, the tensor $A(a)$ depends on the ellipsoid shape and is defined in (3.82). If the inhomogeneity is an elliptical crack opened in the field σ^0, then

$$M^0_{\alpha\beta\lambda\mu} = \frac{2}{3}\pi a_1 a_2 n_{(\alpha} R^0{}_{\beta)(\lambda} n_{\mu)}\ ,\ R^0 = \left(T^0\right)^{-1}. \tag{7.486}$$

where n is the normal to the crack plane, a_1, a_2 are its semiaxes, the tensor T^0 is defined in (3.233)–(3.236).

Let the medium contain a set of point defects, and σ_i^* be the local exciting field that acts on the i-th defect. Then, the stress and strain fields in the medium are presented in the forms that follow from (7.481)–(7.483)

$$\sigma(x) = \sigma^0(x) + \sum_i \int \mathrm{S}(x - x') M_i^0 \sigma_i^* \delta(x' - \xi_i)\, \mathrm{d}x', \tag{7.487}$$

$$\varepsilon(x) = \varepsilon^0(x) + \sum_i \int \mathrm{K}(x - x')\, C^0 M_i^0 \sigma_i^* \delta(x' - \xi_i)\, \mathrm{d}x'. \tag{7.488}$$

Here $M_i^0 \sigma_i^*$ is the coefficient in front of $\delta(x - \xi_i)$ in the multipole series for the density $m_i(x)$ for the i-th inclusion (7.483). The local exciting field σ^* for the k-th inclusion has the form

$$\sigma_k^* = \sigma^0(\xi_k) + \sum_{i \neq k} \int \mathrm{S}(\xi_k - x')\, M_i^0 \sigma_i^* \delta(x' - \xi_i)\, \mathrm{d}x', \tag{7.489}$$

where $\xi_k (k = 1, 2, ...)$ are coordinates of the point defects.

Consider a simple example of two rectilinear cracks with lengths $2l$ located on a straight line; the distances between their centers is r. The external field is uniaxial tension with stress σ^0 along the normal n to the crack line. Let us change the cracks to the equivalent point defects. Because of the symmetry of the problem we have

$$M_1^0 = M_2^0\ ,\quad n_\alpha \sigma^*_{1\alpha\beta} n_\beta = n_\alpha \sigma^*_{2\alpha\beta} n_\beta, \tag{7.490}$$

and from an equation similar to (7.489) we find the normal component of the local exciting field $\sigma_n^* = n_\alpha \sigma^*_{i\alpha\beta} n_\beta$ that acts on every crack. For an isotropic medium we have

$$n_\alpha \mathrm{S}_{\alpha\beta\lambda\mu}(x) M^0_{\lambda\mu\nu\rho} = \frac{l^2}{2x^2} \delta_{\beta(\rho} n_{\nu)}, \tag{7.491}$$

where x is the coordinate along the line of the crack location. Multiplying both side of (7.489) by the vector n and taking into account the previous equation, we find

$$\sigma_n^* = \sigma_n^0 + \frac{l^2}{2r^2}\sigma_n^*\ ,\ \text{or}\ \ \sigma_n^* = \frac{2r^2\sigma_n^0}{2r^2 - l^2}. \tag{7.492}$$

Here $\sigma_n^*(r)$ is the normal component of the stress field that acts on every point defect equivalent to the crack when the distance between them is r. The function $\sigma^*(r)$ is the asymptote of the actual field when $r >> 2l$ in the problem for two cracks located on a line. If $r \leq l/\sqrt{2}$, the solution (7.492) loses physical meaning. Hence, caution is required in employing the point defect model within the framework of the local theory of elasticity. Strictly speaking, if inclusions are replaced by point defects, a characteristic length l is introduced in the continuous medium. The length l is approximately equal to the mean size of the inclusions, and if the distances between the neighboring point defects become smaller than l, one can obtain senseless results. For such distances, the point-defect model can not correctly describe the interactions between inclusions of finite sizes.

Let the positions of point defects be a realization of a random point set homogeneously distributed in space. Let X denote a set of points $\xi_i (i = 1, 2, ...)$ where the point defects are located, and introduce the generalized functions $X(x)$ and $X(x, x')$ by the equations

$$X(x) = \sum_i \delta(x - \xi_i), \quad X(x; x') = \sum_{i \neq k} \delta(x' - \xi_i) \quad \text{if } x = \xi_k. \tag{7.493}$$

By transition to the model of point defects, the equations for the stress and strain fields (7.487), (7.488) take the forms

$$\sigma(x) = \sigma^0(x) + \int \mathrm{S}(x - x') M^0(x') \sigma^*(x') X(x') \, dx', \tag{7.494}$$

$$\varepsilon(x) = \varepsilon^0(x) + \int \mathrm{K}(x - x') C^0 M^0(x') \sigma^*(x') X(x') \, dx', \tag{7.495}$$

where the functions $M^0(x)$ and $\sigma^*(x)$ coincide with M_i^0 and σ_i^* at the points $x = \xi_i$, $i = 1, 2, ...$ Equation (7.489) for $\sigma^*(x)$ can be written in the form

$$\sigma^*(x) = \sigma^0(x) + \int \mathrm{S}(x - x') M^0(x') \sigma^*(x') X(x; x') \, dx', \quad x \in X. \tag{7.496}$$

If the external field σ^0 is constant, the tensor $\sigma^*(x)$ defined at the points $\xi_i \in X$ is a homogeneous random function. Following the EFM (Section 7.3), we suppose that the field σ^* is constant and the same for all the defects. The equation for the tensor σ^* follows after averaging both sides of (7.496) over the ensemble realizations of the sets X and M_i^0 under the condition $x \in X$

$$\sigma^* = \sigma^0 + p \int S(x - x') \bar{M}^0 \Psi(x - x') \, dx' \sigma^*, \tag{7.497}$$

$$\bar{M}^0 = \frac{1}{\langle v \rangle} \langle M^0 \rangle, \quad \Psi(x - x') = \frac{1}{n_0} \langle X(x; x') | x \rangle, \tag{7.498}$$

where n_0 and $p = n_0 \langle v \rangle$ are the numerical and the volume concentrations of the inclusions. The correlation function $\Psi(x)$ in these equations has the properties

$$\Psi(0) = 0, \quad \Psi(x) \to 1 \text{ as } |x| \to \infty. \tag{7.499}$$

Calculating the integral in (7.497) and solving the resulting equation with respect to σ^*, we obtain

$$\sigma^* = \left(I + pD\bar{M}^0\right)^{-1}\sigma^0, \quad D = \int S(x)[1 - \Psi(x)]\mathrm{d}x. \tag{7.500}$$

Averaging the equations for the stress and strain tensors in the medium with the point defects (7.487), (7.488), taking into account that

$$\langle X(x)\rangle = n_0, \tag{7.501}$$

and the definition (7.17) for the action of the operators S and K on constants, we find

$$\langle\sigma\rangle = \sigma^0, \quad \langle\varepsilon\rangle = \varepsilon^0 + p\bar{M}^0\left(I + pD\bar{M}^0\right)^{-1}\sigma^0. \tag{7.502}$$

As a result, the effective compliance tensor B^* of the medium with point defects takes the form

$$\langle\varepsilon\rangle = B^*\langle\sigma\rangle\,, \quad B^* = B^0 + p\bar{M}^0\left(I + pD\bar{M}^0\right)^{-1}. \tag{7.503}$$

Let us consider the tensor B^* in some particular cases.

1. Suppose that there exists a linear transformation α of the x-space that converts the function $\Psi(x)$ into a spherically symmetric function

$$y = \alpha x\,, \quad \Psi\left(\alpha^{-1}y\right) = \Psi\left(|y|\right). \tag{7.504}$$

In this case, the tensor D (7.500) takes the form

$$D = D\left(\alpha\right) = C^0 A\left(\alpha\right) C^0 - C^0, \tag{7.505}$$

where the tensor $A(\alpha)$ is defined in (3.82).

If the point defects model ellipsoidal homogeneous inclusions, the tensor M^0 for has form (7.485), and Equation (7.503) gives the equation for the tensor C^* in the form

$$C^* = C^0 + pP\left(I - pA\left(\alpha\right)P\right)^{-1}, \quad P = \left\langle C^1\left(I + A\left(\alpha\right)C^1\right)^{-1}\right\rangle. \tag{7.506}$$

that coincides with Equation (7.104) for C^* that was obtained for a random set of inclusions of finite sizes.

2. Let the point defects model oriented inhomogeneities (cracks, hard flakes, or fibers). In these cases, the tensor M^0 in (7.494) depends on the inclusion orientation m (7.496). As in Section 7.8, it is possible to obtain the equation for the effective field σ^* that contains the correlation function $\Psi_m(x - x')$

$$\Psi_m\left(x - x'\right) = \frac{1}{n_0}\langle X\left(x; x'\right)|x, m\rangle\,, \tag{7.507}$$

where $\langle\cdot\,|x, m\rangle$ is the mean under the condition that the point x is inside the defect with the orientation m.. If the symmetry of the function $\Psi_m(x)$ coincides with the symmetry of the inclusion, it is possible to obtain an equation for C^* that coincide with that in Sections 7.8–7.14 for composites reinforced with oriented inclusions.

Hence, for stochastic composites, the point defect model with necessary limitations allows us to obtain the same expressions for the effective elastic moduli as the EFM gives for inclusions of finite sizes.

3. Let the point defect model be applied to the calculation of the effective elastic moduli of the medium with a regular lattice of inclusions. For simple lattices and a constant external field σ^0, the local exciting field σ_i^* is the same for all the defects. Let the origin of the coordinate system be taken at an arbitrary lattice node. The effective field σ^* acting on the defect located at this point follows from (7.496) in the form

$$\sigma^* = \sigma^0 + p\int \mathrm{S}(x)\bar{M}^0\Psi(x)\mathrm{d}x\sigma^*, \tag{7.508}$$

$$\bar{M}^0 = \frac{1}{v}M^0, \quad \Psi(x) = \frac{1}{n_0}{\sum_l}'\,\delta\left(\mathbf{x}-\mathbf{l}\right), \quad p = n_0 v\,. \tag{7.509}$$

Here $\mathbf{l}$ is the vector of the lattice, and prime over the sum sign means excluding the term with $\mathbf{l} = \mathbf{0}$. As a result, we go to the expression for σ^* that coincides with (7.316), where the integral D is presented in the form of the following converging series

$$D = -{\sum_l}'\left[\frac{1}{n_0}\mathrm{S}\left(l\right) - \int_{v_l}\mathrm{S}(x)\mathrm{d}x\right] + D^0 + p.v.\int_{v_0}\mathrm{S}(x)\mathrm{d}x. \tag{7.510}$$

Here v_l is an elementary cell which corresponds to the lattice node with the vector $\mathbf{l}$; the tensor D^0 has the form (7.282).

The tensor of the effective elastic moduli of the medium with a lattice of point defects has the form (7.503) where the tensor D is defined in (7.510). Consider the case of an isotropic medium in which spherical inclusions form a cubical lattice. In this case, the tensor C^* has the same form (7.189), except that the coefficient α in this expression is presented in the form of a series similar to (7.510). Numerical summation of this series gives $\alpha = 0,080$.

In the 2D-case, all the constructions are performed as in the 3D-case. Consider a square lattice of circular inclusions in an isotropic plane. In this case, the tensor C^* has form (7.204), where the coefficient α is constant and its value is equal to 0.092.

In Fig. 7.30, the relative effective shear modulus μ_*/μ_0 of an elastic plane with a square lattice of circular holes (line 1) and circular absolutely rigid inclusions (line 2) are presented as functions of the volume concentrations of inclusions p. The solid lines correspond to the exact solution of the problem

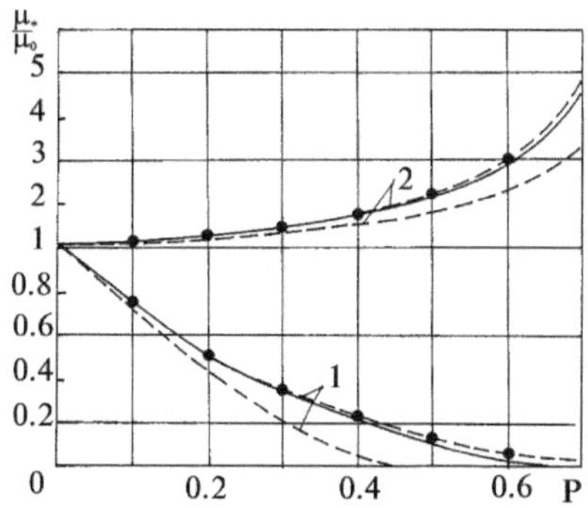

Fig. 7.30. Normalized effective shear modulus versus concentration of inclusions for a medium with a cubic lattice of spherical inclusions.

presented in [57], the dashed lines are obtained for the corresponding point-defects model, and the dashed lines with black dots present the predictions of the EFM for finite inclusions (Section 7.7).

Analysis of the lines in Fig. 7.30 leads the following conclusion. Substitution of inclusions of finite sizes by point defects gives an error less than 10% if the distances between the centers of inclusions are in 1.5–2 times larger than the diameter of the inclusions.

For a regular triangular lattice of circular inclusions, the point-defect model and the effective field method for the lattice of finite inclusions give very close results for the effective elastic constants of the composite.

7.16 Effective elastic properties of hybrid composites

In this section we study the calculation of the overall elastic properties of composite materials that consist of a homogeneous matrix and two different populations of inclusions. The shapes and elastic properties of the inclusions are the same inside every population but different for different populations (three-phase hybrid composites). The applications of known approximate methods to the solution of this problem face a specific difficulty. As shown in [11], [173], [50] (see Section 7.6), the rank four overall elastic moduli tensor of three phase composites obtained in the framework of the Mori–Tanaka method is not symmetric with respect to the first and second pairs of indices. Such symmetry is a general property of elastic moduli tensors that follows from the existence of an overall elastic energy density of the composite; this lack of symmetry is a serious defect of these approximate methods.

A natural generalization of the EFM for three phase composites starts from the assumption that the external field that acts on every inclusion in the composite is different for inclusions of different populations. It is shown that the overall elastic moduli tensor obtained in the framework of such a modification of the EFM has the necessary symmetry with respect to the indices. The final formulas for this tensor contain several specific correlation functions that depend on statistical properties of the spatial distributions of the inclusions of various populations. For the construction of these correlation functions we introduce several probabilistic Boolean type models of random sets of inclusions. The cases of a medium with infinite cylindrical fibers and thin ellipsoidal disks or spherical pores are considered. The corresponding correlation functions are obtained in explicit analytical forms and the overall elastic moduli tensors of these composites are obtained and analyzed. It is shown that the equations of the Mori–Tanaka method (see Section 7.6) may be derived from the equations of this version of the EFM if a general property of the symmetry of cross-correlation functions of different populations of inclusions is violated (see below). Such a violation causes the absence of symmetry of the overall elastic moduli tensor obtained by the Mori–Tanaka method.

7.16.1 Two different populations of inclusions in a homogeneous matrix (hybrid composite)

Consider a homogeneous medium with two different sets (populations) of inclusions. The shapes and elastic properties of the inclusions are the same inside every population but differ for different populations. The natural generalization of the EFM for this case is based on the assumption that the effective field that acts on every inclusion in the composite is different for inclusions of different populations. Thus, we accept the following two hypotheses.

H_1. *The local exciting field $\varepsilon_*(x)$ acting on any inclusion in the composite is constant in the vicinity of this inclusion.*

H_2. *The effective field that acts on any inclusion v_k is the same as for the inclusions of the same population, but it is different for inclusions in different populations.*

Let $V_1(x), V_2(x)$ be the characteristic functions of the regions V_1, V_2 occupied by the inclusions of two different populations

$$V(x) = V_1(x) + V_2(x), \tag{7.511}$$

$$V_1(x) = \sum_i v_i^{(1)}(x), \quad V_2(x) = \sum_j v_j^{(2)}(x). \tag{7.512}$$

Here $v_i^{(1)}(x)$ is the characteristic function of the i-th inclusion of the population 1, $v_j^{(2)}(x)$ is the same for the j-th inclusion of population 2. $C_1^{(1)}$, $C_1^{(2)}$ are the deviations of the elastic moduli inside the inclusions of each population.

The effective fields $\varepsilon_*^{(1)}(x)$ and $\varepsilon_*^{(2)}(x)$ that act on the inclusions of different populations are the following means:

$$\varepsilon_*^{(1)}(x) = \langle \varepsilon_*(x) | x \in V_1 \rangle , \quad \varepsilon_*^{(2)}(x) = \langle \varepsilon_*(x) | x \in V_2 \rangle . \tag{7.513}$$

If hypothesis H_1 holds, the detailed external field $\varepsilon_*(x)$ that acts on each inclusion in the composite satisfies an equation that follows from (7.8)

$$\varepsilon_*(x) + \int K(x-x') C_1^{(1)} \Lambda^{(1)}(x') \varepsilon_*(x') V_1(x;x') dx' +$$

$$+ \int K(x-x') C_1^{(2)} \Lambda^{(2)}(x') \varepsilon_*(x') V_2(x;x') dx' = \varepsilon_0(x), \qquad x \in V, \tag{7.514}$$

where functions $V_1(x;x')$, $V_2(x;x')$ are defined by equations similar to (7.84)

$$V_1(x;x') = \sum_{i \neq k} v_i^{(1)}(x'), \quad \text{if} \ \ x \in v_k^{(1)}, \tag{7.515}$$

$$V_2(x;x') = \sum_{j \neq m} v_j^{(2)}(x'), \quad \text{if} \ \ x \in v_m^{(2)}. \tag{7.516}$$

The functions $\Lambda^{(1)}(x)$, $\Lambda^{(2)}(x)$ have to be built from the solution of the one particle problems for the inclusions in each population. Integral equations of these problems have the forms

$$\varepsilon(x) + \int_{v_k} K(x-x') C_1^{(i)} \varepsilon(x') dx' = \varepsilon_*^{(i)}, \qquad i = 1,2, \tag{7.517}$$

and their solutions may be presented as the following:

$$\varepsilon(x) = \Lambda_k^{(i)}(x) \varepsilon_*^{(i)}, \quad x \in v_k, \quad v_k \in V_i, \quad i = 1,2. \tag{7.518}$$

Here the field $\varepsilon_*^{(i)}$ is assumed to be constant (hypothesis H_1). For ellipsoidal inclusions and their limit forms, $\Lambda_k^{(i)}$ are constant tensors that are the same for the inclusions in a given population.

In order to obtain a closed system of equations for the mean effective fields $\varepsilon_*^{(1)}$ and $\varepsilon_*^{(2)}$, let us average both parts of (7.514) by the conditions $x \in V_1$ and $x \in V_2$. Using (7.513) we obtain the following two equations:

$$\varepsilon_*^{(1)} + p_1 \int K(x-x') T_1 \varepsilon_*^{(1)} \Psi_{11}(x-x') dx' +$$

$$+ p_2 \int K(x-x') T_2 \varepsilon_*^{(2)} \Psi_{12}(x-x') dx' = \varepsilon_0, \tag{7.519}$$

$$\varepsilon_*^{(2)} + p_1 \int K(x-x') T_1 \varepsilon_*^{(1)} \Psi_{21}(x-x') dx' +$$

$$+ p_2 \int K(x-x') T_2 \varepsilon_*^{(2)} \Psi_{22}(x-x') dx' = \varepsilon_0. \tag{7.520}$$

Here we take into account that the mean field $\varepsilon_*^{(1)}$ acts on the inclusions of the first population and the field $\varepsilon_*^{(2)}$ acts on the inclusions of the second population. The correlation functions $\Psi_{ij}(x-x')$ in (7.519) are defined by the equations

$$\Psi_{ij}(x-x') = \frac{1}{p_j}\left\langle V_j(x;x')|x \in V_i\right\rangle, \qquad i,j=1,2, \tag{7.521}$$

where $p_1 = \langle V_1(x)\rangle$, $p_2 = \langle V_2(x)\rangle$ are the volume concentrations of the inclusions of each population;

$$T_i = \left\langle \frac{1}{v}\int_v C_1^{(i)}\Lambda^{(i)}(x)\mathrm{d}x|v \in V_i\right\rangle, \qquad i=1,2. \tag{7.522}$$

After calculating the integrals in (7.519) we obtain the following linear algebraic system for the constant tensors $\varepsilon_*^{(1)}, \varepsilon_*^{(2)}$

$$(I - p_1A_{11}T_1)\varepsilon_*^{(1)} - p_2A_{12}T_2\varepsilon_*^{(2)} = \varepsilon_0, \tag{7.523}$$

$$-p_1A_{21}T_1\varepsilon_*^{(1)} + (I - p_2A_{22}T_2)\varepsilon_*^{(2)} = \varepsilon_0, \tag{7.524}$$

$$A_{ij} = \int K(x)\Phi_{ij}(x)\mathrm{d}x, \qquad \Phi_{ij}(x) = 1 - \Psi_{ij}(x), \qquad i,j=1,2. \tag{7.525}$$

Here the regularization (7.19) was used.

The solution of the system (7.523), (7.524) may be presented in the form

$$\varepsilon_*^{(1)} = \Lambda_*^{(1)}\varepsilon_0, \qquad \varepsilon_*^{(2)} = \Lambda_*^{(2)}\varepsilon_0, \tag{7.526}$$

$$\Lambda_*^{(1)} = \left(I - p_1A_{11}T_1 - p_2A_{12}T_2D_2^{-1}D_1\right)^{-1}, \tag{7.527}$$

$$\Lambda_*^{(2)} = \left(I - p_2A_{22}T_2 - p_1A_{21}T_1D_1^{-1}D_2\right)^{-1}, \tag{7.528}$$

$$D_1 = I - p_1(A_{11} - A_{21})T_1, \qquad D_2 = I - p_2(A_{22} - A_{12})T_2. \tag{7.529}$$

For two populations of inclusions, the mean value of the function $q(x)$ in (7.26) takes the form

$$\langle q(x)\rangle = \left\langle C_1^{(1)}\Lambda^{(1)}(x)\varepsilon_*^{(1)}V_1(x) + C_1^{(2)}\Lambda^{(2)}(x)\varepsilon_*^{(2)}V_2(x)\right\rangle = Q\varepsilon_0, \tag{7.530}$$

$$Q = p_1T_1\Lambda_*^{(1)} + p_2T_2\Lambda_*^{(2)}. \tag{7.531}$$

This equation and (7.30) give the equation for the overall elastic moduli tensor C_* of the composite in the form

$$C_* = C_0 + p_1T_1\Lambda_*^{(1)} + p_2T_2\Lambda_*^{(2)}. \tag{7.532}$$

Let us study the symmetry of this rank four tensor when the inclusions in each population are ellipsoids. For ellipsoids, the tensors T_1, T_2 in (7.526), (7.532) are

$$T_i = C_1^{(i)} \Lambda^{(i)}, \quad \Lambda^{(i)} = (I + A_i C_1^{(i)})^{-1}, \quad A_i = A(a_i), \quad i = 1, 2, \tag{7.533}$$

where a_i is the linear transform that converts ellipsoids in the i-th population into a unit sphere. After expanding the tensor C_* in (7.526), (7.532) into a series with respect to the volume concentrations p_1, p_2, and keeping only the terms of order p_i^2 we get

$$\begin{aligned} C_* = {} & C_0 + p_1 T_1 + p_2 T_2 + p_1^2 T_1 A_{11} T_1 + p_2^2 T_2 A_{22} T_2 \\ & + p_1 p_2 (T_1 A_{12} T_2 + T_2 A_{21} T_1) + \cdots \end{aligned} \tag{7.534}$$

The terms proportional to p_1, p_2, p_1^2, p_2^2 on the right-hand side of this equation have the symmetry of elastic moduli tensors with respect to indices because tensors T_i and A_{ii} have such a symmetry. The last term in (7.534) has such a symmetry if

$$A_{12} = A_{21}. \tag{7.535}$$

Let us consider equation (7.525) for tensors A_{12} and A_{21} in detail. Functions $\Psi_{12}(x)$ and $\Psi_{21}(x)$ in this equation are defined in (7.521) and may be written in the forms

$$\Psi_{12}(x - x') = \frac{1}{p_2} \langle V_2(x; x') | x \in V_1 \rangle = \frac{1}{p_1 p_2} \langle V_2(x') V_1(x) \rangle, \tag{7.536}$$

$$\begin{aligned} \Psi_{21}(x - x') &= \frac{1}{p_1} \langle V_1(x; x') | x \in V_2 \rangle = \\ &= \frac{1}{p_1 p_2} \langle V_1(x') V_2(x) \rangle = \Psi_{12}(x' - x). \end{aligned} \tag{7.537}$$

Here we take into account the fact that if the point x is inside the inclusion of the population V_1, it does not belong to the region V_2, and *vice versa*. This equation yields a general property of the cross-correlation functions Ψ_{12} and Ψ_{21}

$$\Psi_{12}(x) = \Psi_{21}(-x) \tag{7.538}$$

that holds for populations of inclusions that are stationary in space. Because $K(x)$ is an even function of the argument ($K(-x) = K(x)$), from (7.525), (7.538) we obtain

$$\begin{aligned} A_{12} &= \int K(x) \left[1 - \Psi_{12}(x)\right] \mathrm{d}x = \\ &= \int K(x) \left[1 - \Psi_{21}(-x)\right] \mathrm{d}x = \int K(-x) \left[1 - \Psi_{21}(x)\right] \mathrm{d}x = A_{21}. \end{aligned} \tag{7.539}$$

Thus, Equation (7.535) is a consequence of the general property (7.538) of the cross-correlation functions $\Psi_{ij}(x)$. As a result, all the terms on the right-hand side of (7.534) have the symmetry of elastic moduli tensors. It is possible to show that other terms in this expansion have such a symmetry if (7.535) holds. For instance, the terms of the order p_i^3 of this expansion are

$$C_* = \cdots + p_1^3 T_1 A_{11} T_1 A_{11} T_1 + p_1^2 p_2 (T_1 A_{11} T_1 A_3 T_2 + \\ + T_2 A_3 T_1 A_{11} T_1 + T_1 A_3 T_2 A_3 T_1) + p_1 p_2^2 (T_2 A_{22} T_2 A_3 T_1 + \\ + T_1 A_3 T_2 A_{22} T_2 + T_2 A_3 T_1 A_3 T_2) + p_2^3 T_2 A_{22} T_2 A_{22} T_2 + \cdots , \quad (7.540)$$

$$A_3 = A_{12} = A_{21}. \quad (7.541)$$

Direct checking shows that all the terms in this equation are symmetric with respect to pairs of indices.

Let us assume that the following equations hold:

$$A_{11} = A_{21} = A_1, \qquad A_{22} = A_{12} = A_2. \quad (7.542)$$

In that case, Equations (7.526), (7.532) are dramatically simplified, and we obtain the following equation for the overall elastic moduli tensor of the composite ($p_0 = 1 - p_1 - p_2$):

$$C_* = C_0 + (p_1 C_1^{(1)} \Lambda^{(1)} + p_2 C_1^{(2)} \Lambda^{(2)})(p_0 I + p_1 \Lambda^{(1)} + p_2 \Lambda^{(2)})^{-1}. \quad (7.543)$$

This equation coincides with the result of the application of the Mori–Tanaka method to three phase composite materials (see Section (7.6)). As shown in [11] this tensor does not have the symmetry of the elastic moduli tensors (the symmetry with respect to pairs of indices). The reason for such a nonsense is quite clear now: the assumptions (7.542) violate the general property (7.535) of the cross tensors A_{12}, A_{21}, and the penalty for the simplicity of (7.543) is the absence of the necessary symmetry of the overall elastic moduli tensor C_*.

7.16.2 Two-point correlation functions for a hybrid composite with sets of cylindrical and spheroidal inclusions

Consider the three-phase composite studied in [11]: a composite with isotropic matrix (Young's modulus $E_0 = 96.5$ GPa, shear modulus $\mu_0 = 37.1$ GPa), a set of infinite unidirectional isotropic cylindrical fibers ($E^{(1)} = 431$ GPa, $\mu^{(1)} = 172$ GPa) and a set of isotropic spheroids (thin spheroidal disks) ($E^{(2)} = 34.4$ GPa, $\mu^{(2)} = 14.3$ GPa). The directions of the rotation axes of the disks and the axes of the fibers coincide. For the application of Equation (7.526), (7.532) of the EFM to this composite material, we have to calculate tensors $A_{11}, A_{22}, A_{12} = A_{21}$ defined in (7.525). The correlation functions $\Phi_{11}(x)$, $\Phi_{22}(x)$ and $\Phi_{12}(x)$ that are present in this equation depend on the spatial distribution of cylinders and disks in the matrix. Thus we have to introduce probabilistic models of such distributions. In this section, several

Boolean type random models of cylindrical fibers and ellipsoidal inclusions in 3D-space are considered (see Appendix E.3).

We start with one population of a Boolean random set of inclusions. The Boolean model is obtained by implantation of random "grains" v_k on the points $x^{(k)}$ of a Poisson point process of intensity λ (the number of points in a unit volume) [141], [66]:

$$V = \cup_k v_{k,x^{(k)}}. \tag{7.544}$$

For instance, for a population of spheroidal inclusions of the same orientation, the Boolean model is obtained by selecting for v_k a spheroid with semiaxis a_s and l_s (l_s is the semiaxis in the direction of the rotation axis x_3) with the center at point $x^{(k)}$. Random points $x^{(k)}$ are independently and homogeneously distributed in the 3D-space. For this model, the function $\Phi_{ss}(x)$ in (E.20), (E.32) takes the form

$$\Phi_{ss}(x_1, x_2, x_3) = \Phi_s(\rho, x_3) = \\ = 1 - \frac{1}{p_s^2}\left[1 - q_s{}^{1-H(R)}\right]^2 q_s{}^{H(R)}, \tag{7.545}$$

$$R^2 = \left(\frac{\rho}{a_s}\right)^2 + \left(\frac{x_3}{l_s}\right)^2, \quad \rho^2 = x_1^2 + x_2^2 \tag{7.546}$$

Here p_s is the volume concentration of spheroids, $q_s = 1 - p_s$, the function $H(R)$ is defined in (E.27). It is seen that this correlation function has the symmetry of the spheroidal inclusion.

For a boolean population of infinite cylinders of the same radius a_c orientated along the axis x_3, the corresponding correlation function takes the form (E.19), (E.31) [66]

$$\Phi_{cc}(x) = \Phi_c(\rho) = 1 - \frac{1}{p_c^2}\left[1 - q_c{}^{1-h\left(\frac{\rho}{a_c}\right)}\right]^2 q_c{}^{h\left(\frac{\rho}{a_c}\right)}, \tag{7.547}$$

Here p_c is the volume concentration of cylinders, $q_c = 1 - p_c$, and the function h is defined in (E.25). This function clearly has cylindrical symmetry.

Now let us obtain a model of two Boolean populations of inclusions (cylinder and spheroids). In this model the centers of the spheroids are situated in a Poisson set of points $x^{(k)}$ with intensity λ_s in 3D-space and the axes of the cylinders intersect the plane (x_1, x_2) in a set of Poisson points with intensity λ_c. Volume concentrations of cylinders and spheroids are p_c and p_s, respectively. The directions of the rotation axes of spheroids coincide with the directions of the cylinder axes. Because of the independence of the positions of the centers of spheroids and the axes of cylinders there are possibilities of overlaps between inclusions in the same population, as well as overlaps between inclusions in different populations. To specify the model uniquely we have to define a rule of priority, namely to decide which parts of the inclusions will be kept in the intersections. Thus, parts of the inclusion of one population should be deleted from the zones of intersection. There are two possibilities here. Parts of spheroids in the intersections with cylinders may be deleted (cylinder priority) or parts of cylinders may be deleted from such intersections (spheroid priority). Let us consider each of these models.

Cylinder priority. The function $\Phi_{cc}(x)$ (point y and $y+x$ are inside different cylinders) takes the form (E.16), (E.19)

$$\Phi_{cc}(x) = \Phi_{cc}(\rho) = 1 - \frac{1}{p_c^2}\left[1 - q_c^{\,1-h\left(\frac{\rho}{a_c}\right)}\right]^2 q_c^{\,h\left(\frac{\rho}{a_c}\right)} \tag{7.548}$$

and coincides with function Φ_c in (7.547) for one population of cylinders.

Function $\Phi_{ss}(x)$ (point y and $y+x$ inside different spheroids) takes the form that follows from (E.39), (E.20)

$$\Phi_{ss}(x) = \Phi_{ss}(\rho, x_3) = 1 - \frac{1}{p_s^2}\left[1 - q_{cs}^{\,1-H(R)}\right]^2 q_{cs}^{H(R)} q_c^{\,2-h\left(\frac{\rho}{a_c}\right)}, \tag{7.549}$$

$$q_{cs} = 1 - \frac{p_s}{1-p_c}.$$

Function $\Phi_{cs}(x) = \Phi_{sc}(x)$ (point y is inside a cylinder and point $y+x$ is inside a sphere) is

$$\Phi_{cs}(x) = \Phi_{sc}(x) = 1 - \frac{1}{p_c}\left[1 - q_c^{\,1-h\left(\frac{\rho}{a_c}\right)}\right]. \tag{7.550}$$

Because of the intersections with cylinders, the actual volume concentration of spheroids p_{*s} will be less than its original value p_s

$$p_{*s} = p_s(1 - p_c). \tag{7.551}$$

The volume concentration of cylinders p_c does not change in this model.

In Fig. 7.31, a realization of this model for $p_c = p_s = 0.2$, $a_c = a_s = l_s$ is presented. In this figure, a section of such a realization in the plane orthogonal to the axes of the cylinders is shown.

Spheroid priority. In this case, the correlation functions $\Phi_{ij}(x)$ have the forms

$$\Phi_{cc}(x) = 1 - \frac{1}{p_c^2}\left[1 - q_{sc}^{\,1-h\left(\frac{\rho}{a_c}\right)}\right]^2 q_{sc}^{\,h\left(\frac{\rho}{a_c}\right)} q_s^{\,2-H(R)}, \tag{7.552}$$

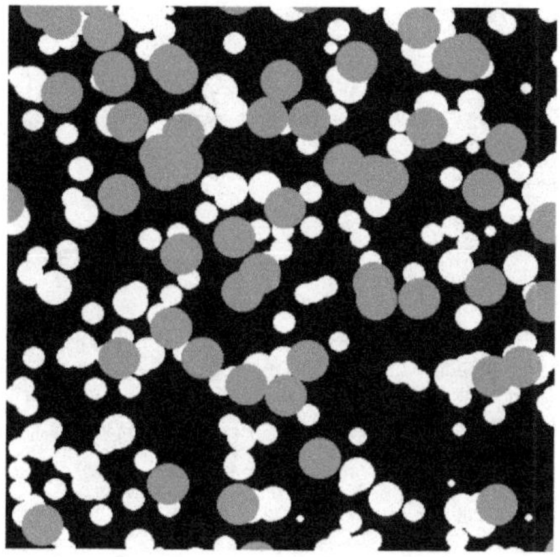

Fig. 7.31. Simulation of a section of a Boolean model of cylinders (grey) and spheres (white) with the cylinder priority $p_c = 0.2$, $p_s = 0.2$.

$$q_{sc} = 1 - \frac{p_c}{1 - p_s}, \tag{7.553}$$

$$\Phi_{ss}(x) = 1 - \frac{1}{p_s^2}\left[1 - q_s{}^{1-H(R)}\right]^2 q_s{}^{H(R)}, \tag{7.554}$$

$$\Phi_{cs}(x) = \Phi_{sc}(x) = 1 - \frac{1}{p_s}\left[1 - q_s{}^{1-H(R)}\right].. \tag{7.555}$$

The actual volume concentration of cylinders p_{*c} differs from its initial concentration p_c because of their intersections with spheroids

$$p_{*c} = p_c(1 - p_s) \tag{7.556}$$

and the volume concentration of spheroids p_s is not changed.

A realization of this model for $p_c = p_s = 0.2$ and $a_c = a_s = l_s$ is presented in Fig. 7.32.

The next probabilistic model of two populations of inclusions is called the "dead leaves" model ([66], [67], [68]). As before, we consider populations of infinite cylinders and spheroids with the same orientations. Two sequences of Poisson point processes for each population are implanted homogeneously in time. In the intersections, we keep parts of inclusions that have appeared later in time and delete parts of all the inclusions that appeared earlier. For this model the correlation functions Φ_{ij} take the forms

$$\Phi_{cc}(x) = 1 - \\ - \frac{2}{p^2}\left(1 - h\left(\frac{\rho}{a_c}\right)\right)\left[\frac{1 - Q(x)}{2 - H_{cs}(x)} - \frac{Q(0) - Q(x)}{1 - H_{cs}(x)}\right], \tag{7.557}$$

$$\Phi_{ss}(x) = 1 - \frac{2}{p^2}(1 - H(R))\left[\frac{1 - Q(x)}{2 - H_{cs}(x)} - \frac{Q(0) - Q(x)}{1 - H_{cs}(x)}\right], \tag{7.558}$$

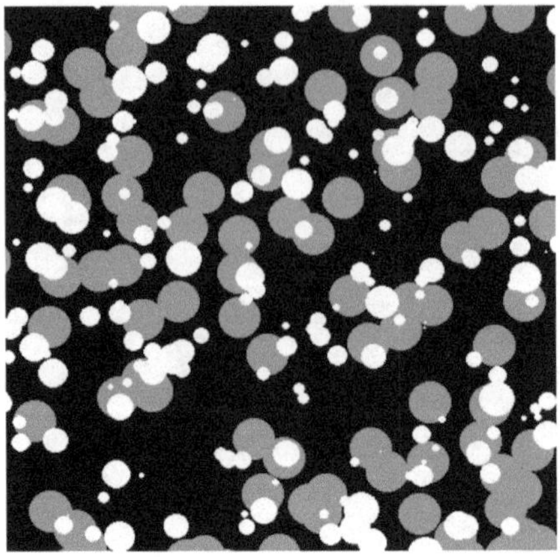

Fig. 7.32. Simulation of a section of a Boolean model of cylinders (grey) and spheres (white) with spheres priority $p_c = 0.2$, $p_s = 0.2$.

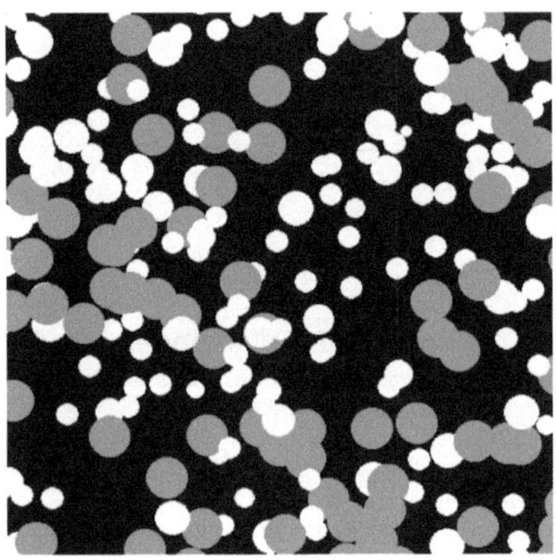

Fig. 7.33. Simulation of a section of a dead leaves model of cylinders (grey) and spheres (white) $p_c = 0.2$, $p_s = 0.2$.

$$\Phi_{cs}(x) = 1 - \\ - \frac{1}{p^2}\left[2 - h\left(\frac{\rho}{a_c}\right) - H(R)\right]\left[\frac{1 - Q(x)}{2 - H_{cs}(x)} - \frac{Q(0) - Q(x)}{1 - H_{cs}(x)}\right].. \tag{7.559}$$

Here $p = p_c + p_s$,

$$Q(x) = (1 - p)^{2 - H_{cs}(x)}, \quad H_{cs}(x) = \frac{p_c}{p} h\left(\frac{\rho}{a_c}\right) + \frac{p_s}{p} H(R). \tag{7.560}$$

A realization of this model for $p_c = p_s = 0.2$, $a_c = a_s = l_s$ is presented in Fig. 7.33.

The next probabilistic model of cylinders and spheroids has no priorities for any type of inclusions. Called the dilution model, it is obtained by implantation of independent inclusions of both populations inside a domain in 3D-space. If the inclusions intersect, the region of intersection is replaced by the matrix phase. As a result, the inclusions do not overlap in this model. The corresponding correlation functions have the following forms:

$$\Phi_{cc}(x) = 1 - \frac{2}{p^2}\left(1 - h\left(\frac{\rho}{a_c}\right)\right)^2 \exp\left[\theta H_{cs}(x)\right], \tag{7.561}$$

$$\Phi_{ss}(x) = 1 - \frac{2}{p^2}\left(1 - H(R)\right)^2 \exp\left[\theta H_{cs}(x)\right], \tag{7.562}$$

$$\Phi_{cs}(x) = 1 - \frac{2}{p^2}\left(1 - h\left(\frac{\rho}{a_c}\right)\right)\left(1 - H(R)\right) \exp\left[\theta H_{cs}(x)\right]. \tag{7.563}$$

Here the parameter θ is connected to the total volume concentration of inclusions $p = p_c + p_s$ by the equation

$$p = \theta \exp(-\theta). \tag{7.564}$$

The maximal value of the concentration p in this model is $p = 0.368$. This value of p corresponds to the maximum of the function on the right-hand side of (7.564). This maximum is achieved at $\theta = 1$. In Fig. 7.34, one can see a realization of the dilution model for $p_c = 0.2$, $p_d = 0.15$, $a_c = a_s = l_s$.

It is important to notice that the various probabilistic models present very different morphologies: the two versions of the Boolean model are not symmetric with respect to spheroids and cylinders; that do not play the same roles as a result of the order involved by the priority rule. On the other hand,

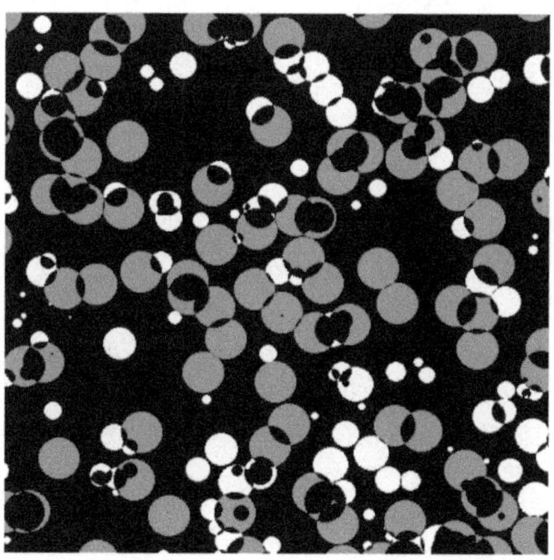

Fig. 7.34. Simulation of a section of a dilution model of cylinders (grey) and spheres (white) $p_c = 0.2$, $p_s = 0.2$.

the dead leaves and the dilution model give the same roles to the two type of inclusions, which can alternatively dominate (according to their date of arrival) for the dead leaves, or which are always erased if overlaps occur for the dilution model.

Note that these probabilistic models are not completely compatible with the main assumptions of the EFM because of the possibility of overlaps of inclusions. Only for small volume concentrations of inclusions the realizations of these models are sets of isolated cylinders and spheroids as implied in the EFM. For high volume concentrations, many inclusions overlap and a typical inhomogeneity in the composite does not have a cylindrical or spheroidal form, but some agglomeration of these forms. Thus, the solutions of the one particle problems that are used in the method and obtained for non overlapping inclusions do not correspond completely to these models. The volume concentrations of inclusions when the assumptions of the EFM are valid, in spite of the overlapping, depend on the relative sizes of cylinders and spheroids. For instance, in the boolean model with cylinder priority the relative number of "defected" spheroids tends to zero if the radii of spheroids become much smaller than the radii of cylinders. Thus, the lower is the ratio a_s/a_c, the more reasons we have to use the boolean model with cylinder priority in the framework of EFM.

7.16.3 Overall elastic moduli of three-phase composites

For the calculation of the overall elastic moduli tensors using (7.532), (7.526) of the EFM let us build the tensors A_{ij} for the probabilistic models of hybrid composites considered in the previous section. For the boolean model of cylinders and spheroids with the cylinder priority, the function $\Phi_{cc}(x)$ has the form (7.548) and depends only on the distance ρ from the axis x_3. Therefore this function has a cylinder symmetry and the corresponding integral in (7.525) may be calculated using (7.100) and takes the form

$$A_{cc} = \int K(x)\Phi_{cc}(x)\mathrm{d}x = A_c. \tag{7.565}$$

$$A_c = -\frac{\lambda_0+\mu_0}{8\mu_0(\lambda_0+2\mu_0)}P_1(n) + \frac{\lambda_0+3\mu_0}{4\mu_0(\lambda_0+2\mu_0)}P_2(n) + \frac{1}{2\mu_0}P_5(n), \tag{7.566}$$

where λ_0, μ_0 are the Lamé constants of the matrix , n is a unit vector along the axis of the cylinder, $P_i(n)$ are the elements of the P-basis.

For the calculation of the integral A_{ss} with the function $\Phi_{ss}(x)$ in (7.549) let us present this function in the form

$$\Phi_{ss}(x) = \Phi_{ss}^{(1)}(\rho) + \Phi_{ss}^{(2)}(x), \tag{7.567}$$

$$\Phi_{ss}^{(1)}(\rho) = 1 - \frac{1}{p_s^2}(1-q_{cs})^2 q_c^{2-h\left(\frac{\rho}{a_c}\right)}, \tag{7.568}$$

$$\Phi_{ss}^{(2)}(x) = \frac{1}{p_s^2}\left[(1-q_{cs})^2 - \left(1-q_{cs}^{1-H(R)}\right)^2 q_{cs}^{H(R)}\right] q_c^{2-h\left(\frac{\rho}{a_c}\right)}. \quad (7.569)$$

Here the function $\Phi_{ss}^{(1)}(\rho)$ depends only on the distance from the x_3 axis, and thus has cylindrical symmetry. The function $\Phi_{ss}^{(2)}(x)$ is equal to zero if $|x| > r_0 = \max(2a_c, 2a_s, 2l_s)$, and

$$\Phi_{ss}^{(1)}(0) = -\frac{p_c}{1-p_c}, \quad \Phi_{ss}^{(2)}(0) = \frac{1}{1-p_c}. \quad (7.570)$$

As a result, the tensor A_{ss} takes the form

$$A_{ss} = \int K(x)\left[\Phi_{ss}^{(1)}(\rho) + \Phi_{ss}^{(2)}(x)\right]\mathrm{d}x = -\frac{p_c}{1-p_c}A_c + \\ + \frac{1}{1-p_c}A_s + \det(a_s) v.p. \int K(a_s y)\Phi_{ss}^{(2)}(a_s y)\mathrm{d}y. \quad (7.571)$$

Here the tensor A_c coincides with (7.566), the tensor $A_s = A(a_s)$ has the form in (7.101) when a_s is the linear transformation that converts the spheroidal inclusion into a unit sphere. The last integral in (7.571) was calculated numerically.

The tensor $A_{cs} = A_{sc}$ for the function $\Phi_{cs}(\rho)$ in (7.550) with a cylinder symmetry takes the form

$$A_{cs} = \int K(x)\Phi_{cs}(\rho)\mathrm{d}x = A_c. \quad (7.572)$$

For the model with spheroid priority, the function $\Phi_{cc}(x)$ in (7.552) may be rewritten in the form

$$\Phi_{cc}(x) = \Phi_{cc}^{(1)}(\rho) + \Phi_{cc}^{(2)}(x), \quad (7.573)$$

$$\Phi_{cc}^{(1)}(\rho) = 1 - \frac{1}{p_c^2}\left[1 - q_{sc}^{1-h\left(\frac{\rho}{a_c}\right)}\right]^2 q_{sc}^{h\left(\frac{\rho}{a_c}\right)} q_s^{\,2}, \quad (7.574)$$

$$\Phi_{cc}^{(2)}(x) = \frac{1}{p_c^2}\left[1 - q_{sc}^{1-h\left(\frac{\rho}{a_c}\right)}\right]^2 q_{sc}^{h\left(\frac{\rho}{a_c}\right)}\left[q_s^{\,2} - q_s^{\,2-H(R)}\right], \quad (7.575)$$

where the function $\Phi_{cc}^{(1)}(\rho)$ tends to zero as $\rho \to \infty$, and $\Phi_{cc}^{(2)}(x) \to 0$ when $|x| \to \infty$,

$$\Phi_{cc}^{(1)}(0) = 1, \quad \Phi_{cc}^{(2)}(0) = 0.$$

As a result, the tensor A_{cc} for the model with the spheroid priority takes the form

$$A_{cc} = \int K(x)\Phi_{cc}(x)\mathrm{d}x = A_c + \det(a_s)\int K(a_s y)\Phi_{cc}^{(2)}(a_s y)\mathrm{d}y, \quad (7.576)$$

where the last integral converges absolutely because the function $\Phi^{(2)}_{cc}(x)$ is equal to zero if $|x| > \max(2a_c, 2a_s, 2l_s)$ and $\Phi^{(2)}_{cc}(0) = 0$.

It follows from (7.554), (7.555) and (7.525) that the tensors A_{ss} and A_{cs} for spheroid priority take the forms

$$A_{ss} = A_{cs} = A_{sc} = A_s. \tag{7.577}$$

Let us consider the case of very thin ellipsoidal disks ($l_s << a_s$). In this case the integrals on the right-hand sides of (7.571), (7.576) have the order of $\frac{l_s}{a_s}$, and may be neglected. Thus, the equations for the tensors A_{ij} in the case of thin disks are essentially simplified, and for cylinder priority we get

$$A_{cc} = A_{cd} = A_{dc} = A_c, \tag{7.578}$$

$$A_{dd} = -\frac{p_c}{1-p_c}A_c + \frac{1}{1-p_c}A_d. \tag{7.579}$$

Here the tensor A_d takes the form

$$A_d = \frac{1}{2\mu_0}P^5(n) + \frac{1}{\lambda_0 + 2\mu_0}P^6(n). \tag{7.580}$$

For disk priority we get

$$A_{cc} = A_c,$$

$$A_{dd} = A_{dc} = A_{cd} = A_d, \tag{7.581}$$

where the tensors A_c and A_d are defined in (7.566), (7.580).

Let us consider the overall elastic moduli tensor of the hybrid composite when inclusions are cylinders and thin spheroids (disks). For cylinder priority, we have from (7.532), (7.526), and (7.578) the following equation for C_*:

$$C_* = C_0 + p_{*c}T_c\Lambda_*^{(c)} + p_{*d}T_d\Lambda_*^{(d)}, \tag{7.582}$$

$$\Lambda_*^{(c)} = \left[I - p_{*c}A_cT_c - p_{*d}A_cT_dD_d^{-1}\right]^{-1}, \tag{7.583}$$

$$\Lambda_*^{(d)} = \left[I - p_{*d}A_{dd}T_d - p_{*c}A_cT_cD_d\right]^{-1}, \tag{7.584}$$

$$D_d = I - p_{*d}(A_{dd} - A_c)T_d, \quad T_c = C_1^c\Lambda^c, \tag{7.585}$$

$$T_d = C_1^d\Lambda^d, \quad p_{*c} = p_c, \quad p_{*d} = p_d(1-p_c). \tag{7.586}$$

For disk priority, the equation for C_* follows from (7.532), (7.540), and (7.581) and takes the form

$$C_* = C_0 + p_{*c}T_c\Lambda_*^{(c)} + p_{*d}T_d\Lambda_*^{(d)}, \tag{7.587}$$

$$\Lambda_*^{(c)} = \left[I - p_{*c}A_cT_c - p_{*d}A_dT_dD_c\right]^{-1}, \tag{7.588}$$

$$\Lambda_*^{(d)} = \left[I - p_{*d}A_dT_d - p_{*c}A_dT_cD_c^{-1}\right]^{-1}, \tag{7.589}$$

$$D_c = I - p_{*c}(A_c - A_d)T_c, \quad p_{*c} = p_c(1-p_d), \quad p_{*d} = p_d. \tag{7.590}$$

In Fig. 7.36, the dependences of Young's moduli E_{*1} in the direction orthogonal to the axes of the cylinders and E_{*3} in the direction of the cylinder axes on the actual volume concentrations of disks p_{*d} are presented for a fixed actual concentration of cylinders $p_{*c} = 0.2$. The bold lines with dots correspond to the results of the EFM for the boolean model with cylinder priority, the bold lines with short horizontal bars present the results of the same method for disk priority.

Note that we can calculate the overall elastic properties of three phase composites using the equations of the two phase theory. One can consider first the matrix with one population of inclusions, and using (3.16) of the EFM for the overall elastic moduli tensor of two phase composites, calculate overall elastic properties of such a heterogeneous medium. After that, one can consider the two-phase system of a matrix with the calculated elastic properties and the second population of inclusions and again apply the formulas of the two-phase theory (sequential approach, Fig. 7.35).

As a result, we get the overall elastic moduli tensor of the hybrid composite that has the necessary symmetry with respect to indices. A thin line

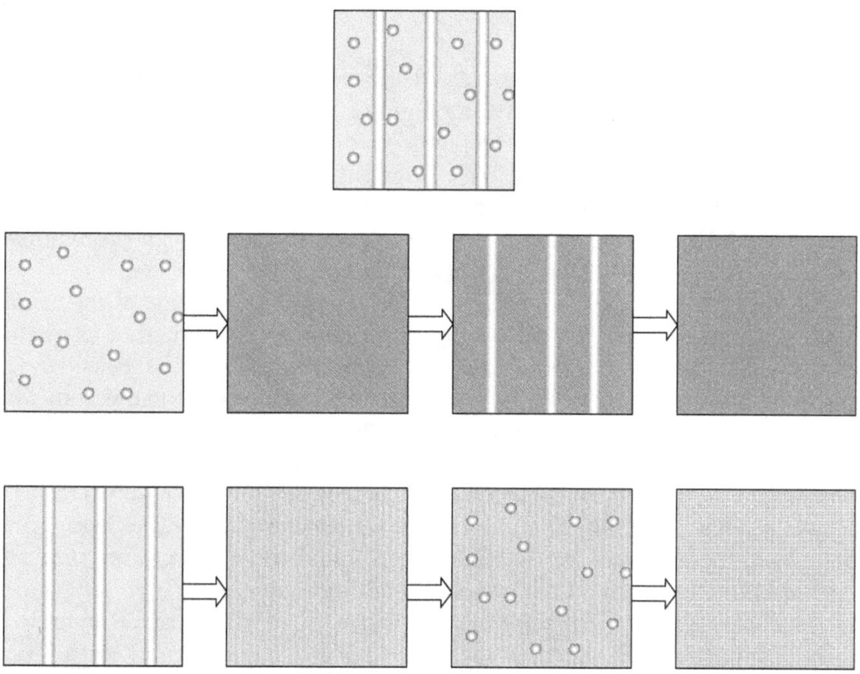

Fig. 7.35. Sequential approach for the calculation of the overall properties of hybrid composites. The final results depends on the order of the application the homogenization equations.

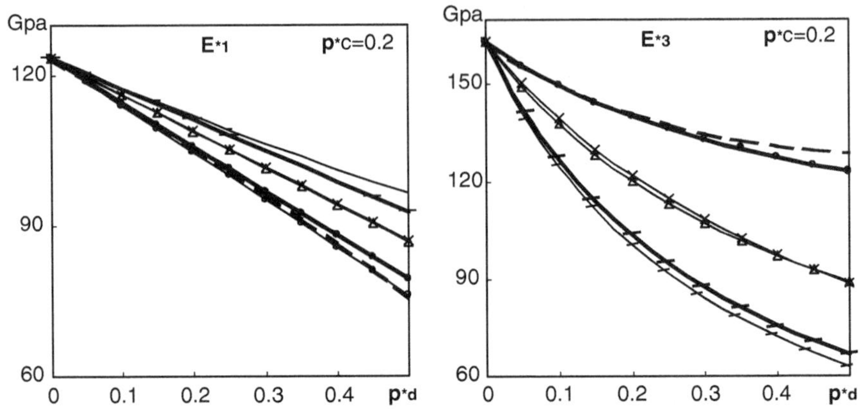

Fig. 7.36. Elastic moduli of the cylinders and disks hybrid composite, E_{*1} is Young's modulus orthogonal to cylinders, E_{*3} is Young's modulus along cylinders. Thin lines with dots correspond to cylinders in the effective medium of matrix and disks, thin lines with short bars correspond to disks in the effective medium matrix with cylinders. Bold lines with dots correspond to the EFM and the boolean model of cylinders and disks with cylinder priority, bold lines with short bars correspond to the EFM and the same model with disk priority, lines with triangles correspond to the EFM and the dead leaves model, lines with crosses are given by the Mori–Tanaka method. Bold lines correspond to the EFM and the dilution model, dashed line is the extrapolation of the dilution model in the region $p > 0.368$.

with dots in Fig. 7.36 corresponds to the sequential approach when cylinders are considered in the effective medium of the matrix with the disks, and the thin line with short bars corresponds to disks in the effective medium of the matrix with cylinders. It is seen from this figure that the result of the sequential approach strongly depends on the order of phase consideration. The sequential approaches and the EFM applied to the boolean probabilistic models with the cylinders or disks priorities practically coincide.

The application of the sequential approach to the calculation of the overall elastic properties of hybrid composites is reasonable if the sizes of the inclusions of different populations are essentially different. Let us consider the sequence (mdc) ((matrix + disks) + cylinders). In this case the medium with disks may be considered as a homogeneous medium with respect to cylindrical inclusions if the radii a_d of disks are much smaller then the radii a_c of the cylinders ($a_d \ll a_c$). Note that because cylinders occupy a part of the volume of the composite, the effective volume concentration p_{*d} of disks in the final three-phase medium will be less then the original concentration p_d that was considered in the first step of the sequential approach, and $p_{*d} = p_d(1 - p_c)$. On the contrary, if we consider the sequence (mcd) ((matrix + cylinders) + disks) the results of this approach is valid, strictly speaking, in the limit $\frac{a_c}{a_d} \to 0$ and the effective volume concentration p_{*c} of cylinders in the hybrid composite is $p_{*c} = p_c(1 - p_d)$.

Let us obtain the tensors A_{cc}, A_{dd}, A_{cd} for the dead leaves model of cylinders and thin disks. From (7.557)–(7.558) and (7.525) we get for this case

$$A_{cc} = A_c, \quad A_{dd} = (1 - g_{dd})A_c + g_{dd}A_d, \tag{7.591}$$

$$A_{cd} = (1 - g_{cd})A_c + g_{cd}A_d, \quad g_{dd} = 2g_{cd}, \tag{7.592}$$

$$g_{cd} = \frac{1}{p}\left[\frac{1-(1-p)^{2-\frac{p_c}{p}}}{2p-p_c} - \frac{1-p}{p_d}\left(1-(1-p)^{\frac{p_d}{p}}\right)\right], \tag{7.593}$$

and $p_d = p_{*d}, p_c = p_{*c}$.

The dependences of Young's moduli E_{*1} and E_{*3} for this model are lines with triangles in Fig. 7.36. These lines are very close to the results of the Mori–Tanaka method that are presented in Fig. 7.36 by the lines with crosses.

For the dilution model of cylinder and disks, the tensors A_{ij} take the forms

$$A_{cc} = A_c, \quad A_{dd} = (1 - g_{dd})A_c + g_{dd}A_d, \quad A_{cd} = A_c, \tag{7.594}$$

$$g_{dd} = \exp\left[p_c\frac{\theta(p)}{p}\right], \quad p = p_c + p_d, \tag{7.595}$$

where the dependence $\theta(p)$ is defined from (7.564), and actual concentrations p_{*c}, p_{*d} of inclusions coincide with p_c, p_d. As mentioned in Section 7.16.2 this model serves only until the volume concentration of disks reaches $p_d = 0.168$ ($p_c = 0.2$), the bold lines in Fig. 7.36 correspond to this solution. In the region $p_c + p_d < 0.368$ the result of the dilution model may be approximated with a sufficient accuracy if the function $\theta(p)$ in (7.564) is approximated by the first terms of its Taylor series

$$\theta(p) = p + p^2 + \frac{3}{2}p^3. \tag{7.596}$$

The dashed lines in Fig. 7.36 correspond to the extrapolation of Young's moduli of the composite into the region $p_d > 0.168$ if the function $\theta(p)$ has the form (7.596).

In Fig. 7.37 the results of all approaches are compared for the composite with infinite cylinders and spherical pores. Young's modulus and Poisson's ratio of the isotropic matrix are $E_0 = 100$GPa, $\nu_0 = 0.4$ and the elastic parameters of cylinders are $E_c = 500$Gpa, $\nu_c = 0.3$. The notations for the different curves are the same as in Fig. 7.37 (the lines with short bars correspond to the boolean model with sphere priority (bold lines) and to the spheres in the effective medium of matrix and cylinders (thin lines)). Note that the integrals in (7.571), (7.576) give a small contribution to the values of the elastic constants, and neglecting these integrals changes the values of the overall elastic moduli by less than about 5%.

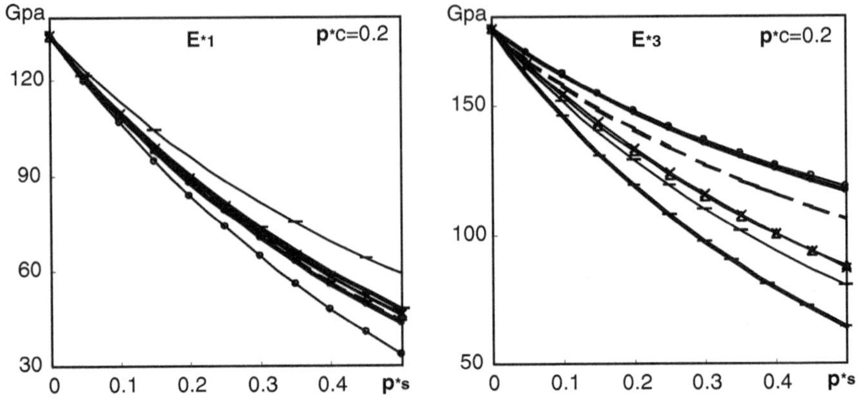

Fig. 7.37. The same dependences as in Fig. 7.36 for cylinder and spherical pore hybrid composite.

As for cylinder-disk composites, the results of the sequential approaches for cylinders-spheres are very close to the EFM method applied to the boolean probabilistic models of cylinders and spheres with cylinder or sphere priorities. The dead leaves model of cylinders and spheres again turns out to be close to the results of the Mori–Tanaka method.

7.17 Conclusions

Among different approximate methods for the solution of the homogenization problem, the EFM has several important advantages. In many cases it gives us simple equations for the overall physical and mechanical properties of composite materials, that are in good agreement with experimental data and with exact solutions for regular composites in a wide range of its microparameters. The method is sufficiently flexible to be applied in various nonstandard situations and to obtain physically reasonable approximate solutions of the homogenization problem for such cases.

In application to two-phase composite materials, the simplest version of the method gives an equation for the overall elastic moduli tensor (7.104) that is not very sensitive to the statistical properties of the random field of inclusions. For instance, if we consider quasispherical inclusions in a homogeneous matrix and restrict ourselves to statistically isotropic sets of inclusions, the final formula of the method (7.108) will be the same for all the boolean type models similar to those considered in Section 7.16. This is connected with the fact that the integral A in (7.100) does not depend on the details of the behavior of the correlation function if the latter has a spherically symmetry $\Psi(x) = \Psi(|x|)$. On the other hand, this fact reflects a physical reality: the overall static elastic properties of composites are not very sensitive to the details of microstructure (e.g., to agglomeration of inclusions, to an order in the positions of neighbor inclusions). Of course there are some phenomena that strongly depend on microstructures and that the EFM cannot predict, for instance, the existence of a percolation threshold for the overall conductivity of a dielectric matrix and conducting inclusions. This phenomenon must be explained in another theoretical framework.

The situation changes dramatically when we consider the influence of microstructure on the elastic properties of hybrid composites. As shown in Section 7.16.4, the overall elastic moduli of hybrid composites with the same volume concentrations and types of inclusions may differ by a factor of two for different isotropic probabilistic models of inclusions distributions. The results of the sequential approach applied in Section 7.16.4 for the calculation of the overall elastic moduli of three-phase composites, may be considered as a good approximation when the sizes of the inclusions of different populations are essentially different. This approach also gives us an essential difference in elastic properties of the studied hybrid composites by exchanging the order of phase implantation. On the other hand, the EFM for the Boolean models with cylinder priority and spheroid priority gives results that are very close to the sequential approaches. This fact may be considered as a verification of essential differences of the elastic properties of hybrid composites with the same volume concentrations but with different types of spatial distributions of the inclusions.

As shown in Section 7.6 the formulas of the Mori–Tanaka method follow from the EFM by some additional assumptions. For three-phase composites, these assumptions violate general symmetry of the cross-correlation functions of different populations of inclusions. As a result, the Mori–Tanaka method gives a wrong symmetry of the overall elastic moduli tensors of three-phase composites. Nevertheless, the numerical values of the elastic moduli predicted by the Mori–Tanaka method turn out to be close to the results of the version of the EFM developed for the dead leaves model of a random set of inclusions.

The main conclusion may be formulated as follows. For the prediction of the overall elastic properties of hybrid composites, we have to use more microstructural information than for two-phase composite materials. The version of the EFM developed in this section allows us to take this information into account. The precision of the predictions of the method strongly depends on the precision of this microstructural information.

7.18 Notes

The effective medium method (version I) was first applied to the calculation of effective elastic properties of polycrystals in [61], [100]. For the calculation of the effective elastic moduli of matrix composite materials, this method was used in [190], [25], [63]. Version II of the EMM was proposed in [95]. Some mistakes of this work were corrected in [191] and [29]. Effective elastic moduli of a medium with cracks were calculated by EMM in [26] and by the DEM in [60]. The EMM was used for the calculation of effective elastic moduli of anisotropic composites in [125] and [139] (DEMM).

The DEMM was first proposed in [23], [24] and used for a medium containing rigid particles in [22] (calculation of viscosity of fluid suspensions) and in [177] (calculation of effective elastic moduli of composites). Later on, the DEMM was exploited in [145], [156], [220]. A detailed discussion of this method is presented in [31]. Effective elastic moduli of a cracked medium were obtained by the differential scheme in [60]. Versions of the effective medium theories were discussed also in [14], [16], [153]..

In implicit forms, the EFM was used for the calculation of effective elastic properties of composites with ellipsoidal inclusions in [207], [208]. An explicit formulation of the EFM in application to the homogenization problem for matrix composites with ellipsoidal inclusions and its limit forms (elliptical cracks and long cylindrical fibers) was used in [70], [115], [71]. In [116], the EFM was used for evaluation of stress concentrations on the inclusions in the matrix composite materials. The results of these works are presented in Section 7.5.

The Mori–Tanaka approximation was first proposed in [151] and then developed in [9], [10] for the calculation of thermal conductivities and elastic properties of multi-phase composites. The connection between the Mori–Tanaka approximation and variational bounds for the effective elastic moduli of composites was analyzed in [157]. Specific features of the Mori–Tanaka method were discussed in [16].

Applications of the EFM to the solution of the homogenization problem for composites with regular lattices of ellipsoidal inclusions were considered in [71], [72]. The results of these works are presented in Sections 7.7.

In [90], the EFM was used to calculate the effective moduli of composites reinforced with thin hard flakes or ribbons, . The elastic moduli of media with cracks were calculated by the EFM [73], [85], [126]. The results of these works are presented in Sections 7.8–7.11.

In [83], the EFM was applied to the calculation of the effective elastic moduli of composites reinforced with short axi-symmetric fibers (Section 7.12).

The coefficients of temperature expansions of composites with ellipsoidal inclusions and concentration of temperature stresses on the inclusions were found by the EFM in [115], [117]. Temperature deformation of the medium containing spherical multilayered inclusions was considered in [89], and the medium reinforced with infinite cylindrical multilayered fibers in [91]. The results of these work are presented in Sections 7.5 and 7.13.

Section 7.16 is based on the work [76], where the problem of modelling of the inclusions by point defects was considered.

Application of the EFM to the solution of the homogenization problem for hybrid composites was performed in [87].

8. Multiparticle interactions in composites

The version of the effective field method considered in Chapter 7 allows us to solve the homogenization problem and to calculate effective elastic and thermoelastic properties of matrix composites with various types of inclusions. But for composites with spherical inclusions that are much harder than the matrix, the predictions of the method deviate essentially from experimental data and results of numerical simulations when the volume concentration of the inclusions exceeds 0.4. The reason for this discrepancy is that the hypotheses of the EFM do not conform to the actual distribution of elastic fields in composites with high volume concentrations of inclusions. In order to correct the predictions of the EFM, its hypotheses should be changed. In this chapter we develop a modification of the EFM accounting for multiparticle interactions between inclusions. This modification allows us to improve the predictions of the method, to calculate higher statistical moments of physical fields in composites, and to describe nonlocal connections between the mean stress and strain fields in regions where these fields change rapidly.

8.1 The effective field method beyond the quasicrystalline approximation

We start with the consideration of a homogeneous elastic medium containing a random set of isolated inclusions. Let us take a typical realization of this set, and consider an arbitrary i-th inclusion that occupies the region v_i. If the local exciting strain field acting on this inclusion is ε_i^*, and the solution of the elasticity problem for an inclusion subjected to this field is known, the strain field inside the inclusion is presented in the form

$$\varepsilon^+(x) = \left(\Lambda^i \varepsilon_i^*\right)(x), \qquad x \in V_i, \quad (i = 1, 2, \ldots), \tag{8.1}$$

where Λ^i is a known linear operator.

The equation for the fields $\varepsilon_i^*(x)$ $(i = 1, 2, ...)$ follows from Equation (7.8) for the strain field $\varepsilon(x)$ in the composite in the form

$$\varepsilon_k^*(x) = \varepsilon^0(x) - \sum_{i \neq k} \int \mathrm{K}\left(x - x'\right) C_i^1\left(x'\right) \left(\Lambda^i \varepsilon_i^*\right)\left(x'\right) v_i\left(x'\right) \mathrm{d}x', \tag{8.2}$$

$$x \in v_k\,, \quad k = 1, 2, \ldots\,.$$

Here $C_i^1(x)$ is the perturbation of the elastic moduli inside the i–th inclusion. If the functions $\varepsilon_k^*(x)$ are found from the solution of this system, the strain and stress fields in the composite are determined from (7.20), (7.21)

$$\varepsilon(x) = \varepsilon^0(x) - \int \mathrm{K}(x - x')\, q(x')\, dx', \tag{8.3}$$

$$\sigma(x) = \sigma^0(x) - \int \mathrm{S}(x - x')\, B^0 q(x')\, dx', \tag{8.4}$$

$$q(x) = \sum_i C_i^1(x) \left(\Lambda^i \varepsilon_i^*\right)(x) V_i(x). \tag{8.5}$$

Let us introduce the field $\varepsilon^*(x)$ that coincides with $\varepsilon_i^*(x)$ when $x \in v_i$, and a linear operator P defined by the equation

$$\left(P\varepsilon^*\right)(x) V(x) = \sum_i C_i^1(x) \left(\Lambda^i \varepsilon_i^*\right)(x) v_i(x), \tag{8.6}$$

$$V(x) = \sum_i v_i(x).$$

As a result, system (8.2) can be written as an equation for the field $\varepsilon^*(x)$ in the region V

$$\varepsilon^*(x) = \varepsilon^0(x) - \int \mathrm{K}(x - x') \left(P\varepsilon^*\right)(x')\, V(x; x')\, dx', \tag{8.7}$$

where the function $V(x; x')$ is defined in (7.84).

If the set of inclusions is random, $\varepsilon^*(x)$ is a random function. In order to construct statistical moments of the function $\varepsilon^*(x)$, we introduce the following hypotheses:

H_1. *The field $\varepsilon^*(x)$ has the same structure in every region v_i occupied by an inclusion.*

In particular, if we accept that in the region v_i the field ε^* is polynomial, the degree of this polynomial is the same for all the inclusions, and only the coefficients may differ from one inclusion to another.

H_2. *The value of the function $\varepsilon^*(x)$ in the region v_i is statistically independent of the physical and geometrical properties of the inclusion that occupies this region.*

Hypothesis H_2 means that the local exciting field acting on an arbitrary inclusion depends weakly on its shape and properties but is determined mainly by some integral characteristics of the random set of inhomogeneities.

Let the inclusions be homogeneous ellipsoids, and the field $\varepsilon^*(x)$ be polynomial. Then, hypothesis H_1 and the polynomial conservation theorem (Section 3.3) imply that the field $\varepsilon(x)$ inside any inclusion is a polynomial of the same degree as the local exciting field $\varepsilon^*(x)$. Particularly, if the field ε^* is constant in the domains v_i, the operator P in (8.7) is multiplication by the function $P(x)$ that is constant in every inclusion

$$(P\varepsilon^*)(x) = P(x)\varepsilon^*(x), \quad x \in V, \tag{8.8}$$

$$P(x) = P_i = C_i^1 \left(I + A(a_i) C_i^1\right)^{-1}, \ \varepsilon^*(x) = \varepsilon_i^*, \ x \in V_i . \tag{8.9}$$

Here the tensor $A(a)$ is defined in (3.82).

Substituting (8.8) into (8.7) we go to the following equation for the field $\varepsilon^*(x)$ in the region V:

$$\varepsilon^*(x) = \varepsilon^0(x) - \int \mathrm{K}(x - x') P^0(x') \varepsilon^*(x') V(x; x') \mathrm{d}x'. \tag{8.10}$$

If the field $\varepsilon^*(x)$ is really constant in every region v_i, the solutions of Equations (8.10) and (8.7) coincide.

Let us suppose that the field $\varepsilon^*(x)$ is linear in every regions v_k, occupied by the inclusions

$$\varepsilon^*_{\alpha\beta}(x) = \varepsilon^*_{k\alpha\beta} + \tau_{k\alpha\beta\lambda}\left(x - x^k\right)_\lambda, \quad x \in V_k, \quad k = 1, 2, \dots . \tag{8.11}$$

Here x^k is the center of the region v_k. Since a linear external field induces a linear field inside an ellipsoidal inhomogeneity, the operator P in (8.6) acts on $\varepsilon^*(x)$ according to the equation ($x \in v_k$):

$$(P\varepsilon^*)_{\alpha\beta}(x) = P^{0(k)}_{\alpha\beta\lambda\mu}\varepsilon^{*(k)}_{\lambda\mu} + P^{1(k)}_{\alpha\beta\lambda\mu\nu\delta}\tau^{(k)}_{\lambda\mu\nu}\left(x - x^k\right)_\delta , \tag{8.12}$$

where the tensor $P^{0(k)}$ is defined in (8.9), and the tensor P_k^1 may be found from the solution of the problem for an isolated ellipsoidal inhomogeneity in a linear external field (Section 3.3). Taking into account that the constant tensors $\varepsilon^{*(k)}$ and $\tau^{(k)}$ in (8.11) are expressed via the coefficients of a linear in v_k field $\varepsilon^*(x)$ by the Equations ($x \in v_k$)

$$\varepsilon^{*(k)}_{\alpha\beta} = \varepsilon^*_{\alpha\beta}(x) - \left[\nabla_\lambda \varepsilon^*_{\alpha\beta}(x)\right](x - \xi_k)_\lambda , \quad \tau^{(k)}_{\alpha\beta\lambda} = \nabla_\lambda \varepsilon^*_{\alpha\beta} , \tag{8.13}$$

we obtain that Equation (8.7) for $\varepsilon^*(x)$ takes the form

$$\begin{aligned}\varepsilon^*_{\alpha\beta}(x) = \varepsilon^0_{\alpha\beta}(x) - \int \mathrm{K}_{\alpha\beta\lambda\mu}(x - x')\left\{P^0_{\lambda\mu\nu\rho}(x')\varepsilon^*_{\nu\rho}(x') + \right. \\ \left. + P^2_{\lambda\mu\nu\rho\tau\delta}(x')\left[\nabla_\nu \varepsilon^*_{\rho\tau}(x')\right] H_\delta(x')\right\} V(x; x')\mathrm{d}x',\end{aligned} \tag{8.14}$$

where the functions $P^2(x)$ and $H_\delta(x)$ are

$$P^2_{\lambda\mu\nu\rho\tau\delta}(x) = P^{1(k)}_{\lambda\mu\nu\rho\tau\delta} - P^{(k)}_{\lambda\mu\nu\rho}\delta_{\tau\delta} , \tag{8.15}$$

$$H_\delta(x) = H_\delta^{(k)}(x) = \left(x - x^k\right)_\delta , \quad x \in V_k , \quad k = 1, 2, \dots . \tag{8.16}$$

Equation (8.14) is an integro-differential equation with respect to $\varepsilon^*(x)$. If the field $\varepsilon^*(x)$ is approximated by a polynomial of the second degree in v_k, one can derive an integro-differential equation for the field $\varepsilon^*(x)$ where the derivatives of the second order of $\varepsilon^*(x)$ are presented. The field $\varepsilon^*(x)$ that satisfies (8.10), (8.15) is called the effective field. Equation (8.10) is called the equation of the zero order, and Equation (8.14) is the equation of the first order for the effective field.

8.2 Mean values of some homogeneous random fields

In this section, conditional means of random functions $V(x)$, $\varepsilon(x)$ and $\sigma(x)$ that appear in the equations of the EFM are considered in detail. Here $V(x)$ is the characteristic function of a random set of ellipsoidal regions homogeneously distributed in space. The elastic strain $\varepsilon(x)$ and stress $\sigma(x)$ fields, as well as the effective field $\varepsilon^*(x)$, are functionals of the random function $V(x)$. For a constant external stress field, all these fields are random functions homogeneous in space.

Let $f(x, V)$ be one of the functions under consideration. The mean $\langle f(x)\rangle$ over the ensemble of realizations of $V(x)$ is the following integral in functional space [55]

$$\langle f(x)\rangle = \int f(x, V)\, \mathrm{d}\mu(V), \tag{8.17}$$

where $\mu(V)$ is the measure in this space that corresponds to the random function $V(x)$. Similarly, the correlation function of the field $f(x)$ is defined by the equation

$$\langle f(x_1) f(x_2)\rangle = \int f(x_1, V) f(x_2, V)\, \mathrm{d}\mu(V), \tag{8.18}$$

and higher statistical moments of the function $f(x)$ are defined by similar equations.

By carrying out the EFM one has to calculate the mean value of a function $f(x)$ under the condition that a point x_1 belongs to the region V occupied by the inclusions. For such a conditional mean, a common notation $\langle f(x)|x_1\rangle$ is used. This conditional mean is defined by the equation [55], [172]

$$\langle f(x)|x_1\rangle = \langle V(x_1)\rangle^{-1} \int f(x) V(x_1)\, \mathrm{d}\mu(V). \tag{8.19}$$

The mean of $f(x)$ under the condition that points $x_1, x_2, ..., x_n$ belong to the region V is defined by the equation

$$\langle f(x)|x_1, x_2, ..., x_n\rangle = \left\langle \prod_{i=1}^{n} V(x_i)\right\rangle^{-1} \int f(x) \prod_{i=1}^{n} V(x_i)\, \mathrm{d}\mu(V).$$

Let $x \in V$ and $x_1 \in V_x$, where the region V_x is defined by the equation

$$V_x = \bigcup_{i \neq k} v_i \quad \text{where} \quad x \in v_k \,. \tag{8.20}$$

The mean of the field $f(x)$ under the conditions $x \in V$ and $x_1 \in V_x$ is denoted by $\langle f(x)|x\,;x_1\rangle$, and the expression for this mean has the form

$$\langle f(x)\,|x;x_1\rangle = \langle V(x)\,V(x;x_1)\rangle^{-1} \int f(x)\,V(x)\,V(x;x_1)\,\mathrm{d}\,\mu(V)\,, \tag{8.21}$$

where the function $V(x;x_1)$ is defined in (7.84).

The definition of the conditional means implies the equations

$$\langle f(x)\,V(x)\rangle = \langle V(x)\rangle\langle f(x)\,|x\rangle\,, \tag{8.22}$$

$$\langle f(x)\,V(x_1;x)\,|x_1\rangle = \langle V(x_1;x)\,|x_1\rangle\langle f(x)\,|x;x_1\rangle \tag{8.23}$$

that were used in Chapter 7 for the solution of the homogenization problem in the framework of the EFM.

Random fields considered in this part of the book are assumed to be ergodic. Thus, the average with respect to the ensemble realizations of $V(x)$ coincides with the average of a fixed typical realization of $f(x)$ over the 3D-space. Let $f_0(x)$ by a realization of $f(x)$ that corresponds to a realization $V_0(x)$ of $V(x)$. Then, for the mean value of the function $f(x)$ we have

$$\langle f(x)\rangle = \lim_{W\to\infty} \frac{1}{W}\int_W f_0(x)\,\mathrm{d}x, \tag{8.24}$$

$$\langle f(x)\,|x\rangle = \langle V_0(x)\rangle^{-1} \lim_{W\to\infty} \frac{1}{W}\int_W f_0(x)\,V_0(x)\,\mathrm{d}x, \tag{8.25}$$

where W is a region in R^3 that occupies the whole space in the limit $W \to \infty$. In what follows, the sign 0 for a fixed realization is omitted.

In many cases, the inclusions in a composite may be approximated by point defects (Section 7.16). By the use of this approximation in the framework of the EFM, there appear conditional means of generalized functions $X(x)$ and $X(x_1;x)$ defined in (7.493). Let X be a random set of points ξ_i $(i = 1, 2, ...)$ in 3D-space, and $X_{x_1 x_2 \ldots x_n}$ be the set X from which the points ξ_j are eliminated if they coincide with fixed points $x_1, x_2, ..., x_n$ of this space. $X(x)$ and $X(x_1,x_2,..., x_n; x)$ are the generalized functions concentrated on the elements (points) of the sets X and $X_{x_1 x_2 \ldots x_n}$

$$X(x) = \sum_{\xi_i \in X} \delta(x - \xi_i)\,, \tag{8.26}$$

$$X(x_1, x_2, ..., x_n; x) = \sum_{\xi_i \in X_{x_1 x_2 \dots x_n}} \delta(x - \xi_i). \tag{8.27}$$

Here $\delta(x)$ is the 3D-Dirac delta-function.

Let us consider the following means of the functions $X(x)$ and $X(x_1; x)$ over the ensemble realizations of the random point set X

$$\langle X(x) \rangle, \quad \langle X(x_1; x) | x_1 \rangle = \frac{\langle X(x_1; x) X(x_1) \rangle}{\langle X(x_1) \rangle}, \tag{8.28}$$

$$\langle X(x_1; x) | x_1; x_2 \rangle = \frac{\langle X(x_1; x) X(x_1; x_2) X(x_1) \rangle}{\langle X(x_1; x_2) X(x_1) \rangle}. \tag{8.29}$$

Here $\langle \cdot | x_1 \rangle$ is the mean under the condition that $x_1 \in X$; $\langle \cdot | x_1; x_2 \rangle$ is the mean under the condition $x_1, x_2 \in X$, $x_1 \neq x_2$. In the general the case, the symbol

$$\langle \cdot | x_1, x_2, ..., x_n; x_{n+1}, ..., x_m \rangle \tag{8.30}$$

is the mean under the condition $x_1, x_2, ..., x_m \in X$, and the sign ";" separates variables that cannot have the same values.

In what follows, the set X is supposed to be ergodic. A method of construction of the means in (8.28), (8.29) is based on applying the ergodic property with the following ensemble averaging of the obtained result. For example, from definition (8.26) of the function $X(x)$ we have

$$\langle X(x) \rangle = \lim_{W \to \infty} \frac{1}{W} \int_W \sum_{\xi_i \in W} \delta(x - \xi_i)\, dx = \lim_{W \to \infty} \frac{N}{W} = n_0. \tag{8.31}$$

Here W is the region in R^3 occupying the whole space in the limit $W \to \infty$, N is the number of the points of a realization of the random set X that are inside W, n_0 is the numerical concentration of the elements of X.

Let us calculate the two-point statistical moment of the function $X(x)$. Using the ergodic property, we obtain

$$\langle X(x) X(x + x_1) \rangle = \lim_{W \to \infty} \frac{1}{W} \int_W \sum_{\xi_i, \xi_j \in W} \delta(x - \xi_i) \delta(x + x_1 - \xi_j)\, dx =$$

$$= \lim_{W \to \infty} \frac{1}{W} \sum_{\xi_i, \xi_j \in W} \delta(x_1 - \xi_j + \xi_i). \tag{8.32}$$

Let us introduce the random vector $\xi_{ij} = \xi_j - \xi_i$ with the distribution density $g_{ij}(x)$. If we average Equation (8.32) once more over the ensemble realizations of X, then for every term in the last sum we obtain

$$\langle \delta(x_1 - \xi_{ij}) \rangle = \int \delta(x_1 - \xi) g_{ij}(\xi)\, d\xi = g_{ij}(x_1), \quad i \neq j, \tag{8.33}$$

$$\langle \delta (x_1 - \xi_{ij}) \rangle = \delta (x_1) , \quad i = j. \tag{8.34}$$

Here we take into account that $\xi_{ii} = 0$ for all i, and thus, $g_{ii}(x) = \delta(x)$. Separating the terms in (8.32) with $i = j$, we can write

$$\langle X (x) X (x + x_1) \rangle = n_0 \delta (x_1) + \lim_{W \to \infty} \frac{1}{W} \sum_{i,j\ (i \neq j)} g_{ij} (x_1). \tag{8.35}$$

Using this result and an evident equation

$$X (x) = X (x_1; x) + \delta (x - x_1) , \quad x_1 \in X, \tag{8.36}$$

we find the equation for the numerator in (8.28):

$$\langle X (x_1; x) X (x_1) \rangle = \lim_{W \to \infty} \frac{1}{W} \sum_{i,j\ (i \neq j)}^{N} g_{ij} (x - x_1). \tag{8.37}$$

Let a random point set X be statistically homogeneous and isotropic, and correlation in positions of the elements of X disappear with increasing distances between them. In this case, when $|x - x_1| \to \infty$, we have

$$\langle X (x_1; x) X (x_1) \rangle \to \langle X (x_1; x) \rangle \langle X (x_1) \rangle = (n_0)^2 , \tag{8.38}$$

and the function $\Psi(x - x_1) = (n_0)^{-2} \langle X(x_1; x) X(x_1) \rangle$ has the properties (compare with (7.97), (7.88))

$$\Psi (x) = \Psi (|x|) ; \quad \Psi (0) = 0; \quad \Psi (x) \to 1 \text{ when } |x| \to \infty . \tag{8.39}$$

Similar functions appeared in Section 7.16 in consideration of a medium with a random set of point defects

Let the points of the set X form a random lattice in space, and vectors of these points ξ_m be defined by the equation

$$\xi_m = m + \rho_m + r, \tag{8.40}$$

where m is the vector of the point of a fixed in space regular lattice, ρ_m are the independent random vectors with the mathematical expectation equal to zero and the distribution density $f(x)$ ($f^*(k)$ is the Fourier transform of $f(x)$), r is a random vector homogeneously distributed in space and the same for all m.

It follows from (8.40) that the characteristic function $g^*_{mn}(k)$ of the random vector $\xi_{mn} = \xi_m - \xi_n = m - n + \rho_m - \rho_n$ has the form

$$g^*_{mn} (k) = f^* (k) f^* (-k) e^{-ik \cdot (m-n)} . \tag{8.41}$$

The function $g_{mn}(x)$ in (8.37) is defined by the following equation:

$$g_{mn} (x) = g (x - m + n) , \quad g (x) = \frac{1}{(2\pi)^3} \int f^* (k) f^* (-k) e^{-ik \cdot x} \mathrm{d}\, k . \tag{8.42}$$

It follows from this equation and (8.28), (8.37), and (8.31) that the expression for the condition mean $\langle X(x_1; x)|x_1\rangle$ takes the form

$$\langle X(x_1; x)|x_1\rangle = \sum_m{}' g(x - x_1 - m), \tag{8.43}$$

where the prime over the sum sign means excluding the term with $m = 0$.

Let us go to the construction of the mean in the numerator in (8.29). Using the ergodic property we find

$$\begin{aligned}
&\langle X(x_1; x_1 + x_2) X(x_1; x_1 + x_3) X(x_1)\rangle = \\
&= \lim_{W\to\infty} \frac{1}{W} \int_W \sum_{\xi_i,\xi_j,\xi_k \in W\ (k,j\neq i)} \delta(x_1 - \xi_i) \times \\
&\times \delta(x_1 + x_2 - \xi_j)\, \delta(x_1 + x_3 - \xi_k)\, \mathrm{d}\, x_1 = \\
&= \lim_{W\to\infty} \frac{1}{W} \sum_{\xi_i,\xi_j,\xi_k \in W\ (k,j\neq i)} \delta(x_2 - \xi_{ij})\, \delta(x_3 - \xi_{ki})\,.
\end{aligned} \tag{8.44}$$

Let us average this equation once more with respect to the ensemble realizations of X. Since ξ_{ij} and ξ_{ki} when $j \neq k$ are independent random vectors, their joint distribution function is $g_{ij}(x)g_{ki}(x)$. Ensemble averaging of every terms of the last sum in (8.44) gives

$$\langle \delta(x_2 - \xi_{ij})\, \delta(x_3 - \xi_{ki})\rangle = \begin{cases} g_{ij}(x_2)\, g_{ki}(x_3)\,, & j \neq k \\ \delta(x_2 - x_3)\, g_{ij}(x_3)\,, & j = k \end{cases}. \tag{8.45}$$

From this equation and (8.44), (8.37) we obtain the expression for the mean in (8.29) in the form

$$\langle X(x_1; x)|x_1; x_2\rangle = \delta(x - x_2) + F(x, x_1, x_2)\,, \tag{8.46}$$

where the function $F(x, x_1, x_2)$ is defined by the equation

$$F(x, x_1, x_2) = \langle X(x_1, x_2; x)|x_1; x_2\rangle, \tag{8.47}$$

and can be presented in the form

$$F(x, x_1, x_2) = \frac{\sum_m{}' g(x - x_1 - m) \sum_{n\neq m}{}' g(x_2 - x_1 - n)}{\sum_m{}' g(x_1 - x_2 - m)}\,. \tag{8.48}$$

Here prime over the sum signs means excluding the term with $m = 0$ ($n = 0$).

For inclusions of finite sizes, the means similar to (8.28), (8.29) are defined by the following equations:

$$\langle V(x)\rangle = p\,, \quad \langle V(x_1;x)\,|x_1\rangle = \frac{\langle V(x_1;x)\,V(x_1)\rangle}{\langle V(x_1)\rangle}, \tag{8.49}$$

$$\langle V(x_1;x)\,|x_1;x_2\rangle = \frac{\langle V(x_1;x)\,V(x_1;x_2)\,V(x_1)\rangle}{\langle V(x_1;x_2)\,V(x_1)\rangle}. \tag{8.50}$$

Here the points x_1 and x_2 are inside different inclusions, p is the volume concentration of inclusions. The means in (8.49) for random and regular sets of inclusions in space were considered in Sections 7.5 and 7.7.

We introduce the region $V_{x_1x_2\ldots x_n}$ by the equation

$$V_{x_1x_2\ldots x_n} = \bigcup_{k\neq i_1,i_2,\ldots,i_n} v_k \quad \text{when} \quad x_1\in v_{i_1}\,,\quad x_2\in v_{i_2}\,,\ \ldots,\quad x_n\in v_{i_n}\,, \tag{8.51}$$

where v_k is the region occupied by the k-th inclusion, and denote by $\bar{V}_{x_1x_2\ldots x_{n-1}x_n}$ the complement of $V_{x_1x_2\ldots x_{n-1}x_n}$ to $V_{x_1x_2\ldots x_{n-1}}$. $\bar{V}_{x_1,x_2,\ldots,x_n}$ is the region v_{i_n} of the i_n-th inclusion which the point x_n belongs to. The characteristic function (with argument x) of the regions $V_{x_1x_2\ldots x_n}$ and $\bar{V}_{x_1x_2\ldots x_n}$ we denote by $V(x_1,x_2,\ldots,x_n;x)$ and $\bar{V}(x_1,x_2,\ldots,x_n;x)$. These definitions imply the equations

$$V(x) = V(x_1;x) + \bar{V}(x_1;x)\;, \tag{8.52}$$

$$V(x_1;x) = V(x_1,x_2;x) + \bar{V}(x_2;x) - \bar{V}(x_2;x)\,\bar{V}(x_1;x)\;. \tag{8.53}$$

If the points x_1 and x_2 are in different inclusions, the last item in the right-hand side of (8.53) disappears. As a result, the mean (8.50) is presented in the form

$$\langle V(x_1;x)\,|x_1;x_2\rangle = \langle\bar{V}(x_2;x)\,|x_1;x_2\rangle + \langle V(x_1,x_2;x\,|x_1;x_2\rangle\;. \tag{8.54}$$

If $|x_1|\to\infty$, the right-hand side of this equation becomes independent of x_1 and takes the forms

$$\langle\bar{V}(x_2;x)\,|x_1;x_2\rangle \to \langle\bar{V}(x_2;x)\,|x_2\rangle = \langle\bar{V}(x_2;x)\,V(x_2)\rangle/\langle V(x_2)\rangle, \tag{8.55}$$

$$\langle V(x_1,x_2;x)\,|x_1;x_2\rangle \to p\Psi(x-x_2)\;, \tag{8.56}$$

where $\Psi(x)$ is defined in (7.88), and the numerator in the right-hand side of (8.55) is presented in the form ($R = x - x_2$)

$$\begin{aligned}&\langle\bar{V}(x_2;x_2+R)\,V(x_2)\rangle\\ &= \lim_{W\to\infty}\frac{1}{W}\int_W \sum_i V_i(x_2+R)\,V_i(x_2)\,\mathrm{d}\,x_2 = pJ(R)\;.\end{aligned} \tag{8.57}$$

Here the function $J(R)$ for ellipsoidal inclusions has form (7.182). Hence,

$$\langle\bar{V}(x_2;x)\,|x_2\rangle = J(x-x_2)\,. \tag{8.58}$$

In the limit $|x_2| \to \infty$, we have

$$\langle \bar{V}(x_2; x)|x_1; x_2\rangle \to 0, \quad \langle V(x_1, x_2; x)|x_1; x_2\rangle \to p\Psi(x - x_1). \tag{8.59}$$

The function

$$\langle V(x_1; x)|x_1; x\rangle = J(x - x_2)\Psi(x_1 - x_2) + \\ + p\Psi(x - x_1)\Psi(x - x_2)\Psi(x_1 - x_2), \tag{8.60}$$

is the mean in (8.50). It is similar to the mean (8.47) in the case of point defects.

8.3 General scheme for constructing multipoint statistical moments

Let us consider multipoint statistical moments of the effective strain field $\varepsilon^*(x)$. We denote by $\varepsilon^{*(n)}(x_1, x_2, ..., x_n)$ the moment of the order n of this field: the mean of the tensor product $\varepsilon^*(x_1) \otimes \varepsilon^*(x_2) \otimes ... \otimes \varepsilon^*(x_n)$ under the condition that points $x_1, x_2, ..., x_n$ belong to the region V occupied by the inclusions. In particular, the first $\varepsilon^{*(1)}(x)$ and the second $\varepsilon^{*(2)}(x_1, x_2)$ moments of the effective field are the following conditional means:

$$\varepsilon^{*(1)}(x) = \langle \varepsilon^*(x)|x\rangle, \tag{8.61}$$

$$\varepsilon^{*(2)}(x_1, x_2) = \langle \varepsilon^*(x_1) \otimes \varepsilon^*(x_2)|x_1, x_2\rangle. \tag{8.62}$$

In this Section, we consider the equation of the zero order for the effective field ε^* (8.10). In order to construct the mean $\varepsilon^{*(1)}(x)$, let us average both sides of (8.7) over the ensemble realizations of the random field of inhomogeneities under the condition $x \in V$

$$\langle \varepsilon^*(x)x\rangle = \varepsilon^0(x) - \int \mathrm{K}(x - x')\langle P(x')\varepsilon^*(x')V(x; x')x\rangle dx'. \tag{8.63}$$

Using the hypothesis H_2 (Section 8.1) about statistical independence of the field $\varepsilon^*(x)$ in the region v_k on the geometrical characteristics and properties of the k-th inclusion, we represent the mean in the integrand (8.63) in the form of the following product:

$$\langle P(x')\varepsilon^*(x')V(x; x')|x\rangle = \langle P(x')V(x; x')|x\rangle\langle \varepsilon^*(x')|x'; x\rangle, \tag{8.64}$$

where $\langle \varepsilon^*(x')|x'; x\rangle$ is the mean under the condition that $x' \in V, x \in V_{x'}$. This mean differs from $\varepsilon^{*(1)}(x)$.

If the properties of the inclusions are independent on their spatial locations, then the first multiplier in the right-hand side of (8.64) is represented in the form

$$\langle P(x')\, V(x;x')\,|x\rangle = \bar{P}\Psi(x,x')\ , \quad \bar{P} = p\langle P(x)|x\rangle\ , \tag{8.65}$$

$$\Psi(x,x') = \frac{\langle V(x;x')\,|x\rangle}{\langle V(x)\rangle}\ , \quad \Psi(x,x') = \Psi(x-x')\ , \tag{8.66}$$

where $\Psi(x,x')$ is a scalar function with properties discussed in Section 6.2. A specific form of the function $\Psi(x,x')$ and more complex means of the function $V(x,x')$ are supposed to be known.

In order to construct the mean $\langle \varepsilon^*(x')|x';x\rangle$ on the right-hand side of (8.64) let us average both sides of (8.10) under the condition $x \in V$, $x_1 \in V_x$ and again use the hypothesis H_2. As the result, we obtain the equations for the means $\langle \varepsilon^*(x)|x\rangle$ and $\langle \varepsilon^*(x)|x;x_1\rangle$ in the following forms:

$$\langle \varepsilon^*(x)|x\rangle = \varepsilon^0(x) - \int \mathrm{K}(x-x')\,\bar{P}\langle \varepsilon^*(x')\,|x';x\rangle \Psi(x,x')\,\mathrm{d}x', \tag{8.67}$$

$$\langle \varepsilon^*(x)|x;x_1\rangle = \varepsilon^0(x) - \\ - \int \mathrm{K}(x-x')\,P\langle \varepsilon^*(x')|x';x,x_1\rangle \langle V(x;x')\,|x;x_1\rangle \mathrm{d}x', \tag{8.68}$$

where $\langle \varepsilon^*(x')\,|x';x,x_1\rangle$ is the mean under the condition $x' \in V$, $x \in V_{x'}$, $x_1 \in V_{x'}$ that differs from $\langle \varepsilon^*(x')|x';x\rangle$, and the mean $\langle V(x;x')|x,x_1\rangle$ is defined in (8.50).

Thus, there appears a chain of equations that connects all the multi-point conditional means of the function $\varepsilon^*(x)$. In order to close this chain, one has to introduce additional assumptions about the statistical properties of the effective field $\varepsilon^*(x)$. A simplest assumption of this kind is called the "quasicrystalline approximation" [113], [217]

$$\langle \varepsilon^*(x)|x;x_1\rangle = \langle \varepsilon^*(x)|x\rangle = \varepsilon^{*(1)}(x)\ . \tag{8.69}$$

It is supposed here that the mean value of the effective field at a point x ($x \in V$) coincides with the mean value of $\varepsilon^*(x)$ by the condition that the point x_1 is inside one of the inclusions. Then, from (8.69) and (8.55) follows a closed equation for the mean value of the effective field $\varepsilon^{*(1)}(x)$

$$\varepsilon^{*(1)}(x) = \varepsilon^0(x) - \int \mathrm{K}^{\Psi}(x-x')\,\bar{P}\varepsilon^{*(1)}(x')\,\mathrm{d}\,x', \tag{8.70}$$

$$\mathrm{K}^{\Psi}(x) = \mathrm{K}(x)\Psi(x). \tag{8.71}$$

When ε^0 and $\varepsilon^{*(1)}$are constants, this equation coincides with Equation (7.92) of the EFM obtained in Chapter 7.

If the external field $\varepsilon^0(x)$ is not constant, the solution of Equation (8.70) takes the form

$$\varepsilon^{*(1)}(x) = \left(\Lambda\varepsilon^0\right)(x)\ , \tag{8.72}$$

where Λ is a pseudodifferential operator with the symbol $\Lambda^*(k)$ defined by the equation

$$\Lambda^*(k) = \left(I + \mathrm{K}^{\Psi *}(k)\bar{P}\right)^{-1} \tag{8.73}$$

Here $\mathrm{K}^{\Psi *}(k)$ is the Fourier transform of the function $\mathrm{K}^{\Psi}(x)$ in (8.70). To develop (8.72), (8.73) the convolution form of Equation (8.70) was taken into account.

The next approximation for $\varepsilon^{*(1)}(x)$ may be found by closing the chain of equations for the condition means of the function $\varepsilon^*(x)$ in (8.21) at the second step by the equation

$$\langle \varepsilon^*(x') | x'; x, x_1 \rangle = \langle \varepsilon^*(x') | x'; x_1 \rangle = \Xi(x', x_1) \ . \tag{8.74}$$

Here the function $\Xi(x, x_1)$ is the mean value of the effective field $\varepsilon^*(x)$ at the point x ($x \in V$) under the condition that $x_1 \in V_x$. This function describes pair interactions between the inclusions in the composite. Clearly that

$$\Xi(x, x_1) \to \varepsilon^{*(1)}(x) \quad \text{when} \quad |x_1| \to \infty. \tag{8.75}$$

The equation for $\Xi(x, x_1)$ follows from (8.68), (8.74) and takes the form

$$\Xi(x, x_1) = \varepsilon^0(x) - \int \mathrm{K}(x - x') P \langle \bar{V}(x_1; x') | x; x_1 \rangle dx' \Xi(x; x_1) - \\ - p \int \mathrm{K}(x - x') P \Xi(x', x_1) F(x', x, x_1) dx', \tag{8.76}$$

$$F(x', x, x_1) = \frac{\langle V(x, x_1; x') | x; x_1 \rangle}{\langle V(x) \rangle} \ . \tag{8.77}$$

Here Equation (8.54) for the mean $\langle V(x_1; x) | x_1; x_2 \rangle$ was taken into account.

Let us go to the construction of the second moment $\varepsilon^{*(2)}(x_1, x_2)$ of the effective field. Multiplying both sides of (8.10) by $\varepsilon^*(x_2)$ and averaging the result under the condition $x_1, x_2 \in V$, we obtain

$$\varepsilon^{*(2)}(x_1, x_2) = \varepsilon^0(x_1) \otimes \langle \varepsilon^*(x_2) | x_2, x_1 \rangle - \\ - \int \mathrm{K}(x_1 - x') \langle P(x') \varepsilon^*(x') \otimes \varepsilon^*(x_2) V(x_1; x') | x_1, x_2 \rangle dx'. \tag{8.78}$$

Due to the hypothesis H_2, the mean under the integral in this equation is represented in the form

$$\langle P(x') \varepsilon^*(x') \otimes \varepsilon^*(x_2) V(x_1, x') | x_1, x_2 \rangle = \\ = P \varepsilon^{*(2)}(x_1, x_2) \langle \bar{V}(x_1; x') | x_1; x_2 \rangle + \\ + pP \langle \varepsilon^*(x') \otimes \varepsilon^*(x_2) | x', x_2; x_1 \rangle F(x', x_1, x_2) \ . \tag{8.79}$$

Using a hypothesis similar to (8.69), (8.74)

$$\langle \varepsilon^* (x') \otimes \varepsilon^* (x_2) | x', x_2; x_1 \rangle = \\ = \langle \varepsilon^* (x') \otimes \varepsilon^* (x_2) | x', x_2 \rangle = \varepsilon^{*(2)} (x', x_2) , \tag{8.80}$$

from (8.78) we obtain a closed equation for $\varepsilon^{*(2)}(x_1, x_2)$

$$\varepsilon^{*(2)} (x_1, x_2) = \varepsilon^0 (x_1) \otimes \Xi (x_1, x_2) - \\ - \int \mathrm{K} (x_1 - x') P \langle \bar{V} (x_1; x') | x_1, x_2 \rangle \mathrm{d}x' \varepsilon^{*(2)} (x_1, x_2) - \\ - \int \mathrm{K} (x_1 - x') \bar{P} \varepsilon^{*(2)} (x', x_2) F (x', x_1, x_2) \, \mathrm{d}x', \tag{8.81}$$

where the function $\Xi(x_1, x_2)$ is the solution of Equation (8.76). Subsequent approximations for $\varepsilon^{*(2)}(x_1, x_2)$ are constructed similarly.

Let us go to the equation for statistical moments of the strain $\varepsilon(x)$ and stress $\sigma(x)$ fields in a composite medium. If the field $\varepsilon^*(x)$ is constant in each inclusion, then due to (8.3), (8.6), and (8.8), the equations for $\varepsilon(x)$ and $\sigma(x)$ take the forms

$$\varepsilon(x) = \varepsilon^0(x) - \int \mathrm{K} (x - x') \, P (x') \, \varepsilon^* (x') \, V (x') \, \mathrm{d}x', \tag{8.82}$$

$$\sigma(x) = \sigma^0(x) - \int \mathrm{S}(x - x') \, B^0 P (x') \, \varepsilon^* (x') \, V (x') \, \mathrm{d}x'. \tag{8.83}$$

Averaging these equations over the ensemble realizations of the random set of inclusions and taking into account the hypothesis H_2

$$\langle P(x) \varepsilon^*(x) V(x) \rangle = \langle P(x) V(x) \rangle \varepsilon^{*(1)}(x) = \bar{P} \varepsilon^{*(1)}(x) , \tag{8.84}$$

we have

$$\langle \varepsilon(x) \rangle = \varepsilon^0(x) - \int \mathrm{K} (x - x') \, \bar{P} \varepsilon^{*(1)} (x') \, \mathrm{d}x', \tag{8.85}$$

$$\langle \sigma(x) \rangle = \sigma^0(x) - \int S (x - x') \, B^0 \bar{P} \varepsilon^{*(1)} (x') \, \mathrm{d}x'. \tag{8.86}$$

Let us derive the equation for the second moment of the strain field $\varepsilon(x)$ in a composite via the the conditional means of the effective field. Due to (8.82), we can write:

$$\langle \varepsilon (x_1) \otimes \varepsilon(x_2) \rangle = \varepsilon^0 (x_1) \otimes \varepsilon^0 (x_2) - \int \mathrm{K} (x_1 - x') \, \bar{P} \varepsilon^{*(1)} (x') \, \mathrm{d}x' \otimes \varepsilon^0 (x_2) - \\ - \varepsilon^0 (x_1) \otimes \int \mathrm{K} (x_2 - x') \, \bar{P} \varepsilon^{*(1)} (x') \, \mathrm{d}x' + \\ + \int \mathrm{K} (x_1 - x') \, \bar{P} \mathrm{d}x' \\ \times \int \mathrm{K} (x_2 - x'') \bar{P} \varepsilon^{*(2)} (x', x'') \, \langle V (x') \, V (x'') \rangle \mathrm{d}x''. \tag{8.87}$$

Here the hypothesis H_2 was used. Similarly, the second moment of the field $\sigma(x)$ is expressed via two first conditional means of the effective field $\varepsilon^*(x)$. Thus, to calculate the first two statistical moments of the fields $\varepsilon(x)$ and $\sigma(x)$ it is necessary to solve Equations (8.70) and (8.81), to determine the means $\varepsilon^{*(1)}(x)$ and $\varepsilon^{*(2)}(x_1, x_2)$, and then, to calculate the integrals in (8.87). The proposed scheme may be used for the calculation of the moments of higher orders for the functions $\varepsilon(x)$ and $\sigma(x)$.

Note that in some cases, instead of the effective strain field $\varepsilon^*(x)$ it is more convenient to consider the effective stress field $\sigma^*(x)$

$$\sigma^*(x) = C^0\varepsilon^*(x). \tag{8.88}$$

The equations for the conditional means of the effective stress field $\sigma^*(x)$ can be obtained by the same way, on the base of the equation similar to (8.10) for $\varepsilon^*(x)$:

$$\sigma^*(x) = \sigma^0(x) + \int S(x-x')\, M^0(x')\, \sigma^*(x')\, V(x;x')\, \mathrm{d}x',$$

$$M^0(x) = -B^0 P(x) B^0, \tag{8.89}$$

and the equations for the stress $\sigma(x)$ and strain $\varepsilon(x)$ tensors in terms of the field $\sigma^*(x)$

$$\sigma(x) = \sigma^0(x) + \int \mathrm{S}(x-x')\, M^0(x')\, \sigma^*(x')\, V(x')\, \mathrm{d}x', \tag{8.90}$$

$$\varepsilon(x) = \varepsilon^0(x) + \int \mathrm{K}(x-x')\, C^0 M^0(x')\, \sigma^*(x')\, V(x')\, \mathrm{d}x'. \tag{8.91}$$

8.4 The operator of the effective properties

Let us introduce the operator C^* that connects the mean values of the stress and strain tensors in a composite

$$\langle \sigma(x)\rangle = (C^*\langle\varepsilon\rangle)(x) = \int C^*(x-x')\langle \varepsilon(x')\rangle \mathrm{d}\, x'. \tag{8.92}$$

Excluding the tensors ε^0 and $\sigma^0 = C^0\varepsilon^0$ from (8.85), (8.86) and taking into account (8.72), (8.73), we obtain the equation for the symbol of the pseudodifferential operator C^* (the Fourier transform of the kernel $C^*(x)$ in (8.92)) in the form

$$C^*(k) = C^0 + \bar{P}^0\left(I - A^*(k)\,\bar{P}^0\right)^{-1}, \tag{8.93}$$

$$A^*(k) = \int \mathrm{K}(x)\,[1-\Psi(x)]e^{ik\cdot x}\mathrm{d}x. \tag{8.94}$$

Note that the connection between the means $\langle\sigma(x)\rangle$ and $\langle\varepsilon(x)\rangle$ is nonlocal because C^* is an operator of the convolution with a generalized function $C^*(x)$ which has singular (proportional to $\delta(x)$) and regular components. An exception is the case of a constant external field. In this case, $\langle\varepsilon\rangle$ and $\langle\sigma\rangle$ are also constant tensors connected by an equation that follows from (8.92)–(8.94)

$$\langle\sigma\rangle = C^*\langle\varepsilon\rangle\,, \quad C^* = C^0 + \bar{P}\left(I - A^0\bar{P}\right)^{-1}, \tag{8.95}$$

$$A^0 = A^*(0) = \int \mathrm{K}(x)\,[1-\Psi(x)]\mathrm{d}x\,. \tag{8.96}$$

This equation for the tensor of the effective elastic moduli of the composite coincides with (7.104). Equations (8.92)–(8.94) are the result of application of the EFM to fast varying external fields.

Let us suppose that the set of inclusions is statistically isotropic ($\Psi(x) = \Psi(|x|)$) and introduce the correlation radius l of the random set of inhomogeneities by the equation

$$l^2 = \int_0^\infty r\,(1-\Psi(r))\,\mathrm{d}r\,. \tag{8.97}$$

If the mean value $\langle\varepsilon(x)\rangle$ of the strain field changes sufficiently slowly in space, the support of the function $\langle\varepsilon(k)\rangle$ (the Fourier transform of the function $\langle\varepsilon(x)\rangle$) is concentrated in the region of the k-space given by the condition $|kl| << 1$. Thus, for the calculation of the mean $\langle\sigma(x)\rangle$ from (8.92), it is possible to expand the symbol $C^*(k)$ of the operator C in Equations (8.93), (8.94) in a series with respect to $|k\cdot x|$ and to keep only two essential terms in this expansion. To construct these terms we replace the function $e^{ik\cdot x}$ in (8.94) by the first two terms of the Taylor series

$$e^{ik\cdot x} = 1 + ik\cdot x - \frac{1}{2}\,(k\cdot x)^2\,. \tag{8.98}$$

As a result, the equation for $A^*(k)$ in (8.94) takes the form

$$A^*_{\alpha\beta\lambda\mu}(k) = A^0_{\alpha\beta\lambda\mu} - l^2 A^1_{\alpha\beta\lambda\mu\nu\rho}k_\nu k_\rho\,, \tag{8.99}$$

$$A^0_{\alpha\beta\lambda\mu} = \int \mathrm{K}_{\alpha\beta\lambda\mu}(x)\,[1-\Psi(x)]\mathrm{d}x\,, \tag{8.100}$$

$$A^1_{\alpha\beta\lambda\mu\nu\rho} = \frac{1}{2}\int_{\Omega_1} \mathrm{K}_{\alpha\beta\lambda\mu}(n)\,n_\nu n_\rho \mathrm{d}\Omega_n\,, \tag{8.101}$$

where $n = x/|x|$, the parameter l is defined in (8.97). From these equations and (8.93) we find the symbol $C^*(k)$ with the accuracy of $(kl)^2$

$$C^*(k) = C^*_0 - (ik)^2 l^2 \bar{P}^1\,\bar{A}^1(m)\,\bar{P}^1\,, \quad C^*_0 = C^0 + \bar{P}^1, \tag{8.102}$$

$$\bar{P}^1 = \bar{P}\left(I - A^0\bar{P}\right)^{-1}, \quad \bar{A}^1_{\alpha\beta\lambda\mu}(m) = A^1_{\alpha\beta\lambda\mu\nu\rho}m_\nu m_\rho, \quad m = \frac{k}{|k|}. \tag{8.103}$$

For an isotropic medium, the tensor A^0 has form (3.133), and tensor $\bar{A}^1(m)$ is defined by the equation ($E^i = E^i(m)$)

$$\bar{A}^1(m) = \frac{1}{105\mu_0}\{(7 - 4\kappa_0)\left(E^1 - 3E^5\right) - \kappa_0\left(2E^2 - 3E^4 - 3E^3\right)\}, \tag{8.104}$$

Let us introduce the average displacement field $\langle u(x)\rangle$ in the composite connected with $\langle\varepsilon(x)\rangle$ by the equation

$$\langle\varepsilon(x)\rangle = \mathrm{def}\langle u(x)\rangle. \tag{8.105}$$

Because of (8.92) and the equilibrium equation for $\langle\sigma(x)\rangle$, the field $\langle u(x)\rangle$ satisfies the equation

$$-\mathrm{div}\, C^*\mathrm{def}\,\langle u\rangle = q. \tag{8.106}$$

Because C^* is a nonlocal operator, (8.106) describes the displacement field $\langle u\rangle$ in some homogeneous medium with spacial dispersion. If the symbol of the operator $C^*$is approximated by the first two terms of the expansion (8.102), Equation (8.106) is transformed into a differential equation with respect to $\langle u\rangle$

$$-C^*_{0\alpha\beta\lambda\mu}\nabla_\alpha\nabla_\lambda\langle u_\mu\rangle - l^2\bar{P}_{\alpha\beta\lambda\mu}\overline{A}^1_{\lambda\mu\nu\rho\tau\delta}\bar{P}^1_{\nu\rho\gamma\pi}\nabla_\alpha\nabla_\pi\nabla_\tau\nabla_\delta\langle u_\gamma\rangle = q_\beta. \tag{8.107}$$

This equation coincides in essence with the system of equations of the moment elasticity theory for a medium with constraint rotation [107], [150]. The correlation radius l of the random set of inclusions plays the role of a parameter of the dimension length that is typical for the moment elasticity theory.

Note that the previous equations do not contain other parameters of the problem, e.g., the inclusion sizes. This fact is the consequence of the assumption that the local external field $\varepsilon^*(x)$ is constant for every inclusion. The dependence of the operator C^* on the inclusion sizes can be taken into account carrying out the EFM on the basic of equation (8.14) of the first order for the effective field $\varepsilon^*(x)$.

8.5 Pair interactions between inclusions

The quasicrystalline approximation (8.69) that was used to derive Equations (8.70) and (7.92) of the EFM is based on neglecting the details of pair interactions between inclusions in the composite. Really, the equation

$$\langle\varepsilon^*(x)\,|x;x_1\rangle = \langle\varepsilon^*(x)\,|x\rangle \tag{8.108}$$

means that presence of an inclusion at the point x_1 in all the realizations does not influence the mean value of the effective field at the point $x \in V$. Let us consider Equations (8.67), (8.68) for the conditional means of the effective field $\varepsilon^*(x)$ but not use hypothesis (8.108)

$$\varepsilon^{*(1)}(x) = \varepsilon^0 - p\int \mathrm{K}(x-x')\, P\Xi_\varepsilon(x',x)\,\Psi(x,x')\,\mathrm{d}x', \tag{8.109}$$

$$\Xi_\varepsilon(x,x_1) = \varepsilon^0 - \\ - \int \mathrm{K}(x-x')\, P\langle \varepsilon^*(x')\,|x',x_1;x\rangle\langle V(x;x')\,|x;x_1\rangle \mathrm{d}x'. \tag{8.110}$$

Here $\Xi_\varepsilon(x,x_1) = \langle \varepsilon^*(x)|x;x_1\rangle$ is the conditional mean that takes into account peculiarities of pair interactions between inclusions. In what follows, we suppose that the random set of inclusions is statistically homogeneous and isotropic in space, and the medium is subjected to a constant external field ε^0. In this case, the function $\Psi(x,x')$ in (8.109) depends only on $|x-x'|$ and $\Xi_\varepsilon(x,x_1)$ on $x-x_1$. If correlations in the inclusion positions decay with increasing the distances between them, the following equations hold:

$$\Xi(x-x_1) \to \varepsilon^{*(1)}, \quad \langle \varepsilon^*(x')\,|x',x_1;x\rangle \to \Xi(x'-x)\ , \tag{8.111}$$

$$\langle V(x;x'|x,x_1)\rangle \to p\Psi(x-x')\ . \tag{8.112}$$

due to decay of interactions between inclusions.

Here $\varepsilon^{*(1)}$ is a constant tensor. Equation (8.109) gives the equation for this tensor in the form

$$\varepsilon^{*(1)} = \varepsilon^0 + pA^0P\varepsilon^{*(1)} - p\int \mathrm{K}(x)\, P\Big[\Xi(x) - \varepsilon^{*(1)}\Big]\Psi(x)\,\mathrm{d}x, \tag{8.113}$$

where the tensor A^0 has the form (3.82), $a_{\alpha\beta} = \delta_{\alpha\beta}$. This equation differs from Equation (7.92) for the effective field by the integral term in the right-hand side containing the function $\Xi_\varepsilon(x)$. In order to find this function, we have to turn to Equations (8.110) and construct the mean $\langle V(x;x')|x;x_1\rangle$ in these equations. For simplicity, we change the inclusions of finite sizes to point defects using the procedure described in Section 7.16. In this case, the functions $V(x)$ and $V(x;x')$ are replaced by the principal terms of their expansion in the series on multipoles concentrated in the inclusion centers ξ_i (2.140)

$$V(x) \cong \sum_i v_i\delta(x-\xi_i);\ \ V(x;x') \cong \sum_{i\neq k} v_i\delta(x'-\xi_i) \ \text{ when } \ x=\xi_k. \tag{8.114}$$

Here v_i is the volume of the i-th inclusion. In this approximation, the mean takes the form

$$\langle V(x;x')\,|x;x_1\rangle \cong \langle v\rangle\langle X(x;x')\,|x;x_1\rangle, \tag{8.115}$$

where the mean in the right-hand side is defined in (8.29), the functions $X(x)$ and $X(x;x')$ have forms (7.493), and $\langle v\rangle$ is the volume of a typical inclusion. From Equation (8.46) we have

$$\langle V(x;x')|x;x_1\rangle \approx \langle v\rangle\delta(x'-x_1)+pF_0(x',x,x_1), \tag{8.116}$$

$$F_0(x',x,x_1)=\frac{1}{n_0}\langle X(x,x_1;x')|x;x_1\rangle, \tag{8.117}$$

where F_0 is a continuous function of the arguments, and

$$F_0(x',x,x_1)\to\Psi(x-x') \quad \text{when} \quad |x_1|\to\infty\,. \tag{8.118}$$

Substituting (8.117) into (8.110) and taking into account (8.113), we come to the equation

$$\Xi(x-x_1)=\varepsilon^{*(1)}-\langle v\rangle \mathrm{K}(x-x_1)P\Xi(x-x_1)-pJ(x,x_1), \tag{8.119}$$

$$\begin{aligned} J(x,x_1)=\int \mathrm{K}(x-x')P\{[\langle\varepsilon^*(x')|x',x;x_1\rangle-\Xi(x-x')]\times \\ \times F_0(x',x,x_1)+\Xi(x-x')[F_0(x',x,x_1)-\Psi(x-x')]\}\mathrm{d}x'. \end{aligned} \tag{8.120}$$

Neglecting in (8.119) the integral term pJ, that has the order p and disappears when $|x_1|\to\infty$, due to (8.112), (8.118), we obtain the function $\Xi(x)$ in the form

$$\Xi(x)=[I+\langle v\rangle\mathrm{K}(x)P]^{-1}\varepsilon^{*(1)}. \tag{8.121}$$

It is possible to come to this equation in the following way. Consider two point defects of the same intensity $\langle v\rangle P^0$ at points ξ_1 and ξ_2 in a homogeneous elastic medium with the tensor of elastic moduli C^0, and let $x=\xi_1-\xi_2$ be the vector connecting these defects. If the medium is subjected to a constant external field $\varepsilon^{*(1)}$, the field $\Xi_\varepsilon(x)$ acting on each defect has the form (8.121). Thus, the function $\Xi(x)$ (8.121) describes the interaction of two point defects in an elastic medium. The existence of surrounding defects is taken into account by the effective external field $\varepsilon^{*(1)}$ acting on these defects.

Substituting (8.121) into (8.113) and solving the obtained equation with respect to $\varepsilon^{*(1)}$, we find

$$\varepsilon^{*(1)}=\left(I-pA^0P+p\mathrm{K}^0\right)^{-1}\varepsilon^0, \tag{8.122}$$

$$\mathrm{K}^0=\int \mathrm{K}(x)P\Big[(I+\langle v\rangle\mathrm{K}(x)P)^{-1}-I\Big]\Psi(x)\,\mathrm{d}\,x\,. \tag{8.123}$$

Because the tensor $\varepsilon^{*(1)}$ in (7.19) is constant, from (8.85), (8.86) we obtain

$$\langle\sigma\rangle=C^*\langle\varepsilon\rangle\,,\quad C^*=C^0+pP\big[I-p(A^0P-\mathrm{K}^0)\big]^{-1}. \tag{8.124}$$

When $\mathrm{K}^0 = 0$, this equation for the tensor of the effective moduli C^* coincides with Equation (7.104) that was obtained by the version of the effective field method developed in Chapter 7.

Let us calculate the tensor K^0 in (8.123). For an isotropic matrix, the tensor P has the form

$$P = P_1 E^2 + P_2 \left(E^1 - \frac{1}{3} E^2 \right), \tag{8.125}$$

where the coefficients P_1, P_2 are defined in (7.130),(7.132) for multilayered spherical inclusions, and in (7.136) for homogeneous inclusions. Introducing the spherical coordinate system ($r = |x|$, $n = x/|x|$) and performing integration over the unit sphere (the vector n) in (8.123), we obtain

$$\mathrm{K}^0 = -\mathrm{k}_1^0 E^2 - \mathrm{k}_2^0 \left(E^1 - \frac{1}{3} E^2 \right), \tag{8.126}$$

$$\mathrm{k}_1^0 = \int_0^\infty \mathrm{K}_1(\xi)\, \Psi(\xi)\, \xi^2 d\,\xi, \quad \mathrm{k}_2^0 = \int_0^\infty \mathrm{K}_2(\xi)\, \Psi(\xi)\, \xi^2 d\,\xi, \tag{8.127}$$

$$\mathrm{K}_1(\xi) = \frac{13}{15} \cdot \frac{f_1^2(\xi)}{6 + 5 f_1(\xi)} + O\left(\xi^{-6}\right), \tag{8.128}$$

$$\mathrm{K}_2(\xi) = \frac{5}{6} \left[\frac{7}{3} \cdot \frac{f_1^2(\xi)}{6 + 5 f_1(\xi)} + \frac{f_2^2(\xi)}{1 + f_2(\xi)} \right] + O\left(\xi^{-6}\right), \tag{8.129}$$

$$f_1(\xi) = \frac{4\kappa_0 - 1}{2\mu_0 \xi^3} P_2^0, \quad f_2(\xi) = \frac{1 - \kappa_0}{2\mu_0 \xi^3} P_2^0, \quad \xi = \frac{r}{a_N}. \tag{8.130}$$

The values of the coefficients k_1^0 and k_2^0 depend on the specific form of the function $\Psi(\xi)$, the properties of which were discussed in Section 7.5. For point defects, the function $\Psi(\xi)$ is a normalized probability density of the defect distribution in space under the condition that the point $x = 0$ is occupied by a defect. If all the inclusions that are modeled by the point defects have the same radius a, their centers cannot be closer than $2a$. Therefore, the simplest approximation of the function $\Psi(x)$ has the form

$$\Psi(r) = H\left(\frac{r}{a} - 2\right) = \begin{cases} 0, & r < 2a \\ 1, & r > 2a \end{cases}. \tag{8.131}$$

Here $H(\xi)$ is the Heaviside function. Strictly speaking, the function $\Psi(r)$ can have such form only in the limit $p \to 0$. When the concentration is finite, $\Psi(r)$ has a maximum at $r = 2a$ and tends to 1 when $r \to \infty$ (Fig. 8.1).

Maximum of $\Psi(r)$ depends on the inclusion concentration and peculiarities of their spatial distribution. Particularly, the maximum of $\Psi(r)$ increases if the inclusions tend to agglomeration.

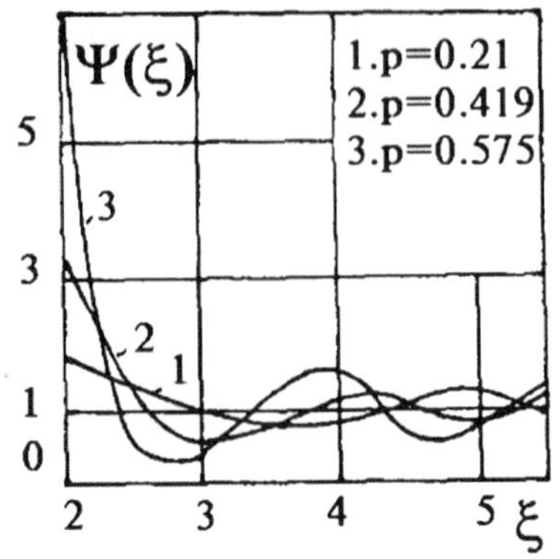

Fig. 8.1. Function $\Psi(r)$.

A reliable approximation of the function $\Psi(r)$ was proposed in [166] and is the solution of the so-called Percus–Yevick equation [218] (see Appendix E). The maximum value of this function is reached at $r = 2a$ and has the form [197]

$$\Psi\left(\xi\right)\big|_{\xi=2} = \frac{2+p}{2\left(1-p\right)^2}\,, \quad \xi = \frac{r}{a}\,. \tag{8.132}$$

The curves in Fig. 8.1 correspond to the numerical solution of the Percus-Yevic equation obtained in [197]. For the calculations of the coefficients k_1^0, k_2^0 in (8.127) we use the approximation of the function $\Psi(\xi)$ suggested in [214]

$$\begin{aligned} &\Psi\left(\xi\right) = 0\,, && \xi < 2\,, \\ &\Psi\left(\xi\right) = 1 + \left[\tfrac{2+p}{2(1-p)^2} - 1\right] \cos\left(\pi\xi\right) \exp\left[2\left(2-\xi\right)\right] && \xi \geq 2\,. \end{aligned} \tag{8.133}$$

From (8.124)–(8.127) follow the equations for the effective bulk K_* and shear μ_* moduli of a composite with spherical inclusions in the form

$$K_* = K_0 + \frac{pP_1}{1 - 3p\left(3a_1^0 P_1 + k_1^0\right)}\,, \tag{8.134}$$

$$\mu_* = \mu_0 + \frac{1}{2} \cdot \frac{pP_2}{1 - p\left(a_2^0 P_2 + k_2^0\right)}\,, \tag{8.135}$$

$$a_1^0 = \frac{1-\kappa_0}{9\mu_0}\,, \quad a_2^0 = \frac{5-2\kappa_0}{15\mu_0}\,, \quad \kappa_0 = \frac{3K_0+\mu_0}{3K_0+4\mu_0}\,. \tag{8.136}$$

Figure 8.2 shows the theoretical and experimental dependence of the effective Young modulus E_* on the inclusion concentration for composites

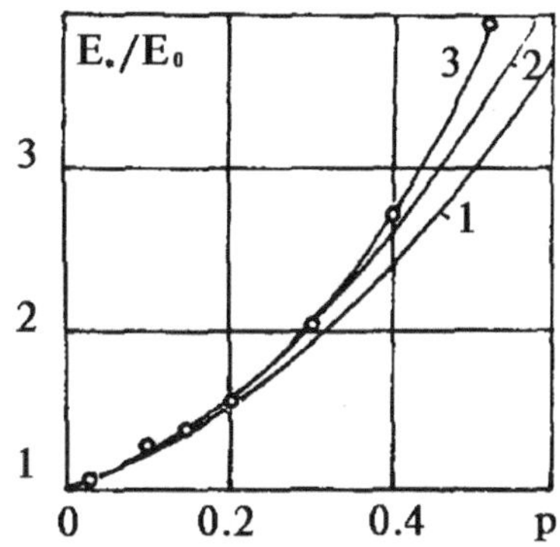

Fig. 8.2. Comparison of the predictions of the EFM and experimental data for the effective Young modulus E_* of the composite with hard spherical inclusions

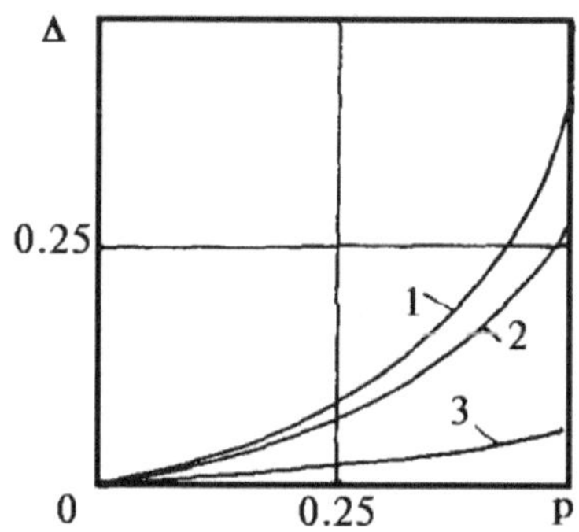

Fig. 8.3. The relative error $\Delta = \left|E_*^e - E_*^T\right| / E_*^e$ of different versions of the EFM in the case of the composite with hard spherical inclusions.

reinforced by spherical inclusions. Line 1 corresponds to the one-particle approximation of the effective field method ($k_1^0 = k_2^0 = 0$). Lines 2 and 3 were obtained from (8.136), (8.127) with $\Psi(\xi)$ in the form (8.131) - 2 and (8.133) - 3. Clear dots correspond to experimental data presented in [192] ($E/E_0 = 28.7$, $\nu = 0.33$, $\nu_0 = 0.394$).

Figure 8.3 presents the relative errors $\Delta = \left|E_*^e - E_*^T\right| / E_*^e$ in the calculation of the effective elastic moduli versus the volume concentration p of inclusions for the composites reinforced by hard spherical inclusions ($E/E_0 = O(\delta)$, $\delta > 10$). Here E_*^e and E_*^T are the experimental and theoretical values of the effective elastic modulus. To construct these curves the experimental data from [175], [192], [200] were used. Line 1 corresponds to the simplest version of the effective field method (Chapter 7), lines 2, 3 are obtained by the EFM accounting pair interactions between inclusions, the function $\Psi(\xi)$ was taken in the forms (8.131) - 2 and (8.133) - 3. The maximum value of the relative error over all the experimental data [175], [192], [200] was taken as Δ.

8.6 Notes

The contents of this chapter is based on the works [75], [78], [77]. A non-local operator for the medium with spherical inclusions was derived in [75]. Stress concentration on the inclusion in the regions of fast changing in fields was considered in [92]. A detailed account of pair interactions between inclusions in the calculation of the effective elastic moduli of matrix composites is considered in [77], [84]. Correlation function of stress field in the composites were considered in [78], [77] in the framework of the point defect model.

9. Thermo- and electroconductive properties of composites

This chapter is devoted to electric and temperature fields in matrix composite materials. Self-consistent methods are used to obtain effective dielectric permittivities, and electro- and thermoconductivities of these materials.

9.1 Integral equations for a medium with isolated inclusions

Many types of physical fields in an inhomogeneous medium may be described by the following system of partial differential equations:

$$\nabla_i J_i(x) = -q(x), \; D_i(x) = C_{ij}(x)E_j(x), \; E_j(x) = \nabla_i \varphi(x), \tag{9.1}$$

Here $E_j(x)$, $D_i(x)$ are the vectors of the intensity and flux of the field, $\varphi(x)$ is the scalar potential of the field, $C_{ij}(x)$ is the tensor of the properties of the medium, q is the density of the sources of the field. For electrostatic problems, D_i is the electrostatic displacement, E_i is the electric field, $C_{ij}(x)$ is the tensor of the dielectric permittivities, φ is the potential of the electric field. For thermostatic problems, D_i is the heat flux, E_i is the gradient of the temperature field, C_{ij} is the tensor of thermoconductivity, and $-\varphi$ is the temperature. For static electroconductivity, E_i is the voltage, D_i is the current, and C_{ij} is the tensor of electroconductivity.

If an infinite homogeneous medium with the property tensor C^0 contains a set of inclusions with the property tensor C, the system of differential equation (9.1) may be reduced to integral equations for the fields $E(x)$ and $D(x)$ inside the inclusions. Let the inclusions occupy a region V with the characteristic function $V(x)$: $V(x) = 1$ if $x \in V$, $V(x) = 0$ if $x \notin V$. The integral equations for the intensity $E_\alpha(x)$ and flux $D_\alpha(x)$ of the field have the forms

$$E_\alpha(x) + \int K_{\alpha\beta}(x - x')C^1_{\beta\lambda}(x')E_\lambda(x')V(x')\mathrm{d}x' = E^0_\alpha(x), \tag{9.2}$$

$$D_\alpha(x) + \int S_{\alpha\beta}(x - x')B^1_{\beta\lambda}(x')D_\lambda(x')V(x')\mathrm{d}x' = D^0_\alpha(x), \tag{9.3}$$

Here $E^0(x)$ and $D^0(x) = C^0E^0(x)$ are the external fields applied to the medium. These equations may be also written in the forms

$$E_\alpha(x) + \int K_{\alpha\beta}(x - x')q_\beta(x')\mathrm{d}x' = E^0_\alpha(x), \tag{9.4}$$

$$D_\alpha(x) + \int S_{\alpha\beta}(x - x')B^0_{\beta\lambda}q_\lambda(x')\mathrm{d}x' = D^0_\alpha(x), \tag{9.5}$$

$$q_\alpha(x) = C^1_{\alpha\beta}(x)E_\beta(x)V(x), \quad B^0 = \left(C^0\right)^{-1}, \tag{9.6}$$

and they are similar to Equations (8.3) and (8.5) for the elastic fields in the medium with inclusions.

Let us average the fields $E_\alpha(x)$ and $D_\alpha(x)$ over the ensemble realizations of a random set of inclusions. As a result, we obtain

$$\langle E_\alpha(x)\rangle + \int K_{\alpha\beta}(x - x')\langle q_\beta(x')\rangle\,\mathrm{d}x' = E^0_\alpha(x), \tag{9.7}$$

$$\langle D_\alpha(x)\rangle + \int S_{\alpha\beta}(x - x')B^0_{\beta\lambda}\langle q_\lambda(x')\rangle\,\mathrm{d}x' = D^0_\alpha(x). \tag{9.8}$$

For a spatially homogeneous random set of inclusions and a constant external field E^0, the mean $\langle q_\beta(x)\rangle$ is constant. The action of the integral operators with the kernels $K_{\alpha\beta}(x)$ and $S_{\alpha\beta}(x)$ on constants depends on which of the fields $E^0_\alpha(x)$ or $D^0_\alpha(x)$ is fixed in the problem. For instance, if the field E^0_α is fixed, the operators K and S act on constants as follows:

$$\int K_{\alpha\beta}(x - x')\mathrm{d}x' = 0, \quad \int S_{\alpha\beta}(x - x')\mathrm{d}x' = -C^0_{\alpha\beta}. \tag{9.9}$$

However, if the field D^0_α is fixed, the result will be different

$$\int K_{\alpha\beta}(x - x')\mathrm{d}x' = C^0_{\alpha\beta}, \quad \int S_{\alpha\beta}(x - x')\mathrm{d}x' = 0. \tag{9.10}$$

The effective properties of the composite do not depend on these conditions, and any of them may be used for the solution of the homogenization problem.

Suppose that the field E^0_α is fixed in the problem. In this case, (9.7) and (9.8) take the forms

$$\langle E_\alpha(x)\rangle = E^0_\alpha,$$

$$\langle D_\alpha(x)\rangle = D^0_\alpha + \langle q_\alpha(x)\rangle = \left[C^0_{\alpha\beta} + p\left\langle C^1_{\alpha\lambda}(x)\Lambda_{\lambda\beta}(x)\right\rangle\right]E^0_\beta(x). \tag{9.11}$$

Here the tensor $\Lambda_{\lambda\beta}(x)$ defines the detailed fields inside the inclusions according to the equation

$$E_\alpha(x) = \Lambda_{\alpha\beta}(x)E^0_\beta. \tag{9.12}$$

Equation (9.11) may be written in the form

$$\langle D_\alpha(x)\rangle = C^*_{\alpha\beta}\langle E_\beta(x)\rangle\,,$$

$$C^*_{\alpha\beta} = C^0_{\alpha\beta} + p\langle P_{\alpha\beta}\rangle\,, \quad P_{\alpha\beta} = \frac{1}{\langle v\rangle}\int\limits_v C^1_{\alpha\lambda}(x)\Lambda_{\lambda\beta}(x)\mathrm{d}x, \tag{9.13}$$

where $C^*_{\alpha\beta}$ is the tensor of the effective properties of the composite.

9.2 The effective medium method

Consider a composite material with isotropic matrix ($C^0_{\alpha\beta} = c_0\delta_{\alpha\beta}$) containing a random set of ellipsoidal inclusions of the same sizes and shapes. Suppose that all the inclusions are isotropic ($C_{\alpha\beta} = c\delta_{\alpha\beta}$) and have the same properties. Similar to Section 7.4, the basic hypothesis of the effective medium method (EMM) is that every inclusion may be considered as an isolated one in a homogeneous medium with the unknown effective properties of the composite. This means that the tensor $P_{\alpha\beta}$ in (9.13) depends on the tensor of the effective properties $C^*_{\alpha\beta}$. Using the solution of the one-particle problem (6.11) we obtain the following self-consistent equation for the tensor $C^*_{\alpha\beta}$

$$C^*_{\alpha\beta} = C^0_{\alpha\beta} + p\left(C_{\alpha\lambda} - C^0_{\alpha\lambda}\right)\left\langle\left[\delta_{\lambda\beta} + A^*_{\alpha\mu}\left(C_{\mu\beta} - C^*_{\mu\beta}\right)\right]^{-1}\right\rangle. \tag{9.14}$$

Here the tensor $A^*_{\alpha\mu}$ has form (6.12), where the tensor $C^0_{\alpha\beta}$ should be replaced by the tensor $C^*_{\alpha\beta}$.

The main difficulty in carrying out this scheme is that, strictly speaking, we cannot predict in advance the symmetry of the effective conductivity tensor $C^*_{\alpha\beta}$ that depends on the spatial distribution of the inclusions. But if this symmetry is known, it is possible to obtain simple equations for the components of the tensor $C^*_{\alpha\beta}$.

Let the set of ellipsoidal inclusions be randomly distributed over the orientations. In this case, the composite material is isotropic ($C^*_{\alpha\beta} = c_*\delta_{\alpha\beta}$) and the effective property tensor C_* satisfies the equation

$$C_* = C_0 + \frac{1}{3}p(C - C_0)\sum_{k=1}^{3}\left[1 + (C - C_*)A^*_k\right]^{-1}, \tag{9.15}$$

where the tensors A^*_k ($k = 1, 2, 3$) are determined in (6.14), and the coefficient C_0 in that equation should to be replaced by C_*. If the inclusions are spheres, $A^*_{\alpha\beta} = (1/3c_*)\delta_{\alpha\beta}$, and Equation (9.15) takes the form

$$c_* = c_0 + \frac{3pc_*(c - c_0)}{2c_* + c}. \tag{9.16}$$

This is a quadratic equation for c_*

$$2c_*^2 - [c_0 - (1-3p)(c - c_0)]\, c_* - cc_0 = 0. \tag{9.17}$$

If $c = 0$ and the inclusions are not conductive, the solution of this equation is

$$c_* = \left(1 - \frac{3}{2}p\right) c_0. \tag{9.18}$$

Hence, c_* becomes zero for $p = 2/3$.

If $c \to \infty$, (highly conductive inclusions), we have

$$c_* = \frac{c_0}{1-3p} \tag{9.19}$$

and c_* tends to infinity as $p \to 1/3$.

Let the inclusions be spheroids with the symmetry axes parallel to the x_3-axis. The material of all the spheroids has the properties C, and their aspect ratio is γ. In this case, the composite is transversally isotropic (with x_1x_2 as the plane of isotropy) and the tensor $C^*_{\alpha\beta}$ has a hexagonal symmetry, i.e., its matrix is diagonal with the components $C^*_{11} = C^*_{22}$, C^*_{33}. The one-particle problem is the problem for a spheroidal inclusion embedded in an infinite transversely isotropic medium, and this problem was considered in Section 6.4. As a result, the unknown components C^*_{11} and C^*_{33} are the roots of the system of algebraic equations

$$C^*_{11} = C_0 + p(C - C_0)\left[1 + A^*_1(C - C^*_{11})\right]^{-1}, \tag{9.20}$$

$$C^*_{33} = C_0 + p(C - C_0)\left[1 + A^*_3(C - C^*_{33})\right]^{-1}, \tag{9.21}$$

where the coefficients A^*_1 and A^*_3 are

$$A^*_1 = \frac{\lambda_*^2}{2C^*_{11}}\left[1 - \frac{\gamma^2 C^*_{33}}{2C^*_{11}}\lambda_* \ln\left(\frac{\lambda_*+1}{\lambda_*-1}\right)\right], \tag{9.22}$$

$$A^*_3 = \frac{(\gamma\lambda_*)^2}{C^*_{11}}\left[\frac{1}{2}\lambda_* \ln\left(\frac{\lambda_*+1}{\lambda_*-1}\right) - 1\right], \quad \lambda_* = \left(1 - \gamma^2\frac{C^*_{33}}{C^*_{11}}\right)^{-1/2}. \tag{9.23}$$

Let us consider the particular case of thin prolate spheroids, corresponding to the limit $\gamma \to 0$ in this equation. From (9.22), (9.23) we have equations for the coefficients A^*_1 and A^*_3

$$A^*_1 \approx \frac{1}{2C^*_{11}}\left\{1 + \gamma^2\frac{C^*_{33}}{C^*_{11}}\left[1 - \ln\left(\frac{2}{\gamma}\sqrt{\frac{C^*_{11}}{C^*_{33}}}\right)\right]\right\}, \tag{9.24}$$

$$A^*_3 \approx \frac{\gamma^2}{C^*_{11}}\left[\ln\left(\frac{2}{\gamma}\sqrt{\frac{C^*_{11}}{C^*_{33}}}\right) - 1\right].$$

If the ratio c/c_0 is neither very small nor very large, the effective coefficients $C^*_{\alpha\beta}$ have the order c_0, and Equations (9.20) are reduced to the following ones:

$$C^*_{11} = c_0 + p(c - c_0)\left[1 + \frac{c - C^*_{11}}{2C^*_{11}}\right]^{-1}, \tag{9.25}$$

$$C^*_{33} = c_0 + p(c - c_0). \tag{9.26}$$

The first of these equations is quadratic

$$(C^*_{11})^2 + (1 - 2p)(c - c_0)C^*_{11} - cc_0 = 0, \tag{9.27}$$

and coincides with the equation for the longitudinal shear modulus μ_* for elastic material reinforced by long parallel fibers.

Suppose that γ tends to zero while p remains finite, and $c \gg c_0$ (a composite with the parallel highly conductive short fibers). From Equations (9.20) we obtain

$$C^*_{11} = \frac{c_0}{1 - 2p}, \tag{9.28}$$

and (9.21) takes the form in this case

$$C^*_{33} = c_0\left\{1 + \frac{p}{\gamma^2(1 - 2p)}\left[\ln\left(\frac{2}{\gamma^2}\sqrt{\frac{c_0}{C^*_{33}(1 - 2p)}}\right) - 1\right]^{-1}\right\}. \tag{9.29}$$

The opposite limiting case $\gamma \to \infty$ corresponds to a composite containing parallel flat disks. In this limit, (9.24) is reduced to the equation

$$A^*_1 \approx \frac{\pi}{4\gamma\sqrt{C^*_{11}C^*_{33}}}, \quad A^*_3 \approx \frac{1}{C^*_{33}}\left[1 - \frac{\pi}{2\gamma}\sqrt{\frac{C^*_{11}}{C^*_{33}}}\right]. \tag{9.30}$$

If C/C_0 is neither very large nor very small, (9.20) and (9.21) take the forms

$$C^*_{11} \approx pc + (1 - p)c_0, \tag{9.31}$$

$$C^*_{33} \approx \left[\frac{p}{c} + \frac{1 - p}{c_0}\right]^{-1}, \tag{9.32}$$

and coincide with the Voigt and Reuss estimates for C^*_{11} and C^*_{33}.

The problem becomes more interesting if we consider the cases of highly conductive and nonconductive disks. First, if $c \gg c_0$, (9.21) is reduced to the equation

$$C^*_{33} = \frac{c_0}{1 - p}, \tag{9.33}$$

and for the coefficient C_{11}^* in (9.20) we obtain

$$C_{11}^* = c_0 + \frac{4p\gamma}{\pi}\sqrt{\frac{C_{11}^* c_0}{1-p}}. \tag{9.34}$$

In the limit $C \to 0$ that corresponds to a matrix containing aligned thin non-conductive inclusions, (9.20) gives the equation

$$C_{11}^* = (1-p)c_0, \tag{9.35}$$

for the coefficient C_{11}^* and according to (9.21), the ratio $\sqrt{c_0/C_{33}^*}$ satisfies the quadratic equation

$$\sqrt{1-p}\frac{c_0}{C_{33}^*} - \frac{2p\gamma}{\pi}\sqrt{\frac{c_0}{C_{33}^*}} - \sqrt{1-p} = 0. \tag{9.36}$$

9.2.1 Differential effective medium method

The differential effective medium method (DEMM) was developed in Section 7.4.1 for elasticity. In this method, a small concentration δq of the inclusions is added to the matrix phase of the composite. The mixture so obtained may be replaced by a homogeneous medium with the effective conductivity $C^* = C^*(\delta q)$. Then, a new small portion of the inclusions of the concentration δq is inserted in the homogeneous effective medium obtained at the first stage; the new medium again is replaced by a homogeneous one with the effective conductivity $C^*(2\delta q)$, and so forth. As a result, the effective conductivity C^* will be a function of the total amount q of inserted particles: $C^* = C^*(q)$. It is essential that at each step of this process, a small concentration of particles ($\delta q \ll 1$) is added into a homogeneous effective medium. Hence, the dilute approximation may be used for calculation of the effective properties of the composite. For ellipsoidal inclusions with conductivity tensor C distributed in a homogeneous matrix (with the conductivity tensor C^0) the dilute approximation has the form

$$C_{\alpha\beta}^* = C_{\alpha\beta}^0 + q\left(C_{\alpha\lambda} - C_{\alpha\lambda}^0\right)\left\langle \Lambda_{\lambda\beta}\right\rangle, \tag{9.37}$$

where tensor $\Lambda_{\lambda\beta}$ is defined in (9.12). This tensor may be calculated explicitly from the solution of the one-particle problem. Averaging in (9.37) is performed over the distribution of the ellipsoidal inclusions over the orientations and sizes. Hence, for the described process, we can write

$$C^*(q+\delta q) = C^*(q) + \delta q\left(C - C^*\right)\left\langle \Lambda^*\right\rangle, \tag{9.38}$$

where the tensor Λ^* is determined in (6.11) for an ellipsoidal inclusion in the infinite medium with the effective conductivity tensor C^*.

In the limit $\delta q \to 0$, we obtain the differential equation for the unknown function $C^*(q)$

$$\frac{\mathrm{d}C^*}{\mathrm{d}q} = (C - C^*) \langle \Lambda^* \rangle, \tag{9.39}$$

that should be solved with the following initial condition

$$C^*(0) = C^0. \tag{9.40}$$

As for the elastic case (see Section 7.4.1) it is possible to show that the total concentration c of the inclusions does not coincide with the volume concentration p of the second phase. For $\delta q \to 0$ these quantities are connected by the equation

$$\mathrm{d}p = (1 - p)\mathrm{d}q, \tag{9.41}$$

and Equation (9.39) is transformed into the following:

$$(1 - p)\frac{\mathrm{d}C^*}{\mathrm{d}p} = (C - C^*(p)) \langle \Lambda^*(C, C^*(p)) \rangle. \tag{9.42}$$

For the composite material containing isotropic spherical inclusions $C_{ij} = c\delta_{ij}$ homogeneously distributed in the matrix, $C^*_{ij} = c_*\delta_{ij}$, and this equation takes the form

$$\frac{\mathrm{d}c_*}{\mathrm{d}p} = \frac{3c_*(c - c_*)}{(1 - p)(c + 2c_*)}, \tag{9.43}$$

with the initial condition

$$c_*\big|_{p=0} = c_0. \tag{9.44}$$

The solution of this equation has the form

$$\left(\frac{c - c_*}{c - c_0}\right)\left(\frac{c_0}{c_*}\right)^{1/3} = 1 - p. \tag{9.45}$$

For highly conductive inclusions, this equation gives us the solution

$$c_* = \frac{c_0}{(1 - p)^3}, \tag{9.46}$$

and for inclusions with zero conductivity ($c = 0$) we have

$$c_* = c_0(1 - p)^{3/2}. \tag{9.47}$$

Hence, in contrast to the "classical" effective medium method, the DEMM gives physically reasonable values for the effective conductivities for the whole range of the volume concentrations of the inclusions.

9.3 The effective field method

The simplest version of the EFM is based on the hypothesis that the local external field E^*_α that acts on each inclusion in the composite is constant and the same for all the inclusions. Using this hypothesis, we can present the field $E_\alpha(x)$ inside each inclusion in the form

$$E_\alpha(x) = \Lambda^E_{\alpha\beta}(x)E^*_\beta, \tag{9.48}$$

where the tensor $\Lambda^E_{\alpha\beta}(x)$ is determined from the solution of the problem for isolated inclusion in the medium with the conductivity tensor C^0 by the action of a constant field E^*_β. The function $q_\alpha(x)$ in (9.6) takes the form

$$q_\alpha(x) = C^1_{\alpha\beta}(x)\Lambda^E_{\beta\lambda}(x)E^*_\lambda V(x), \tag{9.49}$$

where $V(x)$ is the characteristic function of the domain V occupied by the inclusions.

On the other hand, the local exciting field $E^*_\alpha(x)$ at the point $x \in V$ is presented in the form

$$E^*_\alpha(x) = E^0_\alpha - \int K_{\alpha\beta}(x-x')C^1_{\beta\lambda}(x')\Lambda^E_{\lambda\mu}(x')V(x;x')\mathrm{d}x'E^*_\mu, \tag{9.50}$$

where the function $V(x;x')$ is defined in (7.84). Averaging this equation under the condition $x \in V$ and identifying the conditional mean $\langle E^*_\alpha(x)|x\rangle$ with the effective field E^*_α we obtain a closed equation for this vector

$$E^*_\alpha = E^0_\alpha - p\int K_{\alpha\beta}(x-x')P^0_{\beta\lambda}\Psi(x-x')\mathrm{d}x'E^*_\lambda, \tag{9.51}$$

$$P^0_{\alpha\beta} = \frac{1}{\langle v\rangle}\left\langle \int_v C^1_{\alpha\lambda}(x)\Lambda^E_{\lambda\beta}(x)\mathrm{d}x\right\rangle. \tag{9.52}$$

Here the function $\Psi(x-x')$ is defined in (7.88) and has the properties described in Section 7.5.

Taking into account (9.9) we can write

$$\int K_{\alpha\beta}(x-x')\Psi(x-x')\mathrm{d}x' = -\int K_{\alpha\beta}(x)\Phi(x)\mathrm{d}x = K^\Phi_{\alpha\beta}, \tag{9.53}$$

$$\Phi(x) = 1 - \Psi(x). \tag{9.54}$$

The solution of Equation (9.51) has the form

$$E^*_\alpha = \Lambda^*_{\alpha\beta}E^0_\beta, \quad \Lambda^*_{\alpha\beta} = \left(\delta_{\alpha\beta} - pK^\Phi_{\alpha\lambda}P^0_{\lambda\beta}\right)^{-1}, \tag{9.55}$$

and the mean $\langle q_\alpha(x)\rangle$ in (9.49) is transformed into the following:

$$\langle q_\alpha(x)\rangle = \left\langle C^1_{\alpha\lambda}(x)\Lambda^E_{\lambda\mu}(x)V(x)\right\rangle E^*_\mu = \\ = pP_{\alpha\beta}E^0_\beta, \quad P_{\alpha\beta} = P^0_{\alpha\lambda}\Lambda^*_{\lambda\beta}. \tag{9.56}$$

Because of (9.13), the tensor of effective conductive properties is determined by the equation

$$C^*_{\alpha\beta} = C^0_{\alpha\beta} + pP_{\alpha\beta} = C^0_{\alpha\beta} + pP^0_{\alpha\lambda}\left(\delta_{\lambda\beta} - pK^\Phi_{\lambda\mu}P^0_{\mu\beta}\right)^{-1}. \tag{9.57}$$

Let the matrix material be isotropic ($C^0_{\alpha\beta} = c_0\delta_{\alpha\beta}$), and the inclusions be ellipsoids with the same properties and sizes randomly oriented in space. If their centers are distributed homogeneously in space, the function $\Phi(x)$ is spherically symmetric, and tensor $K^\Phi_{\alpha\beta}$ takes the form

$$K^\Phi_{\alpha\beta} = \frac{1}{3c_0}\delta_{\alpha\beta}. \tag{9.58}$$

In this case, the composite is isotropic ($C^*_{\alpha\beta} = c_*\delta_{\alpha\beta}$), and its effective conductivity c_* is defined by the equation

$$c_* = c_0 + p\Lambda_0\left(1 - \frac{p\Lambda_0}{3c_0}\right)^{-1}, \tag{9.59}$$

$$\Lambda_0 = \frac{1}{3}(c - c_0)\sum_{k=1}^{3}\left(1 + (c - c_0)A^0_k\right)^{-1}, \tag{9.60}$$

where the coefficients A^0_k are determined in (6.14).

For spherical inclusions, the expression for C_* takes the form

$$c_* = c_0 + p(c - c_0)\left[1 + (1 - p)\frac{c - c_0}{3c_0}\right]^{-1}. \tag{9.61}$$

For highly conductive inclusions ($c \to \infty$), this equation gives

$$c_* = \frac{1 + 2p}{1 - p}c_0, \tag{9.62}$$

and for nonconductive spheres ($c = 0$) we obtain

$$c_* = \frac{2(1 - p)}{2 + p}c_0. \tag{9.63}$$

Let us consider a composite with spheroidal inclusions of the same orientation. If the symmetry of the function $\Phi(x)$ is defined by a spheroid with the semiaxes $\alpha_1 = \alpha_2 = \alpha$, α_3, $\gamma = \alpha/\alpha_3$, coaxial to inclusion, the tensor K^Φ in (9.53) takes the form

$$K^\Phi_{\alpha\beta} = K^\Phi_1 \theta_{\alpha\beta} + K^\Phi_2 m_\alpha m_\beta, \tag{9.64}$$

$$K^\Phi_1 = \frac{1}{c_0}\left[1 - 2f_0(\gamma)\right], \quad K^\Phi_2 = \frac{1}{c_0} f_0(\gamma), \tag{9.65}$$

where m is the unit vector along the x_3-axis, and the function $f_0(\gamma)$ is defined in (3.92). The composite material is transversely isotropic, and its tensor of the effective conductivity $C^*_{\alpha\beta}$ is

$$C^*_{\alpha\beta} = C^*_{11}\theta_{\alpha\beta} + C^*_{33} m_\alpha m_\beta, \tag{9.66}$$

$$C^*_{11} = c_0 + pP^0_1 \left(1 - pK^\Phi_1 P^0_1\right)^{-1}, \tag{9.67}$$

$$C^*_{33} = c_0 + pP^0_3 \left(1 - pK^\Phi_3 P^0_3\right)^{-1}, \tag{9.68}$$

$$P^0_1 = \left(\frac{1}{c - c_0} + A^0_1\right)^{-1}, \quad P^0_3 = \left(\frac{1}{c - c_0} + A^0_3\right)^{-1}. \tag{9.69}$$

Here the coefficients A^0_1, A^0_3 are determined in (6.15). If the aspect ratio γ has the same order as the inclusion aspect ratio, $K^\Phi_{\alpha\beta} = A^0_{\alpha\beta}$ and (9.69) is transformed into the equation

$$C^*_{11} = c_0 + p\left[\frac{1}{c - c_0} + (1 - p)A^0_1\right]^{-1}, \tag{9.70}$$

$$C^*_{33} = c_0 + p\left[\frac{1}{c - c_0} + (1 - p)A^0_3\right]^{-1}. \tag{9.71}$$

Note that the Mori–Tanaka approach leads to the same equation for the effective conductivity.

In the limit $\gamma \to 0$, we obtain a material containing long parallel fibers in the x_3-direction

$$C^*_{11} = C_0\left[1 + \frac{2p(c - c_0)}{2c_0 + (1 - p)(c - c_0)}\right], \tag{9.72}$$

$$C^*_{33} = c_0 + p(c - c_0). \tag{9.73}$$

The limit $\gamma \to \infty$ corresponds to inclusions in the form of the flat disks. In this case,

$$A^0_1 \approx \frac{\pi}{4c_0\gamma}, \quad A^0_3 \approx \frac{1}{c_0}\left(1 - \frac{\pi}{2\gamma}\right). \tag{9.74}$$

For the highly conducting disks we have

$$C_{11}^* = \left[1 + \frac{4p\gamma}{\pi(1-p)}\right] c_0, \quad C_{33}^* = \frac{c_0}{1-p}, \tag{9.75}$$

and for nonconductive thin inclusions

$$C_{11}^* = (1-p)c_0, \quad C_{33}^* = (1-p)\left(1 + \frac{2p\gamma}{\pi}\right)^{-1}. \tag{9.76}$$

9.3.1 Random set of thin inclusions

Consider an infinite medium with a random set of thin inclusions of low conductivity. The external field D_α^0 applied to the medium is assumed to be constant, and the middle surface Ω_i of the i-th inclusion is an ellipse with the semi-axes $a_1^{(i)}, a_2^{(i)}$ and the normal $m^{(i)}$. The equations for the fields $E_\alpha(x)$ and $D_\alpha(x)$ in such a medium can be represented in the form

$$E_\alpha(x) = E_\alpha^0 - \int K_{\alpha\beta}(x-x')C_0 q_\beta(x')\mathrm{d}x', \tag{9.77}$$

$$D_\alpha(x) = D_\alpha^0 - \int S_{\alpha\beta}(x-x')q_\beta(x')\mathrm{d}x', \tag{9.78}$$

$$q_\alpha(x) = \Lambda_{\alpha\beta}^0(x)D_\beta^* Z(x)\Omega(x), \quad \Omega(x) = \sum_k \Omega_k(x). \tag{9.79}$$

Here $\Omega_k(x)$ is the delta-function concentrated on the middle surface of the k-th inclusion. The function $\Lambda_{\alpha\beta}^0(x)$ is the constant $\Lambda_{\alpha\beta}^0(a_1, a_2)$ determined in (6.101) when $x \in \Omega_k$, and

$$Z(x) = \frac{2(a_1^{(k)})^2}{a_2^{(k)}}\sqrt{1 - \left(\frac{x_1}{a_1^{(k)}}\right)^2 - \left(\frac{x_2}{a_2^{(k)}}\right)^2}, \quad x \in \Omega_k. \tag{9.80}$$

If we introduce the function

$$\Omega(x;x') = \sum_{i\neq j} \Omega_i(x'), \quad \text{when } x \in \Omega_j, \tag{9.81}$$

the equation for the field D_α^* can be written as follows:

$$D_\alpha^* = D_\alpha^0 + \int S_{\alpha\beta}(x-x')\Lambda_{\beta\lambda}^0(x')Z(x')\Omega(x;x')\mathrm{d}x' D_\lambda^*, \quad x \in \Omega. \tag{9.82}$$

The homogenization procedure applied to the Equations (9.77)–(9.82) is similar to that described in Section 7.10 and leads to the following final result:

$$B^*_{\alpha\beta} = B^0_{\alpha\beta} + \langle \Lambda^0_{\alpha\lambda}(m)\rangle \left[\delta_{\lambda\beta} - \langle D_{\lambda\mu}(m)\Lambda^0_{\mu\beta}(m)\rangle\right]^{-1}, \tag{9.83}$$

$$D_{\alpha\beta}(m) = C^0_{\alpha\lambda}\left(A_{\lambda\mu}(m)C^0_{\mu\beta} - \delta_{\lambda\beta}\right), \tag{9.84}$$

$$A_{\alpha\beta}(m) = \int K_{\alpha\beta}(x)\left[1 - \Psi_m(x)\right] \mathrm{d}x. \tag{9.85}$$

Here the function $\Psi_m(x)$ is the same as considered in Section 5.8. If the matrix is isotropic and the symmetry of $\Psi_m(x)$ coincides with the symmetry of the ellipsoid with the semiaxes $\alpha_1, \alpha_2, \alpha_3$ coaxial to an inclusion of the orientation m, then

$$A_{\alpha\beta}(m) = A_1 e^1_\alpha e^1_\beta + A_2 e^2_\alpha e^2_\beta + A_3 m_\alpha m_\beta, \tag{9.86}$$

$$A_k = \frac{\alpha_1\alpha_2\alpha_3}{2c_0}\int\limits_0^\infty \frac{\mathrm{d}\epsilon}{(\alpha_k^2+\epsilon)\sqrt{(\alpha_1^2+\epsilon)(\alpha_1^2+\epsilon)(\alpha_1^2+\epsilon)}}, \qquad k = 1,2,3, \tag{9.87}$$

where e^1_α, e^2_α are the unit vectors of the major axes of the ellipse, the orientation of which is given by the normal m.

The ergodic property allows us to replace ensemble averaging by spatial averaging for a fixed typical realization of the random set of inclusions ($a_1 \geq a_2$)

$$\langle \Lambda^0(m)\rangle = \lim_{W\to\infty}\frac{1}{W}\int\limits_W \Lambda^0(x)Z(x)\Omega(x)\mathrm{d}x =$$

$$= n_0\left\langle v\Lambda^0(a_1,a_2)\right\rangle, \quad v = \frac{4}{3}\pi a_1^3 \tag{9.88}$$

The tensor $v\Lambda^0$ in this equation is averaged over the ensemble distribution of the sizes and orientations of the inclusions, and n_0 is the numerical concentration of the inclusions.

Let thin inclusions in the composite have low conductivity and the same orientation. In this case, Equation (9.83) takes the form

$$B^*_{\alpha\beta} = B_0 e^1_\alpha e^1_\beta + B_0 e^2_\alpha e^2_\beta + B_* m_\alpha m_\beta. \tag{9.89}$$

Hence, the composite medium has orthotropic symmetry. Two of three independent resistance coefficients (in the direction perpendicular of the vector m) coincide with those of the matrix, and the third one is

$$B_* = B_0 + \frac{\tau\Lambda_0(a_1,a_2)}{1 - \tau A_3\Lambda_0(a_1,a_2)}, \tag{9.90}$$

$$\Lambda_0(a_1, a_2) = B_0 \left[\frac{c}{c_0} \frac{a_1}{h\delta} + \frac{E(k)}{1-k} \right], \tag{9.91}$$

where $\delta = a_2/a_1$, $E(k)$ is the elliptic integral of the second kind defined in (3.238) ($k = 1 - (a_2/a_1)^2$).

If thin ellipsoids are homogeneously distributed over the orientations, the composite medium is macroscopically isotropic, and the tensor $B^*_{\alpha\beta}$ is defined by the equation

$$B^*_{\alpha\beta} = B_* \delta_{\alpha\beta}, \quad B_* = B_0 + \frac{\tau}{3} \left(1 - \frac{\tau}{3} \Lambda^0 A_3\right)^{-1}. \tag{9.92}$$

Let us consider the materials with a random set of thin highly conductive inclusions. In this case, the tensor of effective conductivity $C^*_{\alpha\beta}$ takes the form

$$C^*_{\alpha\beta} = C^0_{\alpha\beta} + \langle \Lambda^0_{\alpha\lambda}(m) \rangle \left[\delta_{\lambda\beta} - \langle A_{\lambda\mu}(m) \Lambda^0_{\mu\beta}(m) \rangle \right]^{-1}, \tag{9.93}$$

where the tensor $A_{\mu\lambda}(m)$ is defined in (9.86) and the tensor $\Lambda^0_{\alpha\mu}(m)$ is defined in this case in (6.112).

If highly conductive thin inclusions of the same sizes, properties, and orientations, the tensor $C^*_{\alpha\beta}$ takes the form

$$C^*_{\alpha\beta} = C^*_{11} e^1_\alpha e^1_\beta + C^*_{22} e^2_\alpha e^2_\beta + c_0 m_\alpha m_\beta. \tag{9.94}$$

Hence, in this case the composite medium is also orthotropic. One of the three independent coefficients of the conductivity (in the direction parallel to vector m) coincides with those of the matrix, and two others are determined by the expressions

$$C^*_{11} = c_0 + \frac{\tau \Lambda^0_1(a_1, a_2)}{1 - \tau \Lambda^0_1(a_1, a_2) A_1}, \tag{9.95}$$

$$C^*_{22} = c_0 + \frac{\tau \Lambda^0_2(a_1, a_2)}{1 - \tau \Lambda^0_2(a_1, a_2) A_2}, \tag{9.96}$$

$$\Lambda^0_1(a_1, a_2) = c_0 \left[\frac{c_0}{c} \frac{a_1}{h\delta} + \frac{K(k) - E(k)}{k} \right]^{-1}, \tag{9.97}$$

$$\Lambda^0_2(a_1, a_2) = c_0 \left[\frac{c_0}{c} \frac{a_1}{h\delta} + \frac{E(k) - (1-k)K(k)}{k(1-k)} \right]^{-1}. \tag{9.98}$$

If the inclusions are uniformly distributed over the orientation we have

$$C^*_{\alpha\beta} = c_* \delta_{\alpha\beta}, \quad c_* = c_0 + \frac{\tau}{3} \left(\Lambda^0_1 + \Lambda^0_2 \right) \left[1 - \frac{\tau}{3} \left(\Lambda^0_1 A_1 + \Lambda^0_2 A_2 \right) \right]^{-1}. \tag{9.99}$$

9.4 Dielectric properties of composites with high volume concentrations of inclusions

The EFM based on the quasicrystalline approximation gives simple and physically correct formulas for various effective properties of composite materials. For instance, the formula for the dielectric permittivity of a two-phase composite with spherical inclusions obtained in the framework of the EFM in 2D-case takes the form

$$c_* = c_0 + \frac{2pc_0c_1}{2c_0 + (1-p)c_1}. \tag{9.100}$$

Here c_0 is the dielectric permittivity of the background medium (matrix), $c = c_0 + c_1$ is the same for the material of inclusions, p is the volume concentration of inclusions. Equation (9.100) is in fact the well-known Maxwell–Garnett formula in the 2D-case [111]. This formula and similar formulas for dielectric, elastic, thermoconductive, etc. properties of composites in 2D- and 3D-cases are in agreement with experimental data and numerical solutions for not very high volume concentrations of inclusions. In Fig. 9.1 the dependence $c_*(p)$ corresponding to (9.100) for $c_0 = 1, c = 100$ is presented by a solid line. The line with black dots shows the results of numerical calculations of the effective dielectric permittivities of composites with random sets of circular disks. It can be seen that the numerical results deviate from (9.100) when $p > 0.3$ and the difference grows with p. The same situation may be observed in the case of elastic properties of two-phase composites with spherical inclusions (see, e.g., Fig. 7.4).

Thus, in the region $p > 0.3$ the EFM needs correction. The version of the EFM that allows us to improve the predictions of the method was described

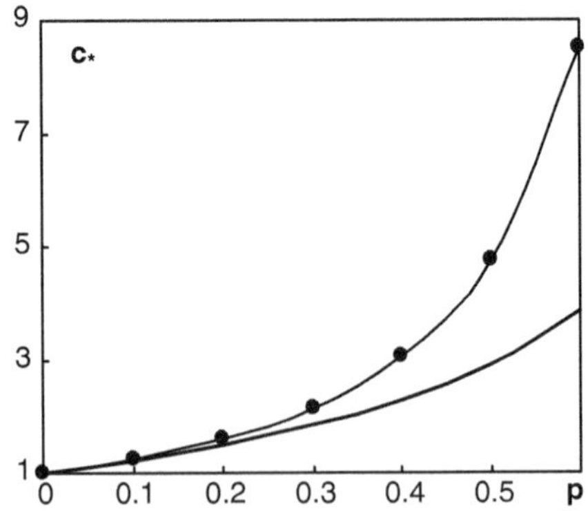

Fig. 9.1. The dependence of the effective dielectric permittivity of two-phase composite c_* on volume concentrations of inclusions p in 2D-case ($c_0 = 1, c = 100$); the solid line corresponds to the EFM based on the quasicrystalline approximation; the line with black dots corresponds to the numerical calculations of c_*.

in Chapter 8. In this version the main unknowns of the problem are multipoint conditional means of the random effective field E_* that acts on the inclusions in the composite. The n-point mean $\langle E_*(x_1)|x_1, x_2, ...x_n\rangle$ is the mean value of the field E_* at point x_1 under the condition that points $x_1, x_2, ..., x_n$ are occupied by different inclusions. The effective properties of the composite are expressed via the first conditional mean $\langle E_*(x_1)|x_1\rangle$. As shown in Chapter 8, the mean $\langle E_*(x_1)|x_1\rangle$ is expressed via all multipoint conditional means of the field E_* by an infinite chain of integral equations. The hypothesis of the quasicrystalline approximation that may be written as follows:

$$\langle E_*(x_1)|x_1, x_2\rangle = \langle E_*(x_1)|x_1\rangle \tag{9.101}$$

closes this chain at the first step. It essentially simplifies the problem by reducing it to the solution of only one integral equation for $\langle E_*(x_1)|x_1\rangle$. The necessary statistical information about the random field of inclusions required in this version of the EFM is a specific two-point correlation function (the probability density that two points x_1 and x_2 are inside different inclusions). Further improvements inside the EFM may be obtained by the assumption that the n-point conditional mean of E_* coincides with the $(n+1)$-point similar mean. This hypothesis closes the mentioned chain of integral equations at the n-th step, and the problem is reduced to the solution of an integral equation whose kernel depends on a specific $(n+1)$-point correlation function of the random set of inclusions.

In this section, a version of the EFM is developed when the chain of integral equation for the conditional means of the effective field E_* is closed at the second step. In this step the integral equation for the two-point mean $\langle E_*(x_1)|x_1, x_2\rangle$ is obtained. This two-point mean characterizes pair interactions between inclusions in the composite medium. The kernel of the integral equation for $\langle E_*(x_1)|x_1, x_2\rangle$ depends on a specific three-point correlation function of a random field of inhomogeneities. The influence of this function on the overall dielectric permittivities of the composites is analyzed in this section.

9.4.1 The EFM in application to two-phase composites (the quasicrystalline approximation)

In previous chapters, a version of the EFM based on the quasicrystalline approximation was developed. Nevertheless, we present here another derivation of the equations of the method that takes account of the mutual positions of the inclusions in the composite. This version of the EFM may be naturally generalized in order to improve the predictions of the method for high volume concentration of inclusions.

Consider an infinite dielectric medium with a homogeneous set of isolated inclusions. For simplicity, we assume that all the inclusions are spheres of unit radius $a = 1$. The integral equation for the electric field $E(x)$ inside the inclusion that occupies an arbitrary region v_k may be written in a form that follows from (9.2)

$$E(x) + \int_{v_k} K(x - x')C_1E(x')\mathrm{d}x' = E_*(x), \qquad x \in v_k, \tag{9.102}$$

where $E_*(x)$ may be interpreted as a local external field that acts on the inclusion v_k

$$E_*(x) = E_0 - \sum_{i \neq k} \int K(x - x')C_1E(x')v_i(x')\mathrm{d}x'. \tag{9.103}$$

It follows from this equation, that the field $E_*(x)$ consists of the applied field E_0 and the sum of disturbances from other inclusions. A general solution of Equation (9.102) may be presented in the form

$$E(x) = (\Lambda^k E_*)(x), \qquad x \in v_k, \tag{9.104}$$

where Λ_k is a linear operator that depends on the dielectric properties and the form of the inclusion v_k. After substituting (9.104) into (9.103) we go to the following integral equation for the field $E_*(x)$

$$E_*(x) = E_0 -$$

$$- \sum_{i \neq k} \int K(x - x')C_1(\Lambda E_*)(x')v_i(x')\mathrm{d}x', \qquad x \in v_k, \quad k = 1, 2, ... \tag{9.105}$$

Equation (9.105) is totally equivalent to the original integral Equation (9.2). Thus, the field $E_*(x)$ may be considered as the main unknown of the problem. If this field is constructed, the electric fields inside any inclusion may be obtained from (9.104), and the field in the matrix is to be built from the original Equation (9.2).

The main hypotheses of the EFM concern the general structure of the field $E_*(x)$, and a simplest version of the method is based on the following hypothesis:

H_1. *The field $E_*(x)$ is constant in the vicinity of every inclusion in the composite medium.*

This hypothesis allows to rewrite (9.104) in the form

$$E(x) = E^k(x) = (\Lambda^k E_*^k)(x) = \Lambda^k(x)E_*^k, \qquad x \in v_k, \quad k = 1, 2, ..., \tag{9.106}$$

where function $\Lambda^k(x)$ is the result of the action of the operator Λ^k on a constant field E_*^k. If v_k are spherical regions, Λ^k is also a constant tensor that is the same for all the inclusions

$$\Lambda^k = \Lambda = (I + A^0C_1)^{-1}. \tag{9.107}$$

Here I is the unit rank two tensor; the tensor A^0 is defined in (6.12).

Equation (9.106) may be considered as an approximation to the polarization fields E^k inside inclusions. In order to obtain the system of equations for the effective fields E_*^k let us substitute (9.106), (9.107) into (9.105) and integrate the result over every subarea v_k (Galerkin's scheme). Finally, we obtain the following system of linear algebraic equations for the fields E_*^k

$$E_*^k + \sum_{i \neq k} \bar{K}(x^k - x^i) P E_*^i = E_0, \qquad k = 1, 2, ..., \tag{9.108}$$

$$P = C_1 \Lambda, \quad \bar{K}(x^k - x^i) = \frac{1}{v} \int_{v_k} \mathrm{d}x \int_{v_i} K(x - x') \mathrm{d}x'. \tag{9.109}$$

Here x^k is the center of the inclusion v_k, v is the volume of an arbitrary inclusion.

Let us introduce two generalized functions concentrated at the centers of the inclusions

$$X(x) = \sum_i \delta(x - x^i), \qquad X(x; x') = \sum_{i \neq k} \delta(x' - x^i) \;\text{ if }\; x = x^k. \tag{9.110}$$

Using these functions, we may write (9.108) in the form

$$E_*(x) = E_0 - \int \bar{K}(x - x') P E_*(x') X(x, x') \mathrm{d}x', \quad x \in X. \tag{9.111}$$

Here X is the set of the centers of inclusions ($X = \{x^1, x^2, ...\}$); the field $E_*(x)$ coincides with E_*^i at point $x = x^i$.

Let us find the first statistical moment of the effective field $E_*(x)$. After averaging (9.109) under the condition $x \in X$ we obtain

$$\langle E_*(x) | x \rangle = E_0 - \int \bar{K}(x - x') P \langle E_*(x') | x', x \rangle \langle X(x; x') | x \rangle \, \mathrm{d}x', \tag{9.112}$$

where $\langle E_*(x) | x \rangle$ is the ensemble averaging of the random field $E_*(x)$ under the condition $x \in X$, $\langle E_*(x') | x', x \rangle$ is the same averaging under the condition that $x, x' \in X$. A closed equation for the mean $\langle E_*(x) | x \rangle$ follows from (9.112) by using an additional assumption

$$\langle E_*(x') | x', x \rangle = \langle E_*(x') | x' \rangle = E_*^{(1)}. \tag{9.113}$$

Equation (9.113) is in fact the hypothesis of quasicrystalline approximation that converts (9.112) into the following equation:

$$\langle E_*(x) | x \rangle = E_0 - \int \bar{K}(x - x') P \langle E_*(x') | x' \rangle \langle X(x; x') | x \rangle \, \mathrm{d}x'. \tag{9.114}$$

Note that for a homogeneous random field of inclusions, the first statistical moment of the effective field $\langle E_*(x')|x'\rangle = E_*^{(1)}$ is constant. The mean $\langle X(x;x')|x\rangle$ in the right-hand side of (9.114) depends only on geometrical properties of the random field of inclusions. If this field is homogeneous and isotropic in space, the mean $\langle X(x;x')|x\rangle$ is a function of only the distance between points x and x' ($|x-x'|$)

$$\langle X(x;x')|x\rangle = n_0 g(|x-x'|). \tag{9.115}$$

Here n_0 is the numerical concentration of inclusions (the number of inclusions in a unit volume). The function $g(x)$ satisfies the following conditions:

$$g(|x|) = 0 \text{ if } |x| < 2; \qquad \lim_{|x|\to\infty} g(|x|) = 1. \tag{9.116}$$

The first of these conditions holds for nonoverlapping sets of spherical inclusions of unit radii, and is a consequence of the definition (9.110) of the function $X(x;x')$. The second condition is correct if the correlation in spatial positions of the inclusions disappears when the distance between their centers tends to infinity.

From (9.114),(9.115) we obtain the following algebraic equation for the mean value of the effective field $E_*^{(1)}$

$$E_*^{(1)} - n_0 A^{\Phi} P E_*^{(1)} = E_0, \tag{9.117}$$

$$A^{\Phi} = -\int \bar{K}(x) g(|x|) \mathrm{d}x = \int \bar{K}(x) \Phi(|x|) \mathrm{d}x, \tag{9.118}$$

$$\Phi(|x|) = 1 - g(|x|). \tag{9.119}$$

Here we take into account the equation $\int \bar{K}(x)\mathrm{d}\, x = 0$ that follows from the regularization formula (9.9). Note that $\Phi(|x|) = 1$ if $|x| < 2$, and $\Phi(|x|) \to 0$ when $|x| \to \infty$ as it follows from (9.116).

The function $\bar{K}(x)$ in (9.109) may be written in the form

$$\bar{K}(x) = \frac{1}{v}\int v_0(x-x')\mathrm{d}x' \int K(x'-x'')v_0(x'')\mathrm{d}x'' =$$

$$= \frac{1}{v}\int K(x-y) f(|y|)\mathrm{d}y, \quad f(|y|) = \int v_0(x+y)v_0(x)\mathrm{d}x, \tag{9.120}$$

where v_0 is the region of the inclusion centered at point $x = 0$. Here $f(|y|)$ is the volume of the intersection of two unit spheres whose centers are separated by the distance y, $f(|y|) = 0$ if $|y| > 2$. Using Parceval's formula: the integral A^{Φ} in (9.118) may be calculated as follows:

$$
\begin{aligned}
A^{\Phi} &= \frac{1}{v}\int\int f(|y|)K(x-y)\Phi(|x|)\mathrm{d}x\mathrm{d}y = \\
&= \frac{1}{(2\pi)^3 v}\int \tilde{f}(|k|)\tilde{K}(k)\tilde{\Phi}(|k|)\mathrm{d}k = \\
&= \frac{1}{2\pi^2 v}A^0\int_0^{\infty}\tilde{f}(|k|)\tilde{\Phi}(|k|)|k|^2\mathrm{d}|k| = \\
&= \frac{4\pi}{v}A^0\int_0^2 f(|y|)|y|^2\mathrm{d}|y| = vA^0,
\end{aligned}
\tag{9.121}
$$

where tensor A^0 is defined in (6.12). This result does not depend on the details of the behavior of the function $\Phi(x)$. After substituting (9.121) into (9.117) we obtain an equation for $E_*^{(1)}$, the solution of which takes the form

$$
E_*^{(1)} = (I - pA^0P)^{-1}E_0, \tag{9.122}
$$

where $p = n_0 v$ is the volume concentration of inclusions.

Let us turn to Equation (9.6) for the function $q(x)$. As follows from (9.106), (9.107) and (9.122), the mean $\langle q(x)\rangle$ takes the form

$$
\langle q(x)\rangle = \langle C_1 E(x)V(x)\rangle = pPE_*^{(1)} = QE_0, \tag{9.123}
$$

$$
Q = pP(I - pA^0P)^{-1}. \tag{9.124}
$$

Equation (9.123) together with (9.13) give us the final formula for the overall dielectric permittivity tensor C_* of the composite in the framework of the EFM and the quasicrystalline approximation (9.113)

$$
C_* = C_0 + pP(I - pA^0P)^{-1}. \tag{9.125}
$$

Taking into account that $P = C_1(I - A^0C_1)^{-1}$ (see (9.107), (9.109)) we may write the last equation in the form

$$
C_* = C_0 + pC_1\left[I + (1-p)A^0C_1\right]^{-1}. \tag{9.126}
$$

9.4.2 The EFM beyond the quasicrystalline approximation

Let us consider again Equation (9.112) for the first statistical moment of the effective field $E_*(x)$. If we do not use the quasicrystalline approximation (9.134), the mean $\langle E_*(x')|x', x\rangle$ in the right-hand side of (9.112) must be constructed from an independent equation. In order to derive the equation for this mean let us average (9.111) under the condition that point x and x_1

occupy the centers of different inclusions. As a result we obtain the equation for the mean $\langle E_*(x)|x,x_1\rangle = E_*^{(2)}(x-x_1)$ in the form

$$E_*^{(2)}(x-x_1) = E_0 - \\ - \int \bar{K}(x-x')P\left\langle E_*(x')|x',x,x_1\right\rangle \left\langle X(x;x')|x,x_1\right\rangle \mathrm{d}x'. \tag{9.127}$$

The mean $\langle E_*(x')|x',x,x_1\rangle$ on the right-hand side of this equation is the mean value of the effective field $E_*(x')$ at the center x' of an inclusion under the condition that points x and x_1 occupy the centers of different inclusions. In order to obtain a closed equation for the mean $E_*^{(2)}(x-x_1)$ let us accept a hypothesis that is similar to (9.113):

$$\langle E_*(x')|x',x,x_1\rangle = \langle E_*(x')|x',x\rangle = E_*^{(2)}(x-x'). \tag{9.128}$$

Otherwise, the mean $\langle E_*(x')|x',x,x_1\rangle$ should be constructed from (9.111) by averaging both parts under the condition that $x,x_1,x_2 \in X$. As a result we obtain an equation for the three-point conditional mean of the effective field, the right-hand side of which depends on the four-point similar mean. Continuing this process we have infinite chain of integral equations that connects all multipoint conditional means of the effective field $E_*(x)$. The hypothesis (9.128) closes this chain at the second step, and gives us the following equation for the mean $E_*^{(2)}(x)$ $(y = x_1 - x, y' = x' - x)$:

$$E_*^{(2)}(y) = E_0 - \\ - \int \bar{K}(y')TE_*^{(2)}(y')\left\langle X(x_1-y;y'-x_1+y)|x_1-y,x_1\right\rangle \mathrm{d}y'. \tag{9.129}$$

$E_*^{(2)}(y)$ is the mean value of the effective field at point $y=0$ occupied by the center of an inclusion under the condition that point y occupies the center of different inclusion. In (9.129) we use the fact that $E_*^{(2)}(y)$ and $\bar{K}(y)$ are even functions of their arguments. For a statistically homogeneous random medium, the conditional mean $\langle X(x_1-y;y'-x_1+y)|x_1-y,x_1\rangle$ in (9.129) is invariant under translation, and the following equation holds:

$$\langle X(x_1-y;y'-x_1+y)|x_1-y,x_1\rangle = \langle X(0;y')|0,y\rangle . \tag{9.130}$$

The last mean may be presented in the form

$$\langle X(0;x')|0,x\rangle = \delta(x-x') + n_0 F(x,x'). \tag{9.131}$$

The first term on the right-hand side of this equation corresponds to the input of the situations in which points x and x' occupy the center of the same inclusions, and $n_0F(x,x')$ is the input of the situations when points x and x' occupy the centers of different inclusions. Using (9.131) one can write (9.129) in the form

$$E_*^{(2)}(x) = E_0 - \bar{K}(x)PE_*^{(2)}(x) - n_0 \int \bar{K}(x')PE_*^{(2)}(x')F(x,x')\mathrm{d}x'. \quad (9.132)$$

If the volume concentration of inclusions p tends to zero, the integral term in this equation disappears, and its solution

$$E_*^{(2)}(x) = (I + \bar{K}(x)P)^{-1}E_0 \quad (9.133)$$

describes the interaction between two isolated inclusions in the homogeneous medium with the properties of the matrix. In this case, $E_*^{(2)}(x)$ is the field that acts on each of two isolated inclusions if the distance between their centers is x, by the application of a constant external field E_0. The influence of the surrounding inclusions on the mean $E_*^{(2)}(x)$ is taken into account by the integral term in (9.132).

As follows from (9.112), the one-point mean $E_*^{(1)}$ is expressed via the two-point mean $E_*^{(2)}(x)$ by the equation

$$E_*^{(1)} = E_0 - n_0 \int \bar{K}(x')PE_*^{(2)}(x')g(x')\mathrm{d}x'. \quad (9.134)$$

Here the function $g(x)$ is defined in (9.115). Using (9.134) one can rewrite (9.132) as follows:

$$(I + \bar{K}(x)P)E_*^{(2)}(x) = E_*^{(1)} - \\ - n_0 \int \bar{K}(x')PE_*^{(2)}(x')\left[F(x,x') - g(x')\right]\mathrm{d}x'. \quad (9.135)$$

The solution of this equation may be found in the form

$$E_{*i}^{(2)}(x) = \Pi_{ij}(x)E_{*j}^{(1)}, \quad (9.136)$$

and the equation for the tensor $\Pi(x)$ follows from (9.136)

$$\left[I + \bar{K}(x)P\right]\Pi(x) = \\ = I - n_0 \int \bar{K}(x')P\Pi(x')\left[F(x,x') - g(x')\right]\mathrm{d}x'. \quad (9.137)$$

Note that if $|x| \to \infty$ the following equations hold:

$$\bar{K}(x) \to 0, \quad F(x,x') \to g(x-x')g(x'), \quad (9.138)$$

$$\left[F(x,x') - g(x')\right] \to \left[g(x-x') - 1\right] = -\Phi(x-x'), \quad (9.139)$$

$$\int \bar{K}(x')P\Pi(x')\,[F(x,x') - g(x')]\,\mathrm{d}x' \to$$

$$\to -\bar{K}(x)\int P\Pi(x')\Phi(x-x')\mathrm{d}x' \to 0. \tag{9.140}$$

Here the properties (9.116) of the two-point correlation function $g(x)$ are used.

As a result, we obtain from (9.137) that

$$\Pi(x) \to I \ \text{ if } \ |x| \to \infty. \tag{9.141}$$

If the solution of Equation (9.137) is found, the one-point mean $E_*^{(1)}$ may be calculated from (9.134)

$$E_*^{(1)} = E_0 - n_0 \int \bar{K}(x')P\Pi(x')g(x')\mathrm{d}x' E_*^{(1)} =$$

$$= E_0 - n_0 \int \bar{K}(x')P(\Pi(x') - I)g(x')\mathrm{d}x' E_*^{(1)} -$$

$$- n_0 \int \bar{K}(x')Pg(x')\mathrm{d}x' E_*^{(1)}. \tag{9.142}$$

Here the first integral on the right-hand side converges because of the property (9.141) of $\Pi(x)$, and the second integral takes the form

$$n_0 \int \bar{K}(x')Pg(x')\mathrm{d}x = n_0 \int \bar{K}(x')P[g(x') - 1]\mathrm{d}x = -pAP \tag{9.143}$$

that is a consequence of (9.110) and (9.102).

The final expression for $E_*^{(1)}$ follows from (9.142), (9.143) in the form

$$E_*^{(1)} = [I - p(AP - K_0)]^{-1}\, E_0, \tag{9.144}$$

$$K_0 = \frac{1}{v}\int \bar{K}(x')P\,[\Pi(x') - I]\,g(x')\mathrm{d}x'. \tag{9.145}$$

Equation (9.145) together with (9.106), (9.107) give us the equation for the mean $\langle q \rangle$

$$\langle q(x)\rangle = \langle C_1 E(x) V(x)\rangle = pPE_*^{(1)} = QE_0,$$

$$Q = pP\,[I - p(AP - K_0)]^{-1}, \tag{9.146}$$

As a result the expression for the effective dielectric permittivity of the composite C_* takes the form

$$C_* = C_0 + pP\,[I - p(AP - K_0)]^{-1}. \tag{9.147}$$

This equation differs from (9.125) by the integral K_0 defined in (9.145). The integral K_0 appears in the equation for the effective dielectric permittivity as a result of a more precise description of interactions between inclusions in the composite medium.

9.4.3 Effective dielectric permittivity in 3D-case

Consider a 3D-composite material that consists of an isotropic matrix and a set of isotropic spherical inclusions of unit radii. In this case $C_0 = c_0 I, C = cI$, where c_0 and c are scalar coefficients of dielectric permittivity of the matrix and inclusions; function $K(x)$ in the integral equation (9.102) takes the form

$$K_{ij}(x) = \frac{1}{4\pi c_0 r^3}\left(\delta_{ij} - 3n_i n_j\right), \quad r = |x|, \quad n = \frac{x}{r}, \tag{9.148}$$

where δ_{ij} are components of the rank two unit tensor I. Tensors $\tilde{K}(k)$, A^0 and P are

$$\tilde{K}_{ij}(k) = \frac{k_i k_j}{c_0 k^2}, \quad A^0_{ij} = \frac{1}{3c_0}\delta_{ij}, \tag{9.149}$$

$$P_{ij} = t\delta_{ij}, \quad t = \frac{3c_0 c_1}{3c_0 + c_1}, \quad k^2 = k_i k_i. \tag{9.150}$$

Direct calculation of the integrals in (9.120) gives for $\bar{K}(x)$ the following equation:

$$\bar{K}(x) = vK(x), \tag{9.151}$$

where $K(x)$ is defined in (9.148).

Equation (9.125) for the effective dielectric permittivity C_* of the composite in the framework of the quasicrystalline approximation takes the form

$$C_* = c_* I, \qquad c_* = c_0 + \frac{3pc_0c_1}{3c_0 + (1-p)c_1}. \tag{9.152}$$

The last equation coincides with the well-known Maxwell–Garnet formula [111].

Consider the two-point conditional mean of the effective field $E_*^{(2)}(x) = \Pi(x)E_*^{(1)}$ in (9.135), (9.136). Equation (9.137) for the tensor $\Pi(x)$ may be written in the form

$$\Pi_0^{-1}(x)\Pi(x) = I - p\int K(x')P\Pi(x')\left[F(x,x') - g(x')\right] \mathrm{d}x', \tag{9.153}$$

$$\Pi_{0ij}(x) = \left[I + \bar{K}(x)P\right]^{-1}_{ij} = p_{01}(r)n_i n_j + p_{02}(r)(\delta_{ij} - n_i n_j), \tag{9.154}$$

$$p_{01}(r) = \left[1 - \frac{2c_1}{r^3(3c_0 + c_1)}\right]^{-1}, \quad p_{02}(r) = \left[1 + \frac{c_1}{r^3(3c_0 + c_1)}\right]^{-1}. \tag{9.155}$$

The function $g(x)$ in (9.153) is defined in (9.115) and is the probability density of finding a center of inclusion at the point x if the point $x = 0$ is the center of another inclusion (two-point correlation function of the center distribution). For a random set of nonoverlapping spheres, the most reliable

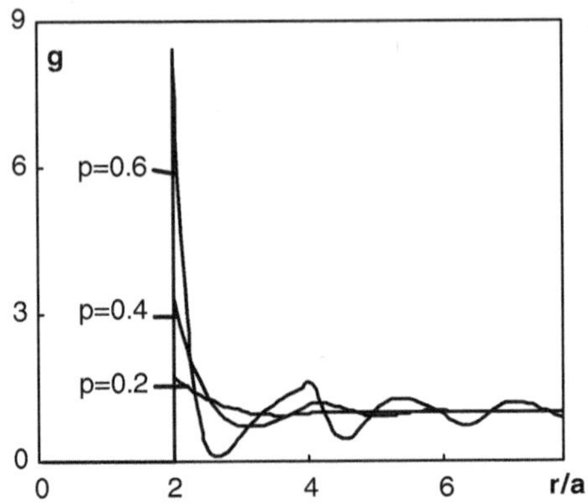

Fig. 9.2. Two-point Percus-Yevick correlation functions of the distribution of centers of inclusions of unit radii.

two-point correlation function is the solution of the so-called Perckus–Yevick equation proposed in the molecular theory of liquids [166] (Appendix E). The Perckus–Yevick correlation function is in agreement with the results of Monte-Carlo simulations of the pair correlation function of the centers of nonoverlapping spheres [7]. The detailed tables of the function $g(x)$ for 2D and 3D-cases may be found in [65]. The graphs of these functions for $p = 0.2, 0.4, 0.6$ (3D-case) are presented in Fig. 9.2.

The function $F(x, x')$ in (9.153) is defined in (9.131). This function is the probability density of finding a center of inclusion at point x' if fixed points $x' = 0$ and $x' = x$ are the centers of different inclusions. The simplest approximation of this function has the form

$$F(x, x') = g(|x'|)g(|x - x'|) \tag{9.156}$$

and is called the Kirkwood superposition approximation (KSA) [3]. Strictly speaking, this approximation of the three-point correlation function $F(x, x')$ is valid only in the limit $|x| \to \infty$.

Many works have been dedicated to the construction of $F(x, x')$ from direct Monte-Carlo simulations of sets of non-overlapping spheres in the 3D-case (see, e.g., [3], [33], [18], and references in these papers). It was shown that the errors of approximation (9.156) are maximal for small values of x ($|x| < 4$) if x' is in the region between two fixed inclusions. Equation (9.156) overestimates the probability of finding a center of an inclusion in this region. The deviation of $F(x, x')$ in (9.156) from the results of Monte-Carlo simulations may reach 40%. In spite of the efforts of many authors, no satisfactory analytical approximations have been found for the function $F(x, x')$ that can serve for wide regions of volume concentrations of inclusions p and distances $|x|$ between the centers of the fixed inclusions. In the calculations, the following approximation to $F(x, x')$ was used

$$F(x,x') = g(|x'|)g(|x-x'|)H(x,x'), \tag{9.157}$$

$$H(x,x') = \frac{1 - a\cos(\theta)\exp[-(\rho + |x| - 4)]}{1 + a\exp[-(\rho + |x| - 4)]}, \tag{9.158}$$

$$\cos(\theta) = \frac{\rho^2 + |x|^2 - \rho_1^2}{2\rho|x|}, \tag{9.159}$$

$$\rho = \min(|x'|, |x-x'|), \quad \rho_1 = \max(|x'|, |x-x'|), \tag{9.160}$$

a is a parameter that depends on the volume concentration of inclusions.

This function reduces the values of the KSA in (9.156) in the region between fixed inclusions. The connections between the parameters ρ, ρ_1, and θ are shown in Fig. 9.3. The graph of the function $F(x,x')$ in the plane that goes through the centers of two fixed inclusions situated at the points $x = 0$ and $x = (6,0,0)$ is presented in Fig. 9.4.

For the solution of Equation (9.153) let us introduce a spherical coordinate system with the origin at point $x' = 0$ (the center of the first fixed inclusion). The direction of the polar axis of this system coincides with the direction of vector x to the center of the second fixed inclusion. In this system, the vector x' has coordinates (r', θ, φ), the vector x has coordinates $(r, 0, 0)$, and the function F does not depends on the polar angle φ in the plane orthogonal to the polar axis: $F(x,x') = F(r,r',\theta)$.

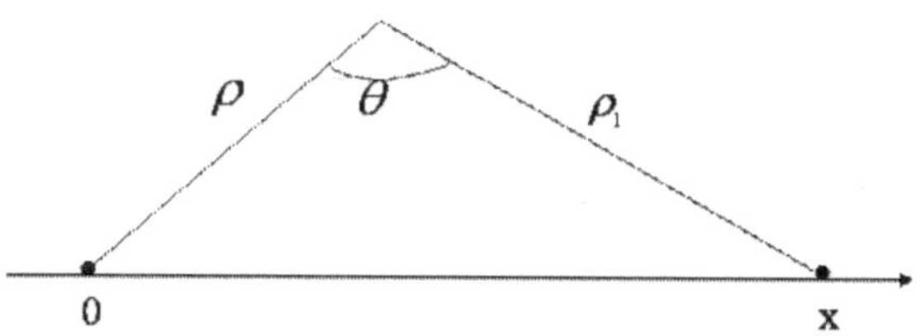

Fig. 9.3. Geometrical parameters ρ, ρ_1 and θ in equation (9.157) for $F(x,x')$.

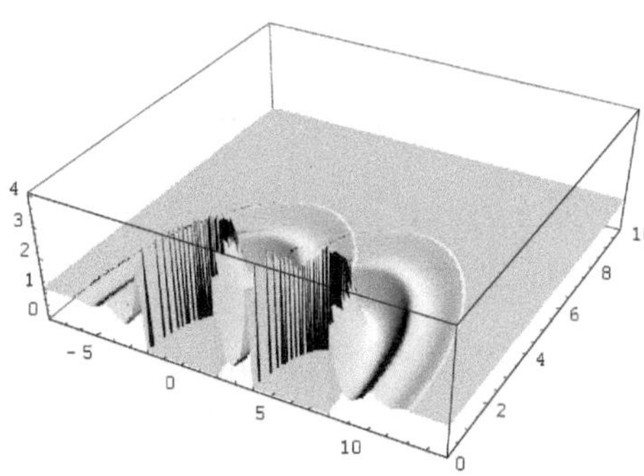

Fig. 9.4. The graph of the function $F(x,x')$ in (9.157) in the plane that goes through the centers of two fixed inclusions with the distance between their centers $|x| = 6$.

The solution of (9.153) may be found in the form

$$\Pi_{ij}(x) = p_1(r)n_in_j + p_2(r)(\delta_{ij} - n_in_j). \tag{9.161}$$

After substituting this equation into (9.153) and integrating over the φ and θ angles we reach the following system of integral equations for the scalar functions $p_1(r)$ and $p_2(r)$:

$$p_1(r) = p_{01}(r)R_1(r), \qquad p_2(r) = p_{02}(r)R_2(r), \tag{9.162}$$

$$R_1(r) = 1 - \frac{pt_0}{4c_0}\int_2^\infty [(p_2(r') - 2p_1(r'))F_1(r,r') - \\ - (p_2(r') + 2p_1(r'))F_2(r,r')]\frac{dr'}{r'}, \tag{9.163}$$

$$R_2(r) = 1 - \frac{pt_0}{8c_0}\int_2^\infty [(3p_2(r') - 2p_1(r'))F_1(r,r') + \\ +(p_2(r') + 2p_1(r'))F_2(r,r')]\frac{dr'}{r'}, \tag{9.164}$$

$$F_1(r,r') = \int_0^\pi [F(r,r',\theta) - g(r')]\sin(\theta)d\theta, \tag{9.165}$$

$$F_2(r,r') = \int_0^\pi [F(r,r',\theta) - g(r')]\cos(2\theta)\sin(\theta)d\theta. \tag{9.166}$$

Here the functions $p_{01}(r), p_{02}(r)$ are defined in (9.155).
After integration over φ and θ, the integral K_0 in (9.145) takes the form

$$K_0 = k_0I, \qquad k_0 = \frac{2c_1}{3c_0 + c_1}\int_2^\infty [p_2(r) - p_1(r)]\, g(r)\frac{dr}{r}. \tag{9.167}$$

This integral converges because $p_2(r) - p_1(r) \to 0$ when $r \to \infty$. As a result, Equation (9.147) for the effective dielectric permittivity C_* of the composite takes the form

$$C_* = c_*I, \qquad c_* = c_0 + \frac{3c_0c_1}{(3c_0 + c_1)(1 + pk_0) - pc_1}. \tag{9.168}$$

Thus, the problem is reduced to the solution of the integral equations (9.162)–(9.166) and the calculation of the integral in (9.167). A subsequent iteration procedure was used for the construction of the solution of these integral equations. As an initial approximation for p_1, p_2 the functions p_{01}, p_{02} in (9.155) were used. It turns out that the number of the iterations necessary depends on the volume concentration of inclusions p and the deviation of dielectric properties inside inclusions. For $c_0 = 1, c = 100$, the number of the iterations does not exceed 10 for $p \leq 0.6$.

Note that if the distance $|x|$ between the centers of the fixed inclusions is sufficiently large, functions F_1 and F_2 in (9.165),(9.166) coincide because $\cos(2\theta) \approx 1$ in the region where $[F(r, r', \theta) - g(r')] \neq 0$, and (9.163),(9.164) take the forms

$$R_1(r) = 1 + p\frac{t_0}{c_0}\int_2^\infty p_1(r')F_1(r, r')\frac{dr'}{r'}, \tag{9.169}$$

$$R_2(r) = 1 - p\frac{t_0}{2c_0}\int_2^\infty p_2(r')F_1(r, r')\frac{dr'}{r'}, \tag{9.170}$$

Because the coefficients p_1, p_2 are positive and $F_1(r, r')$ is mainly negative, coefficient $R_1(r)$ is less than 1, and coefficient $R_2(r)$ is more than 1 for large r. This means that for large r, the coefficients p_1, p_2 are closer to each other and to 1 than the coefficients p_{01}, p_{02}. Note that the coefficients p_{01}, p_{02} describe the interactions between two isolated inclusions in the homogeneous medium. Thus, the presence of surrounding inclusions decreases the rate of interaction between two inclusions in the composite for large r (screen effect). If the distance r is small, the influence of surrounding inclusions on the interaction between two inclusions is more complex and depends on the volume concentration of inclusions p.

The results of the solution of the integral equations (9.163), (9.164) are presented in Fig. 9.5 for $p = 0.3$ and $p = 0.6$. Figure 9.6 presents the dependences of the effective dielectric permittivity of the composites obtained from (9.167), (9.168) on the volume concentrations of inclusions p.

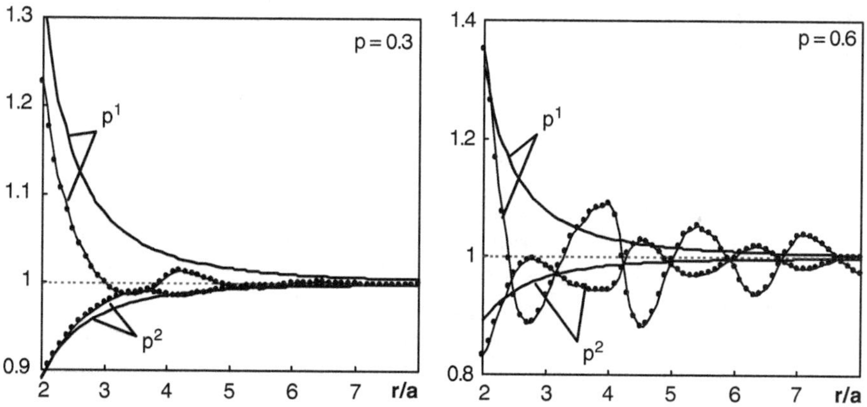

Fig. 9.5. The graphs of the functions $p_1(r), p_2(r)$ describing the interaction between two inclusions in the composite material in the 3D-case. Solid lines describe the interaction between two isolated inclusions in the homogeneous matrix; the dotted lines describe such an interaction in the presence of surrounding inclusions.

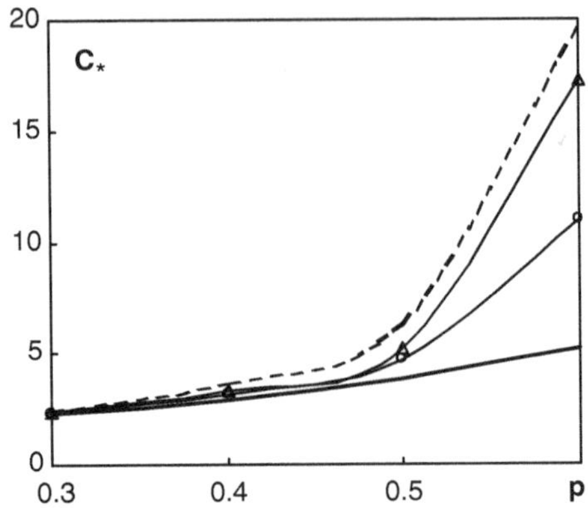

Fig. 9.6. The dependence of the effective dielectric permittivity of a two-phase composite c_* on the volume concentrations of inclusions p ($c_0 = 1, c = 100$) in the 3D-case; the solid line corresponds to the EFM based on the quasicrystalline approximation; the line with circles is the improved version of the EFM with the Kirkwood superposition approximation for the three-point correlation functions of the centers of inclusions; the line with triangles corresponds to the three-point correlation function defined in (9.157) with $a = 0.4$; the dashed line corresponds to the functions $p_1(r) = p_{01}(r), p_2(r) = p_{02}(r)$ in (9.155).

The dielectric permittivities of the matrix and inclusions are $c_0 = 1, c = 100$. The solid line corresponds to the EFM in the framework of the quasicrystalline approximation (9.152), the line with circles corresponds to (9.168) of the improved version of the method when the three-point correlation function $F(x, x')$ is taken in the form of the Kirkwood superposition approximation (9.156). The dashed line corresponds to $\Pi(x) = \Pi_0(x)$ (neglecting the integral term in (9.153)), and the line with triangles corresponds to $\Pi(x)$ obtained from (9.162)–(9.166) for $F(x, x')$ in the form (9.157) with $a = 0.4$.

9.4.4 Interaction between two inclusions in the 2D-case

Consider an infinite homogeneous 2D-medium with two identical inclusions v_1, v_2 that are disks of unit radii the centers of which are separated by a distance z. The electric potential $u(x)$ in the medium satisfies the equation $\Delta u = 0$, where Δ is Laplace operator. If the applied electric field E_0 is orthogonal to the axis connecting the centers of the inclusions $E_0 = E_0^{\perp}$, for the symmetry of the problem, the electric potential $u(x)$ in the medium may be presented in the form ($|E_0^{\perp}| = 1$)

$$u(x) = u^{(1)}(r, \varphi) = r\sin(\varphi) + \sum_{n=1}^{\infty} a_n r^n \sin(n\varphi) \quad \text{if} \quad x \in v_1, \qquad (9.171)$$

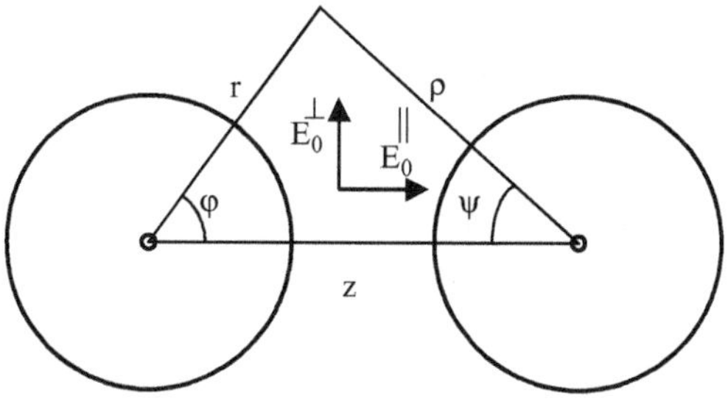

Fig. 9.7. Two isolated inclusions in a constant external field.

$$u(x) = u^-(r,\varphi,\rho,\psi) = r\sin(\varphi) + \\ + \sum_{n=1}^{\infty} b_n \left[r^{-n}\sin(n\varphi) + \rho^{-n}\sin(n\psi)\right] \quad \text{if} \quad \notin v_1 \cup v_2, \tag{9.172}$$

$$u(x) = u^{(2)}(r,\varphi,\rho,\psi) = r\sin(\varphi) + \sum_{n=1}^{\infty} a_n \rho^n \sin(n\psi) \quad \text{if} \quad x \in v_2. \tag{9.173}$$

Here r, φ are polar coordinates of the point x in the system with the origin at the center of the first inclusion and the polar axis directed along the line connecting the centers of the inclusions (see Fig. 9.7), ρ is the length of the vector $\boldsymbol{\rho}$ connecting the center of the second inclusion with the point x, ψ is the angle between vector $\boldsymbol{\rho}$ and the polar axis

$$\rho = \sqrt{r^2 + z^2 - 2rz\cos(\varphi)}, \qquad \psi = Arc\cos\left(\frac{z - r\cos(\varphi)}{\rho}\right). \tag{9.174}$$

The boundary conditions on the border of the first inclusion, namely

$$u^{(1)}(r,\varphi)|_{r=1} = u^-(r,\varphi,\rho,\psi)|_{r=1}, \qquad c\frac{\partial u^{(1)}}{\partial r}|_{r=1} = c_0\frac{\partial u^-}{\partial r}|_{r=1} \tag{9.175}$$

give us the following system of linear equations for the coefficients a_n, b_n in the series (9.171)–(9.173)

$$a_n = b_n + \sum_{m=1}^{\infty} \beta_{nm} b_m, \qquad \sum_{m=1}^{\infty} A_{nm} b_m = f_n, \tag{9.176}$$

$$A_{nm} = (c + c_0)n\delta_{nm} + cn\beta_{nm} + c_0\gamma_{nm}, \tag{9.177}$$

$$f_1 = c_0 - c, \quad f_n = 0 \quad n > 1,$$

$$\beta_{nm} = \frac{1}{\pi}\int_0^{2\pi} \rho_1^{-m}\sin(m\psi_1)\sin(n\varphi)\mathrm{d}\varphi, \tag{9.178}$$

$$\gamma_{nm} = \frac{m}{\pi} \int_0^{2\pi} \rho_1^{-m-2} \left[(1 - z\cos(\varphi))\sin(m\psi_1) - \right.$$
$$\left. -z\sin(\varphi)\cos(m\psi_1)\sin(n\varphi)\right] d\varphi, \tag{9.179}$$

where $\rho_1 = \rho(r,\varphi)|_{r=1}$, $\psi_1 = \psi(r,\varphi)|_{r=1}$.

If the external field acts along the axis connecting the centers of two inclusions ($E_0 = E_0^{||}$) the electric potential in the medium takes the form

$$u(x) = u^{(1)}(r,\varphi) = r\cos(\varphi) + \sum_{n=0}^{\infty} a_n r^n \cos(n\varphi) \quad \text{if} \quad x \in v_1 \tag{9.180}$$

$$u(x) = u^{-}(r,\varphi,\rho,\psi) = r\cos(\varphi)$$
$$+ \sum_{n=1}^{\infty} b_n \left[r^{-n}\cos(n\varphi) - \rho^{-n}\cos(n\psi)\right] \quad \text{if} \quad \notin v_1 \cup v_2 \tag{9.181}$$

$$u(x) = u^{(2)}(r,\varphi,\rho,\psi) = r\cos(\varphi) - \sum_{n=0}^{\infty} a_n \rho^n \cos(n\psi) \quad \text{if} \quad x \in v_2. \tag{9.182}$$

The system of linear equations for the constants a_n, b_n in (9.180) follows from the boundary conditions (9.175) in the form

$$a_0 = \frac{1}{2}\sum_{m=1}^{\infty} \beta_{0m} b_m; \quad a_n = b_n + \sum_{m=1}^{\infty} \beta_{nm} b_m, \quad n > 0; \tag{9.183}$$

$$\sum_{m=1}^{\infty} A_{nm} b_m = f_n, \tag{9.184}$$

$$A_{nm} = (c + c_0) n \delta_{nm} + cn\beta_{nm} + c_0\gamma_{nm}, \tag{9.185}$$
$$f_1 = c_0 - c, \quad f_n = 0 \quad n > 1,$$

$$\beta_{nm} = -\frac{1}{\pi}\int_0^{2\pi} \rho_1^{-m} \cos(m\psi_1)\cos(n\varphi) d\varphi, \tag{9.186}$$

$$\gamma_{nm} = -\frac{m}{\pi} \int_0^{2\pi} \rho_1^{-m-2} \left[(1 - z\cos(\varphi))\cos(m\psi_1) + \right.$$
$$\left. +z\sin(\varphi)\cos(m\psi_1)\cos(n\varphi)\right] d\varphi. \tag{9.187}$$

The mean electric field inside any inclusion takes the form

$$\langle E(x)\rangle_v = \frac{1}{\pi}\int_{v_1} E^{(1)}(x) dx = \frac{1}{\pi}\int_{v_1} \nabla u^{(1)}(x) dx = (1 + a_1) E_0. \tag{9.188}$$

Here ∇ is the vector gradient operator, $\nabla_i = \frac{\partial}{\partial x_i}$. Ten terms in the series (9.171), (9.180) allow us to obtain the coefficient a_1 with an error less than 1%.

9.4.5 Dielectric properties of the composites in 2D-case

Consider the 2D-case when inclusions are circular disks of unit radii. In this case, the function $K(x)$ and the tensors A and P have the forms

$$K_{ij}(x) = \frac{1}{2\pi c_0 r^2}\left(\delta_{ij} - 2n_i n_j\right), \tag{9.189}$$

$$A_{ij} = \frac{1}{2c_0}\delta_{ij}, \quad P_{ij} = t\delta_{ij}, \quad t = \frac{2c_0 c_1}{2c_0 + c_1}. \tag{9.190}$$

The effective permittivity tensor of the composite in the framework of the quasicrystalline approximation (9.126) is

$$C_* = c_* I, \qquad c_* = c_0 + \frac{2pc_0 c_1}{2c_0 + (1-p)c_1}. \tag{9.191}$$

The tensor $\Pi_0(x)$ that describes interaction between two isolated inclusions in the homogeneous matrix takes the form

$$\Pi_{0ij}(x) = p_{01}(r) n_i n_j + p_{02}(r)(\delta_{ij} - n_i n_j), \quad r = |x|, \quad n = \frac{x}{r}, \tag{9.192}$$

$$p_{01}(r) = \left[1 - \frac{c_1}{r^2(2c_0 + c_1)}\right]^{-1}, \quad p_{02}(r) = \left[1 + \frac{c_1}{r^2(2c_0 + c_1)}\right]^{-1}. \tag{9.193}$$

Note that this result was obtained on the assumption that the field acting on every inclusion in the composite is constant (hypothesis H1 of the EFM). Let us compare this result with the exact solution of the problem for two isolated inclusions in a homogeneous matrix. The solution of this problem in the form of Fourier series is presented in the previous section. If the external field E_0 acts along the axis connecting the centers of the inclusions ($E_0 = E_0^{||}$), the mean electric field inside every inclusion in the frame of hypothesis H1 of the EFM takes the form

$$\frac{1}{v}\int_v E(x)\mathrm{d}x = \langle E(x)\rangle_v = \frac{t}{c_1} p_{01}(r) E_0^{||}. \tag{9.194}$$

If the incident field E_0 acts in the direction orthogonal to the axis of the inclusions ($E_0 = E_0^{\perp}$), the mean electric field inside the inclusions takes the form

$$\langle E(x)\rangle_v = \frac{t}{c_1} p_{02}(r) E_0^{\perp}. \tag{9.195}$$

The exact values for the mean electric fields inside the inclusions in these two cases may be written in the forms

$$\langle E(x)\rangle_v = \frac{t}{c_1} p_{e1}(r) E_0^{||}, \qquad \langle E(x)\rangle_v = \frac{t}{c_1} p_{e2}(r) E_0^{\perp}. \tag{9.196}$$

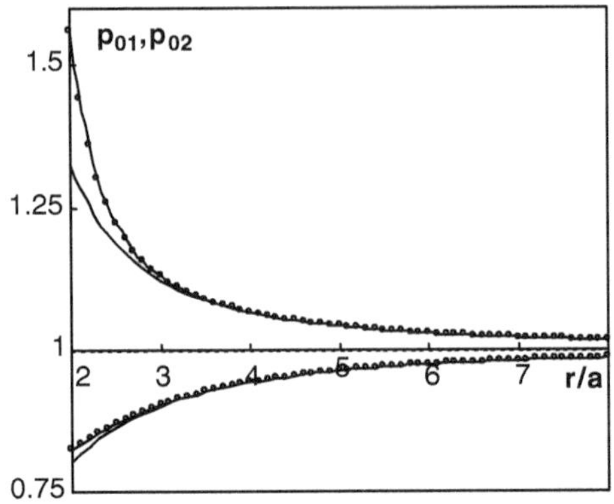

Fig. 9.8. The graphs of the functions $p_{01}(r), p_{02}(r)$ describing the interaction between two inclusions in a homogeneous matrix in the 2D-case. Solid lines correspond to the approximate solutions in (9.192), (9.193), dotted lines are the exact solutions $p_{e1}(r), p_{e2}(r)$.

The graphs of the functions $p_{01}(r), p_{e1}(r)$ and $p_{02}(r), p_{e2}(r)$ are presented in Fig. 9.8. It is seen from these graphs that solution (9.193) is a good approximation to the exact solution of the problem if the mean values of electric fields inside inclusions are evaluated.

The function $\Pi(x)$ that describes the interaction between two inclusions in the composite medium has the form (9.192), and the coefficients $p_1(r), p_2(r)$ in this equation are to be found from the following system of integral equations:

$$p_1(r) = p_{01}(r)R_1(r), \qquad p_2(r) = p_{02}(r)R_2(r), \tag{9.197}$$

$$R_1(r) = 1 - p\frac{t}{c_0}\int_2^\infty [(p_2(r') - p_1(r'))F_1(r,r') - \\ -(p_2(r') + p_1(r'))F_2(r,r')]\frac{\mathrm{d}r'}{r'}, \tag{9.198}$$

$$R_2(r) = 1 - p\frac{t}{c_0}\int_2^\infty [(p_2(r') - p_1(r'))F_1(r,r') + \\ +(p_2(r') + p_1(r'))F_2(r,r')]\frac{\mathrm{d}r'}{r'}, \tag{9.199}$$

$$F_1(r,r') = \int_0^\pi [F(r,r',\varphi) - g(r')]\,\mathrm{d}\varphi, \tag{9.200}$$

$$F_2(r,r') = \int_0^\pi [F(r,r',\varphi) - g(r')]\cos(2\varphi)\mathrm{d}\varphi. \tag{9.201}$$

Here (r', φ) are the polar coordinates of point x' in the system with the origin at the center of the first fixed inclusion ($x' = 0$) and the polar axis directed along the vector x of the centers of the second fixed inclusion, $r = |x|$.

The effective permittivity tensor C_* defined in (4.17) in 2D-case takes the form

$$C_* = c_* I, \qquad c_* = c_0 + \frac{2c_0 c_1}{(2c_0 + c_1)(1 + pk_0) - pc_1}, \tag{9.202}$$

where k_0 is the following integral:

$$k_0 = \frac{t}{2c_0} \int_2^{\infty} [p_2(r) - p_1(r)]\, g(r) \frac{\mathrm{d}r}{r}. \tag{9.203}$$

The dependencies of the effective permittivity c_* on the volume concentrations of inclusions p are presented in Fig. 9.9. The solid line corresponds to the quasicrystalline approximation (9.191). The line with circles corresponds to (9.202) with k_0 obtained from the solution of integral Equations (9.197)–(9.201) and the Kirkwood superposition approximation (9.156) for the function $F(x, x')$. The dashed line is c_* for $\Pi(x) = \Pi_0(x)$, where $\Pi_0(x)$ is defined in (9.192), (9.193). The line with triangles corresponds to the solution of integral Equations (9.197)–(9.201) with $F(x, x')$ in (9.157)–(9.160) and $a = 0.4$. The function $g(x)$ in the 2D-case is taken as the solution of the Perckus–Yevick equation [65].

Vertical bars in Fig. 9.9 are the results of the numerical calculations of the effective dielectric permittivities of the composite in the 2D-case. The

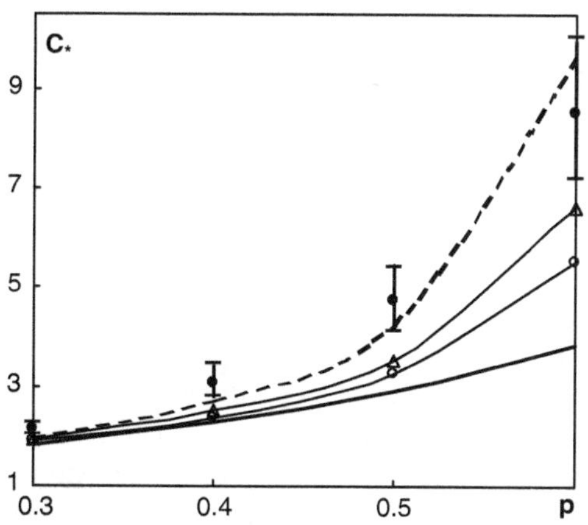

Fig. 9.9. The dependence of the effective dielectric permittivity c_* of a two-phase composite on the volume concentrations of inclusions p ($c_0 = 1, c = 100$) in the 2D-case; the solid line corresponds to the EFM based on the quasicrystalline approximation; the line with light dots is the improved version of the EFM with the Kirkwood superposition approximation for the three-point correlation function of the centers of inclusions; the line with triangles corresponds to the three-point correlation function defined in (9.157) with $a = 0.4$; the dashed line corresponds to the functions $p_1(r) = p_{01}(r), p_2(r) = p_{02}(r)$ in (9.155). The direct bars are the dispersions of the numerical calculations of c_* with the mean values at black dots.

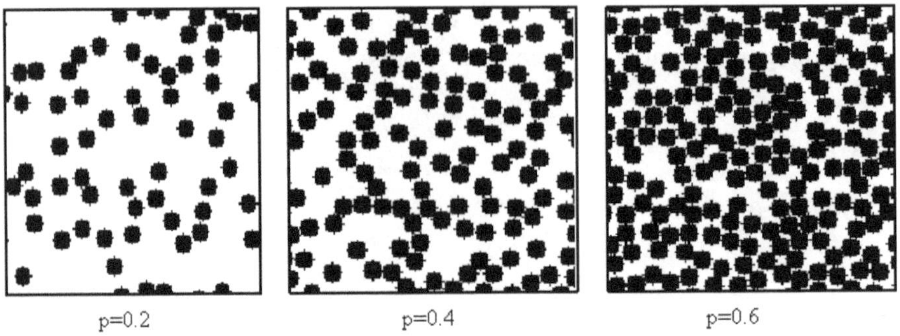

Fig. 9.10. Examples of the realizations of random fields of inclusions in the 2D-case used in the numerical calculations.

method of the calculation is based on the fast Fourier transform algorithm described in [152], [49], [39]. For the construction of the numerical solutions, first a realization of a random set of nonoverlapping disks was generated on a square area. Then, the integral equation (9.2) for the electric field was solved for the obtained realization of inclusions using the mentioned algorithm. Periodic boundary conditions were used in the calculations. The effective dielectric permittivities were obtained by averaging the detailed electric fields and electric displacement fields over the considered area. For every volume concentration of inclusions $p = 0.1, 0.2, ..., 0.6$, ten different realizations of the nonoverlapping inclusions were considered. Typical realizations of random sets of inclusions used in the calculations for $p = 0.2, 0.4, 0.6$ are presented in Fig. 9.9. The lengths of the vertical bars in Fig. 9.10 show the dispersion of the numerical results. The mean values of the dielectric permittivities are black points in this figure.

9.4.6 Discussion and conclusion

The comparison of the prediction of the EFM with the numerical calculation of the effective dielectric parameters in 2D-case (Fig. 9.9) shows that replacing the hypothesis of the quasicrystalline approximation (9.113) with the hypothesis (9.128) improves the predictions of the EFM. This improvement is the result of accounting for pair interactions between inclusions in the composite medium. The use of the three-point correlation function (9.157) with parameter $a = 0.4$ gives the prediction of the method (line with triangles) closer to the numerical data than the use of the Kirkwood superposition approximation (9.156) (line with circles). It is interesting to note that the neglect of the integral term in (9.155) for the two-point conditional mean of the effective field leads to a better coincidence between the predictions of the EFM and the numerical results (dashed line in Fig. 9.9). If the integral term in (9.155) is neglected, the interaction between two inclusions in the composite material coincides with the interaction between two inclusions in the homogeneous matrix. According to the calculations in Sections 9.4.3, 9.4.4 the presence of surrounding inclusions weakens such an interaction and as a result, decreases the values of the predicted effective dielectric constants.

It should be noted that the EFM and its modifications are based on some hypotheses that simplify the solutions of the multiparticle problem. On the other hand, these hypotheses are the main sources of errors in the method. Hypothesis H_1 (Section 9.4.1) simplifies the structure of the local external fields that act on any inclusion in the composite medium. Of course, these fields are not constant as hypothesis H_1 prescribes. But this hypothesis is quite acceptable if the mean polarization fields inside inclusions are the aim of the calculations. Even for two isolated inclusions the approximation of constant local external fields that act on each inclusion is in agreement with the exact solution of the problem (Section 9.4.4). The presence of surrounding inclusions decreases the rate of interactions between two inclusions and smooths the asymmetry of the local effective fields acting on every inclusion. Thus, there are many reasons for using hypothesis H_1.

The hypothesis of the quasicrystalline approximation (9.113) or more complex hypotheses similar to (9.128)

$$E_*^{(n+1)}(x_1, x_2, ..., x_n, x_{n+1}) = E_*^{(n)}(x_1, x_2, ..., x_n), \tag{9.204}$$

where

$$E_*^{(n)}(x_1, x_2, ..., x_n) = \langle E_*(x_1)|x_1, x_2, ..., x_n\rangle\,, \tag{9.205}$$

close the chain of integral equations for multipoint conditional means of the effective field at the n-th equation. Every equation of this chain connects n-point conditional mean of the effective field with $(n+1)$-point similar mean, and has the form

$$E_*^{(n)}(x_1, x_2, ..., x_n) = E_0 - \tag{9.206}$$
$$-\int \bar{K}(x_1 - x')TE_*^{(n+1)}(x', x_1, x_2, ..., x_n)\,\langle X(x_1; x')|x_1, x_2, ..., x_n\rangle\, \mathrm{d}x'.$$

Thus, hypothesis (9.204) gives a closed equation for the n-point conditional mean of the effective field. By closing the chain of the equation for the conditional mean $E_*^{(n)}$ at the first, second, etc., steps one can successively improve the prediction of the effective parameters of the composite materials. The main difficulties of carrying out this procedure are connected with the construction of the $(n+1)$-point correlation function $\langle X(x_1; x')|x_1, x_2, ..., x_n\rangle$ in (9.206), and with the subsequent solution of this equation.

In a number of works, the Monte-Carlo procedure was used for the construction of two and three-point correlation functions of centers of inclusions [3], [33], [18]. The computational cost of this work is rather high, and this cost increases with the order of the correlation functions. The influence of the form of the correlation functions on the effective properties of a composite is essential, as shown in this section. Therefore, simple approximations of multi-point correlation functions similar to the Kirkwood superposition approximation may give significant errors in the calculation of the effective parameters of composites. Thus, the precision of the method depends on the precision of the geometrical information about the microstructure of composites in the form of the multipoint correlation functions.

9.5 Cross-properties relations

Consider a volume V of a two-phase micro inhomogeneous medium, where V_1 and V_2 ($V_1+V_2=V$) are the subregions occupied by the materials of different phases. The solution of the homogenization problems for such a medium gives us various effective properties of the composite. Behind these problems, there is a hidden suggestion that the effective properties, no matter how different in context and interpretation, should be interconnected. Such interconnections are called cross-property relations.

The simplest cross-property relations are due to a mathematical analogy between the equations that govern different phenomena: dielectric and magnetic permittivity, electric conductivity, heat transfer, etc. In these cases, it is sufficient to obtain results for one of these properties, and for the rest, only the interpretations are changed. In some cases however, the mathematical analogy it is not so obvious, and some efforts are needed to clarify it. An example is provided by the heat transfer in a fiber reinforced medium in the direction transverse to the fiber axes, and the shear deformation along the fibers (see Equations (7.433) and (9.72).

Many nontrivial cross-property relations are those where there is no direct mathematical analogy, and the relations result from the underlying internal structure of the medium. Here we present one of the relations of this kind. Consider an inhomogeneous material that consists of the main medium (matrix) and a random set of inclusions with different physical properties. The representative volume V_0 of this material is subjected to surface forces and a constant temperature field T. The constitutive relations for such a medium with incorporated thermal effects can be written in the form

$$\varepsilon_{\alpha\beta}(x) = B_{\alpha\beta\lambda\mu}(x)\sigma_{\lambda\mu}(x) + \alpha_{\alpha\beta}(x)T, \tag{9.207}$$

$$s(x) = \alpha_{\alpha\beta}(x)\sigma_{\alpha\beta}(x) + c_p(x)T. \tag{9.208}$$

Here $B(x)$ is the tensor of elastic compliances, $\alpha(x)$ is the tensor of thermal expansion coefficients, $c_p(x)$ is the heat capacity at constant stresses, $\varepsilon(x)$ and $\sigma(x)$ are the strain and stress tensors, $s(x)$ is the entropy density. $B(x)$, $\alpha(x)$ and $c_p(x)$ are random functions that are B^0, α^0, c_p^0 when a point x is in the matrix and B, α, c_p if x is in an inclusion. Let external forces be applied to the border S of the volume V_0

$$F^0_\alpha(x) = \sigma^0_{\alpha\beta} n_\beta(x). \tag{9.209}$$

Here $\sigma^0_{\alpha\beta}$ is a constant tensor, and $n_\beta(x)$ is the external normal to S. In this case, the mean stress field in V coincides with $\sigma^0_{\alpha\beta}$, as follows from the equation

$$\langle \sigma_{\alpha\beta}(x) \rangle = \frac{1}{V_0} \int\limits_{V_0} \sigma_{\alpha\beta}(x) \mathrm{d}x = \frac{1}{2V_0} \int\limits_S \left(F^0_\alpha x_\beta + F^0_\beta x_\alpha \right) \mathrm{d}S = \sigma^0_{\alpha\beta}. \tag{9.210}$$

Here Gauss' theorem is used.

The functions $B(x), \alpha(x)$ are presented as the following sums:

$$B_{\alpha\beta\lambda\mu}(x) = B^0_{\alpha\beta\lambda\mu} + B^1_{\alpha\beta\lambda\mu} V(x), \quad \alpha_{\alpha\beta}(x) = \alpha^0_{\alpha\beta} + \alpha^1_{\alpha\beta} V(x), \tag{9.211}$$

$$c_p(x) = c^0 + c^1 V(x),$$

$$B^1_{\alpha\beta\lambda\mu} = B_{\alpha\beta\lambda\mu} - B^0_{\alpha\beta\lambda\mu}, \quad \alpha^1_{\alpha\beta} = \alpha_{\alpha\beta} - \alpha^0_{\alpha\beta}, \quad c^1 = c - c^0, \tag{9.212}$$

where $V(x)$ is the characteristic function of the region occupied by the inclusions. After averaging (9.207) and (9.208) over the volume V_0 we obtain

$$\langle \varepsilon_{\alpha\beta} \rangle = B^0_{\alpha\beta\lambda\mu} \langle \sigma_{\lambda\mu} \rangle + p B^1_{\alpha\beta\lambda\mu} \overline{\sigma}_{\lambda\mu} + \langle \alpha_{\alpha\beta} \rangle T, \tag{9.213}$$

$$\langle s \rangle = \alpha^0_{\alpha\beta} \langle \sigma_{\alpha\beta} \rangle + p \alpha^1_{\alpha\beta} \overline{\sigma}_{\alpha\beta} + \langle c_p \rangle T, \tag{9.214}$$

where p is the volume concentration of the inclusion, $\overline{\sigma}_{\alpha\beta}$ is the stress field averaged over the region occupied by the inclusions. Because of the linearity of the problem, the tensor $\overline{\sigma}_{\alpha\beta}$ is a linear function of the mean stress field $\sigma^0_{\alpha\beta} = \langle \sigma_{\alpha\beta} \rangle$ and of the temperature T

$$\overline{\sigma}_{\alpha\beta} = \Lambda_{\alpha\beta\lambda\mu} \langle \sigma_{\lambda\mu} \rangle + \lambda_{\alpha\beta} T, \tag{9.215}$$

where Λ and λ are unknown tensors. Substituting this equation into (9.213) and (9.214) we obtain

$$\langle \varepsilon_{\alpha\beta} \rangle = B^*_{\alpha\beta\lambda\mu} \langle \sigma_{\lambda\mu} \rangle + \alpha^*_{\alpha\beta} T, \tag{9.216}$$

$$\langle s \rangle = \alpha^*_{\alpha\beta} \langle \sigma_{\alpha\beta} \rangle + c^*_p T. \tag{9.217}$$

Here $B^*_{\alpha\beta\lambda\mu}, \alpha^*_{\alpha\beta}$ and c^*_p are the effective thermoelastic characteristics of the inhomogeneous medium. These tensors are defined by the equations

$$\begin{aligned} B^*_{\alpha\beta\lambda\mu} &= B^0_{\alpha\beta\lambda\mu} + p B^1_{\alpha\beta\rho\tau} \Lambda_{\rho\tau\lambda\mu} = \\ &= \langle B_{\alpha\beta\lambda\mu} \rangle + p B^1_{\alpha\beta\rho\tau} \left(\Lambda_{\rho\tau\lambda\mu} - I_{\rho\tau\lambda\mu} \right), \end{aligned} \tag{9.218}$$

$$\begin{aligned} \alpha^*_{\alpha\beta} &= \langle \alpha_{\alpha\beta} \rangle + p B^1_{\alpha\beta\lambda\mu} \lambda_{\lambda\mu} = \\ &= \langle \alpha_{\alpha\beta} \rangle + p \alpha^1_{\lambda\mu} \left(\Lambda_{\lambda\mu\alpha\beta} - I_{\lambda\mu\alpha\beta} \right), \end{aligned} \tag{9.219}$$

$$c^*_p = \langle c_p \rangle + p \alpha^1_{\lambda\mu} \lambda_{\lambda\mu} \tag{9.220}$$

It follows from these equations that the effective constants $\alpha^*_{\alpha\beta}$ and c^*_p are not independent, but expressed via the components of the tensor $B^*_{\alpha\beta\lambda\mu}$. Indeed, from (9.218) and (9.219) we have

$$\alpha^*_{\alpha\beta} - \langle\alpha_{\alpha\beta}\rangle = \alpha^1_{\lambda\mu}\left(B^1_{\lambda\mu\rho\tau}\right)^{-1}\left(B^*_{\rho\tau\alpha\beta} - \langle B_{\rho\tau\alpha\beta}\rangle\right) \tag{9.221}$$

and (9.219) and (9.220) give

$$c^*_p - \langle c_p\rangle = \alpha^1_{\alpha\beta}\left(B^1_{\alpha\beta\lambda\mu}\right)^{-1}\left(\alpha^*_{\lambda\mu} - \langle\alpha_{\lambda\mu}\rangle\right) \tag{9.222}$$

Suppose that both components in the inhomogeneous medium are isotropic with bulk elastic moduli K_0 and K and thermal expansion coefficients α_0, α. Then, Equations (9.221), (9.222) are reduced to the following ones:

$$\alpha_* - \langle\alpha\rangle = \frac{\alpha - \alpha_0}{K^{-1} - K_0^{-1}}\left(\frac{1}{K_*} - \left\langle\frac{1}{K}\right\rangle\right), \tag{9.223}$$

$$c^*_p - \langle c_p\rangle = \frac{9(\alpha - \alpha_0)}{K^{-1} - K_0^{-1}}\left(\alpha_* - \langle\alpha\rangle\right). \tag{9.224}$$

Thus, if the effective bulk modulus of the composite K_* is known, one can calculate its effective coefficient of thermoexpansion α_* from (9.223), and *vice versa*, if the coefficient α_* is known, the bulk modulus may be calculated from the same equation. This is an exact result that does not depend on the microstructure of the composite.

Equation (9.223) was firstly derived in [114]. Almost at the same time, this equation was obtained and generalized in [186] and [178]. Equation (9.224) was obtained in [178]. The background of various cross-property relations was discussed and implemented in [56] and [147]. The approximate theory of cross-property relations was developed in [69], [183], [184].

9.6 Notes

There are many of publications devoted to the solution of the homogenization problems for a field with a scalar potential in a matrix composite. An excellent survey is presented in [138]. In Section 9.2, the results of [213] were used. Conductive properties of composite materials with random sets of thin inclusions (Section 9.3) were considered in [124]. The content of Section 9.4 is based on the work [86].

Self-consistent methods may be used for the evaluation of coupled physical fields in composites. Piezoelectric composites are important examples of materials where coupled fields are to be considered. The constitutive equations of such materials contain, in addition to stress and strain tensors, the vectors of the electric field and electric displacement. In a number of works

([209], [12], [40], [103],[127], [146]), the one-particle problem for an ellipsoidal inclusion in a piezoelectric materials was solved explicitly for constant external fields. These solutions were used in the framework of self-consistent schemes for the calculation of overall electroelastic properties of piezoelectric heterogeneous materials containing random sets of ellipsoidal inclusions. In [27], piezoelectric materials reinforced by piezoelectric fibers were considered, and their effective electroelastic characteristics were calculated by the Mori–Tanaka method. The EMM was used in [210], [42], [43] for piezoelectric materials containing ellipsoidal inclusions. In [118], [129], the EFM was applied to the calculation of the effective properties of composites with an isotropic matrix and spheroidal piezoelectric inclusions. Piezoelectric materials are called piroactive if temperature affects their electric and elastic properties. The effective electro-thermo-elastic characteristics of such materials reinforced by piroactive continuous fibers were obtained in [122], [123] by the effective field method. The same method was used in [128], [119] for piezoelectric polycrystals.

Other coupled problems that have important applications in geomechanics are the problems of poroelasticity. In these problems, the medium contains an interconnected set of pores filled by liquid. This theory was developed in the middle of the last century in [19], [20] and was widely applied to modeling the mechanical properties of porous rocks filled with fluids. These materials may be modelled by a homogeneous matrix containing regions with different poroelastic properties. If these regions are ellipsoids, the solution of the problem for an isolated ellipsoid in a constant external field can be constructed analytically. This solution is the basis for the application of self-consistent methods to this case. The effective medium theories for the multicomponent poroelastic composites were developed in [15]. The effective poroelastic characteristics of such materials were calculated by the effective field method in [120]. A series of cross-property relations (in addition to those considered in Section 9.5) for inhomogeneous poroelastic materials were discovered in [17], [158], [120], [121]. Note, that the mathematical analogy of poroelastic and coupled thermoelastic theories allows one to interpret the results obtained for poroelastic inhomogeneous materials in terms of the theory of thermoelasticity (see, e.g., [158]). Poro-thermo-elastic composites were considered in [121].

A. Special tensor bases of four rank tensors

Rank four tensors with special symmetry are used for the description of elastic properties of solids. For presentation of these tensors and operations with them, it is convenient to introduce special bases. In this Appendix, we present some such bases and give explicit equations for the products and inversions of the tensors that belong to the linear shells of these bases.

A.1 E-basis

Let us introduce six rank four tensors constructed from a unit vector n and the rank two unit tensor $\delta_{\alpha\beta}$.

$$E^1_{\alpha\beta\lambda\mu} = \delta_{\alpha(\lambda}\delta_{\mu)\beta}\,, \quad E^2_{\alpha\beta\lambda\mu} = \delta_{\alpha\beta}\delta_{\lambda\mu}\,, \quad E^3_{\alpha\beta\lambda\mu} = \delta_{\alpha\beta}n_\lambda n_\mu\,, \tag{A.1}$$

$$E^4_{\alpha\beta\lambda\mu} = n_\alpha n_\beta \delta_{\lambda\mu}\,, \quad E^5_{\alpha\beta\lambda\mu} = n_{\alpha)}n_{(\mu}\delta_{\lambda)(\beta}\,, \quad E^6_{\alpha\beta\lambda\mu} = n_\alpha n_\beta n_\lambda n_\mu\,. \tag{A.2}$$

These tensors are symmetric with respect to the first and second pairs of indices, and form a closed algebra with respect to multiplication defined by the equation (convolution over two indices)

$$\left(E^i E^j\right)_{\alpha\beta\lambda\mu} = E^i_{\alpha\beta\rho\tau}E^j_{\rho\tau\lambda\mu}\,. \tag{A.3}$$

The multiplication table of tensors E^i has the form

(A.4)

	E^1	E^2	E^3	E^4	E^5	E^6
E^1	E^1	E^2	E^3	E^4	E^5	E^6
E^2	E^2	$3E^2$	$3E^3$	E^2	E^3	E^3
E^3	E^3	E^2	E^3	E^2	E^3	E^3
E^4	E^4	$3E^4$	$3E^6$	E^4	E^6	E^6
E^5	E^5	E^4	E^6	E^4	$(E^5+E^6)/2$	E^6
E^6	E^6	E^4	E^6	E^4	E^6	E^6

Elastic properties of isotropic materials are described by isotropic tensors symmetric with respect to the first and second pair of indices, as well as over transposition of pairs of indices. The tensors E^1 and E^2 may be used as the basis of such a set of tensors. To construct an inverse tensor with respect to a tensor of this tensor space, it is convenient to introduce the basis that consists of two orthogonal tensors E^2 and $E^1 - \frac{1}{3}E^2$:

$$E^2\left(E^1 - \frac{1}{3}E^2\right) = \left(E^1 - \frac{1}{3}E^2\right)E^2 = 0\,. \tag{A.5}$$

$$E^2E^2 = 3E^2, \quad \left(E^1 - \frac{1}{3}E^2\right)\left(E^1 - \frac{1}{3}E^2\right) = \left(E^1 - \frac{1}{3}E^2\right). \tag{A.6}$$

For an arbitrary tensor A of this subspace, we have the presentation

$$A = a_1E^2 + a_2\left(E^1 - \frac{1}{3}E^2\right), \tag{A.7}$$

where a_1, a_2 are scalar coefficients. From (A.5), (A.6) it is easy to obtain the equation for the inverse tensor $A^{-1}(AA^{-1} = A^{-1}A = E^1)$

$$A^{-1} = \frac{1}{9a_1}E^2 + \frac{1}{a_2}\left(E^1 - \frac{1}{3}E^2\right). \tag{A.8}$$

A.2 *P*-basis

This basis is constructed as (A.1)-(A.2), but projector $\theta_{\alpha\beta}$ on the plane orthogonal to a unit vector m ($|m| = 1$)

$$\theta_{\alpha\beta} = \delta_{\alpha\beta} - m_\alpha m_\beta\,. \tag{A.9}$$

plays the role of rank two unit tensor $\delta_{\alpha\beta}$.

The P-basis consists of the following six four rank four tensors

$$P^1_{\alpha\beta\lambda\mu} = \theta_{\alpha(\lambda}\theta_{\mu)\beta}\,, \quad P^2_{\alpha\beta\lambda\mu} = \theta_{\alpha\beta}\theta_{\lambda\mu}\,, \tag{A.10}$$

$$P^3_{\alpha\beta\lambda\mu} = \theta_{\alpha\beta}m_\lambda m_\mu\,, \quad P^4_{\alpha\beta\lambda\mu} = m_\alpha m_\beta\theta_{\lambda\mu}\,, \tag{A.11}$$

$$P^5_{\alpha\beta\lambda\mu} = m_{\alpha)}m_{(\lambda}\theta_{\mu)(\beta}\,, \quad P^6_{\alpha\beta\lambda\mu} = m_\alpha m_\beta m_\lambda m_\mu\,. \tag{A.12}$$

The elements of the P and E-bases (A.1)-(A.2) are connected by the equations

$$P^j = \alpha^{ji}E^i, \quad E^j = \left(\alpha^{-1}\right)^{ji}P^i, \tag{A.13}$$

where the matrices α and α^{-1} have the following forms:

$$\alpha^{ij} = \begin{bmatrix} 1 & 0 & 0 & 0 & -2 & 1 \\ 0 & 1 & -1 & -1 & 0 & 1 \\ 0 & 0 & 1 & 0 & 0 & -1 \\ 0 & 0 & 0 & 1 & 0 & -1 \\ 0 & 0 & 0 & 0 & 1 & -1 \\ 0 & 0 & 0 & 0 & 0 & 1 \end{bmatrix}, \quad \left(\alpha^{-1}\right)^{ij} = \begin{bmatrix} 1 & 0 & 0 & 0 & 2 & 1 \\ 0 & 1 & 1 & 1 & 0 & 1 \\ 0 & 0 & 1 & 0 & 0 & 1 \\ 0 & 0 & 0 & 1 & 0 & 1 \\ 0 & 0 & 0 & 0 & 1 & 1 \\ 0 & 0 & 0 & 0 & 0 & 1 \end{bmatrix}. \tag{A.14}$$

The six tensors P^i also form a closed algebra with respect to multiplication (A.3). The multiplication table for these tensors has the form

	P^1	P^2	P^3	P^4	P^5	P^6
P^1	P^1	P^2	P^3	0	0	0
P^2	P^2	$2P^2$	$2P^3$	0	0	0
P^3	0	0	0	P^2	0	P^3
P^4	P^4	$2P^4$	$2P^6$	0	0	0
P^5	0	0	0	0	$P^5/2$	0
P^6	0	0	0	P^4	0	P^6

. (A.15)

Note that the tensor $P^1 - \frac{1}{2}P^2$ is orthogonal to all the tensors of the P-basis except P^1, i.e.,

$$P^1\left(P^1 - \frac{1}{2}P^2\right) = \left(P^1 - \frac{1}{2}P^2\right)P^1 = \left(P^1 - \frac{1}{2}P^2\right), \tag{A.16}$$

$$P^i\left(P^1 - \frac{1}{2}P^2\right) = \left(P^1 - \frac{1}{2}P^2\right)P^i = 0\,, \quad i \neq 1\,, \tag{A.17}$$

$$\left(P^1 - \frac{1}{2}P^2\right)\left(P^1 - \frac{1}{2}P^2\right) = \left(P^1 - \frac{1}{2}P^2\right). \tag{A.18}$$

The linear shells of the E- and P-bases coincide.

Let the tensor A in the P-basis is presented in the form

$$A = a_1 P^2 + a_2\left(P^1 - \frac{1}{2}P^2\right) + a_3 P^3 + a_4 P^4 + a_5 P^5 + a_6 P^6, \tag{A.19}$$

where a_i $(i = 1, 2, ..., 6)$ are scalar coefficients. Then, the inverse tensor A^{-1} $(AA^{-1} = A^{-1}A = E^1)$ takes the following form:

$$\begin{aligned} A^{-1} =& \frac{a_6}{2\Delta}P^2 + \frac{1}{a_2}\left(P^1 - \frac{1}{2}P^2\right) - \\ & - \frac{a_3}{\Delta}P^3 - \frac{a_4}{\Delta}P^4 + \frac{4}{a_5}P^5 + \frac{2a_1}{\Delta}P^6, \end{aligned} \tag{A.20}$$

$$\Delta = 2\left(a_1 a_6 - a_3 a_4\right). \tag{A.21}$$

If tensors A and B are presented in the P-basis in the form similar to (A.19) with the coefficients a_i and b_i, the product AB of these tensors is defined by an equation that follows from (A.15) and (A.18)

$$\begin{aligned} AB =& \left(2a_1 b_1 + a_3 b_4\right)P^2 + a_2 b_2\left(P^1 - \frac{1}{2}P^2\right) + \left(2a_1 b_3 + a_3 b_6\right)P^3 + \\ & + \left(2a_4 b_1 + a_6 b_4\right)P^4 + \frac{1}{2}a_5 b_5 P^5 + \left(a_6 b_6 + 2a_4 b_3\right)P^6. \end{aligned} \tag{A.22}$$

A.3 θ-basis

Let $\theta_{\alpha\beta} = \delta_{\alpha\beta} - m_\alpha m_\beta$ be the projector on the plane orthogonal to a unit vector m (A.9), and n be a unit vector in this plane ($m \cdot n = 0$). The elements of the θ-basis are defined by the equations

$$\theta^1_{\alpha\beta\lambda\mu} = \theta_{\alpha(\lambda}\theta_{\mu)\beta}\,, \quad \theta^2_{\alpha\beta\lambda\mu} = \theta_{\alpha\beta}\theta_{\lambda\mu}\,, \quad \theta^3_{\alpha\beta\lambda\mu} = \theta_{\alpha\beta}n_\lambda n_\mu\,, \tag{A.23}$$

$$\theta^4_{\alpha\beta\lambda\mu} = n_\alpha n_\beta \theta_{\lambda\mu}\,, \quad \theta^5_{\alpha\beta\lambda\mu} = \theta_{\alpha)(\lambda}n_{\mu)}n_{(\beta}\,, \quad \theta^6_{\alpha\beta\lambda\mu} = n_\alpha n_\beta n_\lambda n_\mu\,. \tag{A.24}$$

The multiplication table of these tensors has the form

	θ^1	θ^2	θ^3	θ^4	θ^5	θ^6
θ^1	θ^1	θ^2	θ^3	θ^4	θ^5	θ^6
θ^2	θ^2	$2\theta^1$	$2\theta^3$	θ^2	θ^3	θ^3
θ^3	θ^3	θ^2	θ^3	θ^2	θ^3	θ^3
θ^4	θ^4	$2\theta^4$	$2\theta^6$	θ^4	θ^6	θ^6
θ^5	θ^5	θ^4	θ^6	θ^4	$(\theta^5+\theta^6)/2$	θ^6
θ^6	θ^6	θ^4	θ^6	θ^4	θ^6	θ^6

. (A.25)

It is seen from this table that these tensors compose a closed algebra with respect to multiplication (A.3).

A.4 R-basis

This basis consists of the following five tensors:

$$R^1_{\alpha\beta\lambda\mu} = n_\alpha n_\beta n_\lambda n_\mu\,, \quad R^2_{\alpha\beta\lambda\mu} = m_\alpha m_\beta m_\lambda m_\mu\,, \tag{A.26}$$

$$R^3_{\alpha\beta\lambda\mu} = m_\alpha m_\beta n_\lambda n_\mu\,, \quad R^4_{\alpha\beta\lambda\mu} = n_\alpha n_\beta m_\lambda m_\mu\,, \tag{A.27}$$

$$R^5_{\alpha\beta\lambda\mu} = n_{\alpha)}n_{(\lambda}m_{\mu)}m_{(\beta}\,, \tag{A.28}$$

where n and m are two orthogonal unit vectors. Products of the tensors of this basis are presented in the following table:

	R^1	R^2	R^3	R^4	R^5
R^1	R^1	0	0	R^2	0
R^2	0	R^2	R^3	0	0
R^3	R^3	0	0	R^2	0
R^4	0	R^4	R^1	0	0
R^5	0	0	0	0	$R^5/2$

(A.29)

A.5 Averaging the elements of the E-, P-, θ-, and R-bases

For the solution of the homogenization problem, one has to average the elements of bases over a unit sphere in 3D-space, or over a circle in 2D-space. We present the values of the corresponding means.

E-basis.

$$\left\langle E^i(n)\right\rangle = \frac{1}{4\pi}\int\limits_{\Omega_1} E^i(n)\, d\Omega_n\,, \quad i = 1, 2, ..., 6\,, \tag{A.30}$$

Here Ω_1 is the surface of the unit sphere in 3D-space, integration is performed over the vector n on this sphere.

$$\left\langle E^1\right\rangle = E^1, \ \left\langle E^2\right\rangle = E^2, \ \left\langle E^3(n)\right\rangle = \left\langle E^4(n)\right\rangle = \frac{1}{3}E^2, \tag{A.31}$$

$$\left\langle E^5(n)\right\rangle = \frac{1}{3}E^1, \ \left\langle E^6(n)\right\rangle = \frac{1}{15}\left(2E^1 + E^2\right)\,, \tag{A.32}$$

P-basis.

$$\left\langle P^i(m)\right\rangle = \frac{1}{4\pi}\int\limits_{\Omega_1} P^i(m)\, d\Omega_m\,, \quad i = 1, 2, ..., 6\,, \tag{A.33}$$

Here m is the vector on the unit sphere Ω_1.

$$\left\langle P^1(m)\right\rangle = \frac{1}{15}\left(E^2 + 7E^1\right)\,, \ \left\langle P^2(m)\right\rangle = \frac{2}{15}\left(E^1 + 3E^2\right)\,, \tag{A.34}$$

$$\left\langle P^3(m)\right\rangle = \left\langle P^4(m)\right\rangle = \frac{2}{15}\left(2E^2 - E^1\right)\,, \tag{A.35}$$

$$\left\langle P^5(m)\right\rangle = \frac{1}{5}\left(E^1 - \frac{1}{3}E^2\right)\,, \ \left\langle P^6(m)\right\rangle = \frac{1}{15}\left(2E^1 + E^2\right)\,. \tag{A.36}$$

θ-basis. The elements of the θ and R-basis are averaged with respect to the vector n located in the plane orthogonal to the vector m. These means are integrals over the unit circle l_1 in this plane, and they are expressed via the elements of the P-basis as follows:

$$\left\langle \theta^i(m,n)\right\rangle = \frac{1}{2\pi}\int\limits_{l_1} \theta^i(m,n)\, dl_n\,, \quad i = 1, 2, ..., 6\,, \tag{A.37}$$

$$\left\langle \theta^1(m,n)\right\rangle = P^1(m)\,, \ \left\langle \theta^2(m,n)\right\rangle = P^2(m)\,, \tag{A.38}$$

$$\left\langle \theta^3(m,n)\right\rangle = \frac{1}{2}P^2(m)\,, \ \left\langle \theta^4(m,n)\right\rangle = \frac{1}{2}P^2(m)\,, \tag{A.39}$$

$$\left\langle \theta^5(m,n)\right\rangle = \frac{1}{2}P^1(m)\,, \left\langle \theta^6(n)\right\rangle = \frac{1}{2}\left(P^2(m) + 2P^1(m)\right)\,, \tag{A.40}$$

R-basis.

$$\left\langle R^i(m,n)\right\rangle = \frac{1}{2\pi}\int\limits_{l_1} R^i(m,n)\, dl_n\,, \quad i = 1,2,...,5\,, \tag{A.41}$$

$$\left\langle R^1(m,n)\right\rangle = \frac{1}{8}\left(P^2(m) + 2P^1(m)\right)\,, \tag{A.42}$$

$$\left\langle R^2(m,n)\right\rangle = P^6(m)\,, \quad \left\langle R^3(m,n)\right\rangle = \frac{1}{2}P^4(m)\,, \tag{A.43}$$

$$\left\langle R^4(m,n)\right\rangle = \frac{1}{2}P^3(m)\,, \quad \left\langle R^5(m,n)\right\rangle = \frac{1}{2}P^5(m)\,. \tag{A.44}$$

A.6 Tensor bases of rank four tensors in 2D-space

Let us consider the E- and P-bases in 2D-space. The E-basis has form (A.1)-(A.2), but the Greek indices take the values 1, 2. Note that in the 2D-case, not all the tensors $E^i(n)$ are linear independent because the following equation holds:

$$E^1 - E^2 + E^3 + E^4 - 2E^5 = 0. \tag{A.45}$$

Therefore, only five from six tensors $E^i(n)$ (A.1)-(A.2) are linearly independent (any four tensors from the first five, and $E^6(n)$).

For the P-basis, linear dependence of the tensors (A.10)-(A.12) is obvious because the tensors P^1 and P^2 coincide in the 2D-case. The multiplication table for the five linearly independent elements of the P-basis has the form

	P^1	P^3	P^4	P^5	P^6
P^1	P^1	P^3	0	0	0
P^3	0	0	P^1	0	P^3
P^4	P^4	P^6	0	0	0
P^5	0	0	0	$P^5/2$	0
P^6	0	0	P^4	0	P^6

(A.46)

The product of two elements (A and B) of the linear shell of the P-basis

$$A = a_1P^1 + a_3P^3 + a_4P^4 + a_5P^5 + a_6P^6, \tag{A.47}$$

$$B = b_1P^1 + b_3P^3 + b_4P^4 + b_5P^5 + b_6P^6, \tag{A.48}$$

where a_i, b_i are scalar coefficients, takes the following form

$$AB = (a_1b_1 + a_3b_4)\,P^1 + (a_1b_3 + a_3b_6)\,P^3 + \\ + (a_4b_1 + a_6b_4)\,P^4 + \frac{1}{2}a_5b_5P^5 + (a_4b_3 + a_6b_6)\,P^6. \tag{A.49}$$

The tensor A^{-1} inverse with respect to A is defined by the equation

$$A^{-1} = \frac{a_6}{\Delta} P^1 - \frac{a_3}{\Delta} P^3 - \frac{a_4}{\Delta} P^4 + \frac{4}{a_5} P^5 + \frac{a_1}{\Delta} P^6, \tag{A.50}$$

$$\Delta = a_1 a_6 - a_3 a_4 . \tag{A.51}$$

The connection between the elements of the E- and P-basis in 2D-case is determined by the relations:

$$\begin{cases} E^1 = P^1 + 2P^5 + P^6, \;\; E^2 = P^1 + P^3 + P^4 + P^6, \\ E^3 = P^3 + P^6, \;\; E^4 = P^4 + P^6, \\ E^5 = P^5 + P^6, \;\; E^6 = P^6, \end{cases} \tag{A.52}$$

$$\begin{cases} P^1 = E^1 - 2E^5 + E^6, \;\; P^3 = E^3 - E^6, \\ P^4 = E^4 - E^6, \;\; P^5 = E^5 - E^6, \;\; P^6 = E^6. \end{cases} \tag{A.53}$$

The averaging formulas for the E- and P-basis in 2D-case have the forms

$$\langle E^i (n) \rangle = \frac{1}{2\pi} \int\limits_{l_1} E^i (n) \, dl_n , \quad i = 1, 2, ..., 6 , \tag{A.54}$$

$$\langle E^1 \rangle = E^1, \;\; \langle E^2 \rangle = E^2, \;\; \langle E^3 (n) \rangle = \langle E^4 (n) \rangle = \frac{1}{2} E^2, \tag{A.55}$$

$$\langle E^5 (n) \rangle = \frac{1}{2} E^1, \;\; \langle E^6 (n) \rangle = \frac{1}{8} \left(2E^1 + E^2 \right) , \tag{A.56}$$

$$\langle P^i (n) \rangle = \frac{1}{2\pi} \int\limits_{l_1} P^i (n) \, dl_n , \quad i = 1, 3, 4, 5, 6 , \tag{A.57}$$

$$\langle P^1 (n) \rangle = \frac{1}{8} \left(E^2 + 2E^1 \right) , \;\; \langle P^3 (n) \rangle = \langle P^4 (n) \rangle = \frac{1}{8} \left(3E^2 - E^1 \right) , \tag{A.58}$$

$$\langle P^5 (n) \rangle = \frac{1}{8} \left(2E^1 - E^2 \right) , \;\; \langle P^6 (n) \rangle = \frac{1}{8} \left(E^2 + 2E^1 \right) . \tag{A.59}$$

B. Generalized functions connected with the Green function of static elasticity

The Green function $G(x)$ of static elasticity is the solution vanishing at infinity for the equation

$$\left(L^0_{\alpha\beta}G_{\beta\lambda}\right)(x) = \delta_{\alpha\lambda}\delta(x)\ , \quad L^0_{\alpha\beta} = -\nabla_\lambda C^0_{\lambda\alpha\beta\mu}\nabla_\mu\ , \tag{B.1}$$

where $\delta_{\alpha\beta}$ is the Kronecker symbol, $\delta(x)$ is the Dirac delta-function, and C^0 is the tensor of elastic moduli of the medium. This function is also called the Green function for displacements. The Green functions for the strains and stresses are expressed via the second derivatives of $G(x)$. Regularizations of integral operators with kernels that are Green functions, are presented in this Appendix.

B.1 The Green functions of static elasticity in the k-representation

The direct F and inverse F^{-1} Fourier transform of functions in 3D-space are determined by the equations

$$(Ff)(k) = f^*(k) = \int f(x)\exp(ik\cdot x)\,dx\,, \tag{B.2}$$

$$\left(F^{-1}f^*\right)(x) = f(x) = (2\pi)^{-3}\int f^*(k)\exp(-ik\cdot x)\,dk\,. \tag{B.3}$$

The following equation is called the Parseval formula

$$(f,\varphi) = (2\pi)^{-3}(f^*,\varphi^*)\ , \tag{B.4}$$

$$(f,\varphi) = \int f(x)\,\bar{\varphi}(x)\,dx\,. \tag{B.5}$$

Here $f(x)$, $\varphi(x)$ are functions in R^3, $\bar{\varphi}$ is the complex-conjugate function.

The Fourier transform $G^*(k)$ of the Green function follows from (B.1) in the form

$$G^*(k) = \left(L^0(k)\right)^{-1}, \quad L^0_{\alpha\beta}(k) = k_\lambda C^0_{\lambda\alpha\beta\mu}k_\mu \tag{B.6}$$

and is a homogenous function of degree −2.

For an isotropic medium, the function $G^*(k)$ is ($k^2 = k_\alpha k_\alpha$)

$$G^*_{\alpha\beta}(k) = \frac{1}{\mu_0 k^2}\left(\delta_{\alpha\beta} - \varkappa_0 \frac{k_\alpha k_\beta}{k^2}\right), \quad \varkappa_0 = \frac{\lambda_0 + \mu_0}{\lambda_0 + 2\mu_0}. \tag{B.7}$$

The Green function for strains in a medium with sources of external stresses is defined in (2.53)

$$K_{\alpha\beta\lambda\mu}(x) = -[\nabla_\alpha \nabla_\lambda G_{\beta\mu}(x)]_{(\alpha\beta)(\lambda\mu)}. \tag{B.8}$$

The Fourier transform of this function has the form

$$K^*_{\alpha\beta\lambda\mu}(k) = \left[k_\alpha k_\lambda G^*_{\beta\mu}(k)\right]_{(\alpha\beta)(\lambda\mu)}, \tag{B.9}$$

and is a homogenous function of the zero degree with respect to k. For an isotropic medium, the equation for $K^*(k)$ takes the form

$$K^*(k) = \frac{1}{\mu_0}\left[E^5(m) - \varkappa_0 E^6(m)\right], \quad m_\alpha = k_\alpha / |k|, \tag{B.10}$$

where $E^i(m)$ are the elements of the E-basis (A.1), (A.2).

The Green function $S(x)$ for stresses in the medium with dislocation sources is connected with the function $K(x)$ by the equation (2.72)

$$S(x) = C^0 K(x) C^0 - C^0 \delta(x), \tag{B.11}$$

and its Fourier transform is

$$S^*(k) = C^0 K^*(k) C^0 - C^0. \tag{B.12}$$

The function $S^*(k)$, as well as $K^*(k)$, is a homogeneous function of the degree zero. For an isotropic medium, the function $S^*(k)$ takes the form

$$S^*(k) = -2\mu_0\left[P^1(m) + (2\varkappa_0 - 1) P^2(m)\right], \tag{B.13}$$

where $P^i(m)$ are the elements of the P-basis (A.10)–(A.12).

B.2 The Green functions of static elasticity in the x-representation

Let us consider the x-representation of the functions $G(x)$ and $K(x)$ for an isotropic medium. To this end, we use the formulas for the inverse Fourier transform of some generalized functions [21]

$$F^{-1}\left(k^{-2}\right) = \frac{1}{4\pi |x|}, \quad F^{-1}\left(k^{-4}\right) = -\frac{|x|}{8\pi}, \tag{B.14}$$

$$F^{-1}\left(\frac{k_\alpha k_\beta}{k^2}\right) = -\frac{1}{4\pi}\nabla_\alpha\nabla_\beta\frac{1}{|x|} =$$

$$= \frac{1}{4\pi|x|^3}\left(\delta_{\alpha\beta} - 3n_\alpha n_\beta\right), \quad n = \frac{x}{|x|}, \tag{B.15}$$

$$F^{-1}\left(\frac{k_\alpha k_\beta}{k^4}\right) = \frac{1}{8\pi|x|}\left(\delta_{\alpha\beta} - n_\alpha n_\beta\right), \tag{B.16}$$

$$F^{-1}\left(\frac{k\otimes k\otimes k\otimes k\otimes k}{k^4}\right) =$$

$$= \frac{1}{8\pi|x|^3}\left[E^2 + 2E^1 - 3\left(E^3(n) + E^4(n) + 4E^5(n)\right) + 15E^6(n)\right], \tag{B.17}$$

where $E^i(n)$ are the elements of the basis (A.1), (A.2).

These relations and (B.8), (B.10) and (B.12) provide the expressions for the functions $G(x)$, $K(x)$ and $S(x)$ in the forms

$$G_{\alpha\beta}(x) = \frac{1}{8\pi\mu_0}\left[\frac{2}{|x|}\delta_{\alpha\beta} - \varkappa_0\nabla_\alpha\nabla_\beta|x|\right] =$$

$$= \frac{1}{8\pi\mu_0|x|}\left[(2-\varkappa_0)\,\delta_{\alpha\beta} + \varkappa_0 n_\alpha n_\beta\right], \tag{B.18}$$

$$K(x) = \frac{1}{4\pi\mu_0|x|^3}\Big\{E^1 - 3E^5(n) - \frac{\varkappa_0}{2}\big[E^2 + 2E^1 -$$

$$-3\left(E^3(n) + E^4(n) + 4E^5(n)\right) + 15E^6(n)\big]\Big\}, \tag{B.19}$$

$$S(x) =$$

$$= \frac{\mu_0}{2\pi|x|^3}\{(1-\varkappa_0)\left[2\left(E^1 - E^2\right) + 3\left(E^3(n) + E^4(n) - 2E^5(n)\right)\right] +$$

$$+\varkappa_0\left(E^2 + 6E^5(n) - 15E^6(n)\right)\}. \tag{B.20}$$

Here $G(x)$ is a homogeneous function of degree (-1) integrable at $x = 0$, $K(x)$ and $S(x)$ are homogeneous generalized functions of degree (-3). The action of these functions on a smooth finite function $\varphi(x)$ in R^3 is presented by formally divergent integrals. Regularization of generalized functions that are derivatives of regular functionals was proposed in [54] and used below.

Let V be a region with a smooth boundary Ω, and the point $x = 0$ is inside Ω. According to [54] we present the action of the generalized function $K(x)$ on a function $\varphi(x)$ in the form

$$(K,\varphi) = (K^r,\varphi) - A\varphi(0), \tag{B.21}$$

$$(K^r,\varphi) = \int_V K(x)\left[\varphi(x) - \varphi(0)\right]dx + \int_{\bar V} K(x)\,\varphi(x)\,dx, \tag{B.22}$$

$$A = -\int_{\Omega} [\nabla G(x)] n(x)\, d\Omega, \tag{B.23}$$

where $\bar{V}$ is the complement of V in the whole space, $n(x)$ is the external normal to Ω. Here Gauss' theorem and the presentation (B.8) for $K(x)$ were used.

Suppose that the region V has an ellipsoidal form with surface defined by the equation

$$|a^{-1}x| = 1 \tag{B.24}$$

where a is a rank two symmetric tensor, $\det a > 0$.

Using the Parseval's formula (B.4) and Gauss' theorem we present equation (B.23) for A in the form

$$A = \int \nabla G(x) \nabla V(x)\, dx = \frac{1}{(2\pi)^3} \int kG^*(k)\, kV^*(k)\, dk\,. \tag{B.25}$$

Here the relation

$$\nabla V(x) = -n(x)\, \Omega(x) \tag{B.26}$$

was used, $\Omega(x)$ is the delta-function concentrated on the surface Ω that is the boundary of V.

Let us introduce a new variable $y = a^{-1}x$ in the integral (B.25). Then, we have

$$V(ay) = \tilde{V}(|y|)\,, \quad V^*(k) = \det a \tilde{V}^*(|ka|)\,. \tag{B.27}$$

Substituting these equations into (B.25) and replacing k for $a^{-1}k$ we obtain

$$A = \frac{1}{(2\pi)^3} \int K^*(a^{-1}k)\, \tilde{V}^*(|k|)\, dk\,, \tag{B.28}$$

where $K^*(k) = kG^*(k)k$ is a homogeneous function of k of zero degree. Therefore, if $k = |k|m$, where m is a vector on the unit sphere Ω, the equation for A takes the form

$$A = \frac{1}{(2\pi)^3} \int_{\Omega_1} K^*(a^{-1}m)\, d\Omega_m \int_0^\infty \tilde{V}^*(|k|)\, |k|^2 d|k|. \tag{B.29}$$

Because $\tilde{V}(y)$ is the characteristic function of a unit sphere, the following equation holds:

$$\frac{1}{(2\pi)^3} \int \tilde{V}^*(|k|)\, dk = \frac{1}{2\pi^2} \int_0^\infty \tilde{V}^*(|k|)\, |k|^2 d|k| = \tilde{V}(|y|)\,|_{y=0} = 1\,. \tag{B.30}$$

From (B.29) we have finally

$$A = \frac{1}{4\pi} \int\limits_{\Omega_1} K^* \left(a^{-1} m\right) d\Omega_m \, . \tag{B.31}$$

Note that the constant tensor A does not depend on the absolute sizes of the region V because $K^*(k)$ is a homogeneous function of zero degree. As a result, the functional K^r in (B.22) is independent of the absolute sizes of the ellipsoid V. It allows us to go to the limit $V \to 0$ in (B.22), and in this limit, the integral over the volume V disappears because $K(x) \sim |x|^{-3}$, and the second integral in (B.22) tends to its Cauchy principal value. This integral exists because of the existence of the generalized function $K(x)$.

Finally, the action of the generalized function $K(x)$ on an arbitrary finite function $\varphi(x)$ is defined by the equation

$$\int K(x) \varphi(x) dx = A(a) \varphi(0) + (\det a) \, p.v. \int K(ay) \varphi(ay) \, dy \, , \tag{B.32}$$

where $p.v. \int ...dy$ is the Cauchy principal value of the integral; the tensor $A(a)$ is defined in (B.31). The tensor a can be considered as an arbitrary non-singular linear transformation of the x-space.

Similarly, regularization of the generalized function $S(x)$ has the form

$$\int S(x) \varphi(x) dx = D(a) \varphi(0) + (\det a) \, p.v. \int S(ay) \varphi(ay) \, dy \, , \tag{B.33}$$

$$D(a) = \frac{1}{4\pi} \int\limits_{\Omega_1} S^* \left(a^{-1} m\right) d\Omega_m \, . \tag{B.34}$$

From definition (B.12) of the function $S^*(k)$ here follows the equation

$$D(a) = C^0 A(a) C^0 - C^0 \, . \tag{B.35}$$

Note that there is another equation for the tensor $A(a)$ that is more convenient in some cases [202]. For an arbitrary anisotropic medium, the Green function can be represented in the form

$$G_{\alpha\beta}(x) = \frac{1}{|x|} g_{\alpha\beta}(n), \quad n_\alpha = x_\alpha / |x| \tag{B.36}$$

where the function $g_{\alpha\beta}(n)$ is an even function of the unit vector n: $g(n) = g(-n)$. Introducing the local basis $\mathbf{e}^r, \mathbf{e}^\varphi, \mathbf{e}^\varphi$ of the spherical coordinate system (r, θ, φ)

$$\mathbf{e}^r = \frac{\partial \mathbf{x}}{\partial r}, \quad \mathbf{e}^\varphi = \frac{1}{r \sin\theta} \frac{\partial \mathbf{x}}{\partial \varphi}, \quad \mathbf{e}^\theta = \frac{1}{r} \frac{\partial \mathbf{x}}{\partial \theta}, \tag{B.37}$$

we present the Nabla operator in the form

$$\nabla = \mathbf{e}^r \frac{\partial}{\partial r} + \frac{1}{r}\nabla(n), \quad \nabla(n) = \frac{\mathbf{e}^\varphi}{\sin\theta}\frac{\partial}{\partial\varphi} + \mathbf{e}^\theta \frac{\partial}{\partial\theta}. \tag{B.38}$$

As follows from (B.36) and (B.38)

$$\nabla G(x) = \frac{1}{r^2} g^1(n), \quad g^1(n) = \nabla(n) g(n) - g(n)\mathbf{e}^r, \tag{B.39}$$

where the function $g^1(n)$ is also a function of the vector n. In contrast to $g(n)$, $g^1(n)$ is an odd function of n: $g^1(-n) = -g^1(n)$.

Let V be an ellipsoidal region and the origins of the spherical and Cartesian coordinate systems are taken in its center. If $\mathbf{r} = \mathbf{x}' - \mathbf{x}$, we have

$$\int_V \nabla G(x - x')dx' = -\int_0^{r_s} dr \int_{\Omega_1} \left(\nabla(n)g - g\mathbf{e}^r\right) d\Omega_1, \tag{B.40}$$

where r^s is the value of $r = |\mathbf{r}|$ on the ellipsoid surface, and Ω_1 indicates the unit sphere with the center at the point x. Let $\mathbf{a}$ be the rank two tensor

$$\mathbf{a} = \frac{\mathbf{e}^1 \otimes \mathbf{e}^1}{a_1^2} + \frac{\mathbf{e}^2 \otimes \mathbf{e}^2}{a_2^2} + \frac{\mathbf{e}^3 \otimes \mathbf{e}^3}{a_3^2}, \tag{B.41}$$

where $\mathbf{e}^1, \mathbf{e}^2, \mathbf{e}^3$ are the unit vectors along the principal axes of the ellipsoid, and $a_1, a_2 a_3$ are its semiaxes. With the help of this tensor, the ellipsoid surface is defined by the equation

$$x^s_\alpha a_{\alpha\beta} x^s_\beta = 1 \tag{B.42}$$

(x^s_α is the vector x_α when its end is on this surface). Because $\mathbf{r} = \mathbf{x}' - \mathbf{x}$ and $\mathbf{x}^s = \mathbf{r}^s + \mathbf{x}$, (B.42) yields the equation

$$(r^s n^r_\alpha + x_\alpha) a_{\alpha\beta} (r^s n^r_\beta + x_\beta) = 1, \quad n^r_\alpha = \frac{r_\alpha}{|\mathbf{r}|}. \tag{B.43}$$

This is a quadratic equation with respect to r^s that has only one positive root

$$r_s = \left(n^r_\alpha a_{\alpha\beta} n^r_\beta\right)^{-1} \Bigg[- n^r_\alpha a_{\alpha\beta} x_\beta + \sqrt{\left(n^r_\alpha a_{\alpha\beta} n^r_\beta\right)^2 - \left(n^r_\alpha a_{\alpha\beta} n^r_\beta\right)\left(x_\lambda a_{\lambda\mu} x_\mu - 1\right)} \Bigg]. \tag{B.44}$$

Because of (B.42), the equation $x_\lambda a_{\lambda\mu} x_\mu - 1 \leq 0$ holds when $x \in V$. As a result, the expression under the square root in (B.44) is always positive, and r^s is non negative. Equation (B.40) can be written in the form

$$\int_V \boldsymbol{\nabla} G(x-x')dx' = -\int_{\Omega_1} (\boldsymbol{\nabla}(n)g - g\mathbf{e}^r)\, r^s d\Omega_1. \tag{B.45}$$

It follows from (B.44) that r^s is a function on the unit sphere: $r^s = r^s(n)$. This function may be presented as the sum of an odd and an even function with respect to n. Therefore, by substitution r^s from (B.44) into (B.45), the integration of an odd function over the unit sphere is zero, and we finally obtain

$$\int_V \nabla_\mu G_{\beta\lambda}(x-x')dx' = \widehat{A}_{\alpha\mu\beta\lambda} x_\alpha, \quad x \in V \tag{B.46}$$

$$\widehat{A}_{\alpha\mu\beta\lambda} = a_{\alpha\sigma} \int_{\Omega_1} \left(n^r_\rho a_{\rho\tau} n^r_\tau\right)^{-1} n^r_\sigma \left[\nabla_\mu(n^r) g_{\beta\lambda}(n^r) - a^r_\mu g_{\beta\lambda}(n^r)\right] d\Omega_1 \tag{B.47}$$

Here $\widehat{A}_{\alpha\mu\beta\lambda}$ is a tensor with constant components. The tensor A is obtained from $\widehat{A}$ by symmetrization over the indices $\alpha\beta$ and $\lambda\mu$: $A_{\alpha\beta\lambda\mu} = \widehat{A}_{\alpha\mu\beta\lambda}\Big|_{(\alpha\beta)(\lambda\mu)}$. This presentation of the tensor A is used in Section 3.6.

Returning to (B.32) we note that this regularization holds for every function with a Fourier transform that is a homogeneous function of zero degree. Particularly, regularization of the generalized function

$$K^{(m)}_{\alpha_1\alpha_2\ldots\alpha_m\alpha\beta\lambda\mu\mu_1\mu_2\ldots\mu_m}(x) =$$
$$= [\nabla_{\alpha_1}\nabla_{\alpha_2}\ldots\nabla_{\alpha_m} K_{\alpha\beta\lambda\mu}(x)] x_{\mu_1} x_{\mu_2} \ldots x_{\mu_m}, \tag{B.48}$$

where $K(x)$ has the form (B.19), is defined by an expression that is similar to (B.32),

$$\int K^{(m)}(x)\, \varphi(x)\, dx =$$
$$= A^m_m(a)\, \varphi(0) + (\det a)\, p.v. \int K^{(m)}(ay)\, \varphi(ay)\, dy, \tag{B.49}$$

$$A^m_m(a) = \frac{1}{4\pi} \int_{\Omega_1} K^{(m)*}\left(a^{-1}k\right) d\Omega_1, \tag{B.50}$$

where Ω_1 is the surface of the unit sphere in k-space.

The tensor $A^m_m(a)$ may be presented in the form of a volume integral if the function $\varphi(x)$ in (B.49) is the characteristic function of the ellipsoid region given by the equation $|a^{-1}x| \leq 1$. In this case, $V(ay) = \tilde{V}(|y|)$ is the characteristic function of a unit sphere in y-space. A consequence of existing the principal value of the integral in (B.49) is the following equation [149]

$$p.v. \int K^{(m)}(ay)\, V(ay)\, dy = 0\,. \tag{B.51}$$

Therefore, the equation for the tensor $A_m^m(a)$ is presented as the integral over the volume of the considered ellipsoid

$$A_m^m(a) = \int_V K^{(m)}(x)\, dx. \tag{B.52}$$

Let the generalized function $K^1(x)$ be similar to (B.48) but constructed from the scalar Green function $G(x)$, defined in Section 6.1

$$K^{(1)}_{\alpha\beta\lambda\mu}(x) = [\nabla_\lambda K_{\alpha\beta}(x)]x_\mu\,, \quad K_{\alpha\beta} = \nabla_\alpha \nabla_\beta G(x)\,. \tag{B.53}$$

The tensor $A_1^1(a)$ takes the following form:

$$A^1_{1\alpha\beta\lambda\mu}(a) = \int \nabla_\lambda K_{\alpha\beta}(x)\, x_\mu V(x)\, dx. \tag{B.54}$$

Introducing a new variable $y = a^{-1}x$, we transform this integral into an integral over the sphere V_1 of unit radius

$$A^1_{1\alpha\beta\lambda\mu}(a) = |a| \left(a^{-1}\right)_{\lambda\nu} a_{\mu\rho} \int_{V_1} \nabla_\nu K_{\alpha\beta}(ay)\, y_\rho dy\,. \tag{B.55}$$

Gauss' theorem allows this integral to be presented in the form

$$A^1_{1\alpha\beta\lambda\mu}(a) = J^{(1)}_{\alpha\beta\lambda\mu} + J^{(2)}_{\alpha\beta\lambda\mu}\,, \tag{B.56}$$

$$J^{(1)}_{\alpha\beta\lambda\mu} = -|a| \left(a^{-1}\right)_{\lambda\nu} a_{\mu\nu} \int_{V_1} K_{\alpha\beta}(ay)\, dy\,, \tag{B.57}$$

$$J^{(2)}_{\alpha\beta\lambda\mu} = |a| \left(a^{-1}\right)_{\lambda\nu} a_{\mu\rho} \int_{\Omega_1} n_\nu K_{\alpha\beta}(ay)\, n_\rho d\Omega\,. \tag{B.58}$$

The integral $J^{(1)}$ in this expression is calculated as follows:

$$J^{(1)}_{\alpha\beta\lambda\mu} = \delta_{\lambda\mu}|a| \int_{V_1} K_{\alpha\beta}(ay)\, dy = \delta_{\lambda\mu} \int K_{\alpha\beta}(x)\, dx = A_{\alpha\beta}(a)\, \delta_{\lambda\mu}\,, \tag{B.59}$$

where the tensor $A(a)$ is defined in (B.31).

Taking into account the equation

$$\int_0^\infty \exp\left(-y^2/2\right) y dy = 1 \tag{B.60}$$

we transform the integral $J^{(2)}$ in (B.58) into an integral over 3D-space

$$J^{(2)}_{\alpha\beta\lambda\mu} = |a| \left(a^{-1}\right)_{\lambda\nu} a_{\mu\rho} \int y_\nu K_{\alpha\beta}(ay)\, y_\rho e^{-y^2/2} dy \tag{B.61}$$

because $K_{\alpha\beta}(y)$ is a homogeneous function of degree -3.

Using the Parseval's formula, we go to the Fourier transforms of the integrand functions and present $J^{(2)}_{\alpha\beta\lambda\mu}$ in the form

$$J^{(2)}_{\alpha\beta\lambda\mu} = -(2\pi)^{-3/2} \left(a^{-1}\right)_{\lambda\nu} a_{\mu\rho} \int K^*_{\alpha\beta}\left(a^{-1}k\right) \frac{\partial^2}{\partial k_\nu \partial k_\rho} e^{-k^2/2} dk\,. \tag{B.62}$$

Taking into account the equation

$$\int_0^\infty x^{2n} e^{-px^2} dx = \frac{(2n-1)!!}{2(2p)^n} \sqrt{\frac{\pi}{p}}, \tag{B.63}$$

we calculate the integral $J^{(2)}_{\alpha\beta\lambda\mu}$ in the form $(k_\alpha / |k| = m_\alpha)$:

$$J^{(2)}_{\alpha\beta\lambda\mu} = \frac{1}{4\pi} \left(a^{-1}\right)_{\lambda\mu} a_{\mu\rho} \int_{\Omega_1} K^*_{\alpha\beta}\left(a^{-1}k\right) \left(\delta_{\nu\rho} - 3m_\nu m_\rho\right) d\Omega\,. \tag{B.64}$$

Substituting $J^{(1)}$ and $J^{(2)}$ from (B.59), (B.64) into (B.56) we obtain

$$A^1_{1\alpha\beta\lambda\mu} = -\frac{3}{4\pi} \left(a^{-1}\right)_{\lambda\nu} a_{\mu\rho} \int_{\Omega_1} K^*_{\alpha\beta}\left(a^{-1}k\right) m_\nu m_\rho d\Omega\,. \tag{B.65}$$

For a spherical inclusion in an isotropic medium, the tensor A^1_1 takes the form

$$A^1_{1\alpha\beta\lambda\mu} = \frac{1}{5}\left(E^2_{\alpha\beta\lambda\mu} + 3E^1_{\alpha\beta\lambda\mu}\right). \tag{B.66}$$

This equation is used in Section 6.2.

B.3 The Green functions of static elasticity in 2D-case

In the plane case, the k-representation of the Green function has the same form (B.6), (B.9), and (B.12) as in the 3D-case, but the Greek indices take the values 1, 2. The coefficient κ_0 has the form

$$\kappa_0 = \frac{1}{2(1-\nu_0)} \quad \text{(for plane strain state)}, \tag{B.67}$$

$$\kappa_0 = \frac{1+\nu_0}{2} \quad \text{(for plane stress)}. \tag{B.68}$$

In order to obtain the x-representation of the Green tensor we use the equations for the inverse Fourier transform F^{-1}

$$\left(F^{-1}f^*\right)(x) = (2\pi)^{-2} \int f^*(k) \exp\left(-ik \cdot x\right) dk, \tag{B.69}$$

of the functions in (B.6), (B.9) and (B.12) in the 2D-case

$$F^{-1}\left(k^{-2}\right) = -\frac{1}{2\pi} \ln|x|, \quad F^{-1}\left(\frac{k_\alpha k_\beta}{k^2}\right) = \frac{1}{2\pi x^2}\left(\delta_{\alpha\beta} - 2n_\alpha n_\beta\right), \tag{B.70}$$

$$F^{-1}\left(k^{-4}\right) = \frac{x^2}{8\pi} \ln|x|, \quad F^{-1}\left(\frac{k_\alpha k_\beta}{k^4}\right) = -\frac{1}{4\pi}\left(\ln|x|\delta_{\alpha\beta} + n_\alpha n_\beta\right), \tag{B.71}$$

$$F^{-1}\left(\frac{k \otimes k \otimes k \otimes k \otimes k}{k^4}\right) =$$

$$= \frac{1}{4\pi x^2}\left[3E^1 - \left(E^3(n) + E^4(n) + 10E^5(n)\right) + 8E^6(n)\right]. \tag{B.72}$$

Here $E^i(n)$ are the elements of the E-basis in the plane case, $n = x/|x|$. Taking into account these equations, we find that the expressions for the Green functions $G(x)$, $K(x)$, and $S(x)$ take the forms

$$G_{\alpha\beta}(x) = \frac{1}{4\pi\mu_0}\left[(2 - \kappa_0) \ln\frac{1}{|x|}\delta_{\alpha\beta} + \kappa_0 n_\alpha n_\beta\right], \tag{B.73}$$

$$K(x) = \frac{|x|^{-2}}{2\pi\mu_0}\left[E^1 - 2E^5(n) - \right.$$

$$\left. -\frac{\kappa_0}{2}\left(3E^1 - E^3(n) - E^4(n) - 10E^5(n) + 8E^6(n)\right)\right], \tag{B.74}$$

$$S(x) = \frac{\mu_0\kappa_0}{\pi x^2}\left[E^2 + 4E^5(n) - 8E^6(n)\right]. \tag{B.75}$$

The action of the generalized functions $K(x)$ and $S(x)$ on finite functions in R^2 is defined in (B.28), (B.29), where the tensors $A(a)$ and $D(a)$ have the forms

$$A(a) = \frac{1}{2\pi}\int_{l_1} K^*\left(a^{-1}m\right) dl_m, \quad D(a) = \frac{1}{2\pi}\int_{l_1} S^*\left(a^{-1}m\right) dl_m, \tag{B.76}$$

Integration here is performed over the unit circle l_1 in the k-plane.

B.4 Special presentation of the *K*-operator

Let us introduce in the x-space the spherical coordinate system (r, n), $r = |x|, n = x/|x|$ is the vector on the unit sphere Ω_1. We denote by $f^*(s,n)$ the Mellin transform of the tensor function $f(r,n)$ with respect to the variable r [201]

$$f^*(s,n) = \int_0^\infty r^{s-1} f(r,n)\, dr\,, \quad f(r,n) = \frac{1}{2\pi i} \int_{\tau - i\infty}^{\tau + i\infty} r^{-s} f(s,n)\, ds\,. \quad \text{(B.77)}$$

Let us show that the Fourier transform $(Ff)(k)$ of a function $f(r,n)$ is presented in the form $(m = k/|k|)$ [168]:

$$(Ff)(k) = \frac{1}{2\pi i} \int_{\tau - i\infty}^{\tau + i\infty} e^{i\frac{\pi}{2}(p-s)} \Gamma(p-s)\, |k|^{s-p} ds \times$$

$$\times \int_{\Omega_1(p)} f^*(s,n) \left[(n \cdot m) + i0\right]^{s-p} d\Omega_n\,. \quad \text{(B.78)}$$

Here p is the dimension of the $x(k)$-space, in what follows $p = 2, 3$; $\Omega_1(p)$ is the surface of the unit sphere in the pD-space, $\Gamma(s)$ is the Euler Gamma-function. To prove (B.78) we write the Fourier transform of the function $f(r,n)$, using the inversion formula (B.77)

$$(Ff)(k) = \int_0^\infty r^{p-1} dr \int_{\Omega_1(p)} e^{ir|k|(m \cdot n)} d\Omega_n \frac{1}{2\pi i} \int_{\tau - i\infty}^{\tau + i\infty} r^{-s} f^*(s,n)\, ds\,. \quad \text{(B.79)}$$

To justify the change of the order of integration in this equation we introduce the parameter $\varepsilon > 0$ and present $(Ff)(k)$ as follows:

$$(Ff)(k) = \lim_{\varepsilon \to 0} \int_0^\infty r^{p-1} dr \int_{\Omega_1(p)} e^{ir|k|(m \cdot n) - r\varepsilon} d\Omega \frac{1}{2\pi i} \int_{\tau - i\infty}^{\tau + i\infty} r^{-s} f^*(s,n)\, ds =$$

$$= \lim_{\varepsilon \to 0} \frac{1}{2\pi i} \int_{\tau - i\infty}^{\tau + i\infty} ds \int_{\Omega_1(p)} f^*(s,n)\, d\Omega_n \int_0^\infty e^{i|k|r(m \cdot n) - \varepsilon r} r^{p-s-1} dr\,. \quad \text{(B.80)}$$

The internal integral in this equation is calculated explicitly

$$\int_0^\infty e^{i|k|r(m \cdot n) - \varepsilon r} r^{p-s-1} dr = e^{i\frac{\pi}{2}(p-s)} \left[|k|(n \cdot m) + i\varepsilon\right]^{s-p} \Gamma(p-s)\,. \quad \text{(B.81)}$$

Using this equation and (B.80) we find

$$(Ff)(k) = \lim_{\varepsilon\to 0}\frac{1}{2\pi i}\int\limits_{\tau-i\infty}^{\tau+i\infty} e^{i\frac{\pi}{2}(p-s)}\Gamma(p-s)\,|k|^{s-p}ds\times$$
$$\times\int\limits_{\Omega_1(p)} f^8(s,n)\left[(n\cdot m)+i\varepsilon\right]^{s-p}d\Omega_n =$$
$$= \frac{1}{2\pi i}\int\limits_{\tau-i\infty}^{\tau+i\infty} e^{i\frac{\pi}{2}(p-s)}\Gamma(p-s)\,|k|^{s-p}ds\times$$
$$\times\int\limits_{\Omega_1(p)} f^*(s,n)\left[(n\cdot m)+i0\right]^{s-p}d\Omega_n\,, \tag{B.82}$$

where we take into account that when $\varepsilon\to 0$

$$\left[(n\cdot m)+i\varepsilon\right]^{\mu}\to\left[(n\cdot m)+i0\right]^{\mu}\,. \tag{B.83}$$

The right-hand side of this equation is a generalized function on the unit sphere Ω_1. Hence, (B.78) is proved.

Consider now the operator K where symbol $K^*(k)$ has form (B.9). Using the convolution properties we have

$$(Kf)(x) = \int K(x-x')\,f(x')\,dx' = \left[F^{-1}K^*(k)\,Ff\right](x)\,, \tag{B.84}$$

where F is the Fourier transform operator (B.69). Introducing the parameter $\varepsilon>0$ once again and using (B.78) we obtain

$$(Kf)(x) = \lim_{\varepsilon\to+0}(2\pi)^{-p}\int\limits_0^{\infty} d|k|\int\limits_{\Omega_1(p)} e^{-i|k|r(n\cdot m)-r\varepsilon}|k|^{p-1}K^*(m)\,d\Omega_m\times$$
$$\times\frac{1}{2\pi i}\int\limits_{\tau-i\infty}^{\tau+i\infty}\Gamma(p-s)\,e^{i\frac{\pi}{2}(p-s)}|k|^{s-p}ds\int\limits_{\Omega_1(p)} f^*(s,l)\left[(m\cdot l)+i0\right]^{s-p}dl\,. \tag{B.85}$$

Changing the order of integration we go to the equation

$$(Kf)(x) = \lim_{\varepsilon\to 0}\frac{1}{(2\pi)^p}\int\limits_{\tau-i\infty}^{\tau+i\infty} ds\int\limits_{\Omega_1(p)} K^*(m)\,\Gamma(p-s)\,e^{i\frac{\pi}{2}(p-s)}d\Omega_m\times$$
$$\times\int\limits_{\Omega_1(p)} f^*(s,l)\left[(l\cdot m)+i0\right]^{s-p}d\Omega_l\int\limits_0^{\infty} e^{-ir|k|(n\cdot m)-\varepsilon|k|}|k|^{s-1}d|k|\,. \tag{B.86}$$

Using an integral similar to (B.81)

$$\int_0^\infty e^{ir|k|(n\cdot m)-\varepsilon|k|}|k|^{s-1}d|k| = e^{i\frac{\pi}{2}s}\Gamma(s)\left[-r(n\cdot m)+i\varepsilon\right]^{-1}, \tag{B.87}$$

and going to the limit $\varepsilon \to +0$, we obtain finally

$$(Kf)(r,n) = \frac{1}{2\pi i}\int_{\tau-i\infty}^{\tau+i\infty} r^{-s}\left(K_s f^*\right)(s,n)\,ds\ , \tag{B.88}$$

$$\left(K_s f^*\right)(s,n) = \frac{e^{i\frac{\pi}{2}p}}{(2\pi)^p}\Gamma(p-s)\,\Gamma(s)\times$$

$$\times\int_{\Omega_1(p)}\left(-n\cdot m+i0\right)^{-s}K^*(m)\,d\Omega_m\int_{\Omega_1(p)}\left(m\cdot l+i0\right)^{s-p}f^*(s,l)\,d\Omega_l\,. \tag{B.89}$$

This presentation of the singular integral operator is used in Chapter 3 for the solution of the integral equation for the strain field in a medium with spherically symmetric inhomogeneity.

C. Properties of some potentials of static elasticity concentrated on surfaces

C.1 Gauss' and Stokes' integral theorems

Let V be a bounded region in 3D-space with a piece-wise smooth closed surface Ω, and let $n(x)$ be the external (with respect to V) normal to Ω. If $A(x)$ is a tensor function with derivatives that are integrable in R^3, the following relation holds:

$$\int_V \nabla_\alpha A_{\beta\lambda\ldots\mu}(x)\,dx = \int_\Omega n_\alpha(x)\,A_{\beta\lambda\ldots\mu}(x)\,d\Omega\,. \tag{C.1}$$

A special form of this equation is called Gauss' theorem [143]

$$\int \mathrm{div}A(x)\,dx = \int_\Omega n(x)\cdot A(x)\,d\Omega\,. \tag{C.2}$$

Let Ω be an arbitrary Lyapunov surface bounded by a smooth closed contour Γ. The orientation of the surface is defined by the assignment of a continuos vector of normal $n(x)$ to Ω. Suppose that the positive side of Ω is the side of the normal. The positive path-tracing of Γ is such that the positive side of the surface Ω remains on the left-hand side.

Let $\tau(x)$ be the unit vector tangential to Γ and directed along the positive path-tracing of Γ. If $A(x)$ is a tensor function in R^3 with an integrable first derivative, Stokes' theorem for the surface is written in the form [143]

$$\int_\Omega \epsilon_{\alpha\beta\lambda}\nabla_\beta A_{\lambda\mu\ldots\nu}(x)\,n_\alpha(x)\,d\Omega = \int_\Gamma A_{\lambda\mu\ldots\nu}\tau_\lambda(x)\,d\Gamma\,. \tag{C.3}$$

where $\epsilon_{\alpha\beta\lambda}$ is the Levi-Civita tensor. For brevity we write

$$\int_\Omega n(x)\cdot rot\,A(x)\,d\Omega = \int_\Gamma A(x)\cdot d\Gamma,\quad d\Gamma = \tau d\Gamma\,. \tag{C.4}$$

Let us introduce the projection operator on the surface Ω at the point $x \in \Omega$

$$\theta_{\alpha\beta}(x) = \delta_{\alpha\beta} - n_\alpha(x) n_\beta(x) . \tag{C.5}$$

A vector field $a(x)$ belongs to the surface Ω if it satisfies one of the following equations:

$$n_\alpha(x) a_\alpha(x) = 0, \quad \theta_{\alpha\beta}(x) a_\beta(x) = a_\alpha(x) . \tag{C.6}$$

A tensor field $A(x)$ belongs to the surface Ω if the following equation holds:

$$\theta_{\alpha\beta}(x) \theta_{\lambda\mu}(x) \ldots\ldots \theta_{\nu\rho}(x) A_{\beta\mu\ldots\rho}(x) = A_{\alpha\lambda\ldots\nu}(x) . \tag{C.7}$$

The derivative along the surface Ω is defined as follows:

$$\partial_\alpha = \nabla_\alpha - n_\alpha(x) n_\beta(x) \nabla_\beta , \tag{C.8}$$

where ∇ is the Nabla-operator in R^3.

Let us take on the contour Γ (the boundary of Ω) the orthogonal reference basis $n(x)$, $\tau(x)$, $e(x)$ where $n(x)$ is the limit value of the normal $n(x)$ to Ω on Γ, $\tau(x)$ is the unit tangent vector to Γ defining the positive path-tracing, $e(x)$ is the unit vector normal to Γ and located in the plane tangent to Ω at the point $x \in \Gamma$ and directed outside Γ.

For an integrable tensor field $A(x)$ that belongs to the surface Ω and has the first integrable derivative, the following analogue of Gauss' formula (C.2) for a surface holds [143], [206]:

$$\int_\Omega \partial_\alpha A_{\alpha\beta\lambda\ldots\mu}(x) d\Omega = \oint_\Gamma e_\alpha(x) A_{\alpha\beta\lambda\ldots\mu}(x) d\Gamma . \tag{C.9}$$

C.2 Derivatives of the double-layer potential of static elasticity

Let Ω be a closed Lyapunov surface in R^3. We consider the potential

$$\varepsilon_{\alpha\beta}(x) = \int_\Omega \mathrm{K}_{\alpha\beta\lambda\mu}(x - x') C^0_{\lambda\mu\nu\rho} n_\nu(x') b_\rho(x') d\Omega', \tag{C.10}$$

$$\mathrm{K}_{\alpha\beta\lambda\mu}(x) = -[\nabla_\alpha \nabla_\lambda G_{\beta\mu}(x)]_{(\alpha\beta)(\lambda\mu)}, \tag{C.11}$$

where $G(x)$ is the Green function of static elasticity for a homogeneous medium with the elastic moduli tensor C^0. The tensor $\varepsilon_{\alpha\beta}(x)$ in (C.10) is the symmetrized gradient of the vector function $u(x)$ ($\varepsilon(x) = \mathrm{def}u(x)$) that is the double layer potential of static elasticity defined in (2.46), (2.47). For a continuous density $b(x)$, the fields $u(x)$ and $\varepsilon(x)$ are continuous everywhere except on the surface Ω. The jump of the field $u(x)$ across Ω is defined in (2.135).

Consider integral (C.10) when $b = const$. Using Gauss' formula and taking into account that the Green function $G(x)$ satisfies the equation (B.1)

$$\nabla_\alpha C^0_{\alpha\beta\lambda\mu} \nabla_\lambda G_{\mu\nu}(x) = -\delta_{\beta\nu}\delta(x), \tag{C.12}$$

we obtain

$$\begin{aligned} \varepsilon_{\alpha\beta}(x) &= \int_V \nabla'_\nu \mathrm{K}_{\alpha\beta\lambda\mu}(x-x') C^0_{\lambda\mu\nu\rho} dx' b_\rho = \\ &= \int \nabla'_{(\alpha}\delta(x-x') V(x') dx' b_{\beta)} = n_{(\alpha}(x) b_{\beta)} \Omega(x), \end{aligned} \tag{C.13}$$

where $\Omega(x)$ is the delta-function concentrated on the surface Ω; $n(x)$ is the external normal to Ω.

Let $b(x)$ in (C.10) be a linear function

$$b_\alpha(x) = D_{\alpha\beta}(x-x')_\beta, \tag{C.14}$$

where D is a constant tensor. Consider the limiting values of the potential (C.10) when $x \to x_0 \in \Omega$. We denote by Ω_ρ the part of the surface Ω located inside a sphere of a small radius ρ with its center at the point x_0. Let $\bar{\Omega}_\rho$ be the complement Ω_ρ in the whole surface Ω. Introducing the local Cartesian coordinates y_1, y_2, y_3 with the origin at the point x_0 and axis y_3 directed along the normal $n(x_0) = n_0$ we present the integral (C.10) in the form

$$\begin{aligned} \int_\Omega \mathrm{K}(y-y') C^0 n(y') b(y') d\Omega' &= \int_{\Omega_\rho} \mathrm{K}(y-y') C^0 n(y') D(y-y') d\Omega' \\ &+ \int_{\bar{\Omega}_\rho} \mathrm{K}(y-y') C^0 n(y') D(y-y') d\Omega'. \end{aligned} \tag{C.15}$$

The integral over $\bar{\Omega}_\rho$ is a continuous function at the point $y = 0$ $(x = x_0)$.

Let us find the limit of the integral over Ω_ρ when $y \to 0$. Because in what follows $\rho \to 0$, it is possible to consider Ω_ρ as a plane circular disk and $n(y') = n_0$. Introducing the variables $\xi_i = y_i/|y|$, we have

$$\begin{aligned} \lim_{\rho\to 0} \int_{\Omega_\rho} \mathrm{K}(y-y') C^0 n_0 D(y-y') d\Omega' &= \\ = \int \mathrm{K}(\xi-\xi') C^0 n_0 D(\xi-\xi') \delta(n_0\cdot\xi') d\xi, \end{aligned} \tag{C.16}$$

where the integral in the right-hand side is calculated over the whole space R^3. It is taken into account that $\mathrm{K}(x)$ is an even homogeneous function of degree -3.

Let us present the function $n_0\delta(n_0 \cdot \xi')$ in (C.16) in the form

$$n_{0\alpha}\delta\left(n_0 \cdot \xi\right) = \nabla_\alpha H\left(n_0 \cdot \xi'\right), \quad H\left(t\right) = \begin{cases} 1\,, & t > 0 \\ 0\,, & t < 0 \end{cases}, \tag{C.17}$$

and after integrating by parts in (C.16) we find

$$\int \mathrm{K}\left(\xi - \xi'\right) C^0 n_0 D\left(\xi - \xi'\right) \delta\left(n_0 \cdot \xi'\right) d\xi' =$$

$$= -H\left(n_0 \cdot \xi\right) D + \int \mathrm{K}\left(\xi - \xi'\right) H\left(n_0 \cdot \xi'\right) d\xi'^0 D\,. \tag{C.18}$$

Using the convolution properties of the integral on the right-hand side we find

$$\int \mathrm{K}\left(\xi - \xi'\right) H\left(n_0 \cdot \xi'\right) d\xi' = \left(2\pi\right)^{-3} \int \mathrm{K}^*\left(k\right) H^*\left(k\right) e^{-ik\cdot\xi} dk =$$

$$= \frac{1}{2}[\mathrm{K}^*\left(0\right) + \mathrm{sign}\left(n_0 \cdot \xi\right) \mathrm{K}^*\left(n\right)]. \tag{C.19}$$

Here it is taken into account that the Fourier transform of the function $H(n_0 \cdot \xi)$ has the form

$$H^*\left(k\right) = H^*\left(k_1, k_2, k_3\right) = \left(2\pi\right)^2 \delta\left(k_1\right) \delta\left(k_2\right) \left[\pi\delta\left(k_3\right) + ik_3^{-1}\right]. \tag{C.20}$$

Let $J^+(n_0)$ be the limiting value of (C.16) when y tends to zero from the side of the normal n_0, and $J^-(n_0)$ be the same limit when y tends to zero from the opposed side. These limits are independent of the size of the region Ω_ρ and because of (C.18), (C.19) have the form

$$J^+\left(n_0\right) = \frac{1}{2}[\mathrm{K}^*\left(0\right) + \mathrm{K}^*(n_0)]C^0 D - D\,, \tag{C.21}$$

$$J^-\left(n_0\right) = \frac{1}{2}[\mathrm{K}^*\left(0\right) - \mathrm{K}^*(n_0)]C^0 D\,. \tag{C.22}$$

While changing n_0 to $(-n_0)$, the integral (C.16) changes sign, but its limiting values are defined by the same equations (C.21), (C.22). As a result, the equation $J^+(n_0) = -J^-(n_0)$ holds, and that leads to the equation $\mathrm{K}^*(0) = (C^0)^{-1}$ (compare with (2.86)).

Let us find the limiting values of the potential (C.10) with the density (C.14) on Ω. Letting ρ tend to zero in (C.15) and taking into account (C.21), we obtain that when $x \to \Omega$, the limiting values of the potential from the side of the normal (ε^+) and from the opposed side (ε^-) have the forms

$$\varepsilon^{\pm}_{\alpha\beta}(x) = p.v.\int_{\Omega} \mathrm{K}_{\alpha\beta\lambda\mu}(x-x')C^{0}_{\lambda\mu\nu\rho}n_{\nu}(x')\,D_{\rho\delta}(x-x')_{\delta}d\Omega' \pm$$
$$\pm\frac{1}{2}\Lambda_{\alpha\beta\lambda\mu}(n)\,D_{\lambda\mu}\,, \qquad \text{(C.23)}$$

where $p.v.\int ...d\Omega'$ means the Cauchy principal value of the integral that exists because the function $\mathrm{K}(x)$ is even,

$$\Lambda_{\alpha\beta\lambda\mu}(n) = \mathrm{K}^{*}_{\alpha\beta\nu\rho}(n)\,C^{0}_{\nu\rho\lambda\mu} - E^{1}_{\alpha\beta\lambda\mu}\,. \qquad \text{(C.24)}$$

Let us suppose that $b(x)$ in (C.10) is a smooth function, and present the potential $\varepsilon(x)$ in the form

$$\varepsilon_{\alpha\beta}(x) =$$
$$= \int_{\Omega} \mathrm{K}_{\alpha\beta\lambda\mu}(x-x')\,C^{0}_{\lambda\mu\nu\rho}n_{\nu}(x')\,[b_{\rho}(x') - b_{\rho}(x) - \partial_{\tau}b_{\rho}(x)\,(x'-x)_{\tau}]d\Omega' +$$
$$+ \int_{\Omega} \mathrm{K}_{\alpha\beta\lambda\mu}(x-x')\,C^{0}_{\lambda\mu\nu\rho}n_{\nu}(x')\,d\Omega' b_{\rho}(x) +$$
$$+ \int_{\Omega} \mathrm{K}_{\alpha\beta\lambda\mu}(x-x')\,C^{0}_{\lambda\mu\nu\rho}n_{\nu}(x')\,\partial_{\tau}b_{\rho}(x)\,(x'-x)_{\tau}\,d\Omega'. \qquad \text{(C.25)}$$

Here the first integral converges absolutely for all x and is a continuous function in the whole space. When $x \in \Omega$, it may be understood as the Cauchy principal value of the integral and presented in the form of the sum of two integrals with the densities $b(x') - b(x)$ and $\partial b(x)(x'-x)$, correspondingly. Each of these integrals exists in the mentioned sense. The second integral in (C.25) is the generalized function (C.13), and the last integral can be considered as the potential (C.10) with density (C.14). The jumps of this potential on Ω are defined in (C.23). It follows from this equation that the limit values of the potential $\varepsilon(x)$ on Ω have the forms ($x \in \Omega$)

$$\varepsilon^{\pm}_{\alpha\beta}(x) = p.v.\int_{\Omega} \mathrm{K}_{\alpha\beta\lambda\mu}(x-x')\,C^{0}_{\lambda\mu\nu\rho}n_{\nu}(x') \times$$
$$\times [b_{\rho}(x') - b_{\rho}(x)]d\Omega \mp \frac{1}{2}\Lambda_{\alpha\beta\lambda\mu}(n)\,\partial_{\lambda}b_{\mu}(x)\,, \qquad \text{(C.26)}$$

where the tensor $\Lambda(n)$ is defined in (C.24).

Let us consider the potential $\sigma(x)$

$$\sigma_{\alpha\beta}(x) = \int_{\Omega} \mathrm{S}_{\alpha\beta\lambda\mu}(x-x')\,n_{\lambda}(x')\,b_{\mu}(x')\,d\Omega', \qquad \text{(C.27)}$$

$$S_{\alpha\beta\lambda\mu}(x) = C^0_{\alpha\beta\nu\rho} K_{\nu\rho\tau\delta}(x) C^0_{\tau\delta\lambda\mu} - C^0_{\alpha\beta\lambda\mu}\delta(x) , \tag{C.28}$$

where Ω is a closed surface in R^3.

The potentials (C.10) and (C.27) are connected by the equation

$$\sigma_{\alpha\beta}(x) = C^0_{\alpha\beta\lambda\mu}\varepsilon_{\lambda\mu}(x) - C^0_{\alpha\beta\lambda\mu} n_\lambda(x) b_\mu(x) \Omega(x) . \tag{C.29}$$

Therefore, it follows from (C.26) that the limiting values of $\sigma_{\alpha\beta}(x)$ on Ω are determined by the equation

$$\sigma^{\pm}_{\alpha\beta}(x) = p.v. \int_{\Omega} \mathrm{S}_{\alpha\beta\lambda\mu}(x-x') n_\lambda(x') [b_\mu(x') - b_\mu(x)] d\Omega' = \mp\frac{1}{2}\mathrm{S}^*_{\alpha\beta\lambda\mu}(n)\partial_\lambda b_\mu(x) , \quad x \in \Omega , \; . \tag{C.30}$$

$$\mathrm{S}^*(n) = C^0 K^*(n) C^0 - C^0. \tag{C.31}$$

Taking into account Equation (2.58) for $K^*(n)$ we see that the vector $n_\alpha \sigma_{\alpha\beta}(x)$ is continuous on the surface Ω ($n_\alpha \mathrm{S}^*_{\alpha\beta\lambda\mu}(n) = 0$). When $x \in \Omega$, the equation for $n_\alpha(x)\sigma_{\alpha\beta}(x)$ takes the form

$$n_\alpha(x)\sigma_{\alpha\beta}(x) = -p.v. \int_{\Omega} T_{\alpha\beta}(x,x') [b_\beta(x') - b_\beta(x)] d\Omega', \tag{C.32}$$

$$T_{\alpha\beta}(x,x') = -n_\lambda(x) \mathrm{S}_{\lambda\alpha\beta\mu}(x-x') n_\mu(x') . \tag{C.33}$$

If the surface Ω is not closed, the integration in (C.26), (C.30) can be performed over any smooth closed surface that includes Ω, and $b(x) = 0$ outside Ω. Let $\bar{\Omega}$ be the complement of Ω up to an arbitrary closed surface. Taking into account Equation (2.67) for $\mathrm{S}(x)$

$$\mathrm{S}_{\alpha\beta\lambda\mu}(x) = \mathrm{rot}_{\nu\lambda}\mathrm{rot}_{\rho\mu} Z_{\alpha\beta\nu\rho}(x) , \tag{C.34}$$

where $Z(x)$ is the Green tensor for the internal stresses [105], and Stokes' theorem (C.4) we obtain the equation

$$\int_{\bar{\Omega}} \mathrm{S}_{\alpha\beta\lambda\mu}(x-x') n_\mu(x') d\Omega' = \oint_{\Gamma} \mathrm{rot}_{\nu\lambda} Z_{\alpha\beta\nu\mu}(x-x') d\Gamma'_\mu, \tag{C.35}$$

where $d\Gamma_\mu$ is the vector element on the border Γ of $\bar{\Omega}$. Hence, in the case of a nonclosed surface, (C.30) takes the form

$$\sigma^{\pm}_{\alpha\beta}(x) = p.v.\int_{\Omega} \mathrm{S}_{\alpha\beta\lambda\mu}(x-x')\, n_\lambda(x')\,[b_\mu(x') - b_\mu(x)]d\Omega' +$$

$$+\oint_{\Gamma} \mathrm{rot}_{\nu\lambda} Z_{\alpha\beta\nu\mu}(x-x')\, d\Gamma'_\lambda b_\mu(x) \mp \frac{1}{2} S^{*}_{\alpha\beta\lambda\mu}(n)\,\partial_\lambda b_\mu(x)\,. \tag{C.36}$$

From this equation and (C.26), (C.27) it follows that for a nonclosed surface, the limit values of potential (C.10) have the form

$$\varepsilon^{\pm}_{\alpha\beta}(x) = p.v.\int_{\Omega} K_{\alpha\beta\lambda\mu}(x-x')\, C^{0}_{\lambda\mu\nu\rho} n_\nu(x')\,[b_\rho(x') - b_\rho(x)]d\Omega' +$$

$$+ \left(C^{0}\right)^{-1}_{\alpha\beta\lambda\mu} \oint_{\Gamma} \mathrm{rot}_{\nu\delta} Z_{\lambda\mu\nu\tau}(x-x')\, d\Gamma'_\delta b_\tau(x) \mp \frac{1}{2} \Lambda_{\alpha\beta\lambda\mu}(n)\,\partial_\lambda b_\mu(x)\,. \tag{C.37}$$

On the points of a nonclosed surface Ω, the equation for the vector field $n_\alpha(x)\sigma_{\alpha\beta}(x)$, where $\sigma(x)$ is potential (C.27), has the following form:

$$n_\beta(x)\,\sigma_{\beta\alpha}(x) = -p.v.\int_{\Omega} T_{\alpha\beta}(x,x')\,[b_\beta(x') - b_\beta(x)]d\Omega' +$$

$$+\oint_{\Gamma} n_\beta(x)\, \mathrm{rot}_{\nu\lambda} Z_{\alpha\beta\nu\mu}(x-x')\, d\Gamma'_\lambda b_\mu(x)\,. \tag{C.38}$$

C.3 Potentials with densities that are tensors of a surface Ω

Let us consider the potential (C.10)

$$\varepsilon_{\alpha\beta}(x) = \int_{\Omega} K_{\alpha\beta\lambda\mu}(x-x')\, q_{\lambda\mu}(x')\, d\Omega', \tag{C.39}$$

the density $q(x)$ of which is a tensor of the surface Ω:

$$n_\alpha q_{\alpha\beta}(x) = 0\,, \quad \Theta_{\alpha\beta\lambda\mu}(x)\, q_{\lambda\mu}(x) = q_{\alpha\beta}(x)\,, \quad x \in \Omega\,. \tag{C.40}$$

Here the projector $\Theta(x)$ is defined in (4.12)

$$\Theta(x) = \Theta(n) = E^1 - 2E^5(n) + E^6(n)\,, \tag{C.41}$$

$E^i(n)$ are the elements of the E-basis (A.1), (A.2), $n(x)$ is a normal to Ω.
Using Gauss' theorem (C.9) we find

$$\varepsilon_{\alpha\beta} = -\int_{\Omega} \nabla'_{\lambda}\nabla'_{(\alpha}G_{\beta)\mu}(x-x')\Theta_{\lambda\mu\nu\rho}(x')\,q_{\nu\rho}(x')\,d\Omega' =$$

$$= -\int_{\Omega} \partial'_{\lambda}\left[\nabla'_{(\alpha}G_{\beta)\mu}(x-x')\,q_{\lambda\mu}(x')\right]d\Omega' +$$

$$+ \int_{\Omega} \nabla'_{(\alpha}G_{\beta)\mu}(x-x')\,\partial'_{\lambda}q_{\lambda\mu}(x')\,d\Omega' =$$

$$= \int_{\Omega} \nabla'_{(\alpha}G_{\beta)\mu}(x-x')\,\partial'_{\lambda}q_{\lambda\mu}(x')\,d\Omega' +$$

$$- \oint_{\Gamma} \nabla'_{(\alpha}G_{\beta)\mu}(x-x')\,q_{\lambda\mu}(x')\,e_{\lambda}(x')\,d\Gamma', \qquad (C.42)$$

where $e_{\lambda}(x)$ is the normal to Γ, defined in Section 3.3. Here we took into account that

$$\Theta_{\alpha\beta\lambda\mu}(x)\nabla_{\lambda}A_{\mu\ldots\nu}(x) = \Theta_{\alpha\beta\lambda\mu}(x)\partial_{\lambda}A_{\mu\ldots\nu}(x), \qquad (C.43)$$

∂_{α} is the operator of differentiation along the surface (C.8).

The first integral in this relation is the potential (2.108); its limiting values on Ω are defined in (2.132). Hence, the limiting values of the potential (C.39) take the form

$$\varepsilon^{\pm}_{\alpha\beta}(x) = p.v.\int_{\Omega} \nabla'_{(\alpha}G_{\beta)\mu}(x-x')\,\partial'_{\lambda}q_{\lambda\mu}(x')\,d\Omega' +$$

$$- \oint_{\Gamma} \nabla'_{(\alpha}G_{\beta)\mu}(x-x')\,q_{\lambda\mu}(x')\,e_{\lambda}(x')\,d\Gamma' \pm \frac{1}{2}\Lambda^{0}_{\alpha\beta\mu}(n)\,\partial_{\lambda}q_{\lambda\mu}(x)\,, \qquad (C.44)$$

$$\Lambda^{0}_{\alpha\beta\mu}(n) = n_{(\alpha}G^{*}_{\beta)\mu}(n)\,, \quad n = n(x)\,, \quad x \in \Omega\,. \qquad (C.45)$$

The principal value of the integral in this equation can be rewritten in the form

$$p.v.\int_{\Omega} \nabla'_{(\alpha}G_{\beta)\mu}(x-x')\,\partial'_{\lambda}q_{\lambda\mu}(x')\,d\Omega' =$$

$$= p.v.\int_{\Omega} \nabla'_{(\alpha}G_{\beta)\mu}(x-x')\,\partial'_{\lambda}\left[q_{\lambda\mu}(x') - q^{0}_{\lambda\mu}(x,x')\right]d\Omega', \qquad (C.46)$$

where q^{0} is a tensor field on Ω satisfying the equation

$$\partial'_{\lambda}q^{0}_{\lambda\mu}(x,x') = 0, \quad x, x' \in \Omega \qquad (C.47)$$

and coincides with $q(x)$ when $x' = x$.

Using Gauss' theorem for surfaces (C.9), we transform the integral (C.46) into the following one:

$$p.v.\int_{\Omega}\nabla'_{(\alpha}G_{\beta)\mu}(x-x')\,\partial'_{\lambda}q_{\lambda\mu}(x')\,d\Omega'=$$

$$=-p.v.\int_{\Omega}\partial'_{\lambda}\nabla'_{(\alpha}G_{\beta)\mu}(x-x')\left[q_{\lambda\mu}(x')-q^{0}_{\lambda\mu}(x,x')\right]d\Omega'+$$

$$+\oint_{\Gamma}\nabla'_{(\alpha}G_{\beta)\mu}(x-x')\,[q_{\lambda\mu}(x')-q_{\lambda\mu}(x,x')]d\Omega'. \tag{C.48}$$

The regular presentation of the limiting values of the potential (C.39) at a point of the surface Ω follows from this equation in the form

$$\varepsilon^{\pm}_{\alpha\beta}(x)=p.v.\int_{\Omega}K_{\alpha\beta\lambda\mu}(x-x')\left[q_{\lambda\mu}(x')-q^{0}_{\lambda\mu}(x,x')\right]d\Omega'+$$

$$-\oint_{\Gamma}\nabla'_{(\alpha}G_{\beta)\mu}(x-x')\,q^{0}_{\lambda\mu}(x,x')\,e_{\lambda}(x')\,d\Gamma'\pm\frac{1}{2}\Lambda^{0}_{\alpha\beta\mu}(n)\,\partial_{\lambda}q_{\lambda\mu}(x). \tag{C.49}$$

From the definition of the tensor Λ^0 in (C.45) it follows that the "tangential" part of the tensor $\varepsilon_{\alpha\beta}(x)$ $(\Theta_{\alpha\beta\lambda\mu}(x)\varepsilon_{\lambda\mu}(x))$ is continuous across Ω and determined by the equation

$$\Theta_{\alpha\beta\lambda\mu}(x)\,\varepsilon_{\lambda\mu}(x)=p.v.\int_{\Omega}U_{\alpha\beta\lambda\mu}(x,x')\left[q_{\lambda\mu}(x')-q^{0}_{\lambda\mu}(x,x')\right]d\Omega'+$$

$$-\Theta_{\alpha\beta\lambda\mu}(x)\oint_{\Gamma}\nabla_{\lambda}G_{\mu\nu}(x-x')\,q^{0}_{\rho\nu}(x,x')\,e_{\rho}(x')\,d\Gamma, \tag{C.50}$$

$$U_{\alpha\beta\lambda\mu}(x,x')=\Theta_{\alpha\beta\nu\rho}(x)\,K_{\nu\rho\tau\delta}(x-x')\,\Theta_{\tau\delta\lambda\mu}(x'). \tag{C.51}$$

D. Transition through the layers in the problems of thermoelasticity for multilayered inclusions

D.1 Elastic and thermoelastic problems for a spherical multilayered inclusion

The vectors X^i and X^{i+1} of the solutions of the elasticity problem for a layered spherical inclusion are connected at the border a_i between the i-th and $(i+1)$-th layers by Equation (3.354)

$$X^{i+1}(a_i) = F^i + \Gamma^i X^i(a_i) \tag{D.1}$$

Here the transition matrix Γ^i through the border between the i-th and the $(i+1)$-th layers at the point $r = a_i$ $(i = 1, 2, ..., N)$ has dimension (12×12) and is presented in the form ($\oplus$ is the direct (Cartesian) product of the matrices)

$$\Gamma^i = \Gamma_1^i \oplus \Gamma_2^i \oplus \Gamma_2^i \,, \tag{D.2}$$

where the matrices Γ_1^i (8×8) and Γ_2^i (2×2) depend on the elastic moduli of the i-th and $(i+1)$-th layers, and have the forms

$$\Gamma_1^i = \begin{bmatrix} 1 & 0 & 0 & 0 & 0 & 0 & 0 & 0 \\ 0 & 1 & 0 & 0 & 0 & 0 & 0 & 0 \\ \Gamma_{31} & \Gamma_{32} & \Gamma_{33} & 0 & \Gamma_{35} & \Gamma_{36} & 0 & 0 \\ \Gamma_{41} & \Gamma_{42} & 0 & \Gamma_{44} & \Gamma_{45} & \Gamma_{46} & 0 & 0 \\ 0 & 0 & 0 & 0 & 1 & 0 & 0 & 0 \\ 0 & 0 & 0 & 0 & 0 & 1 & 0 & 0 \\ \Gamma_{71} & \Gamma_{72} & 0 & 0 & \Gamma_{75} & \Gamma_{76} & \Gamma_{77} & 0 \\ \Gamma_{81} & \Gamma_{82} & 0 & 0 & \Gamma_{85} & \Gamma_{86} & 0 & \Gamma_{88} \end{bmatrix}, \quad \Gamma_2^i = \begin{bmatrix} 1 & 0 \\ \Gamma_{21} & \Gamma_{22} \end{bmatrix}, \tag{D.3}$$

$$\Gamma_{31} = -\frac{6[\mu]_i}{\mu_{i+1}}, \quad \Gamma_{32} = -\frac{3[\mu]_i}{\mu_{i+1}}, \quad \Gamma_{33} = \Gamma_{44} = \frac{\mu_i}{\mu_{i+1}}, \tag{D.4}$$

$$\Gamma_{35} = \Gamma_{36} = -\frac{4[\mu]_i}{\mu_{i+1}}, \tag{D.5}$$

$$\Gamma_{41} = \frac{48[\mu]_i}{\mu_{i+1}}, \quad \Gamma_{42} = \frac{25[\mu]_i}{\mu_{i+1}}, \quad \Gamma_{44} - \frac{\mu_i}{\mu_{i+1}}, \tag{D.6}$$

$$\Gamma_{45} = \frac{32[\mu]_i}{\mu_{i+1}}\,, \; \Gamma_{46} = \frac{16[\mu]_i}{\mu_{i+1}}\,,$$

$$\Gamma_{71} = \Gamma_{72} = -\frac{6[\mu]_i}{\lambda_{i+1} + 2\mu_{i+1}}\,, \; \Gamma_{75} = -\frac{4[\mu]_i}{\lambda_{i+1} + 2\mu_{i+1}}\,, \tag{D.7}$$

$$\Gamma_{76} = -\frac{[5\lambda + 6\mu]_i}{\lambda_{i+1} + 2\mu_{i+1}}\,, \; \Gamma_{77} = \frac{\lambda_i + 2\mu_i}{\lambda_{i+1} + 2\mu_{i+1}}\,,$$

$$\Gamma_{81} = \frac{48[\mu]_i}{\lambda_{i+1} + 2\mu_{i+1}}\,, \; \Gamma_{82} = \frac{24[\mu]_i}{\lambda_{i+1} + 2\mu_{i+1}}\,, \tag{D.8}$$

$$\Gamma_{85} = \frac{32[\mu]_i}{\lambda_{i+1} + 2\mu_{i+1}}\,, \; \Gamma_{86} = \frac{[25\lambda + 42\mu]_i}{\lambda_{i+1} + 2\mu_{i+1}}\,, \; \Gamma_{88} = \frac{\lambda_i + 2\mu_i}{\lambda_{i+1} + 2\mu_{i+1}}\,, \tag{D.9}$$

$$\Gamma_{21} = -\frac{[3\lambda + 2\mu]_i}{\lambda_{i+1} + 2\mu_{i+1}}\,, \; \Gamma_{22} = \frac{\lambda_1 + 2\mu_i}{\lambda_{i+1} + 2\mu_{i+1}}\,, \tag{D.10}$$

$$[\mu]_i = \mu_{i+1} - \mu_i\,, \; [\lambda]_i = \lambda_{i+1} - \lambda_i\,. \tag{D.11}$$

Here λ_i, μ_i are the Lamé coefficients of the i-th layer ($i = 1, 2, ..., N$), $\lambda_{N+1} = \lambda_0$, $\mu_{N+1} = \mu_0$ are the same for the matrix. The components of the 12-dimensional vector–column F^i in (D.1) have the form

$$F_1^i = F_2^i = F_5^i = F_6^i = F_9^i = F_{11}^i = 0\,, \tag{D.12}$$

$$F_3^i = -\frac{2[\mu]_i}{\mu_{i+1}}\,, \; F_4^i = \frac{16[\mu]_i}{\mu_{i+1}}\,, \tag{D.13}$$

$$F_7^i = -\frac{2[\mu]_i}{\lambda_{i+1} + 2\mu_{i+1}}\,, \; F_8^i = \frac{16[\mu]_i}{\lambda_{i+1} + 2\mu_{i+1}}\,, \tag{D.14}$$

$$F_{10}^i = -\frac{[5\lambda + 6\mu]}{\lambda_{i+1} + 2\mu_{i+1}}\,, \; F_{12}^i = \frac{[\gamma]_i}{\lambda_{i+1} + 2\mu_{i+1}}\,, \tag{D.15}$$

$$[\gamma]_i = \gamma_{i+1} - \gamma_i\,, \; \gamma_i = (3\lambda_i + 2\mu_i)(\alpha_i - \alpha_0)\,, \tag{D.16}$$

where α_i is the thermal expansion coefficient for i-th layer, α_0 is the same for the matrix.

D.2 Elastic and thermoelastic problems for a cylindrical multilayered inclusion

The transition matrix Γ^i through the border between the i-th and $(i+1)$-th layers has dimension (16×16) and is represented in the form

$$\Gamma^i = \Gamma_1^i \oplus \Gamma_2^i \oplus \Gamma_2^i \oplus \Gamma_2^i \oplus \Gamma_2^i\,, \tag{D.17}$$

where the matrices Γ_1^i (8×8) and Γ_2^i (2×2) have the forms

$$\Gamma_1^i = \begin{bmatrix} 1 & 0 & 0 & 0 & 0 & 0 & 0 & 0 \\ 0 & 1 & 0 & 0 & 0 & 0 & 0 & 0 \\ \Gamma_{31} & \Gamma_{32} & \Gamma_{33} & 0 & \Gamma_{35} & \Gamma_{36} & 0 & 0 \\ \Gamma_{41} & \Gamma_{42} & 0 & \Gamma_{44} & \Gamma_{45} & \Gamma_{46} & 0 & 0 \\ 0 & 0 & 0 & 0 & 1 & 0 & 0 & 0 \\ 0 & 0 & 0 & 0 & 0 & 1 & 0 & 0 \\ \Gamma_{71} & \Gamma_{72} & 0 & 0 & \Gamma_{75} & \Gamma_{76} & \Gamma_{77} & 0 \\ \Gamma_{81} & \Gamma_{82} & 0 & 0 & \Gamma_{85} & \Gamma_{86} & 0 & \Gamma_{88} \end{bmatrix}, \tag{D.18}$$

$$\Gamma_{31} = -\frac{4[\mu]_i}{\mu_{i+1}}, \quad \Gamma_{32} = -\frac{2[\mu]_i}{\mu_{i+1}}, \quad \Gamma_{33} = \Gamma_{44} = \frac{\mu_i}{\mu_{i+1}}, \tag{D.19}$$

$$\Gamma_{35} = \Gamma_{36} = -\frac{4[\mu]_i}{\mu_{i+1}}, \tag{D.20}$$

$$\Gamma_{41} = \frac{24[\mu]_i}{\mu_{i+1}}, \quad \Gamma_{42} = \frac{16[\mu]_i}{\mu_{i+1}}, \quad \Gamma_{45} = \frac{24[\mu]_i}{\mu_{i+1}}, \quad \Gamma_{46} = \frac{12[\mu]_i}{\mu_{i+1}}, \tag{D.21}$$

$$\Gamma_{71} = \Gamma_{72} = \Gamma_{75} = -\frac{4[\mu]_i}{\lambda_{i+1} + 2\mu_{i+1}}, \tag{D.22}$$

$$\Gamma_{76} = -\frac{2[2\lambda + 3\mu]_i}{\lambda_{i+1} + 2\mu_{i+1}}, \quad \Gamma_{77} = \frac{\lambda_i + 2\mu_i}{\lambda_{i+1} + 2\mu_{i+1}}, \tag{D.23}$$

$$\Gamma_{81} = \frac{24[\mu]_i}{\lambda_{i+1} + 2\mu_{i+1}}, \quad \Gamma_{82} = \frac{12[\mu]_i}{\lambda_{i+1} + 2\mu_{i+1}}, \quad \Gamma_{85} = \frac{24[\mu]_i}{\lambda_{i+1} + 2\mu_{i+1}}, \tag{D.24}$$

$$\Gamma_{86} = \frac{[16\lambda + 32\mu]_i}{\lambda_{i+1} + 2\mu_{i+1}}, \quad \Gamma_{88} = \frac{\lambda_i + 2\mu_i}{\lambda_{i+1} + 2\mu_{i+1}}, \tag{D.25}$$

$$\Gamma_2^{i+1} = \begin{bmatrix} 1 & 0 \\ \Gamma_{21} & \Gamma_{22} \end{bmatrix}, \quad \Gamma_{21} = -\frac{2[\lambda + \mu]}{\lambda_{i+1} + 2\mu_{i+1}}, \quad \Gamma_{22} = \frac{\lambda_i + 2\mu_i}{\lambda_{i+1} + 2\mu_{i+1}}. \tag{D.26}$$

The components of the 16D-transition vector-column F^i are defined as follows:

$$F_1^i = F_2^i = F_5^i = F_6^i = F_9^i = F_{11}^i = F_{13}^i = F_{15}^i = 0, \tag{D.27}$$

$$F_3^i = -\frac{2[\mu]}{\mu_{i+1}}, \quad F_4^i = \frac{12[\mu]}{\mu_{i+1}}, \tag{D.28}$$

$$F_7^i = -\frac{2[\mu]_i}{\lambda_{i+1} + 2\mu_{i+1}}, \quad F_8^i = \frac{12[\mu]_i}{\lambda_{i+1} + 2\mu_{i+1}}, \tag{D.29}$$

$$F_{10}^i = F_{12}^i = F_{14}^i = -\frac{2[\lambda + \mu]}{\lambda_{i+1} + 2\mu_{i+1}}, \quad F_{16}^i = \frac{[\gamma]_i}{\lambda_{i+1} + 2\mu_{i+1}}, \tag{D.30}$$

$$[\gamma]_i = \gamma_{i+1} - \gamma_i, \quad \gamma_i = (\alpha_i - \alpha_0)(3\lambda_i + 2\mu_i). \tag{D.31}$$

Here λ_i, μ_i are the Lamé parameters and α_i the temperature expansion coefficient of the i-th layer, $\lambda_0, \mu_0, \alpha_0$ are the same parameters for the matrix material.

E. Correlation functions of random sets of spherical inclusions

E.1 The Percus–Yevick correlation function of nonpenetrating sets of spheres in the 3D-case

The Percus–Yevick function $\psi(x)$ is the normalized probability of distribution of the centers of nonoverlapping spheres of radii a (by the conditions that the origin ($x = 0$) is occupied by the center of another sphere. This function is defined by the equations [211]

$$\psi(\zeta) = 1 + \frac{1}{12\pi p\zeta}\int_0^\infty \frac{w^2(s)}{1 - w(s)} \sin\left(\frac{s\zeta}{2}\right) sds \quad \text{if} \quad \zeta \geq 2,$$

$$\psi(\zeta) = 0, \quad \text{if} \quad \zeta < 2, \qquad \zeta = \frac{|x|}{a}. \tag{E.1}$$

The function $w(s)$ in this equation is the following integral

$$w(s) = \frac{24p}{s}\int_0^1 q(t)\sin(ts)tdt, \qquad q(t) = -a_1(1 + p\frac{t^3}{2}) - a_2 t,$$

$$a_1 = \frac{(1+2p)^2}{(1-p)^4}, \quad a_2 = \frac{-3p(2+p)^2}{2(1-p)^4}, \tag{E.2}$$

E.2 The Percus–Yevick correlation function of nonpenetrating sets of spheres in the 2D-case

In the 2D-case, the Percus–Yevick correlation function is presented in the form

$$\psi(\zeta) = 1 + \frac{1}{8p}\int_0^\infty h(x)J_0\left(\frac{\zeta x}{2}\right) xdx, \ |\zeta| \geq 2;$$

$$\psi(\zeta) = 0, \quad |\zeta| < 2. \tag{E.3}$$

Here $J_0(z)$ is the Bessel function of order 0 and the function $h(x)$ is

$$h(x) = \frac{s(x)^2}{1 - s(x)}. \tag{E.4}$$

The function $s(x)$ is the solution of the following nonlinear integral equation

$$s(x) = -8p\frac{J_1(x)}{x} - \frac{1}{2}\int_0^{\infty} \frac{s(t)^2}{1 - s(t)} \left[J_0^2(t) + J_1^2(t)\right] dt. \tag{E.5}$$

E.3 Correlation functions of the Boolean random sets of spheres and cylinders

Let us implant in a finite region W of 3D(2D)-space a number of spheres of the same volume v (inclusions) so that the positions of their centers are independent and homogeneously distributed in W (the Poisson process). If $W \to \infty$, and the region W occupies all the space in this limit, we obtain the Boolean random set of the spheres in this space.

Let us calculate some statistical characteristics of the Boolean random set of inclusions, and start with the probability of a point x to be inside an arbitrary inclusion. This probability coincides with the volume concentration p of the inclusions. For one sphere, the probability not to cover a point x inside W is

$$1 - \frac{v}{W} \tag{E.6}$$

because of a homogeneous distribution of the center of the sphere inside W. If one implants n spheres inside W, the point x will be not covered by the spheres with the probability

$$\left(1 - \frac{v}{W}\right)^n \tag{E.7}$$

because every implantation is independent on the others. Let us introduce a fixed parameter

$$\lambda = \frac{n}{W} \tag{E.8}$$

that is called the intensity of the Poisson process. Thus $n = \lambda W$, and in the limit $W \to \infty$ we obtain

$$q = \lim_{W\to\infty} \left(1 - \frac{v}{W}\right)^{\lambda W} = e^{-\lambda v}, \tag{E.9}$$

and q is the probability that a point x is not covered by the inclusions. As a result, the probability for x to be inside inclusion is

$$p = 1 - q = 1 - e^{-\lambda v}. \tag{E.10}$$

Let us consider the correlation functions of the Boolean random sets of inclusions. Because of independency of the implantations, the probability for two points x and y to be inside inclusions is p^2, where p has form (E.10). In the effective field method, there appears the probability of two points x and y to be covered by different spheres. Firstly, we consider the probability for two points to be covered by one sphere. The latter happens if the center of the sphere is situated in the region of intersection of two spheres which centers are at the points x and y. Let us denote this region as w. Thus, the probability that the point x is covered by an inclusion but by the same time the point y is not covered by the inclusions is $1 - e^{-\lambda(v-w)}$. The same probability but for the point y is also $1 - e^{-\lambda(v-w)}$. As the result, the probability C of the point x and y to be covered by different inclusions is

$$C(x,y) = \left(1 - e^{-\lambda(v-w)}\right) e^{-\lambda w}. \tag{E.11}$$

The multiplier $e^{-\lambda w}$ is the probability that a sphere does not cover both points x and y. Because the region w may be presented in the form

$$w(x,y) = vh(x-y), \tag{E.12}$$

where the function $h(R) = H(R)$ in 3D-case is defined in (E.27), and in 2D-case in (E.25), the function $C(x,y)$ takes the form

$$C(x,y) = C(R) = \left(1 - q^{(1-h(R))}\right)^2 q^{h(R)}, \tag{E.13}$$

$$q = e^{-\lambda v}, \quad R = x - y. \tag{E.14}$$

The specific correlation function $\Phi(R)$ that presents in the equations of the effective field method is connected with the function $C(R)$ by the equation

$$\Phi(R) = 1 - \frac{1}{p^2} C(R). \tag{E.15}$$

E.3.1 Random models of two populations of inclusions

We will restrict the development to spherical and cylindrical inclusions. The change of spheres to spheroids is simply obtained by affine transformations of the coordinates, which can be applied to the two-point probability functions. Various functions $\Phi(x) = 1 - \Psi(x)$ which are required for the calculation of effective properties are deduced from the following probabilities, that are worked out below for some random models of hybrid composites:

$$C_{cc}(x) = P\{y \in C_1, y + x \in C_2\}, \tag{E.16}$$

$$C_{ss}(x) = P\{y \in S_1, y + x \in S_2\}, \tag{E.17}$$

$$C_{cs}(x) = P\{y \in C, y + x \in S\}. \tag{E.18}$$

(probability for the two points x and $x+y$ to belong to two separate cylinders C_1 and C_2, two separate spheres S_1 and S_2, or one sphere S and one cylinder C).

For instance, we have for the function $\Phi_{cc}(x)$, $\Phi_{ss}(x)$, $\Phi_{cs}(x)$

$$\Phi_{cc}(x) = 1 - \Psi_{cc}(x) = 1 - \frac{1}{p_c^2} C_{cc}(x), \tag{E.19}$$

$$\Phi_{ss}(x) = 1 - \Psi_{ss}(x) = 1 - \frac{1}{p_s^2} C_{ss}(x), \tag{E.20}$$

$$\Phi_{cs}(x) = 1 - \Psi_{cs}(x) = 1 - \frac{1}{p_c p_s} C_{cs}(x). \tag{E.21}$$

For the construction of various models combining spheres and parallel cylinders, we will make use of two Poisson point processes: a process in the 3D- space R^3, with intensity (average number of points per unit volume) λ_s, for the implantation of spheres S (with volume V_s), and a process in R^2 (orthogonal to the axis of cylinders) with intensity λ_c for the implantation of discs D (with area A_c), that generates sections of cylinders. We denote $S(y)$, $S(y+x)$, $D(y)$, $D(y+x)$ the spheres located at points x and $x+y$, and the discs located at the projections of x and $x+y$ on the horizontal plane. The complementary set of the set A is denoted as A^c.

The Boolean model is obtained by implantation of spheres and cylinders on the points of the Poisson processes. We first give the priority to cylinders when there is an overlap between a cylinder and spheres. For this model, the probabilities $q_c = 1 - p_c = 1 - P\{y \text{ belongs to cylinders}\}$ and $q_s = 1 - p_s = 1 - P\{y \text{ belongs to spheres}\}$ are given by

$$q_c = \exp(-\lambda_c A_c), \quad q_s = 1 - (1 - q_{cs}) q_c, \quad q_{cs} = \exp(-\lambda_s V_s), \tag{E.22}$$

since $q_c = P\{D(y) \text{ contains no Poisson point of the second process}\}$, and $q_{cs} = P\{S(y) \text{ contains no Poisson point of the first process}\}$. To calculate the probability $C_{cc}(x)$, we have to consider the independent events: $\{D(y) \cap D(y+x)^c \text{ contains at least one Poisson point}\}$, $\{D(y)^c \cap D(y+x) \text{ contains at least one Poisson point}\}$, $\{D(y) \cap D(y+x) \text{ contains no Poisson point}\}$. Therefore, we get, denoting $\rho^2 = x_1^2 + x_2^2$, and using the expression $h(x)$ (7.199) involving the area of the intersection of two discs of radii 1 located at points y and $y+x$:

$$C_{cc}(x) = C_{cc}(\rho) = \left[1 - {q_c}^{1-h(\rho)}\right]^2 {q_c}^{h(\rho)}, \tag{E.23}$$

$$h(\rho) = \frac{2}{\pi}\left[\arccos(\frac{\rho}{2}) - \frac{\rho}{2}\sqrt{1-\frac{\rho}{2}}\right] \quad \text{if} \quad \rho < 2; \tag{E.24}$$

$$h(\rho) = 0 \quad \text{if} \quad \rho \geq 2. \tag{E.25}$$

The probability $C_{ss}(x)$ is derived from the independent events $\{S(y) \cap S(y+x)^c$ contains at least one Poisson point$\}$, $\{S(y)^c \cap S(y+x)$ contains at least one Poisson point$\}$, $\{S(y) \cap S(y+x)$ contains no Poisson point$\}$, $\{D(y) \cup D(y+x)$ contains no Poisson point of the second process$\}$. We get, using the expression $H(\xi)$ involving the volume of the intersection of two spheres of radii 1 located at points y and $y+x$:

$$C_{ss}(x) = \left[1 - {q_{cs}}^{1-H(R)}\right]^2 q_{cs}^{H(R)} {q_c}^{2-h(\rho)}, \; R = |x|, \tag{E.26}$$

$$H(R) = 1 - \frac{3}{4}R + \frac{1}{16}R^3 \; \text{ if } \; R < 2; \quad H(R) = 0 \; \text{ if } \; R \geq 2. \tag{E.27}$$

The probability $C_{cs}(x)$ is calculated from the independent events $\{D(y) \cap D(y+x)^c$ contains at least one Poisson point of the second process$\}$, $\{D(y+x)$ contains no Poisson point of the second process$\}$, $\{S(y+x)$ contains at least one Poisson point$\}$

$$C_{cs}(x) = \left[1 - {q_c}^{1-h(\rho)}\right] q_c(1 - q_s). \tag{E.28}$$

A second version of the model gives the priority to spheres over cylinders. The same proof can be used, after permutation of expressions concerning spheres and cylinders in previous equations. We get now

$$q_s = \exp(-\lambda_s V_s), \quad q_c = 1 - (1 - q_{sc})q_s, \tag{E.29}$$

$$q_{sc} = \exp(-\lambda_c A_c), \tag{E.30}$$

$$C_{cc}(x) = C_{cc}(\rho) = \left[1 - {q_{sc}}^{1-h(\rho)}\right]^2 q_{sc}^{h(\rho)} {q_s}^{2-H(R)}, \tag{E.31}$$

$$C_{ss}(x) = \left[1 - {q_s}^{1-H(R)}\right]^2 {q_s}^{H(R)}, \tag{E.32}$$

$$C_{cs}(x) = \left[1 - {q_s}^{1-H(R)}\right] q_s(1 - q_c). \tag{E.33}$$

Starting as for the Boolean model, the dilution model is obtained by counting the number of grains $N(y) = N_c(y) + N_s(y)$ covering each point y in space, $N_c(y)$ and $N_s(y)$ being the numbers of cylinders and of spheres covering y. For the purpose of the present application, we keep all points y where

$N(y) = 1$.. Since $N(y)$ is a Poisson variable with average $\theta = \lambda_s V_s + \lambda_c A_c = \theta_s + \theta_c$, we have

$$p = P\{N(y) = 1\} = p_c + p_s = \theta \exp(-\theta). \tag{E.34}$$

Similarly, we get the volume fractions p_c and p_s from

$$\begin{aligned} p_c &= P\{D(y) \text{ contains one point, and } S(y) \text{ contains no point}\} \\ &= \theta_c \exp(-\theta), \end{aligned} \tag{E.35}$$

$$\begin{aligned} p_s &= P\{S(y) \text{ contains one point, and } D(y) \text{ contains no point}\} \\ &= \theta_s \exp(-\theta). \end{aligned} \tag{E.36}$$

To calculate $C_{ss}(x)$, we have to consider the probability of the following independent events: $\{S(y) \cap S(y+x)^c$ contains one Poisson point$\}$, $\{S(y)^c \cap S(y+x)$ contains one Poisson point$\}$, $\{S(y) \cap S(y+x)$ contains no Poisson point$\}$, $\{D(y) \cup D(y+x)$ contains no Poisson point of the second process$\}$. Using the independence between the number of Poisson points in separate domains of space, we get

$$C_{ss}(x) = p_s^2 \left(1 - H(R)\right)^2 \exp\left[\theta H_{cs}(x)\right]. \tag{E.37}$$

For $C_{cc}(x)$, we need the following independent events: $\{D(y) \cap D(y+x)^c$ contains one Poisson point$\}$, $\{D(y)^c \cap D(y+x)$ contains one Poisson point$\}$, $\{D(y) \cap D(y+x)$ contains no Poisson point$\}$, $\{S(y) \cup S(y+x)$ contains no Poisson point of the first process$\}$. Therefore,

$$C_{cc}(x) = p_c^2 \left(1 - h(\rho)\right)^2 \exp\left[\theta H_{cs}(x)\right]. \tag{E.38}$$

$C_{cs}(x)$ is given from the following independent events: $\{D(y) \cap D(y+x)^c$ contains one Poisson point of the second process$\}$, $\{D(y+x)$ contains no Poisson point of the second process$\}$, $\{S(y)$ contains no Poisson point of the first process$\}$, $\{S(y)^c \cap S(y+x)$ contains one Poisson point of the first process$\}$. Then,

$$C_{cs}(x) = p_c p_s \left(1 - h(\rho)\right) \left(1 - H(R)\right) \exp\left[\theta H_{cs}(x)\right], \tag{E.39}$$

$$\theta H_{cs}(x) = \lambda_s V_s H(R) + \lambda_c A_c h(\rho) = \frac{p_s}{p} H(R) + \frac{p_c}{p} h(\rho). \tag{E.40}$$

The dead leaves model uses the order of appearance of a sequence of grains implanted according to a space-time Poisson process. From results given in [66], we have

$$p_s = \frac{\lambda_s V_s}{\lambda_s V_s + \lambda_c A_c} \left[1 - \exp\left(-\left(\lambda_s V_s + \lambda_c A_c\right)\right)\right], \tag{E.41}$$

$$p_c = \frac{\lambda_c A_c}{\lambda_s V_s + \lambda_c A_c} \left[1 - \exp\left(-\left(\lambda_s V_s + \lambda_c A_c\right)\right)\right]. \tag{E.42}$$

This is obtained by expressing that during the sequence, point y is covered by some grain (sphere or cylinder, according to the considered phase), and then remains uncovered by new grains, using the Markov property of the sequence. Similarly, after a lengthy proof, we can work out the remaining probabilities,

$$C_{cc}(x) = \frac{2p_c^2}{p^2}(1 - h(\rho))\left[\frac{1 - Q(x)}{2 - H_{cs}(x)} - \frac{Q(0) - Q(x)}{1 - H_{cs}(x)}\right], \tag{E.43}$$

$$C_{ss}(x) = \frac{2p_s^2}{p^2}(1 - H(R))\left[\frac{1 - Q(x)}{2 - H_{cs}(x)} - \frac{Q(0) - Q(x)}{1 - H_{cs}(x)}\right], \tag{E.44}$$

$$C_{cs}(x) = \frac{p_c p_s}{p^2}(2 - h(\rho) - H(R))\left[\frac{1 - Q(x)}{2 - H_{cs}(x)} - \frac{Q(0) - Q(x)}{1 - H_{cs}(x)}\right] \tag{E.45}$$

with $Q(x)$ and $H_{cs}(x)$ given by the equations

$$Q(x) = (1 - p)^{2 - H_{cs}(x)}, \quad H_{cs}(x) = \frac{p_c}{p} h\left(\frac{\rho}{a_c}\right) + \frac{p_s}{p} H(R). \tag{E.46}$$

References

1. Aboudi J (1983) The effective moduli of short-fiber composites. *Int J Solids Struct* 19:693–707
2. Abramowitz M, Stigan I (1964) *Handbook of Mathematical Functions.* Nat Buro Stand, 1046
3. Alder B (1964) Triplet correlation in hard spheres. *Phys Rev Letters* 12: 317–319
4. Aleksandrov V, Mhitaryan S (1983) *Contact Problems for Bodies with Thin Layers and Inclusions.* Nauka, Moscow, 487
5. Andreikiv A (1982) *3D-problems of Crack Theory.* Naukova Dumka, Kiev, 345
6. Andreikiv A, Panasuk V, Stadnik M (1972) Fracture of fragile beams weakened with circular internal cracks. *Strength of Mater* 10:37–41
7. Attard P (1989) Spherically inhomogeneous fluids. I. Percus–Yevick hard spheres: Osmotic coefficients and triplet correlation. *J Chem Phys* 91: 3072–3082
8. Bakhvalov N, Panasenko G (1984) *Homogenization of Periodical Media.* Nauka, Moscow, 352
9. Benveniste Y (1986) On Mori–Tanaka's method in cracked bodies. *Mech Res Commun* 13:193–201
10. Benveniste Y (1987) A new approach to the application of Mori–Tanaka's theory in composite materials. *Mech Materials* 6:147–157
11. Benveniste Y (1991) On diagonal and elastic symmetry of the approximate effective stiffness tensor of heterogeneous media. *J Mech Phys Solids* 39:927–946
12. Benveniste Y (1992) The determination of the elastic and electric fields in a piezoelectric inhomogeneity. *J Appl Phys* 72:1086–1095
13. Beran M (1968) *Statistical Continuum Theories.* Wiley-Interscience, New York, 424
14. Berryman J (1980) Long-wavelength propagation in composite elastic media. II: Ellipsoidal inclusions. *J Acoustic Soc Am* 68:1820–1831
15. Berryman J (2006) Effective medium theories for multicomponent poroelastic composites. *J Eng Mech* May:519–531
16. Berryman J, Berge P (1996) Critique of two explicit schemes for estimating elastic properties of multiphase composites. *Mech Mater* 22:149–164
17. Berryiman J, Milton G (1991) Exact results for generalized Gassmann's equations in composite porous media with two constituents. *Geophys* 56:1950–1960
18. Bildstein B, Kahi G (1994) Triplet correlation function for hard spheres: Computer simulation results. *J Chem Phys* 100:5882–5893
19. Biot M (1941) General theory of three-dimensional consolidation. *J Appl Phys* 12:155–164
20. Biot M (1962) Mechanics of deformation and acoustic propagation in porous media. *J Appl Phys 33*:1482–1498

21. Brichkov Y, Prudnikov A (1977) *Integral Transforms of Generalized Functions.* Nauka, Moscow, 287
22. Brinkman H (1952) The viscosity in concentrated suspension and solutions. *J Chem Phys* 20:571–585
23. Bruggeman D (1935) Berechnung verschiedener physikalischer Konstanten von heterogenen Substanzen. *Ann Phys* 24:636–679
24. Bruggeman D (1937) Berechnung verschiedener physikalischer Konstanten von heterogenen Substanzen. III. Die Elastische Konstanten der Quasiisotropen Mischkoerper aus isotropen Substanzen. *Ann Phys* 29:160–178
25. Budiansky B (1965) On the elastic moduli of some heterogeneous materials. *J Mech Phys Solids* 13:223–234
26. Budiansky B, O'Connell R (1976) Elastic module of a cracked solid. *Int J Solids Struct* 12:81–91
27. Chen T (1994) Micromechanical estimates of the overall thermoelectroelastic moduli of multiphase fibrous composites. *Int J Solids Struct* 31:3099–3111
28. Cherepanov G (1983) *Fracture Mechanics of Composite Materials.* Nauka, Moscow, 296
29. Christensen R, Loo L (1979) Solutions for effective shear properties in three phase and cylinder models. *J Mech Phys Solids* 27:315–330
30. Christensen R (1979) *Mechanics of Composite Materials.* Wiley-Interscience, New York, 384
31. Cleary M, Chen I, Lee S (1987) Self-consistent techniques for heterogeneous solids. *ASCE J Eng Mech* 106:861–887
32. Cox H (1952) The elasticity and strength of paper and other fibrous materials. *Brit J Appl Phys* 3:72–79
33. Curtin W (1990) Triplet direct correlation function in the hard-sphere fluid. *J Chem Phys* 93:1919–1925
34. De Wit R (1970) Linear theory of static disclinations. In: Simmons J, De Wit R, Bullough R (eds) *Fundamental Aspects of Dislocations, Nat Bur Stand* (317) I:651–673
35. De Wit R (1973) Theory of dislocation: II. Continuous and discrete dislocations in anisotropic elasticity. *J Res Nat Bur Stand* 77A:49–100
36. De Wit R (1973) Theory of dislocation: III. Continuous and discrete dislocations in isotropic elasticity. *J Res Nat Bur Stand* 77A:359–368
37. De Wit R (1973) Theory of dislocation: IV. Straight dislocations. *J Res Nat Bur Stand* 77A:607–658
38. Dean G, Lockett F (1973) Determination of the mechanical properties of fiber composites by ultrasonic techniques – Analysis of the test methods for high modulus fibers and composites. *Am Soc Testing Materials* STP N521:326–346
39. Delarue A (2001) Prévision du comportement électromagnétique de matériaux composites á partir de leur mode d'élaboration et leur morphologie, Thése pour obtenir de Docteur de l'Ecole des Mines de Paris, Septembre 2001, 185
40. Dunn M (1994) Electroelastic Green's functions for transversely isotropic media and their application to the solution of inclusion and inhomogeneity problems. *Int J Eng Sci* 32:119–131
41. Dunn M, Wienecke H (1997) Inclusions and inhomogeneities in transversely isotropic piezoelectric media. *Int J Solids Struct* 34(27):3571–3582
42. Dunn M, Taya M (1993) Micromechanical predictions of effective electroelastic moduli of piezoelectric composites. *Int J Solids Struct* 30:161–175
43. Dunn M, Taya M (1993) Analysis of piezoelectric composite materials containing ellipsoidal inhomogeneities. *Proc Royal Soc London* A 443:265–287

44. Elliott, R, Krumhansl J, Leath P (1974) The theory and properties of randomly disordered crystals and related physical system. *Rev Mod Phys* 46: 465–543
45. Eshelby J (1957) The determination of the elastic field of an ellipsoidal inclusion, and related problems. *Proc Royal Soc* A241:376–396
46. Eshelby (1961) Elastic inclusions and inhomogeneities. In: Sneddon I, Hill R (eds) *Progress in Solid Mechanics* 2:88–140
47. Eshelby (1982) The stress on and in a thin inextensible fiber in a stretched elastic medium. *Eng Fracture Mech* 16:453–460
48. Eskin G (1981) *Boundary-value Problems for Elliptic Pseudo-differential Equations.* Math Soc USA, New York, 232
49. Eyre D, Milton G (1999) A fast numerical scheme for computing the response of composites using grid refinement. *Eur Phys J, Appl Phys 6*:41–47
50. Ferrari M (1991) Asymmetry and the high concentration limit of the Mori–Tanaka effective medium theory. *Mech Mater* 11:251–256
51. Filshtinsky L (1974) Interaction of a double-periodical system of rectilinear cracks in an isotropic medium. *Appl Math Mech (PMM)* 38:906–914
52. Gakhov F, Cherskii Yu (1978) *Equations of the Convolution Type.* Nauka, Moscow, 295
53. Garnet J (1904) Colors in metal glasses and metallic films. *Phil Trans Royal Soc London* 203:385–420
54. Gelfand I, Shilov G (1964) *Generalized Functions.* Academic, New York, 470
55. Gichman I, Skorochod A (1977) *Introduction to the Theory of Random Processes.* Nauka, Moscow, 567
56. Grabovsky Yu, Milton G (1998) Exact relations for composites. Towards a complete solution. *Documenta Math*, Extra Volume ICM 1–10.
57. Grigoliuk E, Filshtinsky L (1970) *Perforated Plates and Shells.* Nauka, Moscow, 556
58. Hashin Z, Strikman S (1963) A variational approach to the theory of the elastic behavior of multiphase materials. *J Mech Phys Solids* 11:127–140
59. Hashin Z, Rosen B (1964) The elastic moduli of fiber-reinforced materials. *J Appl Mech* 31:223–232
60. Hashin Z (1988) The differential scheme and its application to cracked materials. *J Mech Phys Solids* 36:719–734
61. Hershey A (1954) The elasticity of an isotropic aggregate of anisotropic cubic crystals. *J Appl Mech* 21:236–240
62. Hill R (1963) Elastic properties of reinforced solids: some theoretical principles. *J Mech Phys Solids* 11:357–372
63. Hill R (1965) Self-consistent mechanics of composite materials. *J Mech Phys Solids* 13:213–222
64. Ilyin A (1989) *Matching of Asymptotic Expansions of the Solutions of Boundary-value Problems.* Nauka, Moscow, 335
65. Ivanov A, Loyko V, Dik V (1998) *Light Propagation through Dense Dispersed Media.* Nauka y Tehnica, Minsk, 192
66. Jeulin D (1987) Random structure analysis and modelling by mathematical morphology. In: Spencer A, Balkema A (eds) *Proc Continuum Media and Discrete Systems* 5, Nottingham: 217–226
67. Jeulin D (1993) Random models for the morphological analysis of powder. *J Microscopy* 172:13–21
68. Jeulin D (1997) Dead leaves models: from space tessellation to random functions. In: Jeulin D (ed) *Proc Int Conf Advances Theory Appl Random Sets, Fontainebleau, 9–11 October 1996.* World Scientific, 137–156

69. Kachanov M, Sevostianov I, Shafiro B (2001) Explicit cross-property correlations for porous materials with anisotropic microstructures. *J Mech Phys Solids* 49:1–25
70. Kanaun S (1975) The method of self-consistent field in the problem of effective elastic properties of composites. *J Appl Mech Techn Phys* 4:194–203
71. Kanaun S (1977) The approximation of self-consistent field for composite elastic media. *J Appl Mech Techn Phys* 2:160–169
72. Kanaun S (1980) Interaction of periodical crack systems in an elastic medium. *Appl Mech* 16:36–42
73. Kanaun S (1980) Poisson's set of cracks in an elastic solid. *Appl Math Mech (PMM)* 44:1129–1139
74. Kanaun S (1981) To the crack problem in 3D anisotropic elastic medium. *Appl Math Mech (PMM)* 45:361–370
75. Kanaun S (1982) Effective field method in the linear static problems of composite media. *Appl Math Mech (PMM)* 46:655–665
76. Kanaun S (1982) On the model of point defects in the mechanics of elastic inhomogeneous media. *Proc Acad Sci USSR, Mech Solids* 4:109–118
77. Kanaun S (1983) Elastic medium with random field of inhomogeneities. In: Kunin I. (ed) *The Theory of Elastic Media with Microstructure*, Vol.2, Chapt.7, Springer, Berlin Heidelberg New York, 165–228
78. Kanaun S (1983) Correlation function of the stress field in an elastic medium with point defects. *Appl Math Mech (PMM)* 47:652–661
79. Kanaun S (1984) On singular models of thin inclusions in an homogeneous elastic medium. *Appl Math Mech (PMM)* 48:81–91
80. Kanaun S (1984) Thin defect in an homogeneous elastic medium. *Proc Acad Sci USSR, Mech Solids* 3:74–83
81. Kanaun S (1987) Static fields in an homogeneous medium disturbed by an inclusion in the form of a curvilinear fiber. *Appl Math Mech (PMM)* 51: 293–304
82. Kanaun S (1988) Equilibrium of a homogeneous elastic medium reinforced with a rectilinear rigid fiber. *Appl Math Mech (PMM)* 52:789–800
83. Kanaun S (1990) Elastic properties of composites reinforced with short axisymmetric fibers. *Appl Mech* 26:56–63
84. Kanaun S (1990) Self-consistent schemes of averaging in the mechanics of matrix composite materials. *Mech Composite Mater* 6:984–994
85. Kanaun S (1993) Effective elastic properties of the medium with an array of thin inclusions. *Revista Mexicana de Fisica* 39:74–83
86. Kanaun S (2003) Dielectric properties of matrix composites materials with high volume concentrations of inclusions (effective field approach). *Int J Eng Sci* 41:1287–1312
87. Kanaun S, Jeulin D (2001) Elastic properties of hybrid composites by the effective field approach. *J Mech Phys Solids* 49:2339–2367
88. Kanaun S, Kudryavtzeva L (1986) Multi-layered spherical inclusions in an homogeneous elastic medium. *Appl Math Mech (PMM)* 50:633–643
89. Kanaun S, Kudryavtzeva L (1987) Thermal stresses in composites with multilayered spherical inclusions. *Proc Acad Sci USSR, Mech Solids* 4:113–121
90. Kanaun S, Kudryavtseva L (1988) Elastic properties of composites reinforced with thin hard inclusions. *Mech Composite Mater* 1:129–136
91. Kanaun S, Kudrjavtzeva L (1989) Thermo elastic properties of composite materials reinforced with unidirectional multi-layered fibers. *Appl Math Mech (PMM)* 53:798–807
92. Kanaun S, Levin V (1984) About the microstresses in composite materials in the regions of fast changing external fields. *Mech Composite Mater* 4:625–629

93. Kanaun S, Levin V (1993) *Effective Field Method in the Mechanics of Composite Materials.* Petrosavodsk University, Petrosavodsk, 600
94. Kanaun S, Levin V (1994) Effective field method on the mechanics of matrix composite materials. In: Markov K (ed) *Advances in Mathematical Modeling of Composite Materials. Series on Advances in Mathematics for Applied Sciences* 15, World Scientific, 1–58
95. Kerner E (1956) The elastic and thermoelastic properties of composite media. *Proc Phys Soc* 69(B):808–813
96. Kit G, Hay M (1989) *The Method of Potentials in 3D thermo-elastic Problems for Bodies with Cracks.* Naukova Dumka, Kiev, 283
97. Kochetkov E., Leont'ev N, Ugodchikov N (1985). Effective thermo-elastic properties of regular fiber reinforced composites. In: Ugodchikov N (ed) *Applied Problems of Strength and Plasticity of Materials.* Gorky State University, Gorky, 29:66–72
98. Kondratiev V (1967) Boundary value problems for regions with conical and angle points. *Proc Moscow Math Soc* 16:209–292
99. Kosevich A (1978) *Dislocations in the Theory of Elasticity.* Naukova Dumka, Kiev, 220
100. Kröner E (1958) Berechnung der elastischen Konstanten des Vielkristalls aus den Konstanten des Einkristalls. *Z Phys* 151:504–518
101. Kröner E (1958) *Kontinuumstheorie der Versetzungen and Eigenspannungen.* Springer, Berlin Göttingen Heidelberg, 179
102. Kröner E (1971) *Statistical Continuum Mechanics.* Springer, Udine, 157
103. Kuhn R, Michelitsch T, Levin V, Vakulenko A (2002) The analogue of the Eshelby tensor for a piezoelectric spheroidal inclusion in the hexagonal electroelastic medium. *ZAMP* 53:584–602
104. Kunez Ya (1987) Axisymmetric torsion of the elastic space with a thin elastic inclusion. *Appl Math Mech (PMM)* 51:638–645
105. Kunin I (1964) Internal stresses in an anisotropic elastic medium. *Appl Math Mech (PMM)* 28:612–622
106. Kunin I (1965) *Methods of Tensor Analysis in the Theory of Dislocations.* US Department of Commerce, Clearing House for Fed Sci Techn Inform, Springfield, VA 22151
107. Kunin I (1980) *The Theory of Elastic Media with Microstructure* I. Springer, Berlin Heilderberg New York, 291
108. Kunin I (1983) *The Theory of Elastic Media with Microstructure* II. Springer, Berlin Heilderberg New York, 272
109. Kunin I, Sosnina E (1971) An ellipsoidal inhomogeneity in an elastic medium. *Proc Acad Sci USSR* 199:571–575
110. Kunin I, Sosnina E (1973) Stress concentration on an ellipsoidal inhomogeneity in an anisotropic elastic medium. *Appl Math Mech (PMM)* 37:306–315
111. Landauer R (1978) Electrical conductivity in inhomogeneous media. In: *Proc Conf on Electrical Transport and Optical Properties of Inhomogeneous Media* (ETOPIM 1) 2 (AIP Conf Proc), 2–46.
112. Lax M (1951) Multiple scattering of waves. *Rev Modern Phys* 23:287–310
113. Lax M (1952) Multiple scattering of waves II. The effective field in dense systems. *Phys Rev* 85:621–629
114. Levin V (1967) On the thermal expansion coefficients of heterogeneous materials. *Proc Acad Sci USSR, Mech Solids* 1:88–93
115. Levin V (1976) On the determination of elastic and thermoelastic constants of composite materials. *Proc Acad Sci USSR, Mech Solids* 6:136–145
116. Levin V (1977) On the stress concentration on inclusions in composite materials. *Appl Math Mech (PMM)* 41:735–743

117. Levin V (1982) On thermoelastic stresses in composite media. *Appl Math Mech (PMM)* 46:502–506
118. Levin V (1996) On the effective properties of piezoactive matrix composite materials. *Appl Math and Mech (PMM)* 60:313–332
119. Levin V (1999) The microstresses evaluation in piezoelectric polycrystals. *Int J Fracture* 97:113–123
120. Levin V, Alvarez-Tostado J (2004) On the effective constants of inhomogeneous poroelastic medium. *Sci Eng Composite Materials* 11:35–46
121. Levin V, Alvarez-Tostado J (2006) Explicit effective constants for an in homogeneous porothermoelastic medium. *Arch Appl Mech* 76:199–214
122. Levin V, Luchaninov A (2001) On the effective properties of thermo-piezoelectric matrix composites. *J Phys D, Appl Phys* 34:1–6
123. Levin V, Luchaninov A (2004) Effective properties of matrix piro-active composites. *Proc Acad Sci USSR, Mech Solids* 5:44–53
124. Levin V, Markov M (2004) Electroconductivity of a medium with thin low-resistivity inclusions. *J Electrostatics* 61:129–145
125. Levin V, Markov M (2005) Elastic properties of inhomogeneous transversely isotropic rocks. *Int J Solids Struct* 42:393–408
126. Levin V, Markov M, Kanaun S (2004) Effective field method for seismic properties of cracked rocks. *J Geophys Research* 109 B08202:1–13
127. Levin V, Michelitsch T, Sevostianov I (2000) Spheroidal inhomogeneity in a transversely isotropic piezoelectric medium. *Arch Appl Mech* 70:673–693
128. Levin V, Rakovskaja M (1997) The effective constants of thermopiezoactive polycrystals. *Proc Acad Sci USSR, Mech Solids* 5:100–109
129. Levin V, Rakovskaja M, Kreer W (1999) The effective thermoelectroelastic properties of microinhomogeneous materials. *Int J Solids Struct* 36:2683–2705
130. Lifshitz I, Gradeskul S, Pastur L (1982) *Introduction to the Theory of Disordered Systems.* Nauka, Moscow, 358
131. Lifshitz I, Rosenzweig L (1947) Green's tensor for an anisotropic infinite elastic medium. *J Exper Teor Phys* 17:83–791
132. Lochin V, Sedov L (1963) Nonlinear tensor functions of several tensor arguments. *Appl Math Mech (PMM)* 27:393–417
133. Lomakin V (1970) *Statistical Problems of the Mechanics of Solids.* Nauka, Moscow, 139
134. Lusis J, Woodhams R, Xanthos M (1973) The effect of flake aspect ratio on the flexural properties of mica reinforced plastics. *Polymer Eng Sci* 13:139–145.
135. Manera M. (1977) Elastic properties of randomly oriented short fiber-glass composites. *J Composite Mater* 11:235–247.
136. Markov K (1981) "One particle" approximation in the mechanics of composite media. In: Brulin O, Hsieh R (eds) *Continuum Models and Discrete Systems.* North-Holand, Amsterdam, 441–448
137. Markov K (2001) Justification of the effective field method in elastostatics of heterogeneous solids. *J Mech Phys Solid* 49:2621–2634
138. Markov K(2000) Elementary micromechanics of heterogeneous solids. In: Markov K, Preziosi L (eds) *Heterogeneous Media: Micromechanics Modeling Methods and Simulations.* Birkhauser, Boston, 1–162.
139. Markov M, Levin V, Mousatov A, Kazatchenko E (2005) Elastic properties of double-porosity rocks using the differential effective medium model. *Geophys Prospecting* 53:733–754
140. Matonis V, Small N (1969) A macroscopic analysis of composites containing layered spherical inclusions. *Polymer Eng Sci* 9:90–99
141. Matheron G (1975) *Random Sets and Integral Geometry*, Wiley & Sons, New York, 425

142. Maz'ya V, Nazarov S, Plamenevskii B (2000) *Asymptotic Theory of Elliptic Boundary Value Problems in Singularly Perturbed Domains*, Birkhauser, Berlin, 460
143. McConnell A (1957) *Application of Tensor Analysis*. Dover, New York, 411
144. McLachlan D, Priou A, Chenerie I, Issac E, Henry F (1992) Modeling of permittivity of composite materials with a general effective medium equation. *J Electromagnetic Waves Appl* 6:1099–1131
145. McLaughlin R (1977) A study of the differential scheme for composite materials. *Int J Eng Sci* 15:237–244
146. Michelitsch T, Levin V (2000) Inclusions and inhomogeneities in electroelastic media with hexagonal symmetry. *Eur Phys J* B 14:527–533
147. Milton G (1997) Composites: A myriad of microstructure independent relations. In: Tatsumi T (ed) *Theoretical and Applied Mechanics*. Elsevier, 443–459
148. Milton G (2002) *The Theory of Composites*. Cambridge University Press, Oxford, 780
149. Mikhlin S (1962) *Multi-dimensional Singular Integrals and Integral Equations*. Phys Math Lit, Moscow, 254
150. Mindlin R, Tiersten H (1962) Effect of couple stresses in linear elasticity. *Arch Rat Mech Analysis* 11:415–448
151. Mori S, Tanaka K (1973) Average stress in matrix and average elastic energy of materials with misfitting inclusions. *Acta Metall*, 21:571–574
152. Moulinec H, Suquet P (1994) A fast numerical method for computing the linear and nonlinear mechanical properties of composites. *Compt Rend, A, Paris* 318:1417–1423
153. Mukerji T, Berryman J, Mavko G, Berge P (1995) Differential effective medium modeling of rock elastic moduli with critical porosity constraints. *Geophys Research Letters* 22:555–558
154. Nikishkov G, Cherepanov G (1984) Tension of the elastic space with an isolated rigid fiber. *Appl Math Mech (PMM)* 48:460–465
155. Niklasson G, Granqvist C (1984) Optical properties and solar selectivity of coevaporated $Co\text{-}Al_2O_3$ system composite films. *J Appl Phys* 55:3382–3410
156. Norris A (1985) A differential scheme for effective moduli of composites. *Mech Mater* 4:1–16
157. Norris A (1989) An examination of the Mori-Tanaka effective medium approximation for multiphase composites. *Trans ASME* 88/56:84–88
158. Norris A (1992) On the correspondence between poroelasticity and thermoelasticity. *J Appl Physics* 71:1138–1141
159. Novozgilov V (1959) *Thin Shell Theory*. Noordhoff, Amsterdam, 376
160. Nye J (1985) *Physical Properties of Crystals, their Representation by Tensors and Matrices*. Clarendon Press, Oxford, 329
161. Okuno K, Woodhams R (1975) Mica reinforced polypropylene. *Polym Eng Sci* 15:308–315
162. Okuno K. (1980) On the flexural modulus of the flake-filled plastics. *Kobunshi Ronbunshu* 37:789–796
163. Panasuk V, Savruk M, Dazishin A (1976) *Stress Distribution around Cracks in Plates and Shells*. Naukova Dumka, Kiev, 443
164. Paris P, Si J (1968) Stress state analysis near cracks. In: Paris P (ed) *Applied Problems of Ductile Fracture*. Mir, Moscow, 64–142
165. Parton V, Perlin P(1981) *Mathematical Methods of the Elasticity Theory*. Nauka, Moscow, 311
166. Percus J, Yevick G (1958) Analysis of classical statistical mechanics by means of collective coordinates. *Phys Rev* 111:1–13

167. Piau M (1979) Attenuation of a plane compressional wave by a random distribution of thin circular cracks. *Int J Eng Sci* 17:151–167
168. Plamenevskii B (1968) On the bounding of singular integrals in weight spaces. *Math Collection* 76:537–592
169. Pobedrya B (1984) *Mechanics of Composite Materials.* Moscow University, Moscow, 336
170. Poddubnyak A (1986) Integral equations of torsion of an elastic body with a thin disk-like inclusion. *App Math Mech (PMM)* 50:644–650
171. Ponte Castañeda P, Willis J (1995) The effect of spatial distribution on the effective behavior of composite materials and cracked media. *J Mech Phys Solids* 43:1919–1951
172. Prochorov Yu, Rosanov Yu (1967) *The Probability Theory (Basic Concepts, Limit Theorems, Random Processes).* Nauka, Moscow, 495
173. Qiu Y, Weng, G (1990) On the application of the Mori-Tanaka theory involving transversely isotropic spheroidal inclusions. *Int J Eng Sci* 28:1121–1137
174. Ramsteiner F (1981) Elastic behavior of unidirectional short-fiber reinforced thermoplastics. *Composites* 12:65–71.
175. Richard T (1975) The mechanical behavior of solid microsphere filled composite. *J Composite Mater* 9:108–113
176. Robinson K (1951) Elastic energy of an ellipsoidal inclusion in an infinite solid. *J Appl Mech* 22:1045–1054
177. Roscoe R (1952) The viscosity of suspension and solution. *J Appl Phys* 3:267–269
178. Rosen B, Hashin Z (1970) Effective thermal expansion coefficient and specific heats of composite materials. *Int J Eng Sci* 8:157–173
179. Sadowsky M, Sternberg E (1949) Stress concentration around a three-axial ellipsoidal cavity. *J Appl Mech* 16:149–157
180. Sanches-Palencia E (1984) *Heterogeneous Media and Oscillation Theory.* Mir, Moscow, 472
181. Segurado J, Llorca J (2002) A numerical approximation to the elastic properties of sphere-reinforced composites. *J Mech Phys Solids* 50:2107–2121
182. Sevostianov I (2003) Explicit relations between elastic and conductive properties of materials containing annular cracks. *Phil Trans Royal Soc London* 361:987–999
183. Sevostianov I, Kachanov M (2002) Explicit cross-property correlations for anisotropic two-phase composite materials. *J Mech Phys Solids* 50:253–282
184. Sevostianov I, Kachanov M (2004) Connection between elastic and conductive properties of microstructures with Hertzian contacts. *Proc Royal Soc London A* 460:1529–1534
185. Sevostianov I, Levin V, Kachanov M (2001) On the modeling and design of piezocomposites with prescribed properties. *Arch Appl Mech* 71:733–747
186. Shapery N (1968) Thermal expansion coefficients of composite materials, based on energy principles. *J Composite Mater* 2:380–404.
187. Simonenko I (1974) Electrostatic problem for an inhomogeneous medium. The case of thin dielectric inclusion with high dielectric constants.-I. *Differential Equations* 10:301–310
188. Simonenko I (1975) Electrostatic problem for an inhomogeneous medium. The case of thin dielectric inclusion with high dielectric constants.-II. *Differential Equations* 11:1870–1878
189. Shermergor T (1977) *The Theory of Elasticity of Microheterogeneous Media.* Nauka, Moskow, 399
190. Skorohod V (1961) Calculation of the effective isotropic moduli of disperse solid system. *Powder Metallurgy* 1:50–55

191. Smith J (1974) Correction and extension of Van der Pole's method for calculating the shear modulus of particulate composite. *J Res Natl Bar Stand* 78A:355–361
192. Smith J (1976) The elastic constants of a particulate filled glassy polymer. *Polymer Eng Sci* 16:394–399
193. Sotkilava O, Cherepanov G (1974) Some problems of the heterogeneous elasticity theory. *J Appl Math Mech (PMM)* 38:537–550
194. Talbot D, Willis J (1982) Variational estimates for dispersion and attenuation of waves in random composites. I. General theory. *Int J Solids Struct* 18: 673–683
195. Talbot D, Willis J (1982) Variational estimates for dispersion and attenuation of waves in random composites. II. Isotropic composites. *Int J Solids Struct* 18:685–698
196. Talbot D, Willis J (1983) Variational estimates for dispersion and attenuation of waves in random composites-III, Fiber-reinforced materials. *Int J Solids Struct* 19:793–811
197. Throop G, Bearman R (1965) Numerical solutions of the Percus–Yevick equation for the hard-sphere potential. *J Chem Phys* 42:2408–2411
198. Torquato S (1999) Exact solutions on physical realizable correlation functions of random media. *J Chem Phys* 111:8332–8337
199. Torquato S (2002) *Random Heterogeneous Materials: Microstructure and Macroscopic Properties.* Springer, New York, 728
200. Tshai O, Cohen L (1967) Elastic properties of filled and porous epoxy composites. *Int J Mech Sci* 9:539–546
201. Ufland Ya (1968) *Integral Transforms in the Problems of the Elasticity Theory.* Nauka, Leningrad, 402
202. Vakulenko A (1998) On the effective properties and strength of ceramic composites. *Proc Acad Sci USSR, Mech Solids* 6:50–71
203. Van der Poel C (1958) On the rheology of concentrated dispersions. *Rheol Acta* B1:198–205.
204. Van Dyke (1964) *Perturbation Methods in Fluid Mechanics.* Academic Press, New York London, 310
205. Vavakin A, Salganik R (1975) Effective characteristics of heterogeneous solids with isolated inhomogeneities. *Proc Acad Sci USSR, Mech Solids* 2:95–107
206. Vekua I (1978) *Tensor Aanalysis and the Theory of Covariants.* Nauka, Moscow, 296
207. Walpole L (1966) On bounds for the overall elastic modules of inhomogeneous systems. I, II. *J Mech Phys Solids* 14:151–301
208. Walpole L (1969) On the overall elastic modules of composite materials. *J Mech Phys Solids* 17:235–251
209. Wang B (1992) Three-dimensional analysis of a flat elliptical crack in a piezoelectric material. *Int J Eng Sci* 30:781–791
210. Wang B (1994) Effective behavior of piezoelectric composites. *Appl Mech Rev* 47:112–121
211. Wertheim M (1963) Exact solution of the Perkus–Yevick integral equation for hard spheres. *Phys Rev Lett* 10:321–323
212. Willis J (1968) The stress field around an elliptical crack in an anisotropic elastic medium. *Int J Eng Sci* 6:253–263
213. Willis J (1977) Bounds and self-consistent estimates for the overall properties of anisotropic composites. *J Mech Phys Solids* 25:185–202
214. Willis J (1978) Variational principles and bounds for overall properties of composites. In: Porovan E (ed) *Continuum models of discrete system. Proc*

Second Int Symp Cont Models Discrete Systems. University Waterloo Press, Waterloo, 185–215
215. Willis J (1980) A polarization approach to the scattering of elastic waves. I. Scattering by a single inclusion, II. Multiple scattering from inclusions. *J Mech Phys Solids* 28:287–327
216. Willis J (1983) Variational characterization of waves in fiber-reinforced materials. In: Hashin Z, Herakovich C (eds) *Mechanics of Composite Materials: Recent Advances.* Pergamon, Oxford, 191–205
217. Willis J. (1984) Some remarks on the application of the QCA to the determination of the overall elastic response of a matrix/ inclusion composite. *J Math Phys* 25:2116–2120
218. Zaiman J (1982) *The Models of Disorder.* Mir, Moskow, 591
219. Zhikov V, Koslov S, Oleinik O (1993) *Homogenization of Differential Operators.* Nauka, Moscow, 398
220. Zimmerman R (1991) Elastic moduli of a solid containing spherical inclusions. *Mech Mater* 12:17–24

Index

Mechanics

SOLID MECHANICS AND ITS APPLICATIONS

Series Editor: G.M.L. Gladwell

Aims and Scope of the Series

The fundamental questions arising in mechanics are: *Why?*, *How?*, and *How much?* The aim of this series is to provide lucid accounts written by authoritative researchers giving vision and insight in answering these questions on the subject of mechanics as it relates to solids. The scope of the series covers the entire spectrum of solid mechanics. Thus it includes the foundation of mechanics; variational formulations; computational mechanics; statics, kinematics and dynamics of rigid and elastic bodies; vibrations of solids and structures; dynamical systems and chaos; the theories of elasticity, plasticity and viscoelasticity; composite materials; rods, beams, shells and membranes; structural control and stability; soils, rocks and geomechanics; fracture; tribology; experimental mechanics; biomechanics and machine design.

1. R.T. Haftka, Z. Gürdal and M.P. Kamat: *Elements of Structural Optimization.* 2nd rev.ed., 1990 ISBN 0-7923-0608-2
2. J.J. Kalker: *Three-Dimensional Elastic Bodies in Rolling Contact.* 1990 ISBN 0-7923-0712-7
3. P. Karasudhi: *Foundations of Solid Mechanics.* 1991 ISBN 0-7923-0772-0
4. *Not published*
5. *Not published.*
6. J.F. Doyle: *Static and Dynamic Analysis of Structures.* With an Emphasis on Mechanics and Computer Matrix Methods. 1991 ISBN 0-7923-1124-8; Pb 0-7923-1208-2
7. O.O. Ochoa and J.N. Reddy: *Finite Element Analysis of Composite Laminates.* ISBN 0-7923-1125-6
8. M.H. Aliabadi and D.P. Rooke: *Numerical Fracture Mechanics.* ISBN 0-7923-1175-2
9. J. Angeles and C.S. López-Cajún: *Optimization of Cam Mechanisms.* 1991 ISBN 0-7923-1355-0
10. D.E. Grierson, A. Franchi and P. Riva (eds.): *Progress in Structural Engineering.* 1991 ISBN 0-7923-1396-8
11. R.T. Haftka and Z. Gürdal: *Elements of Structural Optimization.* 3rd rev. and exp. ed. 1992 ISBN 0-7923-1504-9; Pb 0-7923-1505-7
12. J.R. Barber: *Elasticity.* 1992 ISBN 0-7923-1609-6; Pb 0-7923-1610-X
13. H.S. Tzou and G.L. Anderson (eds.): *Intelligent Structural Systems.* 1992 ISBN 0-7923-1920-6
14. E.E. Gdoutos: *Fracture Mechanics.* An Introduction. 1993 ISBN 0-7923-1932-X
15. J.P. Ward: *Solid Mechanics.* An Introduction. 1992 ISBN 0-7923-1949-4
16. M. Farshad: *Design and Analysis of Shell Structures.* 1992 ISBN 0-7923-1950-8
17. H.S. Tzou and T. Fukuda (eds.): *Precision Sensors, Actuators and Systems.* 1992 ISBN 0-7923-2015-8
18. J.R. Vinson: *The Behavior of Shells Composed of Isotropic and Composite Materials.* 1993 ISBN 0-7923-2113-8
19. H.S. Tzou: *Piezoelectric Shells.* Distributed Sensing and Control of Continua. 1993 ISBN 0-7923-2186-3
20. W. Schiehlen (ed.): *Advanced Multibody System Dynamics.* Simulation and Software Tools. 1993 ISBN 0-7923-2192-8
21. C.-W. Lee: *Vibration Analysis of Rotors.* 1993 ISBN 0-7923-2300-9
22. D.R. Smith: *An Introduction to Continuum Mechanics.* 1993 ISBN 0-7923-2454-4
23. G.M.L. Gladwell: *Inverse Problems in Scattering.* An Introduction. 1993 ISBN 0-7923-2478-1

Mechanics

SOLID MECHANICS AND ITS APPLICATIONS

Series Editor: G.M.L. Gladwell

24. G. Prathap: *The Finite Element Method in Structural Mechanics*. 1993 ISBN 0-7923-2492-7
25. J. Herskovits (ed.): *Advances in Structural Optimization*. 1995 ISBN 0-7923-2510-9
26. M.A. González-Palacios and J. Angeles: *Cam Synthesis*. 1993 ISBN 0-7923-2536-2
27. W.S. Hall: *The Boundary Element Method*. 1993 ISBN 0-7923-2580-X
28. J. Angeles, G. Hommel and P. Kovács (eds.): *Computational Kinematics*. 1993 ISBN 0-7923-2585-0
29. A. Curnier: *Computational Methods in Solid Mechanics*. 1994 ISBN 0-7923-2761-6
30. D.A. Hills and D. Nowell: *Mechanics of Fretting Fatigue*. 1994 ISBN 0-7923-2866-3
31. B. Tabarrok and F.P.J. Rimrott: *Variational Methods and Complementary Formulations in Dynamics*. 1994 ISBN 0-7923-2923-6
32. E.H. Dowell (ed.), E.F. Crawley, H.C. Curtiss Jr., D.A. Peters, R. H. Scanlan and F. Sisto: *A Modern Course in Aeroelasticity*. Third Revised and Enlarged Edition. 1995 ISBN 0-7923-2788-8; Pb: 0-7923-2789-6
33. A. Preumont: *Random Vibration and Spectral Analysis*. 1994 ISBN 0-7923-3036-6
34. J.N. Reddy (ed.): *Mechanics of Composite Materials*. Selected works of Nicholas J. Pagano. 1994 ISBN 0-7923-3041-2
35. A.P.S. Selvadurai (ed.): *Mechanics of Poroelastic Media*. 1996 ISBN 0-7923-3329-2
36. Z. Mróz, D. Weichert, S. Dorosz (eds.): *Inelastic Behaviour of Structures under Variable Loads*. 1995 ISBN 0-7923-3397-7
37. R. Pyrz (ed.): *IUTAM Symposium on Microstructure-Property Interactions in Composite Materials*. Proceedings of the IUTAM Symposium held in Aalborg, Denmark. 1995 ISBN 0-7923-3427-2
38. M.I. Friswell and J.E. Mottershead: *Finite Element Model Updating in Structural Dynamics*. 1995 ISBN 0-7923-3431-0
39. D.F. Parker and A.H. England (eds.): *IUTAM Symposium on Anisotropy, Inhomogeneity and Nonlinearity in Solid Mechanics*. Proceedings of the IUTAM Symposium held in Nottingham, U.K. 1995 ISBN 0-7923-3594-5
40. J.-P. Merlet and B. Ravani (eds.): *Computational Kinematics '95*. 1995 ISBN 0-7923-3673-9
41. L.P. Lebedev, I.I. Vorovich and G.M.L. Gladwell: *Functional Analysis*. Applications in Mechanics and Inverse Problems. 1996 ISBN 0-7923-3849-9
42. J. Menčik: *Mechanics of Components with Treated or Coated Surfaces*. 1996 ISBN 0-7923-3700-X
43. D. Bestle and W. Schiehlen (eds.): *IUTAM Symposium on Optimization of Mechanical Systems*. Proceedings of the IUTAM Symposium held in Stuttgart, Germany. 1996 ISBN 0-7923-3830-8
44. D.A. Hills, P.A. Kelly, D.N. Dai and A.M. Korsunsky: *Solution of Crack Problems*. The Distributed Dislocation Technique. 1996 ISBN 0-7923-3848-0
45. V.A. Squire, R.J. Hosking, A.D. Kerr and P.J. Langhorne: *Moving Loads on Ice Plates*. 1996 ISBN 0-7923-3953-3
46. A. Pineau and A. Zaoui (eds.): *IUTAM Symposium on Micromechanics of Plasticity and Damage of Multiphase Materials*. Proceedings of the IUTAM Symposium held in Sèvres, Paris, France. 1996 ISBN 0-7923-4188-0
47. A. Naess and S. Krenk (eds.): *IUTAM Symposium on Advances in Nonlinear Stochastic Mechanics*. Proceedings of the IUTAM Symposium held in Trondheim, Norway. 1996 ISBN 0-7923-4193-7
48. D. Ieşan and A. Scalia: *Thermoelastic Deformations*. 1996 ISBN 0-7923-4230-5

Mechanics

SOLID MECHANICS AND ITS APPLICATIONS

Series Editor: G.M.L. Gladwell

49. J.R. Willis (ed.): *IUTAM Symposium on Nonlinear Analysis of Fracture*. Proceedings of the IUTAM Symposium held in Cambridge, U.K. 1997 ISBN 0-7923-4378-6
50. A. Preumont: *Vibration Control of Active Structures*. An Introduction. 1997 ISBN 0-7923-4392-1
51. G.P. Cherepanov: *Methods of Fracture Mechanics: Solid Matter Physics*. 1997 ISBN 0-7923-4408-1
52. D.H. van Campen (ed.): *IUTAM Symposium on Interaction between Dynamics and Control in Advanced Mechanical Systems*. Proceedings of the IUTAM Symposium held in Eindhoven, The Netherlands. 1997 ISBN 0-7923-4429-4
53. N.A. Fleck and A.C.F. Cocks (eds.): *IUTAM Symposium on Mechanics of Granular and Porous Materials*. Proceedings of the IUTAM Symposium held in Cambridge, U.K. 1997 ISBN 0-7923-4553-3
54. J. Roorda and N.K. Srivastava (eds.): *Trends in Structural Mechanics*. Theory, Practice, Education. 1997 ISBN 0-7923-4603-3
55. Yu.A. Mitropolskii and N. Van Dao: *Applied Asymptotic Methods in Nonlinear Oscillations*. 1997 ISBN 0-7923-4605-X
56. C. Guedes Soares (ed.): *Probabilistic Methods for Structural Design*. 1997 ISBN 0-7923-4670-X
57. D. François, A. Pineau and A. Zaoui: *Mechanical Behaviour of Materials*. Volume I: Elasticity and Plasticity. 1998 ISBN 0-7923-4894-X
58. D. François, A. Pineau and A. Zaoui: *Mechanical Behaviour of Materials*. Volume II: Viscoplasticity, Damage, Fracture and Contact Mechanics. 1998 ISBN 0-7923-4895-8
59. L.T. Tenek and J. Argyris: *Finite Element Analysis for Composite Structures*. 1998 ISBN 0-7923-4899-0
60. Y.A. Bahei-El-Din and G.J. Dvorak (eds.): *IUTAM Symposium on Transformation Problems in Composite and Active Materials*. Proceedings of the IUTAM Symposium held in Cairo, Egypt. 1998 ISBN 0-7923-5122-3
61. I.G. Goryacheva: *Contact Mechanics in Tribology*. 1998 ISBN 0-7923-5257-2
62. O.T. Bruhns and E. Stein (eds.): *IUTAM Symposium on Micro- and Macrostructural Aspects of Thermoplasticity*. Proceedings of the IUTAM Symposium held in Bochum, Germany. 1999 ISBN 0-7923-5265-3
63. F.C. Moon: *IUTAM Symposium on New Applications of Nonlinear and Chaotic Dynamics in Mechanics*. Proceedings of the IUTAM Symposium held in Ithaca, NY, USA. 1998 ISBN 0-7923-5276-9
64. R. Wang: *IUTAM Symposium on Rheology of Bodies with Defects*. Proceedings of the IUTAM Symposium held in Beijing, China. 1999 ISBN 0-7923-5297-1
65. Yu.I. Dimitrienko: *Thermomechanics of Composites under High Temperatures*. 1999 ISBN 0-7923-4899-0
66. P. Argoul, M. Frémond and Q.S. Nguyen (eds.): *IUTAM Symposium on Variations of Domains and Free-Boundary Problems in Solid Mechanics*. Proceedings of the IUTAM Symposium held in Paris, France. 1999 ISBN 0-7923-5450-8
67. F.J. Fahy and W.G. Price (eds.): *IUTAM Symposium on Statistical Energy Analysis*. Proceedings of the IUTAM Symposium held in Southampton, U.K. 1999 ISBN 0-7923-5457-5
68. H.A. Mang and F.G. Rammerstorfer (eds.): *IUTAM Symposium on Discretization Methods in Structural Mechanics*. Proceedings of the IUTAM Symposium held in Vienna, Austria. 1999 ISBN 0-7923-5591-1

Mechanics

SOLID MECHANICS AND ITS APPLICATIONS

Series Editor: G.M.L. Gladwell

69. P. Pedersen and M.P. Bendsøe (eds.): *IUTAM Symposium on Synthesis in Bio Solid Mechanics.* Proceedings of the IUTAM Symposium held in Copenhagen, Denmark. 1999 ISBN 0-7923-5615-2
70. S.K. Agrawal and B.C. Fabien: *Optimization of Dynamic Systems.* 1999 ISBN 0-7923-5681-0
71. A. Carpinteri: *Nonlinear Crack Models for Nonmetallic Materials.* 1999 ISBN 0-7923-5750-7
72. F. Pfeifer (ed.): *IUTAM Symposium on Unilateral Multibody Contacts.* Proceedings of the IUTAM Symposium held in Munich, Germany. 1999 ISBN 0-7923-6030-3
73. E. Lavendelis and M. Zakrzhevsky (eds.): *IUTAM/IFToMM Symposium on Synthesis of Nonlinear Dynamical Systems.* Proceedings of the IUTAM/IFToMM Symposium held in Riga, Latvia. 2000 ISBN 0-7923-6106-7
74. J.-P. Merlet: *Parallel Robots.* 2000 ISBN 0-7923-6308-6
75. J.T. Pindera: *Techniques of Tomographic Isodyne Stress Analysis.* 2000 ISBN 0-7923-6388-4
76. G.A. Maugin, R. Drouot and F. Sidoroff (eds.): *Continuum Thermomechanics.* The Art and Science of Modelling Material Behaviour. 2000 ISBN 0-7923-6407-4
77. N. Van Dao and E.J. Kreuzer (eds.): *IUTAM Symposium on Recent Developments in Non-linear Oscillations of Mechanical Systems.* 2000 ISBN 0-7923-6470-8
78. S.D. Akbarov and A.N. Guz: *Mechanics of Curved Composites.* 2000 ISBN 0-7923-6477-5
79. M.B. Rubin: *Cosserat Theories: Shells, Rods and Points.* 2000 ISBN 0-7923-6489-9
80. S. Pellegrino and S.D. Guest (eds.): *IUTAM-IASS Symposium on Deployable Structures: Theory and Applications.* Proceedings of the IUTAM-IASS Symposium held in Cambridge, U.K., 6–9 September 1998. 2000 ISBN 0-7923-6516-X
81. A.D. Rosato and D.L. Blackmore (eds.): *IUTAM Symposium on Segregation in Granular Flows.* Proceedings of the IUTAM Symposium held in Cape May, NJ, U.S.A., June 5–10, 1999. 2000 ISBN 0-7923-6547-X
82. A. Lagarde (ed.): *IUTAM Symposium on Advanced Optical Methods and Applications in Solid Mechanics.* Proceedings of the IUTAM Symposium held in Futuroscope, Poitiers, France, August 31–September 4, 1998. 2000 ISBN 0-7923-6604-2
83. D. Weichert and G. Maier (eds.): *Inelastic Analysis of Structures under Variable Loads.* Theory and Engineering Applications. 2000 ISBN 0-7923-6645-X
84. T.-J. Chuang and J.W. Rudnicki (eds.): *Multiscale Deformation and Fracture in Materials and Structures.* The James R. Rice 60th Anniversary Volume. 2001 ISBN 0-7923-6718-9
85. S. Narayanan and R.N. Iyengar (eds.): *IUTAM Symposium on Nonlinearity and Stochastic Structural Dynamics.* Proceedings of the IUTAM Symposium held in Madras, Chennai, India, 4–8 January 1999 ISBN 0-7923-6733-2
86. S. Murakami and N. Ohno (eds.): *IUTAM Symposium on Creep in Structures.* Proceedings of the IUTAM Symposium held in Nagoya, Japan, 3-7 April 2000. 2001 ISBN 0-7923-6737-5
87. W. Ehlers (ed.): *IUTAM Symposium on Theoretical and Numerical Methods in Continuum Mechanics of Porous Materials.* Proceedings of the IUTAM Symposium held at the University of Stuttgart, Germany, September 5-10, 1999. 2001 ISBN 0-7923-6766-9
88. D. Durban, D. Givoli and J.G. Simmonds (eds.): *Advances in the Mechanis of Plates and Shells The Avinoam Libai Anniversary Volume.* 2001 ISBN 0-7923-6785-5
89. U. Gabbert and H.-S. Tzou (eds.): *IUTAM Symposium on Smart Structures and Structonic Systems.* Proceedings of the IUTAM Symposium held in Magdeburg, Germany, 26–29 September 2000. 2001 ISBN 0-7923-6968-8

Mechanics

SOLID MECHANICS AND ITS APPLICATIONS

Series Editor: G.M.L. Gladwell

90. Y. Ivanov, V. Cheshkov and M. Natova: *Polymer Composite Materials – Interface Phenomena & Processes*. 2001 ISBN 0-7923-7008-2
91. R.C. McPhedran, L.C. Botten and N.A. Nicorovici (eds.): *IUTAM Symposium on Mechanical and Electromagnetic Waves in Structured Media*. Proceedings of the IUTAM Symposium held in Sydney, NSW, Australia, 18-22 Januari 1999. 2001 ISBN 0-7923-7038-4
92. D.A. Sotiropoulos (ed.): *IUTAM Symposium on Mechanical Waves for Composite Structures Characterization*. Proceedings of the IUTAM Symposium held in Chania, Crete, Greece, June 14-17, 2000. 2001 ISBN 0-7923-7164-X
93. V.M. Alexandrov and D.A. Pozharskii: *Three-Dimensional Contact Problems*. 2001 ISBN 0-7923-7165-8
94. J.P. Dempsey and H.H. Shen (eds.): *IUTAM Symposium on Scaling Laws in Ice Mechanics and Ice Dynamics*. Proceedings of the IUTAM Symposium held in Fairbanks, Alaska, U.S.A., 13-16 June 2000. 2001 ISBN 1-4020-0171-1
95. U. Kirsch: *Design-Oriented Analysis of Structures*. A Unified Approach. 2002 ISBN 1-4020-0443-5
96. A. Preumont: *Vibration Control of Active Structures*. An Introduction (2nd Edition). 2002 ISBN 1-4020-0496-6
97. B.L. Karihaloo (ed.): *IUTAM Symposium on Analytical and Computational Fracture Mechanics of Non-Homogeneous Materials*. Proceedings of the IUTAM Symposium held in Cardiff, U.K., 18-22 June 2001. 2002 ISBN 1-4020-0510-5
98. S.M. Han and H. Benaroya: *Nonlinear and Stochastic Dynamics of Compliant Offshore Structures*. 2002 ISBN 1-4020-0573-3
99. A.M. Linkov: *Boundary Integral Equations in Elasticity Theory*. 2002 ISBN 1-4020-0574-1
100. L.P. Lebedev, I.I. Vorovich and G.M.L. Gladwell: *Functional Analysis*. Applications in Mechanics and Inverse Problems (2nd Edition). 2002 ISBN 1-4020-0667-5; Pb: 1-4020-0756-6
101. Q.P. Sun (ed.): *IUTAM Symposium on Mechanics of Martensitic Phase Transformation in Solids*. Proceedings of the IUTAM Symposium held in Hong Kong, China, 11-15 June 2001. 2002 ISBN 1-4020-0741-8
102. M.L. Munjal (ed.): *IUTAM Symposium on Designing for Quietness*. Proceedings of the IUTAM Symposium held in Bangkok, India, 12-14 December 2000. 2002 ISBN 1-4020-0765-5
103. J.A.C. Martins and M.D.P. Monteiro Marques (eds.): *Contact Mechanics*. Proceedings of the 3rd Contact Mechanics International Symposium, Praia da Consolação, Peniche, Portugal, 17-21 June 2001. 2002 ISBN 1-4020-0811-2
104. H.R. Drew and S. Pellegrino (eds.): *New Approaches to Structural Mechanics, Shells and Biological Structures*. 2002 ISBN 1-4020-0862-7
105. J.R. Vinson and R.L. Sierakowski: *The Behavior of Structures Composed of Composite Materials*. Second Edition. 2002 ISBN 1-4020-0904-6
106. Not yet published.
107. J.R. Barber: *Elasticity*. Second Edition. 2002 ISBN Hb 1-4020-0964-X; Pb 1-4020-0966-6
108. C. Miehe (ed.): *IUTAM Symposium on Computational Mechanics of Solid Materials at Large Strains*. Proceedings of the IUTAM Symposium held in Stuttgart, Germany, 20-24 August 2001. 2003 ISBN 1-4020-1170-9

Mechanics

SOLID MECHANICS AND ITS APPLICATIONS

Series Editor: G.M.L. Gladwell

109. P. Ståhle and K.G. Sundin (eds.): *IUTAM Symposium on Field Analyses for Determination of Material Parameters – Experimental and Numerical Aspects.* Proceedings of the IUTAM Symposium held in Abisko National Park, Kiruna, Sweden, July 31 – August 4, 2000. 2003 ISBN 1-4020-1283-7
110. N. Sri Namachchivaya and Y.K. Lin (eds.): *IUTAM Symposium on Nonlinear Stochastic Dynamics.* Proceedings of the IUTAM Symposium held in Monticello, IL, USA, 26 – 30 August, 2000. 2003 ISBN 1-4020-1471-6
111. H. Sobieckzky (ed.): *IUTAM Symposium Transsonicum IV.* Proceedings of the IUTAM Symposium held in Göttingen, Germany, 2–6 September 2002, 2003 ISBN 1-4020-1608-5
112. J.-C. Samin and P. Fisette: *Symbolic Modeling of Multibody Systems.* 2003 ISBN 1-4020-1629-8
113. A.B. Movchan (ed.): *IUTAM Symposium on Asymptotics, Singularities and Homogenisation in Problems of Mechanics.* Proceedings of the IUTAM Symposium held in Liverpool, United Kingdom, 8-11 July 2002. 2003 ISBN 1-4020-1780-4
114. S. Ahzi, M. Cherkaoui, M.A. Khaleel, H.M. Zbib, M.A. Zikry and B. LaMatina (eds.): *IUTAM Symposium on Multiscale Modeling and Characterization of Elastic-Inelastic Behavior of Engineering Materials.* Proceedings of the IUTAM Symposium held in Marrakech, Morocco, 20-25 October 2002. 2004 ISBN 1-4020-1861-4
115. H. Kitagawa and Y. Shibutani (eds.): *IUTAM Symposium on Mesoscopic Dynamics of Fracture Process and Materials Strength.* Proceedings of the IUTAM Symposium held in Osaka, Japan, 6-11 July 2003. Volume in celebration of Professor Kitagawa's retirement. 2004 ISBN 1-4020-2037-6
116. E.H. Dowell, R.L. Clark, D. Cox, H.C. Curtiss, Jr., K.C. Hall, D.A. Peters, R.H. Scanlan, E. Simiu, F. Sisto and D. Tang: *A Modern Course in Aeroelasticity.* 4th Edition, 2004 ISBN 1-4020-2039-2
117. T. Burczyński and A. Osyczka (eds.): *IUTAM Symposium on Evolutionary Methods in Mechanics.* Proceedings of the IUTAM Symposium held in Cracow, Poland, 24-27 September 2002. 2004 ISBN 1-4020-2266-2
118. D. Ieşan: *Thermoelastic Models of Continua.* 2004 ISBN 1-4020-2309-X
119. G.M.L. Gladwell: *Inverse Problems in Vibration.* Second Edition. 2004 ISBN 1-4020-2670-6
120. J.R. Vinson: *Plate and Panel Structures of Isotropic, Composite and Piezoelectric Materials, Including Sandwich Construction.* 2005 ISBN 1-4020-3110-6
121. *Forthcoming*
122. G. Rega and F. Vestroni (eds.): *IUTAM Symposium on Chaotic Dynamics and Control of Systems and Processes in Mechanics.* Proceedings of the IUTAM Symposium held in Rome, Italy, 8–13 June 2003. 2005 ISBN 1-4020-3267-6
123. E.E. Gdoutos: *Fracture Mechanics. An Introduction. 2nd edition.* 2005 ISBN 1-4020-3267-6
124. M.D. Gilchrist (ed.): *IUTAM Symposium on Impact Biomechanics from Fundamental Insights to Applications.* 2005 ISBN 1-4020-3795-3
125. J.M. Huyghe, P.A.C. Raats and S. C. Cowin (eds.): *IUTAM Symposium on Physicochemical and Electromechanical Interactions in Porous Media.* 2005 ISBN 1-4020-3864-X
126. H. Ding, W. Chen and L. Zhang: *Elasticity of Transversely Isotropic Materials.* 2005 ISBN 1-4020-4033-4
127. W. Yang (ed): *IUTAM Symposium on Mechanics and Reliability of Actuating Materials.* Proceedings of the IUTAM Symposium held in Beijing, China, 1–3 September 2004. 2005 ISBN 1-4020-4131-6

Mechanics

SOLID MECHANICS AND ITS APPLICATIONS

Series Editor: G.M.L. Gladwell

128. J.-P. Merlet: *Parallel Robots*. 2006 ISBN 1-4020-4132-2
129. G.E.A. Meier and K.R. Sreenivasan (eds.): *IUTAM Symposium on One Hundred Years of Boundary Layer Research*. Proceedings of the IUTAM Symposium held at DLR-Göttingen, Germany, August 12–14, 2004. 2006 ISBN 1-4020-4149-7
130. H. Ulbrich and W. Günthner (eds.): *IUTAM Symposium on Vibration Control of Nonlinear Mechanisms and Structures*. 2006 ISBN 1-4020-4160-8
131. L. Librescu and O. Song: *Thin-Walled Composite Beams. Theory and Application*. 2006 ISBN 1-4020-3457-1
132. G. Ben-Dor, A. Dubinsky and T. Elperin: *Applied High-Speed Plate Penetration Dynamics*. 2006 ISBN 1-4020-3452-0
133. X. Markenscoff and A. Gupta (eds.): *Collected Works of J. D. Eshelby*. Mechanics of Defects and Inhomogeneities. 2006 ISBN 1-4020-4416-X
134. R.W. Snidle and H.P. Evans (eds.): *IUTAM Symposium on Elastohydrodynamics and Microelastohydrodynamics*. Proceedings of the IUTAM Symposium held in Cardiff, UK, 1–3 September, 2004. 2006 ISBN 1-4020-4532-8
135. T. Sadowski (ed.): *IUTAM Symposium on Multiscale Modelling of Damage and Fracture Processes in Composite Materials*. Proceedings of the IUTAM Symposium held in Kazimierz Dolny, Poland, 23–27 May 2005. 2006 ISBN 1-4020-4565-4
136. A. Preumont: *Mechatronics. Dynamics of Electromechanical and Piezoelectric Systems*. 2006 ISBN 1-4020-4695-2
137. M.P. Bendsøe, N. Olhoff and O. Sigmund (eds.): *IUTAM Symposium on Topological Design Optimization of Structures, Machines and Materials*. Status and Perspectives. 2006 ISBN 1-4020-4729-0
138. A. Klarbring: *Models of Mechanics*. 2006 ISBN 1-4020-4834-3
139. H.D. Bui: *Fracture Mechanics*. Inverse Problems and Solutions. 2006 ISBN 1-4020-4836-X
140. M. Pandey, W.-C. Xie and L. Xu (eds.): *Advances in Engineering Structures, Mechanics and Construction*. Proceedings of an International Conference on Advances in Engineering Structures, Mechanics & Construction, held in Waterloo, Ontario, Canada, May 14–17, 2006. 2006 ISBN 1-4020-4890-4
141. G.Q. Zhang, W.D. van Driel and X. J. Fan: *Mechanics of Microelectronics*. 2006 ISBN 1-4020-4934-X
142. Q.P. Sun and P. Tong (eds.): *IUTAM Symposium on Size Effects on Material and Structural Behavior at Micron- and Nano-Scales*. Proceedings of the IUTAM Symposium held in Hong Kong, China, 31 May–4 June, 2004. 2006 ISBN 1-4020-4945-5
143. A.P. Mouritz and A.G. Gibson: *Fire Properties of Polymer Composite Materials*. 2006 ISBN 1-4020-5355-X
144. Y.L. Bai, Q.S. Zheng and Y.G. Wei (eds.): *IUTAM Symposium on Mechanical Behavior and Micro-Mechanics of Nanostructured Materials*. Proceedings of the IUTAM Symposium held in Beijing, China, 27–30 June 2005. 2007 ISBN 1-4020-5623-0
145. L.P. Pook: *Metal Fatigue*. What It Is, Why It Matters. 2007 ISBN 1-4020-5596-6
146. H.I. Ling, L. Callisto, D. Leshchinsky and J. Koseki (eds.): *Soil Stress-Strain Behavior: Measurement, Modeling and Analysis*. A Collection of Papers of the Geotechnical Symposium in Rome, March 16–17, 2006. 2007 ISBN 978-1-4020-6145-5
147. A.S. Kravchuk and P.J. Neittaanmäki: *Variational and Quasi-Variational Inequalities in Mechanics*. 2007 ISBN 978-1-4020-6376-3

Mechanics

SOLID MECHANICS AND ITS APPLICATIONS

Series Editor: G.M.L. Gladwell

148. S.K. Kanaun and V. Levin: *Self-Consistent Methods for Composites*. Vol. 1: Static Problems. 2008 ISBN 978-1-4020-6663-4
149. G. Gogu: *Structural Synthesis of Parallel Robots*. Part 1: Methodology. 2008 ISBN 978-1-4020-5102-9

FSC
www.fsc.org
MIX
Papier | Fördert
gute Waldnutzung
FSC® C083411

Kolibri 360 GmbH
Ferdinand-Jühlke-Straße 7
99095 Erfurt, Deutschland
produktsicherheit@kolibri360.de